International Conference on Innovation in Engineering and Vocational Education 2015

IOP Conference Series: Materials Science and Engineering
Volume 128

Bandung, Indonesia
14 November 2015

Editors:

Ade Gafar Abdullah
Budi Mulyanti
Asep Bayu Dani Nandiyanto

ISBN: 978-1-5108-2448-5
ISSN: 1757-8981

Printed from e-media with permission by:

Curran Associates, Inc.
57 Morehouse Lane
Red Hook, NY 12571

Some format issues inherent in the e-media version may also appear in this print version.

Printed by Curran Associates, Inc. (2016)

For permission requests, please contact the Institute of Physics
at the address below.

Institute of Physics
Dirac House, Temple Back
Bristol BS1 6BE UK

Phone: 44 1 17 929 7481
Fax: 44 1 17 920 0979

techtracking@iop.org

Additional copies of this publication are available from:

Curran Associates, Inc.
57 Morehouse Lane
Red Hook, NY 12571 USA
Phone: 845-758-0400
Fax: 845-758-2633
Email: curran@proceedings.com
Web: www.proceedings.com

Table of contents

Volume 128

International Conference on Innovation in Engineering and Vocational Education
14 November 2015, Bandung, Indonesia

Accepted papers received: 7 April 2016
Published online: 24 May 2016

Preface

Papers

IOP Conf. Series: Materials Science and Engineering **128** (2016) 011001 doi:10.1088/1757-899X/128/1/011001

PREFACE

The International Conference on Innovation in Engineering and Vocational Education (ICIEVE) is a conference held by the Vocational Technology Education Study Programs, School of Postgraduates Studies, Universitas Pendidikan Indonesia. This conference is a wonderful example of creativity that provides opportunities for the delegates to exchange new ideas and application experiences, to establish research networks and to find global partners for future collaboration.

The theme of ICIEVE 2015 is "Sustainable Development for Engineering and Vocational Education", and we provide a platform for scientists, scholars, engineers, industrial professionals, and researchers to exchange, share and discuss their innovation, experience, research works and problem solving techniques in all issues in engineering and vocational education. There are 229 papers from various topics including engineering and technology innovation, engineering education, vocational education and training, and innovation in teaching and learning. More than 54 papers are selected to go through a strict peer reviewed process, and these papers will be published in conference proceedings.

We would like to thank the organizing committee and the members of reviewers for their kind assistance in reviewing the papers. We would also extend our best gratitude for Professor Hiroyuki Iida and Associate Professor Agus Setiawan for their invaluable contribution and worthwhile ideas shared in the conference.

The Editors

Dr. Ade Gafar Abdullah

Dr. Budi Mulyanti

Dr. Eng. Asep Bayu Dani Nandiyanto

CONFERENCE PHOTOGRAPHS

CONFERENCE
THEME
SUSTAINABLE DEVELOPMENT FOR ENGINEERING & VOCATIONAL EDUCATION

IOP Conf. Series: Materials Science and Engineering **128** (2016) 011001 doi:10.1088/1757-899X/128/1/011001

LISTS OF COMMITTEES

INTERNATIONAL ADVISORY BOARDS:

1. Prof. Hiroyuki Iida (JAIST, Japan)
2. Prof. Erica Smith (Federation University, Australia)
3. Assoc. Prof. Tetsu Kubota (Hiroshima University, Japan)
4. Prof. Thomas Schroder (Germany)
5. Dr. Luisa Brotas (London Metropolitan University, UK)
6. Assoc. Prof. Sirilak Hanvatananukul (RAVTE, Rajamangala University of Technology Thanyaburi, Thailand)
7. Prof. Lilia Halim (UKM, Malaysia)
8. Dr. Asnul Dahar Mingat (UTM, Malaysia)
9. Assoc. Prof. Muhammad Sukri Saud (UTM, Malaysia)
10. Prof. Ramlee bin Mustapha(UPSI, Malaysia)
11. Dr. Mohd. Sattar bin Rasul (UKM, Malaysia)
12. Dr. Mohd. Nazeri (IPGK Perlis, Malaysia)
13. Dr. Phil. Dadang Kurnia (GIZ, Germany)
14. Dr. Joachim Dittrich (ITB, Germany)
15. Dr. Sigit Dwiananto Arifwidodo (KASETSART University, Thailand)

SCIENTIFIC COMMITEE:

1. Prof. M. Syaom Barliana – Universitas Pendidikan Indonesia
2. Prof. Liliasari – Universitas Pendidikan Indonesia
3. Prof. Nuryani Rustaman – Universitas Pendidikan Indonesia
4. Prof. Asep Kadarohman – Universitas Pendidikan Indonesia
5. Dr. Eng. Agus Setiawan – Universitas Pendidikan Indonesia
6. Dr. Ida Hamidah – Universitas Pendidikan Indonesia
7. Dr. Eng. Asep Bayu Dani Nandiyanto - Universitas Pendidikan Indonesia
8. Prof. Bambang Riyanto – Institut Teknologi Bandung
9. Prof. Andi Isra Mahyuddin – Institut Teknologi Bandung
10. Prof. Armein Z. Langi – Institut Teknologi Bandung
11. Prof. Khairurrijal – Institut Teknologi Bandung
12. Prof. Zaki Suud – Institut Teknologi Bandung
13. Prof. Mikrajuddin Abdullah – Institut Teknologi Bandung
14. Prof. Widjaja Martokusumo – Institut Teknologi Bandung
15. Prof. Setiawan Sabana – Institut Teknologi Bandung
16. Prof. Heri Hermansyah – Universitas Indonesia
17. Prof. Ucok Hasibuan– Universitas Indonesia
18. Dr. Heru Purnomo– Universitas Indonesia
19. Dr. Ariadne Juwono– Universitas Indonesia
20. Dr. Kuwat Triyana – Universitas Gadjah Mada
21. Prof. Ali Khomsan – Institut Pertanian Bogor
22. Dr. Ika Amalia Kartika - Institut Pertanian Bogor
23. Prof. Luthfiyah Nurlaela – Universitas Negeri Surabaya
24. Prof. Ivan Hanafi – Universitas Negeri Jakarta

25. Dr. Wagiran – Universitas Negeri Yogyakarta
26. Dr. Putu Sudira – Universitas Negeri Yogyakarta
27. Prof. Ali Saukah – Universitas Negeri Malang
28. Prof. Jasruddin Daud Malago – Universitas Negeri Makasar
29. Prof. Mahfud Arifin – Universitas Padjajaran
30. Prof. Dian Fiantis – Universitas Andalas

ORGANIZING COMMITEE:

1. Dr. Budi Mulyanti - Conference Chair
2. Dr. Johar Maknun
3. Dr. Ade Gafar Abdullah
4. Dr. Kamin Sumardi
5. Dr. Eng. Beta Paramita
6. Dr. Ana
7. Dr. Isma Widiaty
8. Tutin Aryanti, Ph. D
9. Dr. Dadang Lukman Hakim
10. Dr. Yatti Sugiarti

IOP Conf. Series: Materials Science and Engineering **128** (2016) 011002 doi:10.1088/1757-899X/128/1/011002

Peer review statement

All papers published in this volume of *IOP Conference Series: Materials Science and Engineering* have been peer reviewed through processes administered by the proceedings Editors. Reviews were conducted by expert referees to the professional and scientific standards expected of a proceedings journal published by IOP Publishing.

Evaluation of expert system application based on usability aspects

C P C Munaiseche* and O E S Liando

Department of Information Technology and Communication Education
Faculty of Engineering, Universitas Negeri Manado, Tondano 95618, Indonesia

*Corresponding author: cindymunaiseche@gmail.com

Abstract. Usability usually defined as a point of human acceptance to a product or a system based on understands and right reaction to an interface. The performance of web application has been influence by the quality of the interface of that web to supporting information transfer process. Preferably, before the applications of expert systems were installed in the operational environment, these applications must be evaluated first by usability testing. This research aimed to measure the usability of the expert system application using tasks as interaction media. This study uses an expert system application to diagnose skin disease in human using questionnaire method which utilize the tasks as interaction media in measuring the usability. Certain tasks were executed by the participants in observing usability value of the application. The usability aspects observed were learnability, efficiency, memorability, errors, and satisfaction. Each questionnaire question represent aspects of usability. The results present the usability value for each aspect and the total average merit for all the five-usability aspect was 4.28, this indicated that the tested expert system application is in the range excellent for the usability level, so the application can be implemented as the operated system by user. The main contribution of the study is the research became the first step in using task model in the usability evaluation for the expert system application software.

1. Introduction

From the point of view of people who need to use any interactive software system, usability is the most important aspect of the system. Usability is a quality attribute that assesses how easy user interfaces are to use. One way to evaluate the interface known as usability testing.

Badre provides a definition of usability testing as follows: "Usability testing has traditionally meant testing for efficiency, ease of learning, and the ability to remember how to perform interactive tasks without difficulty or errors."[1]. The ISO 9241 standard defines usability as ''the extent to Which a product can be used by specified users to achieve specified goals with effectiveness, efficiency And satisfaction in a specified context of use'' [2].

Usability tests are necessary to obtain interfaces according to the needs and user satisfaction in using the system.

Clairmont defines formal usability testing as, "The observation and analysis of user behavior while user use a product prototype to achieve a goal" [3]. Covey defines as: "a structured, exploratory observation of clearly defined aspects of the behavior of an individual performing one or more design at the tasks" [4].

From the description above can be concluded that the formal usability testing method is to observe the use of systems or prototypes by the user, by giving the user a series of tasks that have been made by researchers.

Any researchers have used the aspects of usability in their testing for some application software. In previous studies, Ardito et. al, has evaluated the usability of e-learning application. Usability dimension for e-learning platform namely: presentation, hypermediality, application proactivity and user activity [5]. Chrisna et. al, has evaluated the usability of Business Intelligence (BI) application at a coal mining organization. For usability criteria, Chrisna confirms the importance of efficiency, effect, learnability, helpfulness and control but he adds another criteria, namely: information architecture, learnability and operability which relates to reporting format, data quality, accessibility and processing speed as required in the mining context [6]. Shafinah et. al, has evaluated the usability of Decision Support Systems (DSSs). Eight usability factors (efficiency, understandability, operability, attractiveness, error prevention, learnability, accuracy and effectiveness) have been identified for the evaluation process [7].

This study discussed the system evaluation for an Expert System (ES) application based on usability aspects. Five usability factors have been identified for the evaluation process, namely: learnability, efficiency, memorability, errors, and satisfaction. This evaluation aimed to determine how much the ease of an interface can be used by the user so that the system can be received in the operational environment.

The main contribution is the research became the first step in using task model in the purpose of usability measurement of expert system application software. The advantage of this research is the usability evaluation criteria for expert system applications presented as guidelines for another researchers in their study using other expert system applications

2. Literature Review

2.1 Basic Concepts of Expert System

Expert system is a piece of software programmed using Artificial Intelligence (AI) techniques. Such systems use databases of expert knowledge to offer advice or make decisions in such areas as medical diagnosis and trading on the stock exchange. Expert systems development requires knowledge acquisition from people, involving both knowledge engineers and application domain experts in specialist interactions with computing systems. Expert systems may be used to provide support and advice to a user of any complex information system and hence to improve the human-computer interface [8].

2.2 Usability Testing.

Usability ensures that interactive products are easy to learn, effective to use and enjoyable from the user's perspective. This culminates in the goals of effectiveness, effiency, safety, utility, learnability and memorability [9]. The measurement of usability depends on the users, their needs (goals) and the context – three variables that are inconsistent and unstable in themselves [10].

There are some experts who provide the definition and quality components of usability, including [11]:

1) Dix et al, (2004): The system can help users to solve their problems is a system that:

- Useful: a system which functions as desired by the user.
- Usable: a system that is easy to operate
- Used: a system that motivates the user to use, appealing to the user, fun, and etc.

IOP Conf. Series: Materials Science and Engineering **128** (2016) 012001 doi:10.1088/1757-899X/128/1/012001

2) Jakob Nielsen (2003): Usability is a quality attribute that assesses how easy user interfaces are to use. The word "usability" also refers to methods for improving ease-of-use during the design process. Usability is defined by 5 quality components [12]:
   1. Learnability: How easy is it for users to accomplish basic tasks the first time they encounter the interface?
   2. Efficiency: Once users have learned the design or application, how quickly can they perform tasks? (system has the simple steps (fewer steps) to support the users in performing their work, example: a single button on the automatic answering machine, one-click shopping: Amazon.com)
   3. Memorability: When users return to the design or application after a period of not using it, how easily can they reestablish proficiency?
   4. Errors: How many errors do users make, how severe are these errors, and how easily can they recover from the errors?
   5. Satisfaction: How pleasant is it to use the design or application?

3) Palmer (2002): Viewing other quality attributes of usability, namely: the download time, navigabilitas, interactivity, responsiveness, content quality.

3. Expert System Application Software

The application software that used in usability testing is the expert system application to diagnose skin disease in human. Some interfaces from the system are shown in Figure 1, Figure 2 and Figure 3.

Figure 1. The main page interface (Home Menu)

The main page interface of the application (Figure 1) consist of five Menus (Home, Daftar Penyakit, Konsultasi, Login Pakar dan Kontak).

Interface of admin page (Figure 2) consist of eight Menus (Input Penyakit, Input Gejala, Input Relasi, Ubah Penyakit, Ubah Gejala, Laporan Penyakit, Laporan Gejala dan Logout).

Figure 3 displays Analysis Result of Skin Disease after the user answered the questions on the system.

Figure 2. Interface of Admin Menu

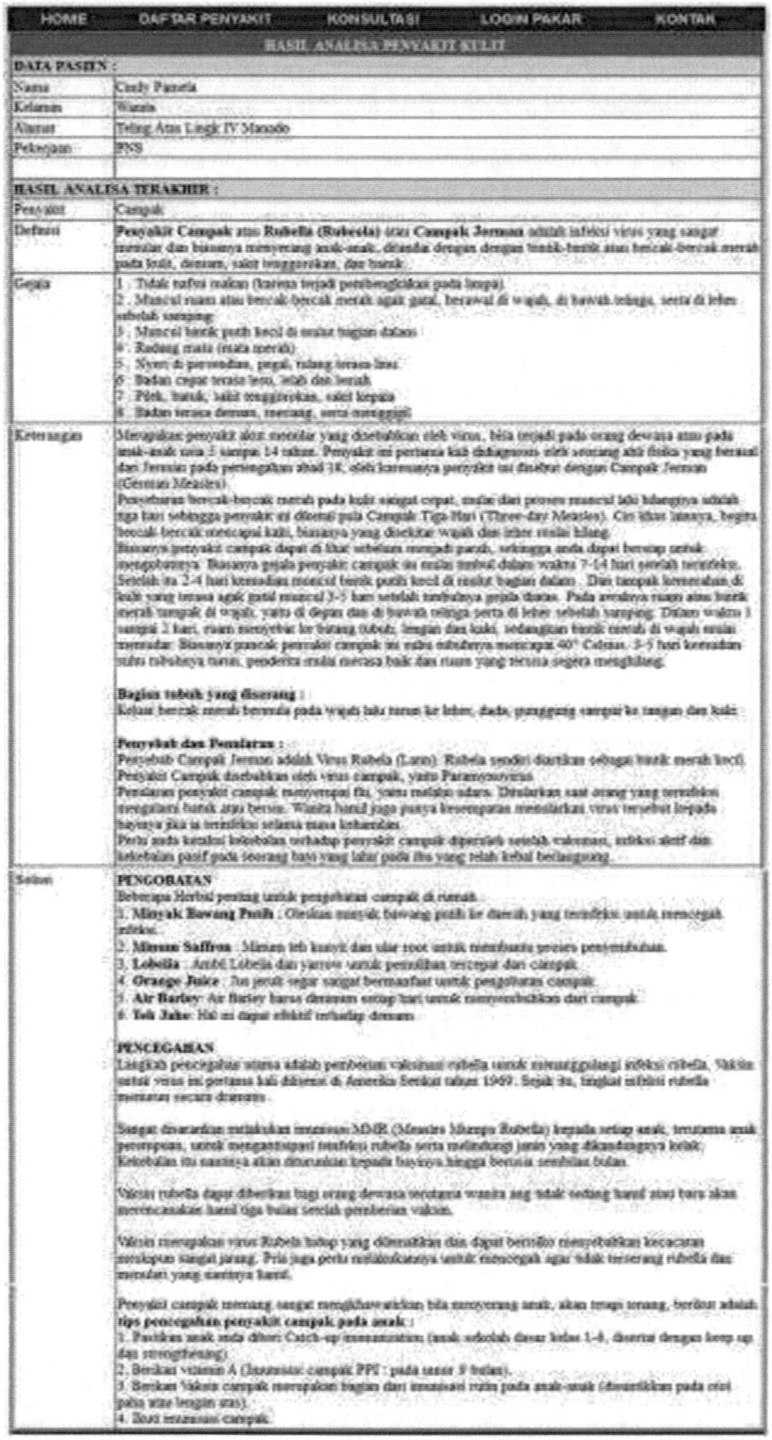

Figure 3. Result of Skin Disease Diagnosis

IOP Conf. Series: Materials Science and Engineering **128** (2016) 012001 doi:10.1088/1757-899X/128/1/012001

4. Research Method

The methodology in this study is shown in Figure 4.

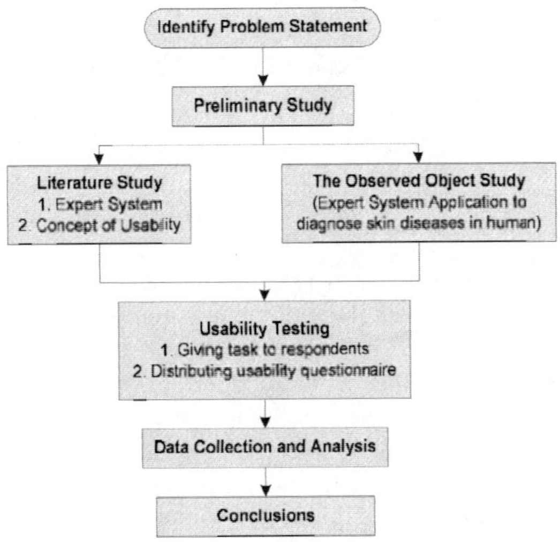

Figure 4. Procedure of Research

4.1. Identifying Problems Statement

Firstly, the issues related to usability especially focused on Expert System application were identified.

4.2. Preliminary Study

Consist of two parts: First, study of literature about expert system and concept of usability. Second, study of the observed object, in this case Expert System application to diagnose skin disease in humans. Some of interfaces are shown in Figure 1, Figure 2 and Figure 3.

4.3. Usability Testing

In this stage, the respondents were asked to perform the tasks. Questionnaires were later carried out and they were to be filled in and completed by the respondents based on experience of each respondent when performing the tasks. This stage consist of two activities: giving task to respondents and distributing usability questionnaire.

4.3.1 Providing the set of tasks that have been prepared by the researcher

The tasks were given to 30 respondents. The respondent were chosen among students of PTIK (Pendidikan Teknologi Informasi dan Komunikasi) and medical personnel who have knowledge in using the web browser features or having a computer science background. The tasks are used as a interaction media in the measurement of usability [13].

Table 1. Tasks of usability testing

Number	Tasks
1.	Log in as a admin, log out, and then log in again.
2.	Sign up as a patient in "Konsultasi" Menu
3.	Answer the displayed questions on the system
4.	Add, modify and delete the data of type and symptoms of the skin disease on admin page.
5.	Search information about the names and symptoms of skin diseases

The set of usability testing tasks are displayed in Table 1. There are five tasks. Each task in Table 1 can be explained as follows:

Task1. Loginas aadmin,logout,andthenloginagain.

Users are required to login as admin, starting from finding and filling the login form to log into the admin page, then asked to looking for the logout button, and then login again. The task is considered completed if the user has login and logout of the system.

Task2. Signupasapatient inKonsultasi Menu

Users are required to register as a patient in "Konsultasi" menu. The task is considered completed when the user has filled the data and then click "Daftar" button.

Task3. Answer thedisplayedquestionsonthesystem

Task 3 relates to task 2. Once the user clicks the "Daftar" button on the task 2, patients as the user can directly answer option true (YES) or False (NO) of the questions displayed by the system, regarding the symptoms felt by patient. The task is considered completed if the system has shown the type of disease suffered by the patient, the symptoms, disease information and solutions, as seen in the Skin Diseases Analysis Result (Figure 3).

Task4. Add,modifyanddeletethedataof typeandsymptomsoftheskindisease onadminpage

The users are asked to enter the admin page and search the forms of adding, editing and deleting type and symptoms of skin diseases. The task is considered completed if the user has been succeeded to add, edit and delete data of type and symptoms of skin diseases.

Task5. Searchinformationabout thenamesandsymptomsofskindiseases

The users are asked to find information about the name list and the symptons of skin diseases. The task is considered completed if the user has found the information.

4.3.2 Distributing the usability questionnaire

After the task was given to respondents, the next step distribute the questionnaire to get an assessment of the thirty respondents to know how much the interface of expert system aplication is understood, learned, remembered, and etc.

The questionnaires were designed using Likert scale values 1 to 5 (1 = very difficult; 2 = difficult; 3 = fairly easy; 4 = easy; 5 = very easy). The questionnaire to evaluate the usability of Expert System application is presented in Table 2. Respondents fill out the questionnaire in Table 2 based on their experience when performing the tasks (Subsubsection 4.3.1).

Table 2. Questionnaire Questions

No.	Question	Scale				
		VD	D	FE	E	VE
		1	2	3	4	5
1.	From the initial interface (Home Menu), could you recognize that this is an application of expert systems to diagnose skin diseases in humans?					
2.	Could you register as a patient via the Consultation Menu?					
3.	Could you answer the questions about the symptoms of disease deployed by the system after registering as a patient via the Consultation Menu?					
4.	Could you obtain the final results regarding the type of disease, symptoms, description of disease, prevention and treatment offered by the system?					
5.	Are the letters easy to read?					
6.	Are the picture symbols easy to understood?					
7.	Does the color design of this application comfortable viewing?					
8.	Could you find the login form?					
9.	Could you access the information on each page?					
10.	Could you recall menus and page view once you exit the application?					

Note: VD = **Very Difficult**; D = **Difficult**; FE = **Fairly Easy**; E = **Easy**; VE = **Very Easy**

The questionnaire consist of 10 questions. Each question has separate section to represent each aspect of usability. Each question of the questionnaire aimed to demonstrate the usability levels according to user acceptance, which will be assessed on the scale values 1 to 5 [14]. Table 3 shows the contributions of questionnaire question for each aspect of usability

Table 3. Plot of Usability Aspects

No.	Question	Aspect of Usability				
		Learn	Eff	Mem	Err	Sat
1.	From the initial interface (Home Menu), could you recognize that this is an application of expert systems to diagnose skin diseases in humans?	■				■
2.	Could you register as a patient via the Consultation Menu?	■	■	■		■
3.	Could you answer the questions about the symptoms of disease displayed by the system after registering as a patient via the Consultation Menu?	■	■			■
4.	Could you obtain the final results regarding the type of disease, symptoms, description of disease, prevention and treatment offered by the system?	■	■			■
5.	Are the letters easy to read?				■	■
6.	Are the picture symbols easy to understood?				■	■
7.	Does the color design of this application comfortable viewing?					■
8.	Could you find the login form?		■	■		■
9.	Could you access the information on each page?		■			■
10.	Could you recall menus and page view once you exit the application?			■		■

Note: Learn = **Learnability**; Eff = **Efficiency**; Mem = **Memorability**; Err = **Errors**; Sat = **Satisfaction**

Table 4. Questionnaire Answers Percentage

QQ	VD	D	FE	E	VE
	1	2	3	4	5
1.	0%	0%	10%	30%	60%
2.	0%	0%	7%	63%	30%
3.	0%	0%	7%	26%	67%
4.	0%	0%	4%	63%	33%
5.	0%	0%	10%	63%	27%
6.	0%	0%	7%	33%	60%
7.	0%	3%	10%	60%	27%
8.	0%	0%	10%	27%	63%
9.	0%	0%	7%	63%	30%
10.	0%	3%	17%	60%	20%

Note: QQ = *Questions of Questionnaire*; VD = Very Difficult; D = Difficult; FE = Fairly Easy; E = Easy; VE = Very Easy

As can be seen from Table 3, question number 1 represents learnability and satisfaction aspects of usability. Question number 2 represents learnability, efficiency, memorability and satisfaction aspects of usability. Question number 3 and number 4 represent learnability, efficiency and satisfaction aspects of usability. Question number 5 and number 6 represent errors and satisfaction aspects of usability, etc.

It can be said that **Learnability** aspect was represented by questions number 1, 2, 3, 4; **Efficiency** aspect was represented by question number 2, 3, 4, 8, 9; **Memorability** by question number 2, 8, 10; **Errors** by question number 5 and 6; and **Satisfaction** by question number 1 until 10

4.4. Data Collection and Analysis

After filled by the user, the questionnaires were collected and then make the recapitulation based on questionnaire results. Table 4 displays the questionnaires recapitulation that shown in percentage.

As can be seen from Table 4, for question number 1, there were 60 % of respondents answered very easy, 30 % of respondents answered easy, and 10% of respondents answered fairly easy, there was none answered difficult or very difficult.

For the data analysis, the Likert scale values in questionnaires are interpreted to corresponding merit 0.00 until 5.00. Table 5 shows the interpretation from Likert Scale values to the corresponding merit for question 1 to 10. Usability average merit and usability level were depicted in Table 6 [7].

Table 5. Data Interpretation Process

QQ	VD	D	FE	E	VE	User Acceptance Value
	1	2	3	4	5	
1.	0.00	0.00	0.3	1.20	3.00	**4.50**
2.	0.00	0.00	0.21	2.52	1.5	**3.93**
3.	0.00	0.00	0.21	1.04	3.35	**4.60**
4.	0.00	0.00	0.12	2.52	1.65	**4.29**
5.	0.00	0.00	0.3	2.52	1.35	**4.17**
6.	0.00	0.00	0.21	1.32	3.00	**4.53**
7.	0.00	0.06	0.3	2.4	1.35	**4.11**
8.	0.00	0.00	0.3	1.08	3.15	**4.53**
9.	0.00	0.00	0.21	2.52	1.50	**4.23**
10.	0.00	0.06	0.51	2.4	1.00	**3.97**

Note: QQ = *Questions of Questionnaire*; VD = Very Difficult; D = Difficult; FE = Fairly Easy; E = Easy; VE = Very Easy

Table 6. Usability average merit and usability level

Average Merit, A	Usability Level
0.00 ≤ A ≤ 1.00	Bad
1.00 < A ≤ 2.00	Poor
2.00 < A ≤ 3.00	Moderate
3.00 < A ≤ 4.00	Good
4.00 < A ≤ 5.00	Excellent

From the data interpretation process in Table 5, the user acceptance value for question number 1 (QQ1) was 4.50 and number 2 (QQ2) was 3.93, its mean that the usability value of attribute "Ease of interface skin disease diagnosis expert system recognized" was 4.50 and the usability value of attribute "Ease of patients registration through the "Konsultasi Menu" was 3.93, etc. These were shown in Table 7.

4.5. Conclusions

From the analysis, will be drawn a conclusion about usability evaluation of the expert system application based on usability aspects and how much the user acceptance value of the application sotware (usability value and usability level) before being installed in the operational environment.

5. Result and Discussion

5.1 Result

There were two results of usability testing on the expert system application found: **the usability value for each attribute** (each question of the questionnaire) and **the usability value for each aspect of usability.** Both of them were shown in Table 7 and Table 8.

Table 7. Usability value and usability level for each attribute

No.	Attribute	Usability Value	Usability Level
1.	Ease of expert system application interface for diagnosing skin disease recognized	**4.50**	**Excellent**
2.	Ease of patient registration through the Consultation Menu	**3.93**	**Good**
3.	Ease of patient answer the displayed questions about the symptoms of the disease after registering on the Consultation Menu	**4.60**	**Excellent**
4.	Ease of final analysis results from diseases to be found	**4.29**	**Excellent**
5.	Ease of letters on the application can be read	**4.17**	**Excellent**
6.	Ease of picture symbols understood.	**4.53**	**Excellent**
7.	Color design comfortable.	**4.11**	**Excellent**
8.	Ease the login form is found	**4.53**	**Excellent**
9.	Ease of access to information on every page.	**4.23**	**Excellent**
10.	Ease of recal menus and pages display	**3.97**	**Good**

Table 8. Usability value for each aspect of usability

Number	Aspect of Usability	Usability Value
1.	Learnability	4.33
2.	Efficiency	4.31
3.	Memorability	4.14
4.	Errors	4.35
5.	Satisfaction	4.27

5.2 Discussion

There were two results found: **the usability value for each attribute** (each question of the questionnaire) and **the usability value for each aspect of usability**

5.2.1 The usability value and level for each attribute

Each question of the questionnaire aimed to demonstrate the usability levels according to user acceptance value (usability value) [14]

As can be seen from Table 7, the user acceptance value or usability value for the attribute "Ease of interface skin disease diagnosis expert system recognized" was 4.50, according to Table 6, the attribute number 1 is in the range of **excellent** for the corresponding usability level.

The user acceptance value or usability value for the attribute "Ease of patients registration through the "Konsultasi Menu" was 3.93, according to Table 6, the attribute number 2 is in the range of **good** for the corresponding usability level.

The user acceptance value or usability value for the attribute "Ease of patients answer the displayed questions about the symptoms of the disease after registering on "Konsultasi" Menu" was 4.60, according to Table 6, the attribute number 3 is in the range of **excellent** for the corresponding usability level.

The attribute number 4 until 9 is in the range of **excellent** for the corresponding usability level, and number 10 is in the range of **good**.

The usability value and usability level for each attribute that explained above was shown in Table 7.

5.2.2 The usability value for each aspect of usability

According to Table 3, Learnability aspect was represented by questions number 1, 2, 3, 4. Usability values of question number 1, 2, 3, 4 were 4.5, 3.93, 4.60, 4.29 (Table 7), the average merit for these question numbers was 4.33, and this indicated that usability value of ***Learnability*** aspect was **4.33**

Efficiency aspect was represented by questions number 2, 3, 4, 8, 9. Usability values of question number 2, 3, 4, 8, 9 were 3.93, 4.60, 4.29, 4.53, 4.23 , the average merit for these question numbers was 4.31, and this indicated that usability value of ***Efficiency*** aspect was **4.31**

Memorability aspect was represented by questions number 2, 8, 10. Usability values of question number 2, 8, 10 were 3.93, 4.53, 3.97 , the average merit for these question numbers was 4.14, and this indicated that usability value of ***Memorability*** aspect was **4.14**

Errors aspect was represented by questions number 5 and 6. Usability values of question number 5 and 6 were 4.17 and 4.35, the average merit for these question numbers was 4.35, and this indicated that usability value of ***Errors*** aspect was **4.35**

Satisfaction aspect was represented by questions number 1 until 10 . Usability values of question number 1 until 10 was depictedin Table 5, the average merit for all question numbers was 4.27, and this indicated that usability value of ***Satisfaction*** aspect was **4.27**.

The highest average merit was scored by the errors \.35). This was followed by the aspect of learnability (4.33), efficiency (4,31), satisfaction (4.27) and memorability (4.14). The total average merit for all the five-usability aspect was 4.28 and this indicated that the tested expert system application is in the range of **excellent** for the corresponding usability level. Usability value for each aspect of usability was shown in Table 8.

6. Conclussion

The software evaluation result of this application indicates that the user acceptance value or usability value of Learnability aspect was 4.33, Efficiency aspect was 4.31, Memorability aspect was 4.14, Errorrs aspect was 4.35 and Satisfaction aspect was 4.27.

The results also present the usability level for each factor and the total average merit for all the five-usability factor was 4.28, this indicated that the tested expert system application is in the range **excellent** for the usability level, so the application can be implemented as the operated system by user.

References

[1] Badre AN. Shaping Web Usability: Interaction Design in Context. Boston: Addison-Wesley. 2002.

[2] ISO. 9241-210: 2010. Ergonomics of Human System Interaction-Part 210: Human-Centred Design for Interactive Systems. International Standardization Organization (ISO). Switzerland, 2009.

[3] Clairmont, Michelle et al. Testing for usability in the design of a new information gateway. Paper presented at Living the Future 2, Tucson, Arizona, 1998 April. Available from: http://dizzy.library.arizona.edu/library/teams/access9798/lftpaper.htm

[4] Covey, DT. Usage and usability assessment: library practices and concerns. 2002. Available from: http://www.clir.org/pubs/reports/pub105/contents.html

[5] Ardito C, Costabile MF, et al. An Approach to Usability Evaluation of E-Learning Applications. Univ Access Inf Soc (2006) 4: 270-283, 2005 Dec 8. Available from: http://www.qou.edu/arabic/researchProgram/eLearningResearchs/anApproach.pdf

[6] Chrisna J, Judy VB, Jan M. Usability Evaluation for Business Intelligence Applications:A User Support Perspective. *South African Computer Journal*, no.53, pp.32-44, 2014 August

[7] Shafinah K, Selamat MH, Abdullah R, Nik Muhammad AM, Awang Nur AG. System Evaluation for A Decission Support System. *Information Technology Journal* 9 (5): 889-898, 2010, ISSN 1812-5638.*Asian Network for Scientific Information* (ANSI Net). Available from: http://scialert.net/qredirect.php?doi=itj.2010.889.898&linkid=pdf

[8] Brian R. Gaines. Designing Expert System for Usability. Available from: http://citeseerx.ist.psu.edu/viewdoc/download?doi=10.1.1.414.583&rep=rep1&type=pdf

[9] Rogers Y, Sharp H, Preece J. Interaction Design: Beyond Human-Computer Interaction. John Wiley Sons, 2011

[10] Gebus S, Leivis K. Knowledge Acquisition for Decision Support Systems on An Electronic Assembly Line. *Expert Systems with Applications*, vol. 36, no. 1, pp. 93-101, 2009.

[11] Paulus Insap S. Bahan Kuliah Teknik Komputer Interaktif. Jurusan Teknik Elektro dan Teknologi Informasi, Fakultas Teknik Universitas Gadjah Mada. 2010.

[12] Jakob Nielsen. Usability 101: Introduction to Usability. 2014 Jan 2012. Available from: http://www.useit.com/alertbox/20030825.html

[13] Sastramihardja et al. Usability Measurement By Means Task Model In User Center Software Development. 2008.

[14] Wingnjosoebroto S, Sudiarno A, Harenda D. Designing Web Interface Prototype Based On Usability Aspects (*Case Studi: Ergonomic and Work System Design Laboratory*). 2009. Available from: http://digilib.its.ac.id/public/ITS-Undergraduate-8877-2504100030-Paper.pdf.

[15] Andrew Dillon. The evaluation of Software Usability. 2015 November 21. Available from: https://www.ischool.utexas.edu/~adillon/BookChapters/Dillon-evaluation%20of%20software%20usab.htm

Life cycle assessment of energy and CO_2 emissions for residential buildings in Jakarta, Indonesia.

U Surahman[1 *], T Kubota[2] and A Wijaya[3]
[1]Universitas Pendidikan Indonesia, Jl. Dr. Setiabudhi No. 207 Bandung 40154, Indonesia.
[2]Graduate School for International Development and Cooperation, Hiroshima University, 1-5-1 Kagamiyama Higashi-Hiroshima, 739-8529 Hiroshima, Japan.
[3]Indonesia University of Persada, Jl. Salemba Raya No. 10 Jakarta 10440, Indonesia.

[*]Corresponding author. Tel.: +62-89634730153; fax: +62-22-424-6956.
E-mail address: usep@upi.edu (Usep Surahman).

Abstract. In order to develop low energy and low carbon residential buildings, it is important to understand their detailed energy profiles. This study provides the results of life cycle assessment of energy and CO_2 emissions for residential buildings in Jakarta, Indonesia. A survey was conducted in the city in 2012 to obtain both material inventory and household energy consumption data within the selected residential buildings (n=300), which are classified into three categories, namely simple, medium and luxurious houses. The results showed that the average embodied energy of simple, medium and luxurious houses was 58.5, 201.0, and 559.5 GJ, respectively. It was found that total embodied energy of each house can be explained by its total floor area alone with high accuracy in respective house categories. Meanwhile, it was seen that operational energy usage patterns varied largely among house categories as well as households especially in the simple and medium houses. The energy consumption for cooling was found to be the most significant factor of the increase in operational energy from simple to luxurious houses. Further, in the life cycle energy, the operational energy accounted for much larger proportions of about 86-92% than embodied energy regardless of the house categories. The life cycle CO_2 emissions for medium and luxurious houses were larger than that of simple houses by 2 and 6 times on average. In the simple houses, cooking was the largest contributor to the CO_2 emissions (25%), while the emissions caused by cooling increased largely with the house category and became the largest contributors in the medium (26%) and luxurious houses (41%).

1. Introduction

The ultimate purpose of this study is to propose low energy and low carbon residential buildings in major cities of Indonesia. Over the last few decades, Indonesia has been experiencing rapid urbanization and population growth. The total population increased from 97.1 million in 1970 to 237.6 million in 2010 [1]. As a consequence, the needs for living areas increased faster and enormous number of residential buildings have been developed especially in major cities, such as Jakarta. Due to the tremendous urbanization seen in the major cities, the present nationwide final energy consumption in Indonesia became about 14 times larger than that of 1970s [2]. This increasing consumption of energy will result in serious energy scarcity in major cities and cause further threat to the global warming.

A building consumes various natural resources, including water, materials and energy, and releases many pollutants during its life cycle [3]. Thus, any comprehensive assessment of building energy consumption and its environmental impacts, such as life cycle assessment (LCA), is essential in order to achieve a low energy and low carbon building. Several LCA methods for buildings were developed and are commonly used in many parts of the world, particularly in developed nations [3, 4, 5]. However, there are relatively few LCA studies for buildings in developing countries to date [6,7,8,9]. This is mainly because of relatively poor data availability of building, economy and environment, which are necessary for LCA analyses.

A few LCA studies were conducted in Indonesia [6,8,9], but these studies focused only on planned houses and apartments. In Indonesia, unplanned houses are typical of residences in major cities rather than the said planned houses, as discussed in the section 2.1. Nevertheless, there are few LCA studies for these unplanned houses to date. The present authors conducted a preliminary study in Bandung, Indonesia in 2011 to investigate life cycle energy and CO_2 emissions for unplanned houses [7].

This study provides the results of life cycle assessment of energy and CO_2 emissions for residential buildings comprising both planned and unplanned houses in Jakarta, Indonesia. A survey was conducted in the city from September to October in 2012 to obtain both material inventory and household energy consumption data in these buildings. This paper analyzes the life cycle energy and CO_2 emissions among different categories of houses through an input-output (I-O) analysis-based method.

2. Methodology

2.1. Case study houses

Jakarta, the capital city of Indonesia, was selected as a case study city. Its population increased from 4.4 million in 1970 to 9.6 million in 2010 [10]. The city has a hot and humid climate and the average rainfall intensity throughout the year ranges from 117.2-388.8 mm per month with the daily average outdoor relative humidity of 70-81%. The monthly average temperature varies at 27.1-28.9 ºC throughout the year [11].

Residential buildings (landed houses) in Jakarta can be grouped into two types, namely planned and unplanned houses. Originally, the unplanned houses, called '*Kampungs*'. These dwellings settled in unplanned and overcrowded urban villages. Moreover, some of them were squatters on public land and located in strategic parts of the city without being provided with basic urban infrastructure and services properly [12]. Moreover, these planned and unplanned houses can be further classified into three house categories based on its construction cost and lot size, namely simple, medium, and luxurious houses (figure 1). These houses have an approximate lifespan of 20, 35, and 50 years on average, respectively [13].

Figure 1. Views of sample residential buildings. (a) Simple house; (b) Medium house; (c) Luxurious house.

A total of 300 residential buildings were selected in the survey (table 1). As shown, the average household size was about 4-5 persons with a small variation between the three categories. The monthly average household income was also investigated by a multiple-choice question. As expected, the average income increases with house category from simple to luxurious houses. As shown, the total floor area also increases with house category. The largest percentage of total floor area was less than 50 m^2 (71%) for simple houses, 50 to 99 m^2 (51%) for medium houses and 100 to 300 m^2 (83%) for luxurious houses. The major building materials used were found to be almost the same among the above categories, though slight differences could be seen in terms of materials for floor and roof.

Table 1. Brief profile of sample houses.

	Simple houses	Medium houses	Luxurious houses	Whole sample
Sample size (Unplanned/ Planned)	125 (125/0)	115 (75/40)	60 (30/30)	300 (230/70)
Age (%)				
<40	32.0	17.4	10.0	22.0
40-49	32.0	37.4	30.0	33.7
50-60	23.2	30.4	50.0	32.3
>60	12.8	14.8	10.0	12.0
Household size (persons)	4.3	4.5	5.4	4.6
Monthly income (%)				
< 100 (USD)	4.8	1.7	1.6	4.0
100-400	76.8	59.1	20.0	58.7
500-1000	16.8	31.3	38.3	28.0
>1000	1.6	7.9	40.1	9.3
Total floor area (%)				
<50 (m^2)	71.2	9.6	0.0	57.7
50 - 99	20.0	51.3	0.0	15.7
100 - 300	8.8	36.5	83.3	23.3
> 300	0.0	2.6	16.7	3.3

2.2. Life cycle assessment (LCA)

LCA generally involves six phases, namely design, material production, construction, operation, maintenance, and demolition phases. However, design, construction and demolition phases were not considered in this paper. This was due to very limited possibilities to consume energy in the above three phases since most of the residential buildings in Jakarta are constructed and demolished by manual labour. Thus, the energy consumption and materials used during the above phases are considered negligible.

The design records such as building drawing are required for the analysis of embodied energy of building materials. These data can normally be obtained from the local authorities or developers, etc. Nevertheless, in the case of Jakarta, these data were available for planned houses and most of the unplanned luxurious houses only. The other houses including most of the simple and unplanned medium houses were not constructed in the formal way in practice (they are normally constructed by non-professional neighbors), and therefore the said design records could not be obtained. Thus, the actual on-site measurements by using laser distance meters and tape measures were conducted for simple and unplanned medium houses instead in order to acquire the data (figure 2a).

Meanwhile, the detailed household energy consumption data are necessary for the analysis of operational energy. Since the household energy consumption data were not available in Jakarta, the detailed interviews and measurement of appliance capacity by using watt checkers (MWC01, OSAKI) were conducted in order to obtain the data along with the above survey on building material inventory (figure 2b). The material inventory data for refurbishment were also obtained during the same interviews.

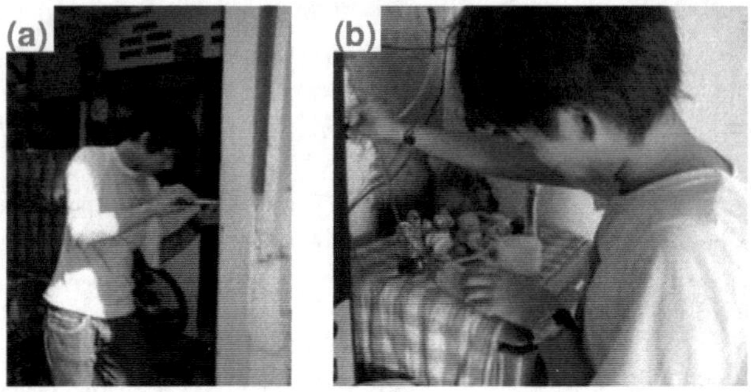

Figure 2. On-site measurements; (a) Building material survey; (b) Household energy consumption survey.

Previous studies showed that there are three main methods commonly used for analysis of energy and environmental impacts, namely process-based, economic input-output (I-O) analysis-based, and hybrid-based methods [14]. Although it is impossible to trace all the processes unlike the process-based or the hybrid-based methods, this study adopted the I-O analysis-based method to calculate the embodied energy of households and estimate their CO_2 emissions, which consistently followed the method described by Nansai et al. [15]. The latest Indonesian nationwide I-O table published in 2005 [16] consisting of 175 by 175 sectors was used for calculating the embodied energy and CO_2 emissions.

Energy used in a building is divided into two categories. The first is embodied energy for raw material extraction, material production, transportation and building construction, and maintenance. The second is energy consumption during the operation phase. The total energy of the above two categories are expressed as life cycle energy in this paper, which is given by the following equation.

$$LCE = EE_i + EE_{rec} + (OE \text{ x } lifespan) \quad (1)$$

Where LCE is the life cycle energy (GJ), EE_i is the initial embodied energy of materials (GJ), EE_{rec} is the recurrent embodied energy of materials (maintenance) (GJ), OE is the annual operational energy (GJ/year) and *lifespan* is the time period of a building's life (year).

The choice of energy resource is also important as type of fuel is crucial for the CO_2 emissions. This means that minimizing the final or purchased energy (secondary energy) does not automatically minimize the use of natural resource or the life cycle CO_2 emissions of a building. Thus, energy should be measured in the form of primary energy especially for a life cycle CO_2 assessment [17]. For the above reason, both embodied and operational energy were measured in the form of primary energy in this paper.

3. Results and discussion

3.1. Embodied energy

Figure 3 shows the average volume of each material used in respective houses, including new, reused, recycled, and maintenance materials. As shown, the sand accounted for the largest proportions in both simple and luxurious houses (28-33%), followed by stone (26-32%), clay brick (18-19%), wood (7-12%), and cement (7-9%), etc. In the case of medium houses, the stone was the largest contributor (33%), followed by sand (31%), clay brick (14%), etc. The steel contributed for about 0.2 to 0.4%. The reused and recycled materials accounted for a small amount in total (6%) for simple and medium houses. The maintenance materials were found to be 4% for medium houses and 10% for luxurious houses, respectively.

The total embodied energy was obtained by combining initial and maintenance embodied energy for respective houses (figure 4a). A shown, the average embodied energy increases with the house category from simple to luxurious houses sharply. For instance, the average value of luxurious houses

(559.5 GJ) was about 3 times larger than that of medium houses (201.0 GJ) and about 10 times than that of simple houses (58.5 GJ). This is mainly due to the differences of average building sizes among house categories (see table 1). The similar pattern was occurred for the average embodied energy per total floor area and per person. As shown in figure 4b, the embodied energy per floor area increases with house category from simple to luxurious houses. Since the total floor area was found to be the most significant explanatory variable to the embodied energy in three house categories as described before, the above per-floor area unit values can be used in predicting the embodied energy of each house category. On the other hand, as shown in figure 4c, the proportion of average embodied energy per household size showed the similar patterns with that of the original embodied energy (see figure. 4). This is mainly due to the small variation of household size among different house categories (see table 1).

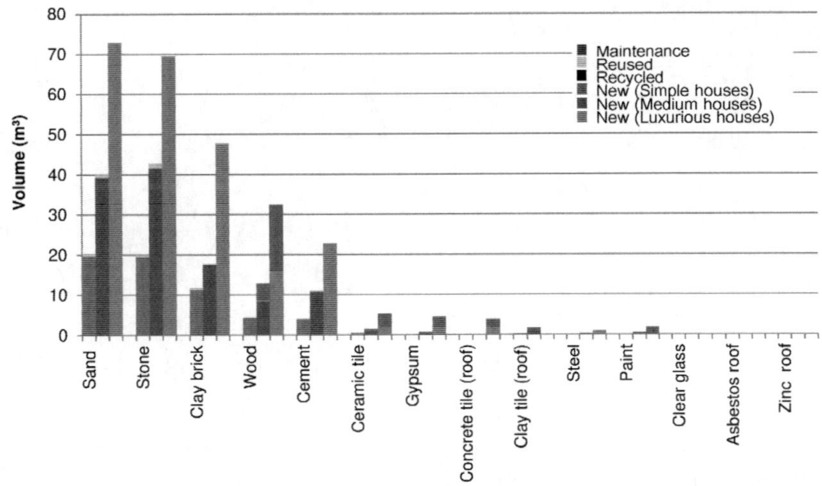

Figure 3. Proportion of building material volume for respective house categories.

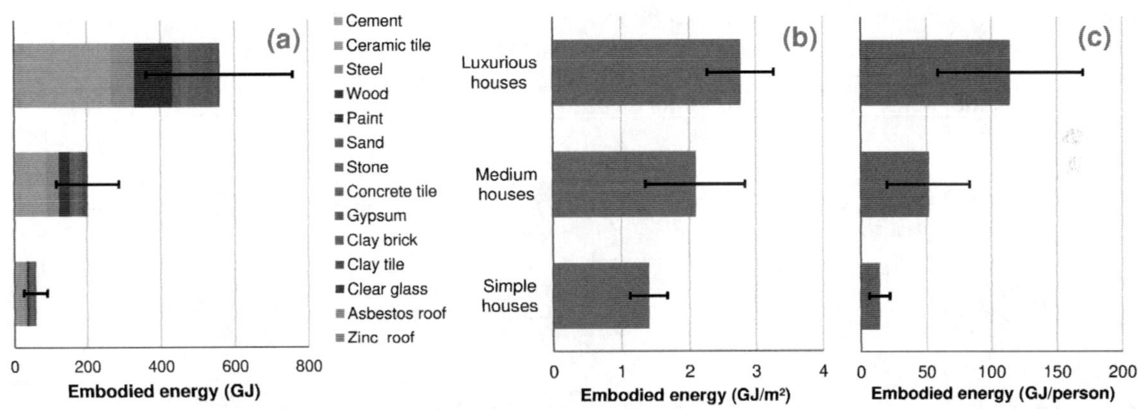

Figure 4. Average embodied energy for respective house categories. (a) Total embodied energy; (b) Per floor area; (c) Per person.

3.2. Operational energy

As describe before, the ownership level and usage time of every appliance were investigated, while gas (LPG) and kerosene consumption were also surveyed. Figure 5 shows the average ownership levels of respective appliances in three house categories. As shown, for instance, in terms of the lighting appliances, the Compact Fluorescent Lamp (CFL) was well penetrated through the households and its ownership level gave more than 90% in all the house categories. Major appliances such as rice cooker, refrigerator, television and electric iron recorded high ownership levels of about

80-100% similarly in the three house categories. In contrast, several appliances showed different ownership levels among them. This includes the blender, water dispenser, air-conditioner, laptop PC, and hair blower. In terms of the above appliances, the ownership level increases with house category from simple to luxurious houses, respectively. For instance, the ownership level of air-conditioners was found to be 6%, 30% and 90% for simple, medium and luxurious houses.
The annual average energy consumption including electricity, gas and kerosene was then calculated in

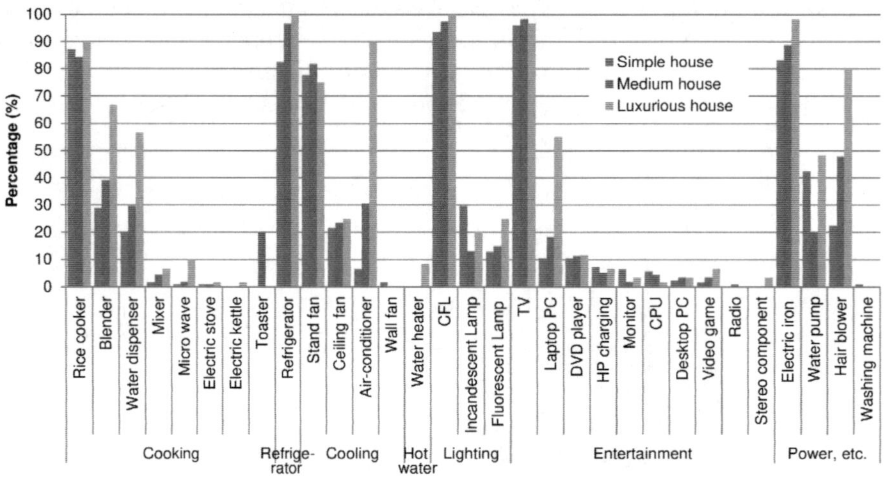

Figure 5. Ownership level of appliances.

primary energy terms (figure 6a). As shown, the annual average energy consumption recorded about 32.5 GJ for simple houses, 42.3 GJ for medium houses and 76.5 GJ for luxurious houses, respectively. The energy consumption for cooking accounted for the largest percentages in the simple (36%) and medium houses (29%), while that for cooling was the largest contributor in the luxurious houses (40%). (see figure. 6a). The annual energy consumption per person increases with house category from simple to luxurious houses sharply (figure. 6c). This is mainly due to the difference of ownership and usage levels of cooling appliances, especially air-conditioners, as described before. In contrast, the unit energy consumption per floor area decreases with house category from simple to luxurious. This is because the number of occupants per floor area is largely different among three house categories.

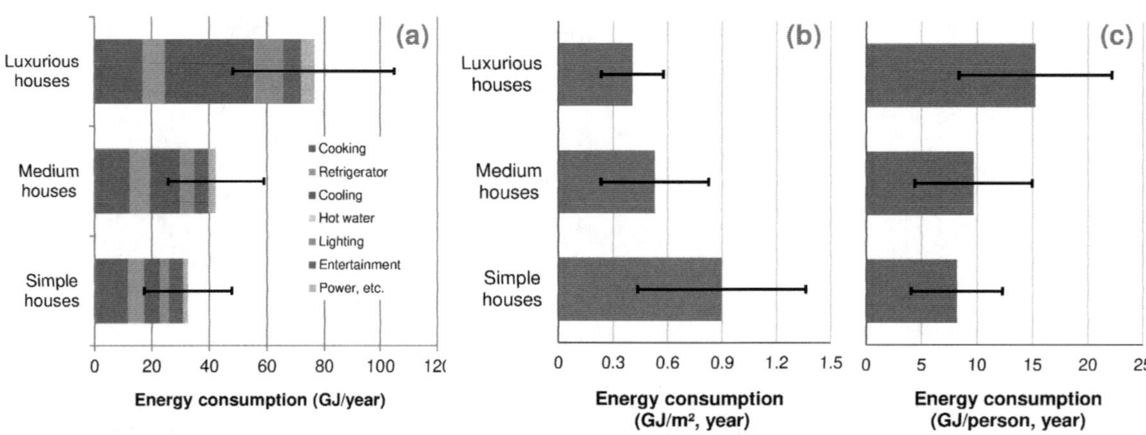

Figure 6. Annual average energy consumption for respective house categories.
(a) Annual average energy consumption; (b) Per floor area; (c) Per person.

We assumed that annual energy consumption patterns would continue uniformly for the whole life-span of a building in this study. Thus, the total operational energy during the whole life-span was simply estimated through multiplying the said annual average energy consumption by its life span for respective house categories. The percentages of electricity gave higher values (83-87%) than those of gas (13-16%) particularly in the medium and luxurious houses. The average operational energy was measured at 650.4 GJ for simple houses, 1479.5 GJ for medium houses and 3826.8 GJ for luxurious houses, respectively.

3.3. Life cycle energy and CO_2 emissions

The life cycle energy was obtained by combining embodied energy and operational energy for respective houses (figure 7a). As shown in figure 7a, the operational energy accounted for much larger proportions of about 86-92% than embodied energy regardless of the house categories. The total life cycle energy was measured at 708.9, 1680.4 and 4386.3 GJ for simple, medium and luxurious houses, respectively. As shown in figure 7b, the CO_2 emissions during operation phase were larger than the embodied CO_2 emissions by 12 to 19 times in three house categories. The estimated total life cycle CO_2 emissions were 117.0, 280.2 and 743.0 tons CO_2-eq for simple, medium, and luxurious houses, respectively. The life cycle CO_2 emissions for medium and luxurious houses were larger than that of simple houses by 2 and 6 times on average.

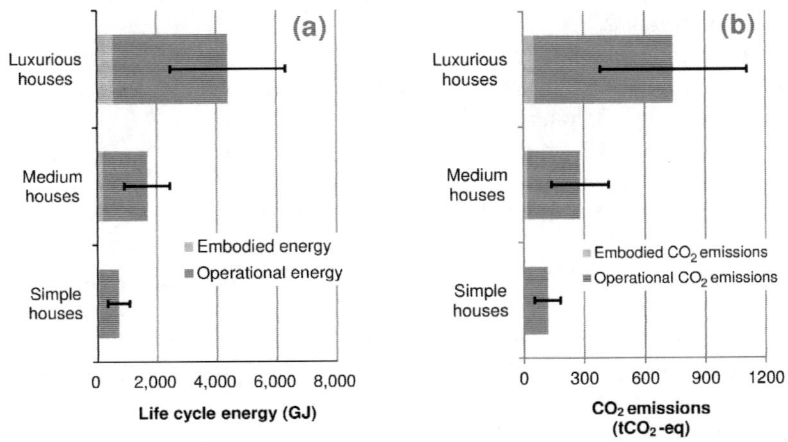

Figure 11. Average life cycle energy and CO_2 emissions for respective house categories. (a) Average life cycle energy; (b) Average life cycle CO_2 emissions

4. Conclusion

This paper provided the detailed profiles of life cycle energy and CO_2 emissions in urban houses of Jakarta.

(1) The results showed that the average embodied energy for the simple, medium and luxurious houses was found to be 58.5, 201.0 and 559.5 GJ, respectively. It was found that the total embodied energy of each house can be explained by its total floor area alone with high accuracy in respective house categories. The average embodied energy per total floor area was 1.4 GJ/m^2 for simple houses, 2.1 GJ/m^2 for medium houses and 2.8 GJ/m^2 for luxurious houses, respectively.

(2) It was seen that operational energy usage patterns varied largely among house categories as well as households especially in the simple and medium houses. The annual average operational energy for simple, medium and luxurious houses was measured at 32.5, 42.3 and 76.5 GJ/year, respectively. The energy consumption for cooling was found to be the most significant factor of the increase in operational energy from simple to luxurious houses.

(3) In the life cycle energy, the operational energy accounted for much larger proportions of about 86-92% than embodied energy regardless of the house categories. The total life cycle energy was found to be 709.0, 1680.4 and 4386.3 GJ for simple, medium and luxurious houses, respectively.

(5) The number of air-conditioners in residential buildings is expected to rise further in the near future

as household income increases further. It is important to find out the means to reduce usage of air-conditioning particularly for emerging middle-class houses in order to achieve low energy and low carbon residential buildings in the hot-humid climate of Jakarta.

Acknowledgements
This research was supported by a JSPS Grant-in-Aid for Young Scientist (B) (No. 23760551). We also would like to thank Mr. Yohei Ito and students of UPI and ITB who kindly supported our survey.

References

[1] Indonesia 2010 Population Census of Indonesia. Statistical Centre of Indonesia, Jakarta.
[2] Indonesia 2011 Handbook of Energy and Economic Statistics of Indonesia. Ministry of Energy and Mineral Resources of Indonesia, Jakarta.
[3] Crawford R H 2011. *Life Cycle Assessment in the Built Environment*. (New York: Spon Press).
[4] Monahan J, Powel J C, 2011 An embodied carbon and energy analysis of modern methods of construction in housing: A case study using a life cycle assessment framework. *Energy and Buildings.* 43, p. 179–188.
[5] Mithraratne N, Vale B, 2004 Life cycle analysis for New Zealand houses, *Building and Environment.* 39, p. 483-492.
[6] Kurdi Z, 2006 Determining Factors of CO_2 Emission in Housing and Settlement of Urban Area in Indonesia. Ministry of Public Work of Indonesia and National Institute for Land, Infrastructure and Management of Japan, Bandung.
[7] Surahman U, Kubota T, 2013 Life cycle energy and CO_2 emissions of residential buildings in Bandung, Indonesia. *Advanced Materials Research*. 689, p. 54-59.
[8] Utama A, Gheewala S H, 2009 Indonesian high rise buildings: A life cycle energy assessment, *Energy and Buildings*. 41, p. 1263–1268.
[9] Utama A, Gheewala S H, 2008 Life cycle energy of single landed houses in Indonesia. Energy and Buildings. 40, p. 1911–1916.
[10] Jakarta 2010a Jakarta in figures. Statistical Centre of Jakarta, Jakarta.
[11] Jakarta 2010b Climatology Data in Jakarta from 2001-2010. Climatology and Meteorology Station of Kemayoran, Jakarta.
[12] World Bank 1995 Indonesia Impact Evaluation Report, Enhancing the Quality of Life in Urban Indonesia: The Legacy of Kampung Improvement Program. Report No. 14747-IND. Operation Evaluation Department, the World Bank, Washington, D.C., United States of America.
[13] Indonesia 2002 Planning Procedures for Developing Earthquake Resistant on Buildings. Department of Public Work (PU), Indonesia.
[14] Dixit M K, Fernandez J L, Lavy S, Culp C H, 2010 Identification of parameters for embodied energy measurement: A literature review. *Energy and Buildings*. 42, p. 1238-1247.
[15] Nansai K, Moriguchi Y, Tohno S, 2002 Embodied Energy and Emission Intensity Data for Japan Using Input-Output Table (3EID) -Inventory Data for LCA-, National Institute for Environmental Studies, Japan.
[16] Indonesia 2005 Input-Output Table of Indonesia. Statistic Centre of Indonesia, Jakarta.
[17] Gustavsson L, Joelsson A, 2010 Life cycle primary energy analysis of residential building. *Energy and Buildings*, 42, p. 210-220.

Numerical simulation of marine currents in the Bunaken Strait, North Sulawesi, Indonesia

P T D Rompas* and J D I Manongko
Faculty of Engineering, Universitas Negeri Manado, Tondano, North Sulawesi, 95618, Indonesia

*Corresponding author: parabelem_rompas@yahoo.com

Abstract. This study intended for the generation of hydroelectric power at suitable area of the strait in order to provide electric current to a close environment. The project uses a three-dimensional model of taking flow into account the variation of hydrostatic pressure in the liquid vertical layers. We brought back to a two-dimensional calculation using the shallow water equations. The objectives of the study are getting simultaneous obtaining the velocities of currents by the component of velocities and distributions of the kinetic energy from the numerical results. The Bunaken strait is 5280 m width for an average depth of 130 m. Numerical calculation is simulated using horizontal meshes of 60 side meters. The numerical solutions obtained by using a time step of one second. It found that there was no great difference between 2D and 3D numerical simulations because the effect of flow velocity in the vertical direction is very small. The numerical results have shown that the average current velocities when low and high tide currents are 1.46 m/s and 0.85 m/s respectively. The kinetic energy ranged from 0.01 to 2.54 kW/m^2 when low and high tide in the Bunaken strait area at discharge of 1 Sv, whereas at discharge 2 Sv, 0.11-17.40 kW/m^2 and 0.11-2.77 kW/m^2 (when low and high tide currents). These results can used in the design of turbines for power generation marine currents in the Bunaken strait at depths below 60 meters.

1. Introduction

Construction of electrical power plants in Indonesia is urgently needed to overcome the shortage of electrical energy to date. One of the components for the construction of the power plant is marine current turbine [8, 9]. Marine current turbine designs requiring variable of current velocity which is directly proportional to the kinetic energy. Many ways to get the data of current velocity, one way is by numerical model [2]. A numerical model of marine currents in Bunaken strait used a semi-implicit finite difference method for the numerical solution of three-dimensional shallow water flows. Several numerical methods with solution of shallow water equations used in practical applications [1, 2, 3, 5]. Several existing numerical model for two and three dimensional shallow water flow simulations based on an alternating direction implicit ADI method. In semi-implicit method, only the *barotropic* pressure gradient in the momentum equations and the velocity divergence in the continuity equation taken implicitly. Each time step a linear five-diagonal system solved in new the water surface elevations for the entire domain are the unknowns. The model is generally explicit with the exception that the vertical eddy viscosity terms discretized implicitly. In the model formulation, the governing system of equations split into an external and an internal mode [2]. Momentum exchanges between vertical layers expressed in a set of tri-diagonal matrix equations relating the discrete horizontal velocities in each vertical level to the gradient of the water surface elevations [5].

This paper is more majoring to study the velocities of current and know the availability of kinetic energy in the Bunaken Strait. In addition, intended for the installation of marine current turbine in the place more adapted strait in order to provide electrical current to the close environment.

The objectives of the study are getting simultaneous obtaining the velocities of currents by the component of velocities and distributions of the kinetic energy from the numerical results.

2. Experimental Setup

2.1 Mathematical Model

The model of three-dimensional equations that developed from the Navier-Stokes equations after turbulent averaging and under assumption that the pressure is hydrostatic [2, 3], the equations as follow:

$$\frac{\partial u}{\partial t}+u\frac{\partial u}{\partial x}+v\frac{\partial u}{\partial y}+w\frac{\partial u}{\partial z}=-g\frac{\partial \eta}{\partial x}+\mu\left(\frac{\partial^2 u}{\partial x^2}+\frac{\partial^2 u}{\partial y^2}\right)+\frac{\partial}{\partial z}\left(\upsilon_t\frac{\partial u}{\partial z}\right)+f.v \quad (1)$$

$$\frac{\partial v}{\partial t}+u\frac{\partial v}{\partial x}+v\frac{\partial v}{\partial y}+w\frac{\partial v}{\partial z}=-g\frac{\partial \eta}{\partial y}+\mu\left(\frac{\partial^2 v}{\partial x^2}+\frac{\partial^2 v}{\partial y^2}\right)+\frac{\partial}{\partial z}\left(\upsilon_t\frac{\partial v}{\partial z}\right)+f.u \quad (2)$$

$$\frac{\partial u}{\partial x}+\frac{\partial v}{\partial y}+\frac{\partial w}{\partial z}=0 \quad (3)$$

where $u(x,y,z,t)$, $v(x,y,z,t)$, and $w(x,y,,z,t)$ are the velocity components in the directions represent x, y, and z respectively,
$\eta(x,y,t)$ is the free surface, t is the time, μ and υ_t are the eddy viscosity coefficients of horizontal and vertical respectively,
f is the Coriolis parameter, assumed to be constant,
And g is the constant gravitational acceleration.

The current computing power does allow taken into the account direct one by using the Reynolds Average Navier-Stokes Equations (RANS) [1].

2.1.1. Turbulence model.

A formula for turbulent viscosity is the standard form as defined:

$$\upsilon_t=\sqrt{\left(l_h^4\left[2\left(\frac{\partial u}{\partial x}\right)^2+2\left(\frac{\partial v}{\partial y}\right)^2+\left(\frac{\partial v}{\partial x}+\frac{\partial u}{\partial y}\right)^2\right]+l_v^4\left[\left(\frac{\partial u}{\partial z}\right)^2+\left(\frac{\partial v}{\partial z}\right)^2\right]\right)} \quad (4)$$

Where: $l_v = \kappa(z-z_b)$, for $(z-z_b)/h < \lambda/\kappa$;
$l_v = \lambda h$, for $\lambda/\kappa < (z-z_b)/h < 1$, for $l_h = \beta\ l_v$, for the horizontal length scale is larger, κ is the von Karman's constant ($\kappa = 0.41$), λ is a constant ($\lambda = 0.09$); $(z-z_b)$ is the distance from the wall, h is the boundary layer thickness assumed to be equal to the water depth,
l_v and l_h are the vertical and horizontal length scales,
In addition, the constant β has to be determined from comparison with experiment.

2.1.2. Boundary condition.

Some types of boundary conditions are required as the boundary conditions at the free surface are specified by the prescribed wind stresses of directions x and y, and a slip boundary $\partial u/\partial z=\partial v/\partial z=0$. At the bottom stress can be related to the turbulent law of the wall, a drag coefficient associated with using a Chezy formula [2]. Velocity on a solid wall is a no-slip condition, and on the open boundary, we used principally two condition, the first is Neumann method and the second is a condition radiation which derived from Orlanski's algorithm that developed by Treguier *et al.* [4].

IOP Conf. Series: Materials Science and Engineering **128** (2016) 012003 doi:10.1088/1757-899X/128/1/012003

The kinetic energy in the Bunaken strait (P_A), we can be obtaining from equation [6]:

$$P_A = \frac{P}{A} = \frac{1}{2}\rho\, v^3 10^{-3} \tag{5}$$

Where P_A in kW/m^2, v is the velocity resultant of marine current (m/s) and ρ is seawater density (kg/m^3).

2.2 Numerical Model

Semi-implicit finite difference method for the numerical solution of the three-dimensional Equation 1 and Equation 2 used by Casulli & Cheng [2], and Stansby [3] in the computation of shallow water flows. The free surface flow equations can derive in which the gradient of surface elevation in the momentum equations and the velocity can discretized implicitly. The convective, Coriolis and horizontal viscosity terms in the momentum equations discretized explicitly, but in order to eliminate a stability condition due to the vertical eddy viscosity, the vertical mixing terms discretized implicitly.

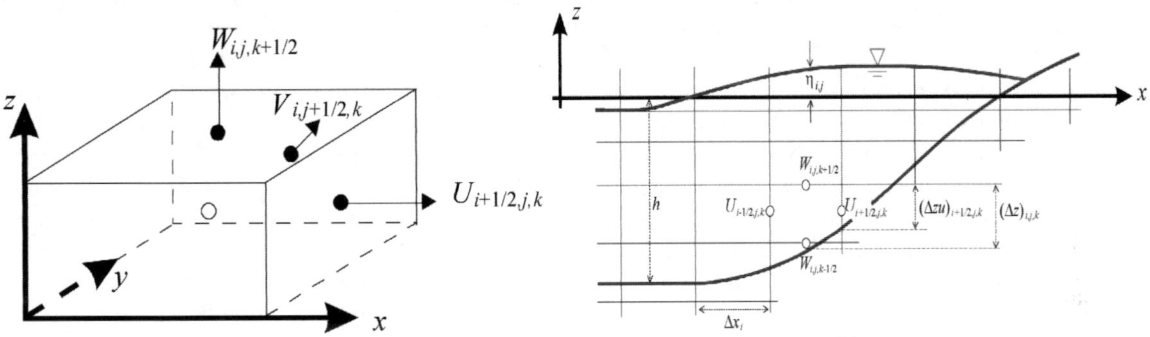

(a) (b)

Figure 1. Schematic diagram of computational mesh and notations

Figure 1 shown that a spatial mesh which consists of rectangular cells of length Δx, width Δy and height Δz_k is introduced. Each cell is numbered at its centre with indices *i, j* and *k*. Figure 1 (a) show that the discrete *u*-velocity is then defined at half-integer *i, j* and *k*; *v* is defined at integers *i, k,* and half-integer *j*; *w* is defined at integers *i, j,* and half-integer *k*. Then figure 1 (b) show that η is defined at integers *i* and *j*. The water depth *h(x,y)* is specified at the *u* and *v* horizontal points. So that a general semi-implicit discretization of the momentum equations in Equation 1 and Equation 2 can be written in the more compact matrix form as

$$\mathbf{A}^n_{i+1/2,j}\mathbf{U}^{n+1}_{i+1/2,j} = \mathbf{G}^n_{i+1/2,j} - g\frac{\Delta t}{\Delta x}\left(\eta^{n+1}_{i+1,j} - \eta^{n+1}_{i,j}\right)\mathbf{\Delta Z}^n_{i+1/2,j} \tag{6}$$

$$\mathbf{A}^n_{i,j+1/2}\mathbf{V}^{n+1}_{i,j+1/2} = \mathbf{G}^n_{i,j+1/2} - g\frac{\Delta t}{\Delta y}\left(\eta^{n+1}_{i,j+1} - \eta^{n+1}_{i,j}\right)\mathbf{\Delta Z}^n_{i,j+1/2} \tag{7}$$

Equation 6 and Equation 7 are linear tri-diagonal systems. For determine the free surface $\eta^{n+1}_{i,j}$ can be written in the matrix notation form

$$\eta^{n+1}_{i,j,} = \eta^{n+1}_{i,j,} - \frac{\Delta t}{\Delta x}\left[\left(\mathbf{\Delta Z}_{i+1/2,j}\right)^{\mathrm{T}}\mathbf{U}^{n+1}_{i+1/2,j} - \left(\mathbf{\Delta Z}_{i-1/2,j}\right)^{\mathrm{T}}\mathbf{U}^{n+1}_{i-1/2,j}\right] - \frac{\Delta t}{\Delta y}\left[\left(\mathbf{\Delta Z}_{i,j+1/2}\right)^{\mathrm{T}}\mathbf{V}^{n+1}_{i,j+1/2} - \left(\mathbf{\Delta Z}_{i,j-1/2}\right)^{\mathrm{T}}\mathbf{V}^{n+1}_{i,j-1/2}\right] \tag{8}$$

The available energy that investigated in this study is the kinetic energy (kW/m^2). The first, we will back at the equation of the kinetic energy which is equation of the marine current power in the Bunaken strait can be discretized from Equation 5 becomes:

$$P_A = \frac{P}{A} = \frac{1}{2}\rho\ (\mathrm{v}_{i,j,k}^{n+1})^3 10^{-3} \qquad (9)$$

3. Results and Discussion

3.1. The Domain Presentation of the Bunaken Strait

The Bunaken strait is located between the Pacific ocean and the Sulawesi sea (Celebes sea) whose area is approximately 200 km^2 (Figure 2), with a minimum width between Bunaken island and Sulawesi island about 5.28 km and the average depth of 130 m.

The three-dimensional current circulation in the Bunaken strait is simulated using the present model with 174 x 318 x 4 finite difference meshes of equal $\Delta x = \Delta y = 60$ m and $\Delta x = 20$ m. The numerical solutions used an integration time $\Delta t = 1$ sec and inlet volume transports (discharges) are 1 Sv to 2 Sv (1 Sv = 10^6 m^{3}s^{-1}). Figure 3 illustrates the bathymetry (a) and the meshes (b), the 3D of the Bunaken strait used for numerical simulation. The water depth distributions show the complex areas.

Currents in the Bunaken strait closely related with ocean currents and Indonesian throughflow. According to Brown *et al.* [7] that the global wind system would be if the Earth were completely covered with water, and the atmospheric circulation transports heat from low to high latitudes and the same is true in the oceans, where surface currents warmed in low latitudes carry heat polewards, while currents cooled at high latitudes flow equatorwards. Marine currents in the Bunaken strait consist of two i.e. high tide currents and low tide currents. In the morning, occurred high tide currents where input currents from from Maluku Sea (see right of Figure 2). The currents then go out to Manado bay and then they go out to Sulawesi Sea. In the evening, occurred low tides currents that are inversed on obtaining of high tide currents. The currents obtain from Sulawesi Sea and Manado bay. The currents then go inside pass in the left of Bunaken strait. Then the currents go out to the right of Bunaken strait and then go out to Maluku Sea.

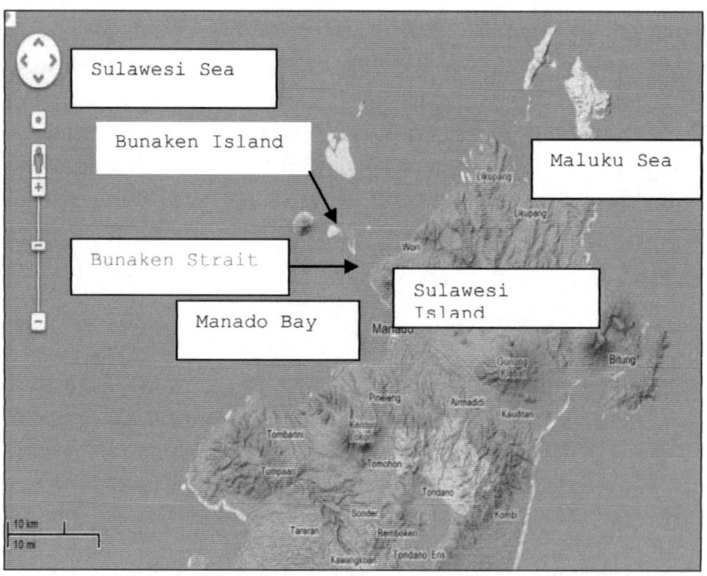

Figure 2. Location of the Bunaken strait in North Sulawesi, Indonesia.

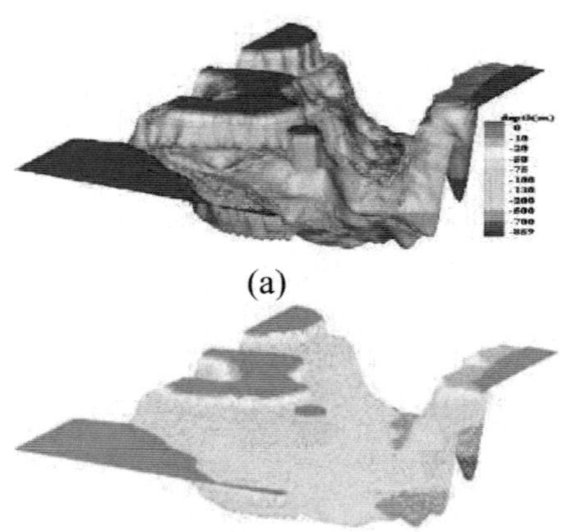

Figure 3. The 3D of the Bunaken strait with water depth: (a) Bathymetry and (b) Meshes.

In 2D simulations, we made two type of simulations with two variations of discharge. In first simulation, we have conducted when low tide current where each simulation has considerate with constant discharge inside. In second simulation, when high tide currents with condition discharge same as in first simulation. Parameter of discharge varies from 1 to 2 Sv with classifications are 1 Sv and 2 Sv.

Like in 2D simulations, 3D simulations we also have made two type of simulations with two variations of discharge. In 3D simulations, there are 4 layers of seawater column which each layer depth of 20 m (Δz), respectively. Whereas in 2D simulations only one layer of seawater column which layer depth is maximum depth.

3.2. Numerical Simulation of Marine Currents in the Bunaken Strait

The current velocity is a key factor in the design of a marine current power plant, since it sets the limits for both the power output as well as the forces acting on the turbine and support structures [8]. Figure 4 shows simulated of velocity component distributions at seawater column when low (a) and high (b) tide currents. Generally, when low tide currents which water enters from left side section and then flows go to section of right side. A small part to top side section which previous rotate form two eddies like elliptic diameter at centre. The average velocity at enter of Bunaken strait is 1.46 m/s (figure 4a). On the contrary, when high tide current (figure 4b), current enters from right side section go to section of left side and a small part to top side section which previous happened eddy is very small at center east area near Bunaken island. The average velocity at enter of Bunaken strait is 0.85 m/s.

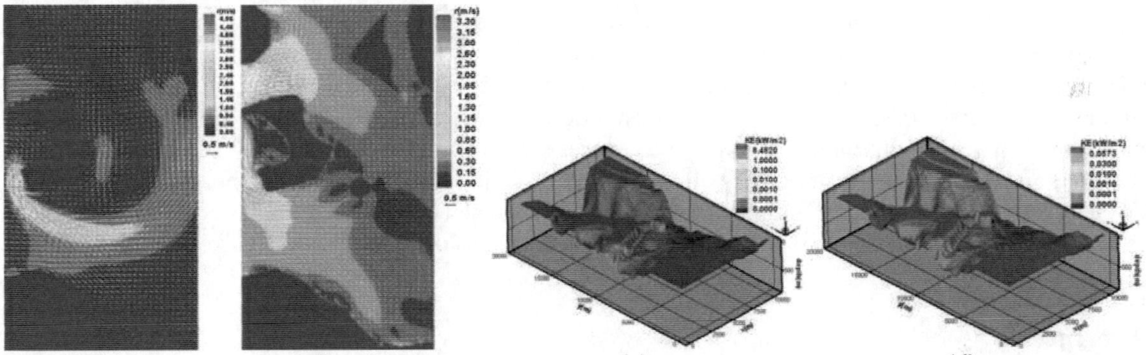

(a) (b) (c) (d)

Figure 4. Simulated of velocity component (2D) and kinetic energy (3D) distributions at seawater column when the tide currents of low (a and c) and high (b and d) at discharge of 1 Sv.

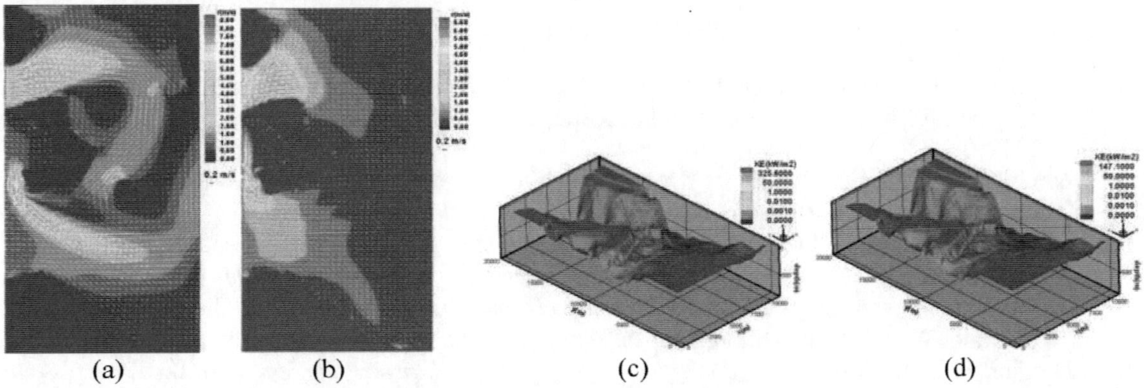

(a) (b) (c) (d)

Figure 5. Simulated of velocity component (2D) and kinetic energy (3D) distributions at seawater column when the tide currents of low (a and c) and high (b and d) at discharge of 2 Sv.

Simulated (2D) of velocity component distributions at seawater column when low (a) and high (b) tide currents shown Figure 5. Water flows when tide currents but the average velocities are different. When low tide current and high tide current at enter of Bunaken strait are 2.6 m/s and 2.0 m/s respectively.

The distributions of the kinetic energy when the tide currents of low (c) and high (d) in the Bunaken strait at discharge of 1 and 2 Sv respectively showed in figure 4 and figure 5. Discharge influence to the kinetic energy is very big where ever greater of discharge then ever greater also kinetic energy. Discharge of 1 Sv shows that there are about 0.05-2.54 kW/m^2 (when low tide currents) and 0.01-0.38 kW/m^2 (when high tide currents) kinetic energies, whereas at discharge of 2 Sv, 0.41-17.40 kW/m^2 (when low tide currents) and 0.11-2.77 kW/m^2 (when high tide currents) available in the Bunaken strait.

We can see generally that the values of current velocities and kinetic energy are bigger when low tide currents. That's because the cross sectional area of water that entered the Bunaken strait is smaller when low tide currents than when high tide currents.

4. Conclusions

Numerical simulation of marine currents in the Bunaken strait, North Sulawesi, Indonesia has presented. The values of the kinetic energy obtained by calculations could be enabling to choose a suitable place for installing the turbines adapted well for a future undersea electricity power plant in the Bunaken strait. The numerical results have shown that the values of current velocities and kinetic energy are bigger when low tide currents. At the discharge of 1 Sv, the average current velocities are 1.46 m/s (when low tide currents) and 0.85 m/s (when high tide currents), the distributions of the kinetic energy are 0.05-2.54 kW/m^2 (when low tide currents) and 0.01-0.38 kW/m^2 (when high tide currents). Whereas at discharge of 2 Sv, the average current velocities are 2.6 m/s (when low tide currents) and 2.00 m/s (when high tide currents), the distributions of the kinetic energy are 0.41-17.40 kW/m^2 (when low tide currents) and 0.11-2.77 kW/m^2 (when high tide currents). These results can used in the design of turbines for power generation marine currents in the Bunaken strait at depths below 60 meters.

References

[1] Broomans P 2003 *Numerical Accuracy in Solution of the Shallow-Water Equations* (Master thesis, TU Delft & WL, Delft Hydraulics)

[2] Casulli V and Cheng R T 1992 Semi-implicit finite difference methods for three-dimensional shallow water flow *International Journal for Numerical Methods in Fluids* **15** 629-648

[3] Stansby P K 1997 Semi-implicit finite volume shallow-water flow and solute transport solver with k-ε turbulence model *International Journal for Numerical Methods in Fluids* **25** 285-313

[4] Treguier A M, Barnier B, and De Miranda A P 2001 An eddy-permitting model of the Atlantic circulation: Evaluating open boundary condition *J. Geophy. Res. Oceans* **106** (C10) (22115-22129) 1-23

[5] Zarrati A R and Jin Y C 2004 Development of a generalized multi-layer model for 3-D simulation of free surface flows *Int. J. Numer. Meth. Fluids* **46** 1049-1067

[6] Luquet R, Bellevre D, Fréchou D, Perdon P, and Guinard P 2013 Design and model testing of an optimized ducted marine current turbine *International Journal of Marine Energy* **2** 61-80

[7] Brown E, Colling A, and Park D 2001 *Ocean circulation* (second edition, The open university, England: Walton Hall, Milton Keynes, MK7 6AA and Butterworth-Heinemann)

[8] Thomas K 2007 *Low Speed Energy Conversion from Marine Currents* (Ph.D. Thesis, Acta Universitatis Upsaliensis, Uppsala: Digital comprehensive summaries of Uppsala Dissertations from the Faculty of Science and Technology)

[9] Fraenkel P L 1999 *New Development in Tidal and Wave Power Technologies*

(Presented at Towards a renewable future: silver jubilee conference, Brighton, UK)

Acknowledgments
The authors wish to express their appreciation to Dirlitabmas, Dirjendikti, Ministry of Education and Culture of the Republic of Indonesia which had finance all of the research activities, and Rector of Manado State University, Indonesia who has proposed research grant.

IOP Conf. Series: Materials Science and Engineering **128** (2016) 012004 doi:10.1088/1757-899X/128/1/012004

The voltage profile improvement using static var compensator (SVC) in power system transmission

G M Ramdan*, Y Mulyadi and Hasbullah

Department of Electrical Engineering Education, Universitas Pendidikan Indonesia

*Corresponding author: ramdhan.gia@gmail.com

Abstract. In transmission system named 'Subsistem Bandung Selatan dan New Ujungberung' there are the voltage drop which relatively high and the voltage profile at the receiving ends below 0.95 p.u. Therefore, this research proposed a method to improve the voltage profile in the transmission system using one of Flexible Alternating Current Transmission System (FACTS) technology which is Static Var Compensator (SVC) and 'Subsistem Bandung Selatan and New Ujungberung' as the object. This research aims to get the voltage profile in 'Subsistem Bandung Selatan dan New Ujungberung' before and after connected to SVC and to set optimal location and rating of SVC to maintain the voltage profile at the system that has desire range (0.95 p.u – 1.05 p.u). To get the result in accordance with these objects, Newton –Raphson power flow solution is applied to the system. The result of Newton- Raphson power flow solution of the system shows the voltage profile before connecting to SVC are averagely 140.95 kV or 0.94 p.u while after connecting to SVC are 145.28 kV or 0.97 p.u. The SVC installation is connected to 'Bandung Utara I' as the weakest bus, and the SVC rating is -250 Mvar to 300 Mvar.

1. Introduction

In electric power system, significantly voltage change is affected by the load variation and the network topology changes. Voltage can drop considerably and even to a voltage collapse when the network is operating under heavy loading. When voltage drop highly, it will affect the phase voltage at the receiving end to become low, that case can affect performance of the equipment and possibly cause them damage [1]. Whereas when voltage collapse happens, it can operate under-voltage relay and other protection system, leading to extensive disconnection of load and thus affecting consumer loss. On the other hand, when the load level in the system is low, over-voltage can arise due to Ferranti effect. Capacitive over-compensation and over-excitation of synchronous machines can also occur [2].

Controlling voltage regulation is needed to inject or absorb reactive power to the network. Power reactive supports the buses which has voltage level outside acceptable limits. It can be improved by power reactive compensator (inject or absorb). FACTS (Flexible Alternating Current Transmission System) technology is suggested by Narian G. Hingorani from Electrical Power Research Institute (EPRI). It is one the advanced compensator for reactive power support. SVC, widely used member of FACTS family, supports to maintain voltage profile when connected at the weakest bus by injection of current [3]. Static Var Compensator (SVC) is shunt connected static generators and/or absorber whose outputs are varied so as to control specific parameter of the electric power system. The term "static" is used to indicate that SVC, unlike synchronous compensator, have no moving or rotating main component. Thus SVC consists of Static Var Generator (SVG) or absorber device and a suitable control device [1]. Identification of bus which is connected by SVC is using power flow analysis.

1.1. Newton-Raphson Power Flow Solution

Figure 1 shows a typical bus of a power system network. For large power system, the NewtonRaphson method is found to be more efficient and practical. The number of iteration required to obtain a solution is independent of the system size. Since in the power flow problem real power and voltage magnitude are specified for the voltage-controlled buses, the power flow equation is formulated in polar form. For the typical bus of the power system shown in figure 1, the current entering bus i is given by (1).

$$I_i = \sum_{j=1}^{n} Y_{ij} V_j \quad (1)$$

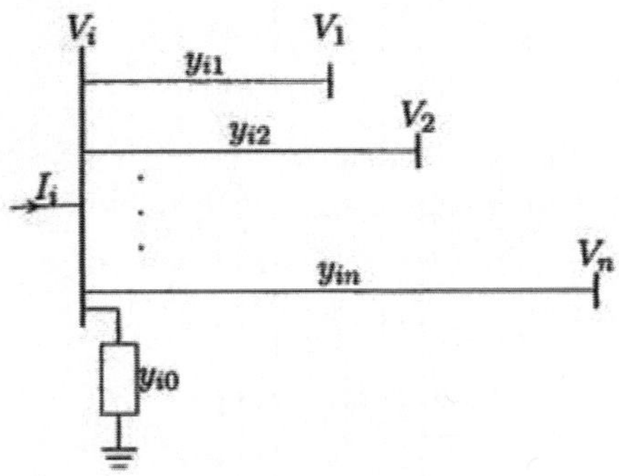

Figure 1. A typical bus of the power system.

In the above equation, j includes bus i. Expressing this equation in polar form, we have

$$I_i = \sum_{j=1}^{n} |Y_{ij}|\, |V_j| \angle \theta_{ij} + \delta_j \quad (2)$$

The complex power at bus i is

$$P_i - jQ_i = V_i^* I_i \quad (3)$$

Substituting from (2) for I_i in (3),

$$P_i - jQ_i = |V_i| \angle{-\delta_i} \sum_{j=1}^{n} |Y_{ij}|\, |V_j| \angle \theta_{ij} + \delta_j \quad (4)$$

Separating the real and imaginary parts,

$$P_i = \sum_{j=1}^{n} |V_i|\, |V_j| |Y_{ij}| \cos(\theta_{ij} - \delta_i - \delta_j) \quad (5)$$

$$Q_i = -\sum_{j=1}^{n} |V_i|\, |V_j| |Y_{ij}| \sin(\theta_{ij} - \delta_i + \delta_j) \quad (6)$$

Equation (5) and (6) constitute a set of nonlinear algebraic equation in term of the independent variables, voltage magnitude in per unit, and phase angle in radians. We have two equation for each load bus, given by (5) and (6), and one equation for each voltage-controlled bus, given by (5). Expanding (5) and (6) in Taylor's series about the initial estimate and neglecting all higher order terms produce results in the following set of linear equation.

$$\begin{bmatrix} \Delta P_2^{(k)} \\ \vdots \\ \Delta P_n^{(k)} \\ \hline \Delta Q_2^{(k)} \\ \vdots \\ \Delta Q_n^{(k)} \end{bmatrix} = \left[\begin{array}{ccc|ccc} \frac{\partial P_2}{\partial \delta_2}^{(k)} & \cdots & \frac{\partial P_2}{\partial \delta_n}^{(k)} & \frac{\partial P_2}{\partial |V_2|}^{(k)} & \cdots & \frac{\partial P_2}{\partial |V_n|}^{(k)} \\ \vdots & \ddots & \vdots & \vdots & \ddots & \vdots \\ \frac{\partial P_n}{\partial \delta_2}^{(k)} & \cdots & \frac{\partial P_n}{\partial \delta_n}^{(k)} & \frac{\partial P_n}{\partial |V_2|}^{(k)} & \cdots & \frac{\partial P_n}{\partial |V_n|}^{(k)} \\ \hline \frac{\partial Q_2}{\partial \delta_2}^{(k)} & \cdots & \frac{\partial Q_2}{\partial \delta_n}^{(k)} & \frac{\partial Q_2}{\partial |V_2|}^{(k)} & \cdots & \frac{\partial Q_2}{\partial |V_n|}^{(k)} \\ \vdots & \ddots & \vdots & \vdots & \ddots & \vdots \\ \frac{\partial Q_n}{\partial \delta_2}^{(k)} & \cdots & \frac{\partial Q_n}{\partial \delta_n}^{(k)} & \frac{\partial Q_n}{\partial |V_2|}^{(k)} & \cdots & \frac{\partial Q_n}{\partial |V_n|}^{(k)} \end{array}\right] \begin{bmatrix} \Delta \delta_2^{(k)} \\ \vdots \\ \Delta \delta_n^{(k)} \\ \hline \Delta |V_2^{(k)}| \\ \vdots \\ \Delta |V_n^{(k)}| \end{bmatrix}$$

As the equation above, bus 1 is assumed as the slack bus. The Jacobian matrix gives the linearized relationship between small changes in voltage angle $\Delta\delta_i^{(k)}$ and voltage magnitude $\Delta\,|^{(k)}|$ with the small changes in real and reactive power $\Delta P_i^{(k)}$ and $\Delta Q_i^{(k)}$. Elements of the Jacobian matrix are the partial derivatives of (5) and (6). In short form, it can be written as

$$\begin{bmatrix} \Delta P \\ \Delta Q \end{bmatrix} = \begin{bmatrix} J_1 & J_2 \\ J_3 & J_4 \end{bmatrix} \begin{bmatrix} \Delta \delta \\ \Delta |V| \end{bmatrix} \tag{7}$$

For voltage-controlled buses, the voltage magnitude are known. Therefore, if m buses of the system are voltage-controlled, m equation involving ΔQ and ΔV and the corresponding columns of the Jacobian matrix are eliminated. Accordingly, there are $n-1$ real power constrains and $n-1-m$ reactive power constraints, and the Jacobian matrix is of order $(2n-2-m) \times (2n-2-m)$. $\boldsymbol{J_1}$ is of the order $(n-1) \times (n-1)$, $\boldsymbol{J_2}$ is of the order $(n-1) \times (n-1-m)$, $\boldsymbol{J_3}$ is of the order $(n-1-m) \times (n-1)$, and $\boldsymbol{J_4}$ is of the order $(n-1-m) \times (n-1-m)$.

The diagonal and the off-diagonal elements of $\boldsymbol{J_1}$ are

$$\frac{\partial P_i}{\partial \delta_i} = \sum_{j \neq i} |V_i||V_j||Y_{ij}| \operatorname{sim}(\theta_{ij} - \delta_i + \delta_j) \tag{8}$$

$$\frac{\partial P_i}{\partial \delta_i} = -|V_i||V_j||Y_{ij}| \sin(\theta_{ij} - \delta_i + \delta_j) \qquad j \neq i \tag{9}$$

The diagonal and the off-diagonal elements of $\boldsymbol{J_2}$ are

$$\frac{\partial P_i}{\partial |V_i|} = 2|V_i||Y_{ii}| \cos\theta_{ii} + \sum_{i \neq i} |V_j|\,|Y_{ij}| \cos(\theta_{ij} - \delta_i + \delta_j) \tag{10}$$

$$\frac{\partial P_i}{\partial |V_j|} = |V_i||Y_{ij}| \cos(\theta_{ij} - \delta_i + \delta_j) \qquad j \neq i \tag{11}$$

The diagonal and the off-diagonal elements of $\boldsymbol{J_3}$ are

$$\frac{\partial Q_i}{\partial \delta_i} = \sum_{j \neq i} |V_i|\,|V_j||Y_{ij}| \cos(\theta_{ij} - \delta_i + \delta_j) \tag{12}$$

$$\frac{\partial P_i}{\partial \delta_j} = -|V_i||V_j||Y_{ij}| \cos(\theta_{ij} - \delta_i + \delta_j) \qquad j \neq i \tag{13}$$

The diagonal and the off-diagonal elements of $\boldsymbol{J_4}$ are

$$\frac{\partial Q_i}{\partial |V_i|} = -2|V_i||Y_{ii}|\sin\theta_{ii} + \sum_{j\neq i}|V_j|\,|Y_{ij}|\sin(\theta_{ij} - \delta_i + \delta_j) \tag{14}$$

$$\frac{\partial Q_i}{\partial |V_j|} = -|V_i||Y_{ij}|\sin(\theta_{ij} - \delta_i + \delta_j) \qquad j \neq i \tag{15}$$

The terms $\Delta P_i^{(k)}$ and $\Delta Q_i^{(k)}$ are different between the schedule and calculated values, known as the power residuals, given by

$$\Delta P_i^{(k)} = P_i^{sch} - P_i^{(k)} \tag{16}$$

$$\Delta Q_i^{(k)} = Q_i^{sch} - Q_i^{(k)} \tag{17}$$

The new estimates for voltages are

$$\delta_i^{(k+1)} = \delta_i^{(k)} + \Delta\delta_i^{(k)} \tag{18}$$

$$|^{(k+1)}| = |V_i^{(k)}| + \Delta\,|V_i^{(k)}| \tag{19}$$

1.2. Flexible Alternating Current Transmission System (FACTS)
The FACTS is a concept based on power-electronic controller, which enhance the value of transmission networks by increasing the use of their capacity. As these controllers operates very fast, they enlarge the safe operating limits of a transmission system without risking stability. Needless to say, the era of FACTS is trigged by the development of new solid-state electrical switching device. Gradually, the use of the FACTS has given a development of new controllable system.

1.3. Static Var Compensator (SVC)
Static Var Compensator (SVC) is one of FACTS controllers, which can control one or more variables in a power system [7]. In its simplest form, the SVC behaves of a Thyristor Controlled Reactor (TCR) in parallel with a bank of capacitor [8], the SVC configuration is shown in figure 2. The working principle of the SVC is set on the thyristor firing angle. Thyristor firing angle would regulate reactive power output of the SVC. The magnitude voltage of the system is output to the controller which will adjust the thyristor firing angle. Therefore the SVC will compensate the reactive power according to the system requirements [9].

To analyse the SVC reactive power compensation of the power system, SVC can be modeled in several ways as follows [10]:

1. SVC firing angle model, that is the SVC modelling of SVC's equivalent reactance X_{SVC}, which constitutes a function of a changing firing angle α, it consists by the parallel combination of a Thyristor Controlled Reactor (TCR) equivalent admittance and a fixed capacitive susceptance as shown in figure 2(a). This model provides information on the SVC firing angel required to achieve a given level of compensation.
2. SVC total susceptance model. A changing susceptance B_{SVC} represents the fundamental frequency equivalent susceptance of all shunt modules making up the SVC. As shown in figure 2(b).

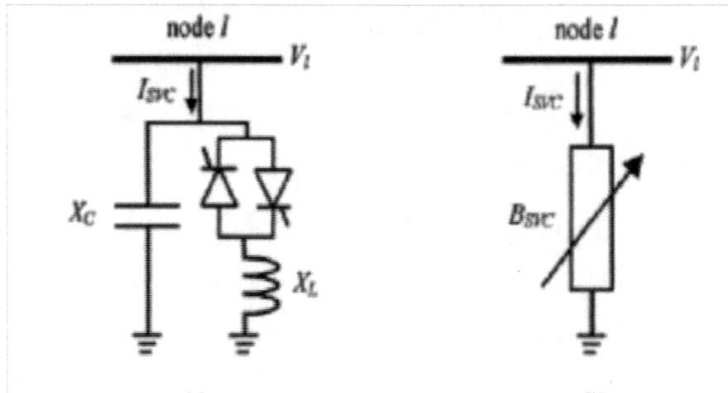

Figure 2. SVC modelling [10].

The most popular configuration for SVC is a combination of fixed capacitor (FC) and Thyristor Controlled Reactor (TCR). SVC modelling as variable sources shown in figure 2(b), we can set the maximum and minimum limits on reactive power output as following the inductive and capacitive susceptance available and voltage reference (V_{reff}). This limitation can be written by (20) and (21).

$$Q_{max} = B_{ind} \times V_{ref}{}^2 \tag{20}$$

$$Q_{min} = B_{cap} \times V_{ref}{}^2 \tag{21}$$

Where: $B_{cap} = \frac{1}{X_c}$ and $B_{ind} = \frac{1}{X_L}$

X_L = Inductive reactance

X_C = Capacitive reactance

whereas SVC can be expressed as a function of the firing angle α, shown in equation (22), as follows:

$$B_{svc} = B_{cap} - (\alpha) \tag{22}$$

Reactive power generated by SVC is calculated by an equation (22) are

$$Q_{svc} = -V_1{}^2 \times B_{svc} \tag{23}$$

with the balance of reactive power and voltage magnitude on bus k at the voltage range in accordance with the maximum and minimum limits of SVC's susceptance. Meanwhile, from figure 2(b), the current is supplied by SVC is expressed by (24)

$$I_{svc} = jB_{svc}V_1 \tag{24}$$

3. Power reactive injection model that is SVC can be used to compensate inductive reactive power or capacitive reactive power to the system. The power flow analysis, SVC can be modeled as an ideal reactive power injection on bus i as (25)

$$\Delta Q_i = Q_{svc} \tag{25}$$

2. Research Methodology

This research is located in PT. PLN (PERSERO) Area Pengaturan Beban (APB) Jawa barat. The data which is used in this research are single line diagram of 'Subsistem Bandung Selatan 150 kV dan New

Ujungberung 150 kV', power generation and loading, and line parameters. All data which is collected is the data on October 22, 2014.

To set the optimal location and rating of SVC that is connecting the SVC to the weakest bus (smallest voltage profile) to get voltage profile at each buses in the system in the acceptable range which is 0.94 p.u – 1.05 p.u [4]. After getting a suitable capacity, then the SVC is connecting to the other bus one by one, and record the voltage magnitude in the system at any different SVC position.

3. Results and Discussion

The results of Newton-Raphson power flow solution have got optimal location and rating of SVC in order to maintain the voltage magnitude in the system at range 0.95. p.u – 1.05 p.u. Optimal rating of SVC is 250 Mvar for reactive power capacitive and 300 Mvar for inductive power reactive (or -250 Mvar and 300 Mvar). Optimal location of SVC is the SVC installation at bus 'Bandung Utara I' because it shows the greatest of voltage profile increase. 'Bandung Utara I' is the bus which has weakest voltage profile before installment of the SVC. The voltage profile at the system before and after connected to SVC are shown by Table 1.

Table 1. Voltage profile increase before and after installment of the SVC.

No.	Id Bus	Rating (kV)	Operating (kV)	
			Without SVC	With SVC
1	Cibereum I	150	140.40	142.53
2	Cibereum II	150	140.40	142.53
3	Bandung Selatan I	150	144.22	146.30
4	Bandung Selatan II	150	144.22	146.30
5	Wayang Windu I	150	142.45	145.34
6	Wayang Windu II	150	142.45	145.34
7	Cigereleng I	150	141.10	143.22
8	Panasia I	150	143.91	146.00
9	Panasia II	150	143.91	146.00
10	Kiaracondong I	150	138.69	143.91
11	Kiaracondong II	150	138.69	143.91
12	Kamojang I	150	141.59	145.24
13	Kamojang II	150	141.59	145.24
14	Darajat Swasta	150	141.92	145.58
15	Darajat I	150	141.92	145.58
16	Ujungberung I	150	137.82	144.59
17	Ujungberung II	150	137.82	144.59
18	Rancakasumba I	150	140.64	143.89
19	rancakasumba II	150	140.64	143.89
20	Cikasungka I	150	140.95	144.65
21	Cikasungka II	150	140.95	144.65
22	Dagopakar I	150	136.25	146.52
23	Dagopakar II	150	136.25	146.52
24	New Ujungberung I	150	145.32	148.00
25	New Ujungberung II	150	145.32	148.00
26	Bandung Utara I	150	134.97	148.64
27	Rancaekek	150	141.42	145.62

In table 1 there are voltage profile increase at all the buses in the system after SVC is connected to the bus 'Bandung Utara I' and the voltage profile at the entire bus is at 0.95 p.u – 1.05 p.u. the voltage profile which marked with red color means that the operating voltage is below 0.95 p.u or below 142.5 kV.

4. Conclusion

150 kV transmission system region II 'Jawa Barat Subsistem Bandung Selatan and New Ujungberung' have the low voltage profile at the some buses, there are 21 buses with the voltage magnitude below 0.95 p.u. After connecting SVC to the system, it can enhance the voltage profile at the acceptable value (0.95 p.u – 1.05 p.u) with the smallest voltage profile at bus 'Cibereum I' and 'Cibereum II" are 142.53 kV, and the highest voltage profile at bus 'Bandung Utara I'. Optimal location and SVC rating in order to maintain the voltage profile at entire buses that are connecting the -250 Mvar – 300 Mvar SVC rating to bus 'Bandung Utara I'.

References

[1] Kundur, P. (1994). Power system stability and control. New York: McGraw-Hill.A reference
[2] Perez, H.A., Acha, E. & Fuerte, C.R., (2000). Advance SVC Models for Newton-Raphson Load Flow and Newton Optimal Power Flow Studies. IEEE Transaction on Power System, 15 (1), pp 129-136.
[3] Katira, M.J. & Porate, K.B. (2009). Load Flow Analysis of 132/11 kV Distribution Sub Station Using Static Var Compensator for Voltage Enhancement – A Case Study, TENCON 2009 (pp 15). 2009 IEEE Region 10 Conference.
[4] Area, G., & Sullivan, D. (2005). Design and Application of a Static VAR Compensator for Voltage Support in the Dublin, Georgia Area, 1–8.
[5] Saadat, H. (1999). Power System Analysis, New York: McGraw-Hill.
[6] Mathur, R.M. & Varma, R.K. (2002). Thyristor-based facts controllers for electrical transmission system. New York: IEEE Press & Wiley-Interscience.
[7] Biswas, M., & Das, K. K. (2011). Voltage Level Improving by Using Static VAR Compensator (SVC), 11(5).
[8] Acha, E. dkk (2004). FACTS modelling and simulation in power network. West Sussex: John Wiley & sons, LTD.
[9] Suyono, H., & Hasanah, N. (2013). Perbaikan Profil Tegangan pada Feeder Harapan Baru Lima (H5) Area Samarinda untuk Pengurangan Susut Energi, 7(2), 131–136.
[10] Anwar, S., Suyono, H. & Soekotjo, H., (2012). Optimasi Penempatan SVC untuk Memperbaiki Profil Tegangan dengan Menggunakan Algoritma Genetika. Jurnal Elektro ELTEK Vol. 3, No. 1.
[11] Liliana. & Syahputra, I., (2014). Penempatan SVC (Static Var Compensator) pada jaringan Distribusi dengan ETAP 7.5.0. Jurnal sains, Teknologi dan Industri, 12(1), pp 1-8.

Coordination Hydrothermal Interconnection Java-Bali Using Simulated Annealing

B Wicaksono, A G Abdullah and W S Saputra

Faculty of Technology and Vocational Education, UniversitasPendidikan Indonesia JalanSetiabudhi 229, Bandung 40154, Indonesia

Email: bagus.wicaksono@student.upi.edu

Abstract. Hydrothermal power plant coordination aims to minimize the total cost of operating system that is represented by fuel costand constraints during optimization. To perform the optimization, there are several methods that can be used. Simulated Annealing (SA) is a method that can be used to solve the optimization problems. This method was inspired by annealing or cooling process in the manufacture of materials composed of crystals. The basic principle of hydrothermal power plant coordination includes the use of hydro power plants to support basic load while thermal power plants were used to support the remaining load. This study used two hydro power plant units and six thermal power plant units with 25 buses by calculating transmission losses and considering power limits in each power plant unit aided by MATLAB software during the process. Hydrothermal power plant coordination using simulated annealing plants showed that a total cost of generation for 24 hours is $ 13,288,508.01.

Keywords : Coordination hydrothermal, Simulated Annealing, Transmission losses

1. Introduction

Nowadays, scheduling of hydro and thermal power plants in the most economical way is the most important thing of modern power systems to meet the needs of energy market [1]. The main goal of hydrothermal power plants optimization is to minimize the total cost of operating system that is represented by cost of fuel, hydro power plant constraints and optimization of thermal power plants during the time of optimization [2]. Scheduling of hydrothermal power plants have a basic concept, that is to maximize hydro power plants and minimize thermal power plant production, thus ensuring reliability and power supply economically.

Over the last few years, there are several methods used by researchers to solve the optimization problem, such as interactive fuzzy satisfying method, improved genetic algorithm (IGA), modified NSGA-II, differential evolution, simulated annealing (SA), particle swarm optimization technique, improved quantum-behaved particle swarm optimization, differential evolution with adaptive Cauchy mutation, bacterial foraging algorithm, mixed integer programming, and so on [3]. SA algorithm was introduced by Metropolis in 1953 which is analogized on annealing (cooling) process applied in the manufacture of materials composed of crystals. Metropolis used several variables in scheduling the refrigerant to control the search process. Simulated annealing algorithm in performing economic dispatch approach requires some parameters in order to produce solutions in the scheduling of power plant units [4] [5].

In this study, the results of optimization using simulated annealing optimization is expected to give an optimal result in the coordination of the Java-Bali 500 kV hydrothermal power plant. The result of this study is expected to be a reference for certain parties in performing the operation of electrical power generation, in order to obtain an optimal level of power generation economics.

2. Hydrothermal Coordination

Optimized scheduling of hydrothermal power plants involve the setting of hydro and thermal power plants to meet the load demand and constraints of the power generation system. These problems can be solved by knowing the discharge of hydroand thermal power plants to minimize the total cost of fuel as well as to overcome the existing constraints [6]. Hydrothermal power plant scheduling, from economic perspective, is the most important thing in modern power system to increase competition in power market [1].

2.1. Objective Function

$$F_T = \sum_{i=1}^{T}\sum_{j=1}^{N} F_j(P_S(j,t)) \tag{1}$$

Where:

F_T : Total production cost function
P_S (j,t) : Power of thermal generator of unit j in period t
F_j (P_S (j,t)) : Production cost for P_S (j,t)
N : Thermal power plant unit number
T : Period time number

Function of electrical production cost is explained as follows:

$$F_j(P_S(j,t)) = a_j + b_j P_S(j,t) + c_j P_S^2(j,t) \tag{2}$$

Where a_j, b_j, c_j are coefficients of cost of thermal power plant unit.

2.2 Hydraulic Model

Input-output characteristics of hydro-electric power plant units can be formulated using the flow of water and output power such as the modelling by Glimn-Kirchmayer:

$$q_j(t) = K\psi(h_j(t))\phi(P_{Hj}(t)) \tag{3}$$

Where:

q_j (t) : Discharge of reservoir j inperiod t
K : Comparison value
h_j (t) : Head effectiveness of j in period t
P_{Hj} (t) : Power of hydro generator of unit j in period t
$\psi\phi$: Independent function, can be found using:

$$K\psi\left(h_j(t)\right) = a_0 + a_1 h_j(t) + a_2 h_j^2(t) \tag{4}$$

$$\phi\left(P_{Hj}(t)\right) = b_0 + b_1 P_{Hj}(t) + b_2 P_{Hj}^2(t) \tag{5}$$

where a_0,a_1,a_2,b_0,b_1,b_2 are hydro coefficient models.

2.3 Constraint

The coordination of hydrothermal plants have constraints as follows:

- Load Balance

$$\sum_{i=1}^{M} P_H(i,t) + \sum_{j=1}^{N} P_S(j,t) = P_D(t) - P_L(t) = 0 \tag{6}$$

IOP Conf. Series: Materials Science and Engineering **128** (2016) 012005 doi:10.1088/1757-899X/128/1/012005

- Power plant capacity

$$P_S(j)^{min} \leq P_S(j,t) \leq P_S(j)^{max} \tag{7}$$
$$P_H(i)^{min} \leq P_H(i,t) \leq P_H(i)^{max} \tag{8}$$

3. Simulated Annealing

Simulated Annealing (SA) uses the concept of local search or neighborhood search at each iteration of SA to perform the optimal layout search with minimum cost. SA algorithm can be used to find the optimal value of the problem of combination of unit commitment optimization. The linkage of physics regarding the simulated annealing method with optimization, which can be translated into atomic energy, is analogized as objective cost function and final ground state and is stated as a minimum result of cost function. Besides, the determination of the temperature becomes the main parameters in the optimization process [8][9]. Simulated annealing approach used for global optimization aims to discover the approximate global solution. Simulated annealing is a development of Monte Carlo method in his research on states and frozen states in a system. The resulting solution in simulated annealing is obtained from initialization parameters that are randomly generated. The solution is solved by selecting one position and moving it to another location. If the new solution is better than the previous one, the solution is stored. Where as if it is not, it can be stored with probability setting by reducing the parameters of temperature, commonly known as Boltzmann probability that is expressed as $e^{-(s_n-s_b/T)}$, where s_n is the current value and sb is the previous value, T is temperature parameter whose initialization value is 100. Subsequently, the temperature parameter is reduced in the next iteration, thus possibility reduction is acceptable. The iteration process is complete when the criteria is achieved and a solution is found [10]. If depicted in the flowchart would like Figure 1.

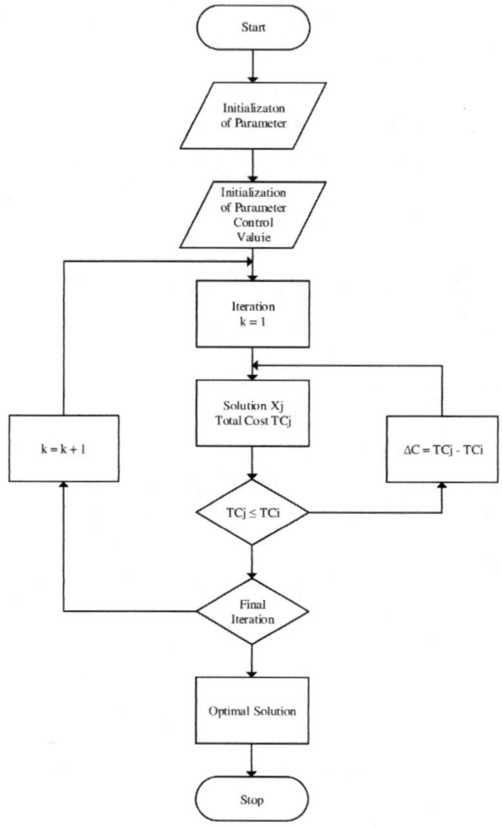

Figure 1. Flow chart Simulated Annealing

4. Simulation Results

Characteristics of hydro power plants in this study can be seen in the table below:

Table 1. Comparison of Input and Output Characteristics

No	Hydro power plant	Comparison of Input and output characteristics (m^3/hour)
1	Saguling	Q = 72077.5 + 335.68 P
2	Cirata	Q = 234703.20 + 563.83 P

While the thermal power plants have the following characteristics:

Table 2. Generation Cost Characteristic

Power plant	Generation cost characteristic ($/h)
Suralaya	$275.17 + 34.21P - 0.00149P^2$
Muaratawar	$624.93 + 178.86P - 0.0512P^2$
TanjungJati	$348.55 + 27.25P - 0.00188P^2$
Gresik	$538.37 + 52.83P - 0.00026P^2$
Paiton	$-212.50 + 151.78P - 0.00835P^2$
Grati	$-119.39 + 45.64P - 0.00606P^2$

In Java-Bali 500 kV interconnection system, there are 25 buses with transmission losses as follows:

$$B = \begin{bmatrix} 0.0157 & 0.0089 & -0.0093 & -0.0202 & -0.0241 & -0.0104 \\ 0.0089 & 0.0319 & -0.0106 & -0.0422 & -0.0481 & -0.0093 \\ -0.0093 & -0.0106 & 0.0342 & 0.0155 & 0.0178 & 0.0089 \\ -0.0202 & -0.0422 & 0.0155 & 0.1253 & 0.1037 & 0.0145 \\ -0.0241 & -0.0481 & 0.0178 & 0.1037 & 0.1544 & 0.0257 \\ -0.0104 & -0.00930 & 0.0089 & 0.0145 & 0.0257 & 0.0268 \end{bmatrix}$$

$$B0 = [0.00021 \quad 0.00039 \quad -0.000099 \quad -0.0008 \quad -0.0009 \quad -0.000039]$$

$$\boldsymbol{B00 = 0.0000032896}$$

After the characteristics of each plant were known, the calculation using simulated annealing could be performed with the aid of MATLAB software, so the results of hydrothermal plant optimization can be seen in the following table:

Table 3. Results of Hydrothermal Coordination

Time	PD	MW							
		Saguling	Cirata	Suralaya	MuaraTawar	TanjungJati	Gresik	Paiton	Grati
01.00	7783.32	105.97	0	1600	750	2630.24	900	1400	150
02.00	7368.35	105.97	0	1600	750	2630.24	900	1400	150
03.00	7229.87	105.97	0	1600	750	2630.24	900	1400	150
04.00	7068.49	118.44	0	1874	750	2700	900	1400	150
05.00	8075.23	118.44	0	1874	750	2700	900	1400	150
06.00	8309.28	118.44	0	1874	750	2700	900	1400	150
07.00	8063.53	135.17	83.67	2957.45	750	2700	900	1400	150
08.00	9009.28	135.17	83.67	2957.45	750	2700	900	1400	150
09.00	9953.89	135.17	83.67	2957.45	750	2700	900	1400	150
10.00	10193.6	255.91	62.67	3400	779.43	2700	1220.56	1726.05	152.33
11.00	10556.99	255.91	62.67	3400	779.43	2700	1220.56	1726.05	152.33
12.00	9844.09	255.91	62.67	3400	779.43	2700	1220.56	1726.05	152.33
13.00	10439.09	250.987	76	3400	1018.44	2700	1144.77	2006.67	151.02
14.00	10798.41	250.987	76	3400	1018.44	2700	1144.77	2006.67	151.02
15.00	10681.73	250.987	76	3400	1018.44	2700	1144.77	2006.67	151.02
16.00	10482.43	242.123	102	3400	1005.81	2700	1215.59	1768.69	224.21
17.00	10118.5	242.123	102	3400	1005.81	2700	1215.59	1768.69	224.21
18.00	11083.13	242.123	102	3400	1005.81	2700	1215.59	1768.69	224.21
19.00	11014.04	301.19	155.67	3400	1070.29	2700	1217.13	1904.04	251.98
20.00	11002.32	301.19	155.67	3400	1070.29	2700	1217.13	1904.04	251.98
21.00	10669.44	301.19	155.67	3400	1070.29	2700	1217.13	1904.04	251.98
22.00	9824.48	93.1733	63	3400	836.91	2700	902.854	1569.16	157.84
23.00	9644.4	93.1733	63	3400	836.91	2700	902.854	1569.16	157.84
24.00	9481.54	93.1733	63	3400	836.91	2700	902.854	1569.16	157.84

Coordination results of hydro thermal plants used simulated annealing on 9 September 2013 for 24 hours by taking into account the transmission losses, resulting in a total generation cost of $ 13,288,508.01 of 2 hydro power plant units and 6 thermal power plant units. In 2 early periods or at 1:00 to 6:00,Cirata hydro power plant unit was not operated because its water debit, if the plant was operated, would have produced output power below the minimum generation limit of Cirata unit. The most expensive operating costs occurred in the period of 7 or at 19.00 - 21.00 because this time was the peak load of the day, so all power plants were operated, with hydro power plant producing 301.19 MW of power for Saguling unit and 155.67 MW for Cirata unit. For thermal power plants in Suralaya and Tanjung Jati unit, the maximum power produced was 3400 MW and 2700 MW, while for the other units it was 1070.29 MW for Muara Tawar power plant unit, 1217.13 MW for Gresik power plant unit, 1904.04 MW for Paiton power plant unit and 157.84 for Grati power plant unit.

5. Conclusion

Concerning the load distribution of hydro and thermal power plants where hydro power plants were used to support basic load while thermal power plants were used to support the remaining load that could not be supported by hydro power plants, the optimization result using simulated annealing showed that the total operating cost obtained was $ 13,288,508.01.

References

[1] S. Cherian, “Hydro Thermal Scheduling using Particle Swarm Optimization,” in *Transmission and Distribution Conference and Exposition, 2008. T&D. IEEE/PES*, 2008, pp. 1–5.

[2] B. Monte, S. Soares, and S. Member, “Fuzzy Inference Systems Approach for Long Term Hydrothermal Scheduling,” *Power Syst. Conf. Expo. 2009. PSCE ’09. IEEE/PES*, pp. 1–7, 2009.

[3] J. Xingwen and Z. Jianzhong, “Hybrid DE-TLBO Algorithm for Solving Short Term Hydro-thermal Optimal Scheduling with Incommensurable Objectives,” *Control Conf. (CCC), 2013 32nd Chinese*, no. 1, pp. 2474–2479, 2013.

[4] Z. Chen and P. Luo, “QISA: Incorporating quantum computation into Simulated Annealing for optimization problems,” *2011 IEEE Congr. Evol. Comput. CEC 2011*, pp. 2480–2487, 2011.

[5] N. Encinas, N. Encinas, C. Alvarez, D. Alfonso, D. Alfonso, a. Perez-Navarro, C. Lvarez, F. Garcıa-Franco, and a Perez-, “Energy market segmentation for distributed energy resources implementation purposes,” *Gener. Transm. Distrib. IET*, vol. 1, no. 2, p. 324, 2007.

[6] K. Dasgupta, “Short-Term Hydrothermal Scheduling Using Particle Swarm Optimization with Constriction Factor and Inertia Weight Approach,” *Autom. Control. Energy Syst. (ACES), 2014 First Int. Conf.*, pp. 1–6, 2014.

[7] I. A. Farhat and S. Member, “Fixed-Head Hydro-Thermal Scheduling Using a Modified Bacterial Foraging Algorithm,” *Electr. Power Energy Conf. (EPEC), 2010 IEEE*, vol. 10, pp. 1 – 6, 2010.

[8] K. P. Dahal, G. M. Burt, J. R. McDonald, and S. J. Galloway, “GA/SA-based hybrid techniques for the scheduling of generator maintenance in power systems,” *Proc. 2000 Congr. Evol. Comput. CEC00 (Cat. No.00TH8512)*, vol. 1, 2000.

[9] D. Simopoulos and S. Kavatza, “Consideration of ramp rate constraints in unit commitment using simulated annealing,” *2005 IEEE Russ. Power Tech, PowerTech*, 2005.

[10] M. Bilbao and E. Alba, "Simulated annealing for optimization of wind farm annual profit," *2009 2nd Int. Symp. Logist. Ind. Informatics, LINDI 2009*, vol. 0, no. 2, pp. 1–5, 2009.

IOP Conf. Series: Materials Science and Engineering **128** (2016) 012006 doi:10.1088/1757-899X/128/1/012006

Designing a SCADA system simulator for fast breeder reactor

E Nugraha*, A G Abdullah and D L Hakim

Electrical Engineering Department
Faculty of Technology and Vocational Education, Universitas Pendidikan Indonesia
Jalan Setiabudhi 229, Bandung 40154, Indonesia

*Corresponding author: eka.nugraha@student.upi.edu

Abstract. SCADA (Supervisory Control and Data Acquisition) system simulator is a Human Machine Interface-based software that is able to visualize the process of a plant. This study describes the results of the process of designing a SCADA system simulator that aims to facilitate the operator in monitoring, controlling, handling the alarm, accessing historical data and historical trend in Nuclear Power Plant (NPP) type Fast Breeder Reactor (FBR). This research used simulation to simulate NPP type FBR Kalpakkam in India. This simulator was developed using Wonderware Intouch software 10 and is equipped with main menu, plant overview, area graphics, control display, set point display, alarm system, real-time trending, historical trending and security system. This simulator can properly simulate the principle of energy flow and energy conversion process on NPP type FBR. This SCADA system simulator can be used as training media for NPP type FBR prospective operators.

1. Introduction

Nuclear Power Plant (NPP) type Fast Breeder Reactor (FBR) uses Metal Oxide Fuel (Uranium and Plutonium) as fuel. Plutonium is placed in the center of the reactor core, and then its outer side is surrounded by U-238. This Uranium-238 absorbs neutrons coming from the fission reaction in the central part of the reactor, thus it is transformed into Pu-239. The production of Pu-239 is known as fuel breeding. Since there is no moderator in the reactor to lower the neutron energy, this reactor is called fast breeder. The coolant used in this process is liquid metal, sodium (Na), which is not moderate and resistant to extreme temperatures inside the reactor [1]–[4].

Supervisory Control and Data Acquisition (SCADA) is a communication system and control system used to monitor, control operation and maintain infrastructure on the network and can acquire data [5]. SCADA is widely used in large and medium-sized systems, so there are many types of software in the development of SCADA [6]. SCADA software functions as a link between HMI (Human Machine Interface) and MTU (Master Terminal Unit).

This study refers to previous research, about Role of Animated Human Machine Interface in Nuclear Power Plant Simulation [7] and Development Strategies of an Intelligent Human Machine Interface for Next Generation Nuclear Power Plants [8]. The Advantages of this research process simulator can visualize the flow energy nuclear power plant such as the type FBR actual conditions in the field.

Nowadays, SCADA system is inseparable from the software application that is used to view, monitor and control industrial processes. Many types of HMI software are used in the development of SCADA. With the recent developments of industrial computing systems and Human Machine Interface (HMI) software packages, engineers and researchers that develop electric power systems encountered the opportunity to produce simulator tools that can be used to process the electric power system [9].

Wonderware Intouch is a Human Machine Interface (HMI) program that meets all the needs of an integrated process, namely visualization, data access, history, event handling, alarm logging, reporting and analysis tools. Human Machine Interface is an equipment that serves to present the process data to the operator and provide input control in various formats on the monitor, including graphics, schemes, windows, screen touch pull down menu and so on. Using HMI, the operator can monitor and control the industrial processes. [10]–[12]. Wondeware intouch consists of three main constituent components, including; Intouch application manager, Intouch windowmaker, Intouch window viewer.

The Application of SCADA system is intended to simulate the dynamic real time data of Nuclear Power Plant (NPP) type Fast Breeder Reactor (FBR), and the operation is aimed for training. The increasing demand for trained power plant operators has created awareness that simulator is the most effective tool for training. An actual nuclear power plant cannot be used for such extensive training, such as start-up, shutdown, damage, incident, etc., because of security considerations and the reactor availability [7]. Therefore, a simulator that can be used as training material for power plant operators is needed.

2. Research and Methodology

This Research is a design research that produced products in the form of SCADA system simulator visualization for NPP type FBR. The designing of this simulator was conducted using Wonderware Intouch software. In order To make it easier to understand the steps involved in this study, the research procedure is shown on Figure 1.

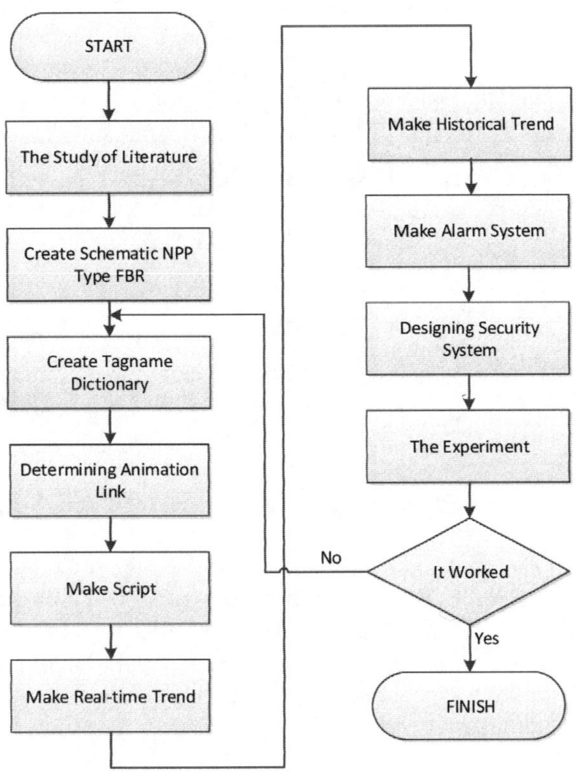

Figure 1. Flowchart Designing a SCADA System Simulator for NPP type FBR

The First step in doing this research was conducting literature study from various reliable sources, such as journals from International Institute of Electrical and Electronics Engineers (IEEE), technical reports from International Atomic Energy Agency (IAEA), technical data from Indira Gandhi Centre for Atomic Research, etc. The second step was creating schematic by inputting the components existing at NPP type FBR. The third step was creating tagname dictionary that was given to each component. The fourth step was determining the animation link on each component in order to identify the working principle of the link, so that when it is runtimed it will look alive like the real situation in the field.

After determining the animation link, the fifth step was creating a script that functioned to execute commands and logical operations based on the criteria for each component. The sixth step was creating real-time trend to display data in real time. The seventh step was creating historical trend to graphically display stored past tagname data. The eighth step was creating alarm systems that will give a warning in an unsafe process conditions on the system. The ninth step was creating security system that aims to keep the system safe from operator's working fault which is not in accordance with the setpoint system that had been made.

The next step is performing test to the system using intouch windowviewer, starting from the process of energy flow at NPP type FBR, real-time trend, historical trend alarm system and security system in order to know wheter these components work as expected or not. If there is a function that does not work, then go back to step three, which is giving tagname and, follow the next steps until the system works as expected.

3. Results and Discussion

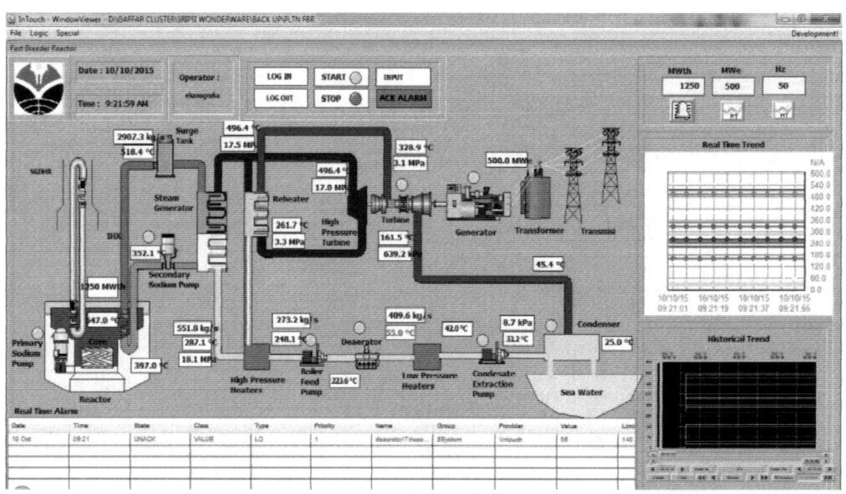

Figure 2. Visualization of Human Machine Interface SCADA System Simulator for NPP type FBR.

3.1. The Process of Electric Energy Generation in NPP type FBR

Nuclear Power Plant type Fast Breeder Reactor has a capacity of 500MWe pool type that utilizes sodium as a heat transfer medium. Nuclear Power Plant type Fast Breeder Reactor contains three main systems of heat transfer, namely primary sodium system, secondary sodium system and steam system [13]. The heat coming from the reactor core is circulated by primary sodium pump, then the heat is transferred to the steam generator through Intermediate Heat Exchangers, the hot sodium from the steam generator is recirculated to the reactor using secondary sodium pump. The steam from the steam generator is generated from the combination of hot sodium and sea water that has previously been through the heating stage, starting from condensate extraction pump, low pressure heater, deaerator, boiler feed pump and the high pressure heater. The steam generated from the steam generator is subsequently used to rotate the high pressure turbine, then the rest of the steam is heated back through the reheater to rotate the intermediate pressure and low pressure turbines which are mounted coaxially with the generator, thereby producing electrical energy. Steam from the rest of the turbine is condensed back into water in the condenser. Condensate water flows back into the steam generator through the heating process starting from the condensate extraction pump and so on.

3.2. Designing a SCADA System Simulator NPP type Fast Breeder Reactor

Design SCADA system for NPP type FBR have relatively many indicators that can facilitate the process of monitoring and controlling the generation process. Figure 2 shows the visualization design of the SCADA system simulator fo NPP type FBR.

SCADA system simulator is equipped with a security system that aims to safeguard the system security so that it cannot be run by those who do not possess access rights to the system. To be able to operate the system, the operator must LOG IN by inserting username and password via the LOG IN button. START button is used to begin running the system, and INPUT button is to display the working parameters for each component.

When the START button is pressed, the system will begin to visualize the generation process at NPP type FBR. Components that are working are marked with green light indicator. In addition, there is also a visualization of blink-blink animation on the components. Working parameters, such as temperature, pressure and mass flow rate of the component will appear or change when the flow of energy passes through the components. The generation process at NPP type FBR consists of several types of energy flow. First, the sodium in the reactor, intermediate heat exchanger, steam generator and pipe are colored red, beige and brick red. Second, the pipe water that comes from sea water is colored light blue. Third, the steam generated from the steam generator which will rotate the turbine is colored purple.

Real-time trend is a graph that displays the working parameters of the components in real-time according to the specified time interval. The maximum time range displayed in a trend is 60 seconds and the update rate is every one second. Historical trend is a graph of past (historical) data that occur at a specific time. Real-time alarm is a warning if the process condition is abnormal or beyond the parameters that have been determined. If the alarm works, there will be a red colored description on the alarm box. To normalize the real-time alarm, the operator can press the ACK ALARM button, so that the information displayed on the alarm box will turn black, indicating that the alarm that occurred has been acknowledged by the operator.

4. Conclusions

The components of Human Machine Interface (HMI) for NPP type FBR can be designed using the symbol factory in Intouch Window Maker in Wonderware Intouch software, and the results of other supporting software. The system is simulated using Intouch Window Viewer. This system can visualize the energy flow, can display energy conversion at NPP type FBR, and is equipped with manual book for each component, real-time trend, historical trend, real-time alarm and security system. Therefore, SCADA system simulator that had been designed can be used to function as instructional and training media operators prior to actual implementation in SCADA system and the process of electric energy generation at NPP type FBR.

5. Acknowledgments

The authors would like to express gratitude to Electrical Power System Research Group of Electrical Engineering Education Department, Indonesia University of Education.

References

[1] M. R. Prusty, J. Chakraborty, H. Seetha, T. Jayanthi, and K. Velusamy, "Fuzzy Logic Based Transient Identification System for Operator Guidance Using Prototype Fast Breeder Reactor Operator Training Simulator," *IEEE*, pp. 1259–1264, 2014.

[2] S. Holm, "Nuclear reactor theory," *Nuclear Physics*, vol. 10. p. 627, 1959.

[3] N. Jasmine, R. Nawlakha, H. Seetha, T. Jayanthi, S. A. V. S. Murty, and P. Swaminathan, "Simulation of control logics for plant transition state for PFBR Operator Training Simulator," in *International Conference on Recent Trends in Information Technology, ICRTIT 2011*, 2011, pp. 380–384.

[4] P. Parimalam, A. Shanmugam, A. S. Raj, N. Murali, and S. A. V. S. Murty, "Convenient and Elegant HCI features of PFBR Operator Consoles for Safe operation," *IEEE Int. Conf. Intell. Hum. Comput. Interact.*, pp. 1–6, 2012.

[5] J. Wu, Y. Cheng, and N. N. Schulz, "Overview of Real-Time Database Management System Design for Power System SCADA System," *Proc. IEEE South East Conf.*, pp. 62–66, 2006.

[6] I. Morsi, M. El Deeb, and A. El Zawawi, "SCADA/HMI development for a multi stage desalination plant," *IEEE Comput. Soc.*, pp. 67–71, 2009.

[7] S. Bindu, T. Jayanthi, and S. a. V. SatyaMurty, "Role of animated Human Machine Interface in nuclear power plant," *Int. J. Simul. Model.*, vol. 10, no. 1, pp. 5–16, 2011.

[8] S. S. Choi, J. K. Park, J. H. Hong, H. G. Kim, S. H. Chang, and K. S. Kang, "Development Strategies of an Intelligent Human-Machine Interface for Next Generation Nuclear Power Plants," *IEEE Trans. Nucl. Sci.*, vol. 43, no. 3, pp. 2096–2114, 1996.

[9] K. W. Darwish, a. R. Al Ali, and R. Dhaouadi, "Virtual SCADA simulation system for power substation," *Innov. 4th Int. Conf. Innov. Inf. Technol. IIT*, pp. 322–326, 2008.

[10] C. . Patil, H. . Sonawane, and K. . Patil, "Overview of SCADA Application in Thermal," *Int. J. Adv. Electron. Commun. Syst.*, 2014.

[11] W. C. W. Chunlei, F. L. F. Lan, and D. Y. D. Yiqi, "A Simulation Environment for SCADA Security Analysis and Assessment," *IEEE Int. Conf. Meas. Technol. Mechatronics Autom.*, vol. 1, 2010.

[12] A. M. Zaw and H. M. Tun, "Design and Implementation of SCADA System Based Power Distribution for Primary Substation (Monitoring System)," *Int. J. Sience, Eng. Technol. Res.*, vol. 3, no. 5, pp. 1542–1546, 2014.

[13] R. Gajapathy, K. Velusamy, P. Selvaraj, P. Chellapandi, S. C. Chetal, and T. Sundararajan, "Thermal hydraulic investigations of intermediate heat exchanger in a pool-type fast breeder reactor," *Nucl. Eng. Des.*, vol. 238, no. 7, pp. 1577–1591, 2008.

Short term load forecasting of anomalous load using hybrid soft computing methods

S A Rasyid*, A G Abdullah and Y Mulyadi

Faculty of Technology and Vocational Education, Universitas Pendidikan Indonesia
Jalan Setiabudhi 229, Bandung 40154, Indonesia

*Corresponding author: sopian.al@student.upi.edu

Abstract. Load forecast accuracy will have an impact on the generation cost is more economical.The use of electrical energy by consumers on holiday, show the tendency of the load patterns are not identical, it is different from the pattern of the load on a normal day. It is then defined as a anomalous load. In this paper, the method of hybrid ANN-Particle Swarm proposed to improve the accuracy of anomalous load forecasting that often occur on holidays. The proposed methodology has been used to forecast the half-hourly electricity demand for power systems in the Indonesia National Electricity Market in West Java region. Experiments were conducted by testing various of learning rate and learning data input. Performance of this methodology will be validated with real data from the national of electricity company. The result of observations show that the proposed formula is very effective to short-term load forecasting in the case of anomalous load. Hybrid ANN-Swarm Particle relatively simple and easy as a analysis tool by engineers.

1. Introduction

Short Term Load Forecasting is a type of electrical load forecasting conducted through the analysis of electrical load patterns (weekday electrical load, weekend electrical load or anomalous electrical load) within a short time series [1][2][3]. The results of short term load forecasting have a very important role in improving the performance of problem solving--the problems on power systems such as economic dispatch, unit commitment, power flow analysis, maintenance scheduling, etc [4].

The main objective of short term load forecasting is to obtain accurate prediction results. Therefore, to gain highly accurate results, various methods were developed that support the forecasting process. At the beginning of its development, the methods frequently used by the forecasters were conventional methods, such as automatic regressive moving average, exponential smoothing models, stochastic processes, linear regression models, etc [5].

Conventional methods have limitations in processing inaccurate or incomplete input data, resulting in predictions that were not adequately good. Nowadays, the methods used by the forecaster are no longer conventional since the forecasters tend to shift to artificial intelligence-based soft computing, such as fuzzy logic, expert systems, artificial neural networks and so forth [6][7].

Nowadays one of the methods often used in short term load forecasting is artificial neural network with Back Propagation algorithm. The strength of this algorithm lies on its learning process where any error obtained at each iteration will be tracked back as feedback which is then studied, so there is improvement in result accuracy (through weights and biases improvement) in the next iteration. To improve the pattern recognition accuracy resulted by Back Propagation, another algorithm which has the strength of optimization was added, that was Particle Swarm Optimization. The hybrid of these two algorithms becomes an optimal combination for the forecasting process. Based on this background, the authors raised this issue into a research entitled "Short Term Load Forecasting of Anomalous load using Hybrid Soft Computing Methods."

2. Methods

The data of electrical load output during national holidays and collective leaves was based on the data in the Indonesia National Electricity Market in West Java region. This data was plotted to determine the characteristic patterns of electrical load during those days. Analysis of characteristic patterns produced the results about the days which showed normal load pattern and the days which showed anomalous load pattern. The algorithm used was a hybrid of back propagation - swarm particle. The steps of short term anomalous load forecasting using this algorithm can be seen on the following flowchart:

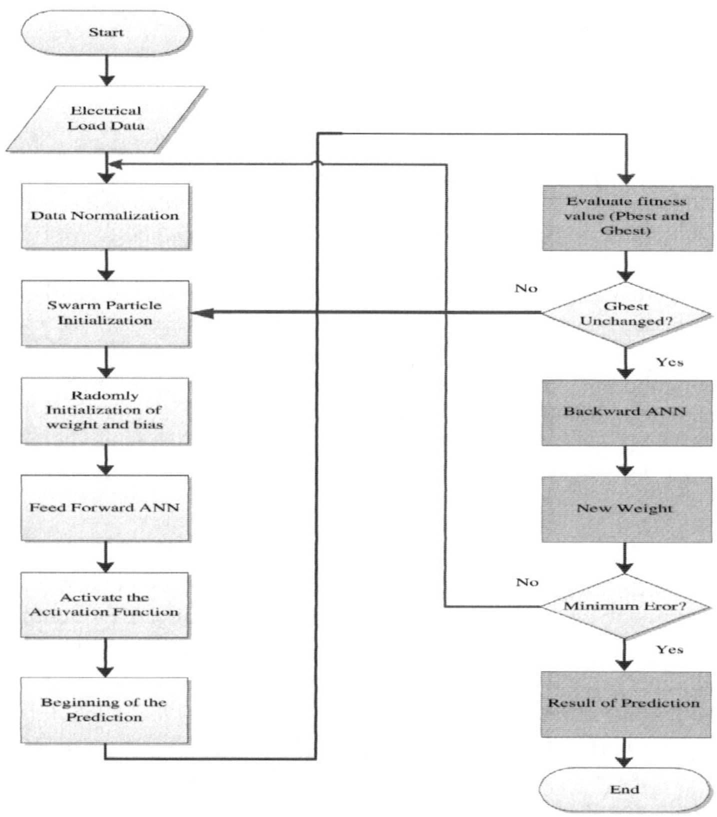

Figure 1. Flowchart of Back Propagation – Swarm Particle Hybrid Algorithm

The above flowchart is explained more clearly [8][9][10] The normalization of electrical load data during national holidays, determination of the type of Back Propagation, the number of neurons on hidden layer and initialization of initial weights and biases randomly between 0 and 1 and then, Initialization of initial position and velocity of particles, number of particles, upper limit and lower limit of search space and number of iterations that would be used. After that, initialing position and velocity of particles were determined randomly with a range between 0 and 1 obtained through Back Propagation. The position was its weight and bias, while the velocity was the value which affected the position

change at each iteration. So, the amount of particles that would be deployed was determined (N). In addition, the upper limit and lower limit range, the position and velocity of particles and number of iterations used on Swarm Particle were specified.
Forward phase of Back Propagation was conducted:
The data was spread on input layer towards the hidden layer by multiplying each data on input layer neuron unit with the weights that connected neurons, then the multiplication results were added and the biases were also added. Each signal was activated using segmoid activation function.

$$z\_in_j = vo_j + \sum_{i=1}^{n} x_i . v_{ij} \quad (1)$$

$$z_j = f(z\_in_j) = \frac{1}{1+\exp^{(z\_in_j)}} \quad (2)$$

The signal that had been activated was sent to the output layer by multiplying the signal on hidden layer with the weights connecting neurons to the output layer, then the biases were added up to the multiplication results. Then, Neurons on the output layer were activated in order to get the output.

$$y\_in_k = wo_j + \sum_{j=1}^{p} z_j . w \quad (3)$$

$$y_k = f(y\_in_k) \quad (4)$$

The output was compared with the target data to obtain fitness. Fitness value was obtained from MAPE (Mean Absolute Percentage Error), and then Particle Swarm Optimization on back propagation was performed by evaluating the fitness value of each particle by determining Pbest (Pbest is the function best value/smallest value for each particle along iteration) and Gbest (Gbest is the function best value/smallest value for the overall particles along the iteration). The position and velocity of particles were updated using particle velocity equation, and this update was affected by inertia coefficient.

$$v(t+1) = w(t) * v(t) + c_1 * rand_1(t) * (Pbest(t) - x(t)) + c_2 * rand_2(t) * (Gbest(t) - x(t)) \quad (5)$$

To find the best particles based on Pbest and Gbest after the position and velocity were updated, the fitness value of each iteration was calculated. So, the results of Pbest and best Gbest were saved as weights and continuous optimization was performed until Gbest experienced convergence and until the maximum iteration, then the algorithm was shifted into backward phase of Back Propagation.Then the algorithm diverted to backward phase of the Back Propagation. The input parameters of Back Propagation algorithm were set using the weights and biases obtained from Swarm Particle. The parameters of learning rate (α) and the maximum value of the epoch were defined and to discover the average MAPE, the objective function at each iteration was calculated. Last Step, weights and biases were fixed, steps 22 to 23 were repeated until the maximum value of iterations was reached, the output was denormalized, and then MAPE of the denormalized output was calculated.

3. Results and Discussion

Results on figure 2 shows that the prediction result using BP-SP Hybrid algorithm without any optimization can still produce a fairly good accuracy, with the average accuracy value obtained in the initial prediction reached the value of 98%. Although the highest error at one point reached 7%, error value of initial prediction could be minimized by finding the factors/parameters that can be optimized, the optimization factors can be obtained by conducting trials testing on the value of the initial parameter setting so that the best prediction results with the highest accuracy is obtained.

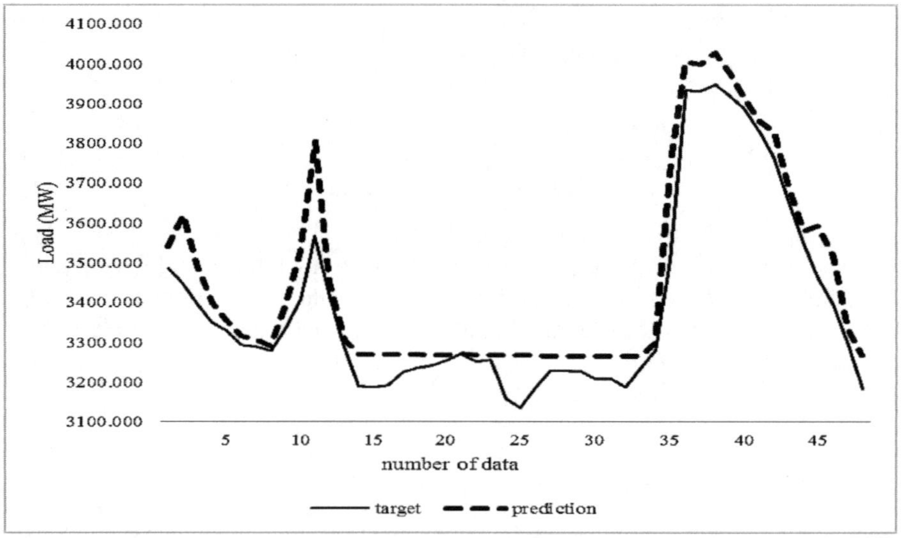

Figure 2. Curve of prediction result compared with the target

90 experiments were conducted with a fixed value setting, which was obtained from the optimization parameter testing in previous discussion. Error range obtained during the experiments were 0.77 to 2.875, thus 1 trial that showed the smallest error from these 90 experiments was sampled, that is the trial with the error value of 0.77 with the curve shown in figure 3.

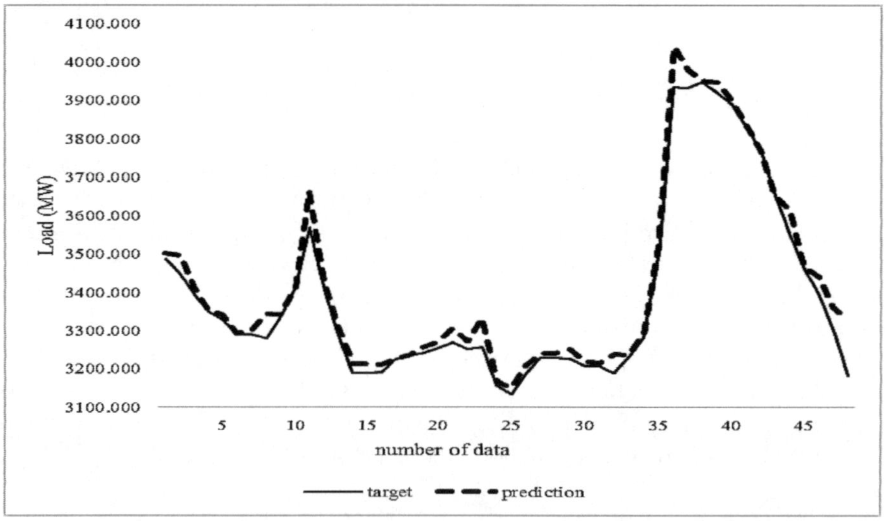

Figure 3. Prediction results (optimization) curve compared with target curve

4. Conclusions

Based on the discussions and findings obtained in this research, there are several things that can be concluded regarding predetermined problem scope. The first conclusion, regarding the pattern characteristics, it was discovered that electrical load conditions during national holidays can be defined as anomalous load condition if the load condition is compared with the load patterns showed during weekdays. Second, regarding short term load forecast process using back propagation - swarm particle hybrid algorithm, it can be concluded that this algorithm is an algorithm that has a very good accuracy, and this conclusion is based on the percentage of prediction accuracy. Compared with targets, this algorithm demonstrated the value of 98 126%. Third, the results of the prediction results of back propagation - swarm particle hybrid algorithm can be optimized by finding the correct input parameter values suitable to produce low error prediction, and in this study it can concluded that the parameters of the optimization were achieved by setting 10 data as the number of input data. In addition, the number of hidden layer should also be set on 1 layer and the learning rate should be set on the value of 0.6. These values of the optimization settings can improve the prediction accuracy of 1,104%.

5. Acknowledgments

The authors would like to express gratitude to Electrical Power System Research Group of Electrical Engineering Education Department, Universitas Pendidikan Indonesia

References

[1] Chaojun G, Yang D, Jirutitijaroen P, Walsh W M and Reindl T 2014 Spatial Load Forecasting with Communication Failure Using Time-Forward Kriging *IEEE Trans. Power Syst* **29** 2875–2882

[2] Shayeghi H, Shayanfar H A and Azimi G 2010 A Hybrid Particle Swarm Optimization Back Propagation Algorithm for Short Term Load Forecasting. *Int. J. Tech. Phys. Probl. Eng.* **2**, 12–22

[3] Choobeh A K 2013 Short Term Load Forecasting for Shiraz Region Using Adaptive Back Propagation Neural Network. *IEEE Int. Conf.* **On page** 3–5.

[4] Rahman S and Bhatnagar R 1988 An Expert System based Algorithm for Short Term Load Forecast *IEEE Trans. Power Syst* **3,** 392–399

[5] Fan S, Member S and Hyndman R J 2012 Short-Term Load Forecasting Based on a Semi-Parametric Additive Model *IEEE Trans. Power Syst* **27,** 134–141

[6] Arora S and Taylor J W 2013 Short-term forecasting of anomalous load using rule-based triple seasonal methods *IEEE Trans. Power Syst* **28** 3235–3242

[7] Paparoditis E and Sapatinas T 2013 Short-Term Load Forecasting: The Similar Shape Functional Time-Series Predictor *IEEE Trans. Power Syst* **28** 3818–3825

[8] Abdullah A G, Suranegara G M and Hakim D L 2014 Hybrid PSO-ANN Application for Improved Accuracy of Short Term Load Forecasting *wseas Trans. power Syst* **9,** 446–451

[9] He Y and Xu Q 2012 Short term Power Load Forecasting Based on Self-adapting PSO-BP Neural Network Model *Inst. Electr. Electron. Eng* **1** 1096–1099

[10] Sun W E I and Zou Y 2007 Short Term Load Forecasting based on BP Neural Network trainde by PSO *Inst. Electr. Electron. Eng* **1** 19–22

IOP Conf. Series: Materials Science and Engineering **128** (2016) 012008 doi:10.1088/1757-899X/128/1/012008

Naturally surveilled space: the design of a male drug rehabilitation center

A R Permana*, T Aryanti and F Rahmanullah

Department of Architectural Education, Universitas Pendidikan Indonesia, Bandung, Indonesia

*Corresponding author: akbarradityap@gmail.com

Abstract. The increase of drug addicts in Indonesia has not been supported by adequate facilities, both quantitatively and qualitatively. Despite being treated in a rehabilitation center, drug addicts may still use drugs surreptitiously and put themselves in danger. Architectural design may contribute to this either positively or negatively. This article elaborates a therapeutic design of a male rehabilitation center in the borderland of Bandung city, Indonesia. Employing the notion of natural surveillance, the rehabilitation center is designed to allow continual control over attendees without them feeling suppressed. The center design uses the behavioral approach to consider both attendees' physical and psychological comforts, as well as their security. Building masses are designed in a way that forms an inward orientation and are laid out circularly according to the therapy processes that attendees must undertake. Moreover, rooms are planned differently in response to attendees' unique conditions and restrictive physical requirements, such as their restriction on lighting and requirement of water for treatment. The landscape uses shady trees and vegetations as natural borders to demarcate the private zone, where attendees live, from the public area, where visitors may enter. The design is intended to provide a model for a responsive drug rehabilitation center that facilitates drug addicts' recovery.

1. Drug rehabilitation center and the spatial problems

A year ago, 35 drug addicts who were being rehabilitated in Lido Rehabilitation Center in Bogor were reported escaping from the center. They defeated the security officers and forcefully opened the main gate of the center. The patients held an after-prayer meeting in the mosque to arrange the escape. It was suspected that they did so because of the environmental high pressure. In many other rehabilitation centers, patients are still able to consume drugs surreptitiously. There are three possible reasons underlying these conditions. First, the center is not designed to provide rooms for patients according to their addiction characteristics. This has resulted in discomfort because of patients' sensitivity differences. Unclassified spaces may also allow bullying to happen among patients. Second, the center is designed like a jail, in which the victims are treated as inmates and thus live under high pressure. Third, the design of the center's public spaces has posited patients to be out of surveillance and thus obtain the opportunity to arrange the escape.

Data launched by Indonesia's Ministry of Health reveals that there has been an increasing number of the drug victims in Indonesia within the last five years [1]. The cases involve people of all ages, including children and teenagers, and both sexes. Data shows that the highest number of drug users is male, 29 years old or older. The condition is worsened by the fact that death caused by drug abuse is

40 people daily, particularly because only 10% of users obtain rehabilitation therapy. This has not conformed the policy made by the National Narcotics Board (Law No. 35 of 2009 on Narcotics, Article 54), in accordance with the United Nations Office on Drugs and Crime, according to which drug victims and addicts have to be rehabilitated, instead of being jailed [2]. Rehabilitation is a recovery process that drug victims and drug addicts should undertake in order to cease their dependency on psychotropic substances. Rehabilitation process consists of medical rehabilitation method, therapeutic community rehabilitation method, and religious rehabilitation method. Medical rehabilitation method is a hospital-based recovery method, which involves medical treatment and pharmacotherapy in order to reduce patients' symptoms and behavior changes. In therapeutic community rehabilitation method, a patient is encouraged to change her/his self-perception and to find her/himself. Patients are treated using behavioral approach through a reward-and-punishment system and group approach. The religious rehabilitation method employs spiritual and cognitive approach in treating patients.

Existing rehabilitation centers mostly have not served their optimum functions. Our preliminary studies show that they still lack in security system, resulting in patients' escape and bullying. Just like Lido's case, 35 patients could runaway because they have a chance to do so. They were hold their preparation to escape inside Lido's area. It is showed that some rehabilitation centers still have blind spots. In 2004, Prime Manager in Lido stated that more than 97% of patients who were escape in the middle way of treatment process were come back in some months later and restart the program from the beginning phase again[3]. Those problems showed that rehabilitation centers still lack with surveillance and also a comfortable environment to live in. The low supervision even allows hidden drug transactions and drug consumptions by patients. It also provides patients with the opportunity for hurting themselves due to drug withdrawal. Most rehabilitation centers also have not considered patients' specificities and this has resulted in their discomfort. This article explores the application of a therapeutic design to a male rehabilitation center in Mt. Manglayang hillside. It provides a model of a behavior-sensitive design for drug addicts' recovery.

2. Design method

We apply behavioral approach in the design of the male rehabilitation center. Prior to the design, we completed studies on the classification of drug consumers, drug addicts' behaviors, and their spatial needs. We conducted interviews with psychiatrists, counselors, psychologists, and medical practitioners on drug characteristics and the addicts' spatial behaviors. We also perform participant observations in a number of drug rehabilitation centers, such as Lido Therapy and Rehabilitation Unit in Bogor, FAN Campus Bogor, Palma Asylum in Cisarua, and Putera Parmadi Social Rehabilitation Center in Lembang. We observed the activities, the spatial arrangement and the use of spaces in the centers. We also inspected users' behaviors, including their motivations, social interactions, privacy, comfort, and character building. Data obtained is used to formulate the design concepts.

3. The design concepts and the design of a therapeutic space

The natural surveillance in the design of this male rehabilitation center is developed based on Jeremy Bentham's idea of panopticon (1787) and Oscar Newman's natural surveillance. Bentham's panopticon is a surveilling system embedded in the architectural space of prison and other correctional facilities. Bentham proposed the use of panopticon to allow surveillance on all inmates through the use of a surveilling tower at the center of the prison. Even in the absence of the guard, inmates always feel that they are being surveilled due to the presence of the surveilling tower [4]. The main idea of the panopticon is the efficiency and optimation of one-direction control from the tower to the cells.

In a similar way to the panopticon system, Newman's natural surveillance controls crime in public areas. Newman identifies that most crimes in public spaces occur in the visually deprived spots, where activities rarely take place. Therefore he proposes improvements in "surveillance capacity," which he defines as the ability to observe the public areas and to make a person who enters the areas continually feel being observed. Such an area will discourage crime, reduce inhabitants' fear and anxieties, and improve their feeling of being secure. Moreover, this will encourage inhabitants' use of the area and in turn will improve its security as a result of the intensive use. Newman has found that such a spatial characteristic is found in a public area that is surrounded by building openings or close to public circulations [5].

We propose the combination of both surveillance systems in a therapeutic rehabilitation space. The design concepts comprise site zoning, site circulation, building masses, interior design, and landscaping.

3.1. Site zoning

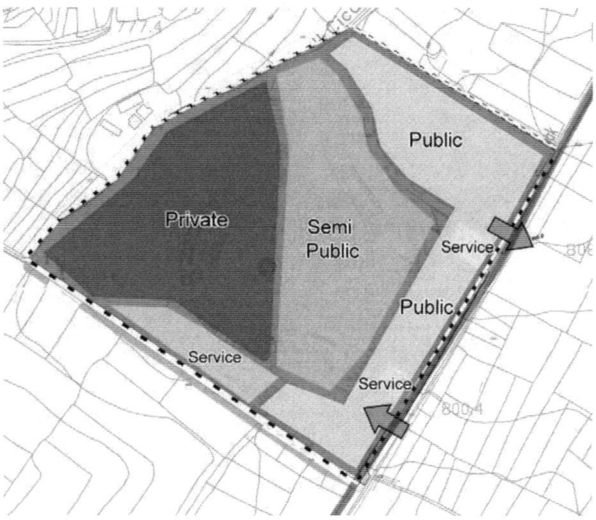

Fig. 1. Site zoning.

A rehabilitation center's spatial arrangement deals with who can or cannot enter to guarantee that the patients are safe from outsiders, other patients, and themselves. Thus, the site is divided into zones according to their security level, which also organizes users according to their accessibility (Fig. 1). The private zone is located at the innermost area where the psychological rehabilitation unit, the social rehabilitation unit, the detoxification facility and sport facilities are situated. This zone is restricted to patients and staffs. Low sound pollution and nice scenery around the building in this zone also help patients' healing process. Public area is located at the front of the site to which visitors enter. This zone includes a parking area for visitors. The semipublic zone, comprising a general service unit, an emergency unit, and an outpatient unit, is located between the public zone and the private zone. In order to limit access to the inner site, service areas are also arranged on this side. An inner courtyard, in the middle of which stands a control tower, separates the semipublic zone from the private zone. The control tower generates a sense of surveillance among patients.

3.2. Site circulation

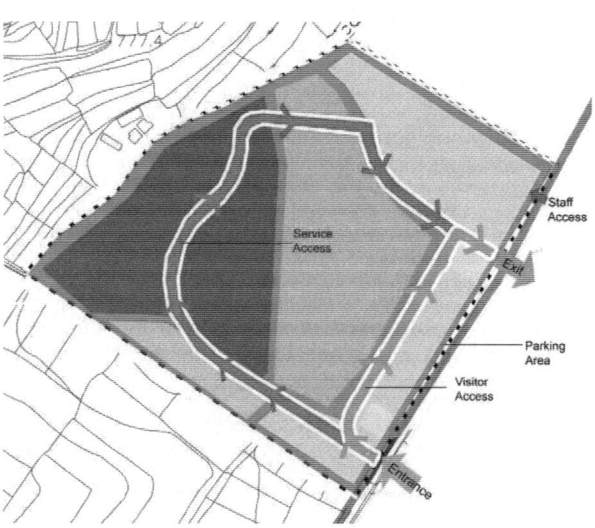

Fig. 2. Site circulation.

The site circulation considers the maximum surveillance on the site. Therefore, there are only a single access for the public and one for the service (Fig. 2).

3.3. Building masses

The shape of the building masses is a transformation a circle (Fig. 3a). The circle's round character generates an inward orientation (Fig. 3b), so it can increase the surveillance, which is good for rehabilitation center. Masses must not have any dead space because it can increase the patients' risk for using drugs inside the rehabilitation center. Furthermore, the building masses are laid according to the rehabilitation procedure as follow: the general service, the emergency unit, the detoxification unit, the psychological rehabilitation unit, and the social rehabilitation unit (fig. 3c). Developed from a circle, the units are arranged in a circular way, which in addition to facilitating flowing activity transitions also enabling maximum security and surveillance system. The open visual access and unisolated social space allow constant observation, which is essential to a secure environment [6].

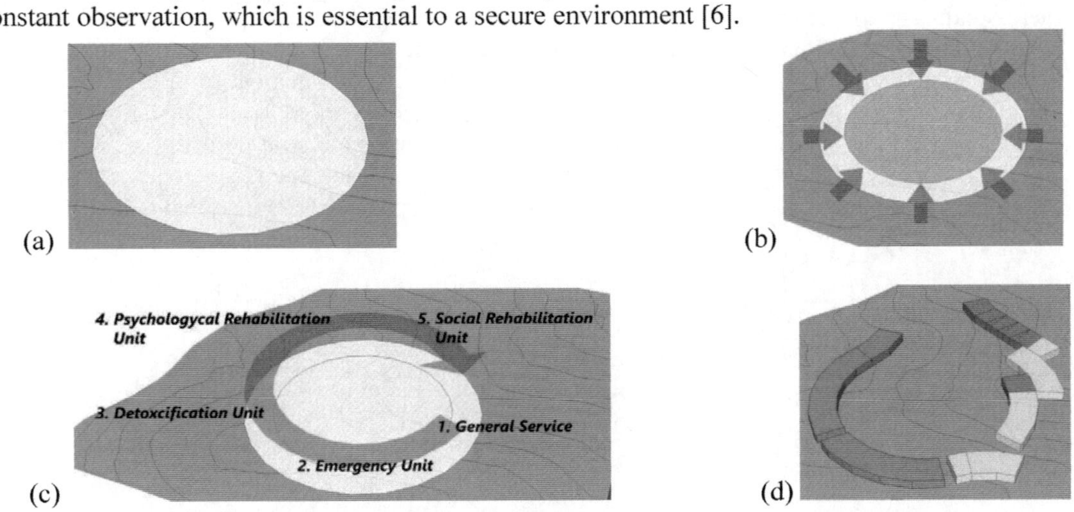

Fig. 3. (a) circle shape, (b) inward orientation, (c) treatment process, (d) building masses.

Different from other centers, in which patients' activities are located on the upper floors, while public activities are on the ground floor, our design proposes that all activities are accommodated on the ground floor in order to minimize accidents. We also separate rehabilitation stages into different buildings to facilitate maximum building function (fig. 3d).

3.4. Interior design

The interior of the buildings reflects the reward system that is applied to all patients. Those who successfully complete a step of therapy are advanced to the next step in another building unit whose facilities are better and more comfortable. This system is applied as a way to motivate patients to recover from addiction.

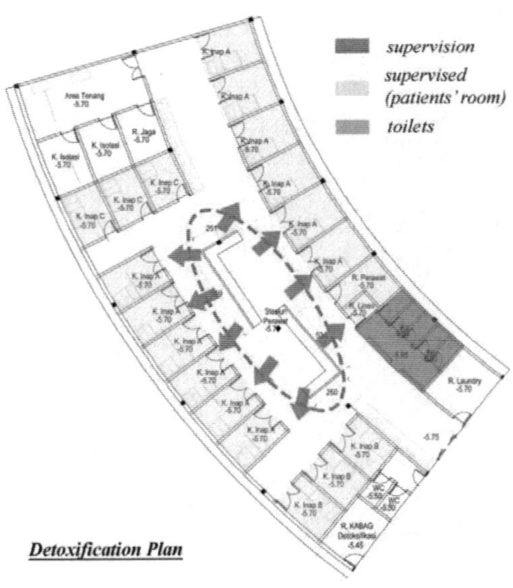

At the detoxification phase, patients are not able to control themselves, thus, they are assigned an isolated, yet under observation, room (Fig. 4).

Fig. 4. (left) surveillance at the detoxification phase, (right) the isolated patient uses glasses for constant observation by staffs.

In detoxification phase, patients also need special requirements for their rooms. Patients' drug's effects are divided into three types of rooms. First, small rooms with acoustic wall for patients with sensitive hearing and difficult to concentrate. Second, rooms with acoustic wall and no windows to accommodate patients with sensitive sight and sensitive hearing. Last, wider rooms with acoustic wall for patients with sensitive earing and hyperactive.

At the psychological rehabilitation phase, patients are able to control their own mind, but still unsure with what they're doing. At this stage, they need more encouragement to release themselves from drug dependency. Thus the design of the facility emphasizes more on social interaction. The facility is designed to allow more caring and motivating interactions among patients as shown in the provision of a number of lounges and common bedrooms. By sharing the rooms where private and public activities take place, patients are expected to grow encouragement among them (Fig. 5).

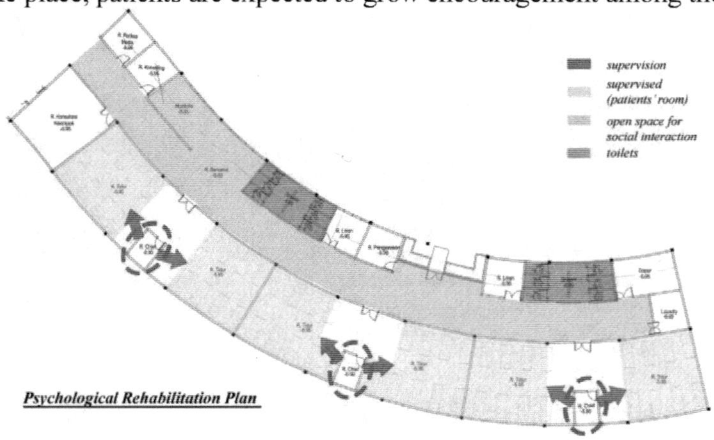

Fig. 5. Surveillance at the psychological rehabilitation phase.

At the social rehabilitation phase, patients have already obtained their self-confidence so that they can interact with the outside world, yet still under supervision. Rooms in this facility allow more privacy for patients as found in bedrooms and toilets (Fig. 6).

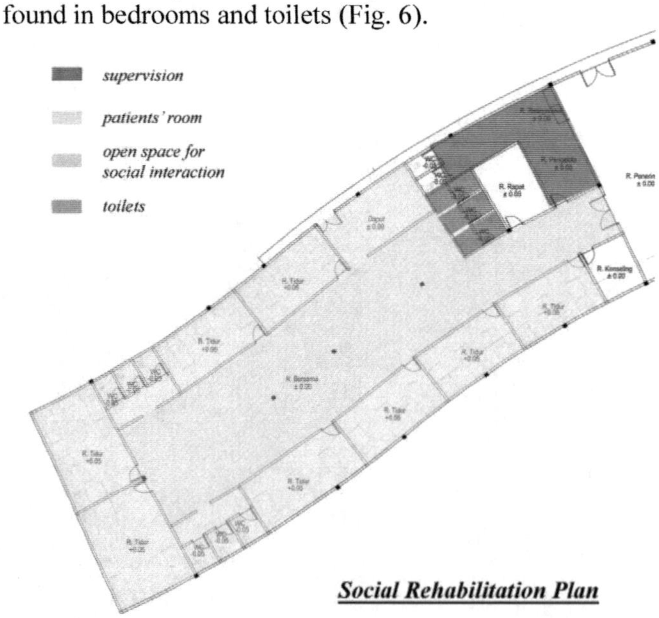

Fig. 6. Privacy for patients at the social rehabilitation phase.

Most penal facility design limits inmates' and patients' freedom of movement through the use of trellis. However, such a design has led to inhabitants' frustration and aggression [7]. In most rehabilitation centers, the supervision room is located at the middle of each floor level. This design instead proposes that the room is positioned in front of patient's rooms. This will enable a better control to patients, in addition to facilitating easier access when emergency situations occur. In the psychological rehabilitation unit, the supervision room is located in common rooms (Fig. 6a). Moreover, public spaces such as prayer room and dining room are not equipped with a supervision room but are designed as open spaces. Such a design generates a sense of surveillance among patients. In the detoxification building, the supervision room is placed outside the patients' room, which uses glass wall (Fig. 6b).

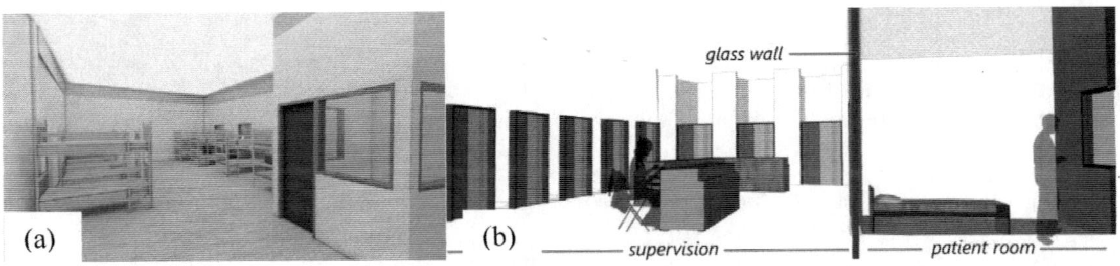

Fig. 6. (a) Supervision room is placed inside the patients' room; (b) Supervision room is placed outside the patients' room.

3.5. Landscaping

Vegetations, consisting of trees and shrubs, are planted surrounding buildings in order to form subtle borders. They also function to demarcate the public, semipublic, and private areas in the center. Vegetations and greenery create relaxing environment that is significant in a therapeutic place. Additionally, trees also serve to give direction to users and visitors. The landscape also includes water and color consideration in the design. Water is a critical element in drug addiction therapy.

Fig. 7. (a) Vegetations to create relaxing environment. (b) vegetations as subtle borders.

4. Conclusion

Compared to most rehabilitation centers, the design that we propose considers patients' unique characteristics according to their phase of therapy. The site and building zoning is designed to enable maximum security and surveillance. So is the interior. Walls may create extra pressure to patients. To avoid such an architectural expression, the buildings are laid in circular arrangement. Thus while forming an inward orientation toward the inner court, the buildings themselves function as fences that restrict patients' movements within the buildings while creating a system of natural surveillance to them. Each of rehabilitation phase has different interior design aspects, such as room arrangement and material, to accommodate patients' unique characteristics. Landscape uses water and vegetations as part of treatments method and also used to create subtle borders to divide zones in the center.

References

[1] Kementerian Kesehatan RI. 2014. *Pengguna Narkoba Dapat Dicegah dan Dapat Direhabilitasi.* Buletin Jendela Data dan Informasi Kesehatan. Semester I 26 Juni 2014. Jakarta.

[2] Handayani, F. 2010. *Hubungan Antara Kekuatan Karakter Dengan Resiliensi Residen NARKOBA di Unit Pelaksanaan Teknis (UPT) Terapi dan Rehabilitasi Badan Narkotika Nasional Lido*, pp. 11. Skripsi. Jakarta: Fakultas Psikologi. Universitas Islam Negeri Syarif Hidayatulah.

[3] Undang-Undang No. 35 Tahun 2009 Tentang Narkotika

[4] Bentham J. 1791. *Panopticon or the Inspection House*. London: Mews-Gate.

[5] Newman O. 1973. Natural Surveillance In *Defensible Space: Crime Prevention Through Urban Design*, pp. 78-101. New York: Collier Books

[6] Katyal NK. 2002. Architecture as Crime Control. *Yale Law Journal* 111: 1039-139

[7] Mazuch R, Rona S. Dec 2005. Creating Healing Environments: Humanistic Architecture and Therapeutic Design. *Journal of Public Mental Health* 4: 48-52

Preliminary study: kinetics of oil extraction from sandalwood by microwave-assisted hydrodistillation

H S Kusuma* and M Mahfud
Department of Chemical Engineering, Institut Teknologi Sepuluh Nopember, Surabaya, Indonesia

*Corresponding author: mahfud@chem-eng.its.ac.id

Abstract. Sandalwood and its oil, is one of the oldest known perfume materials and has a long history (more than 4000 years) of use as mentioned in Sanskrit manuscripts. Sandalwood oil plays an important role as an export commodity in many countries and its widely used in the food, perfumery and pharmaceuticals industries. The aim of this study is to know and verify the kinetics and mechanism of microwave-assisted hydrodistillation of sandalwood based on a second-order model. In this study, microwave-assisted hydrodistillation is used to extract essential oils from sandalwood. The extraction was carried out in ten extraction cycles of 15 min to 2.5 hours. The initial extraction rate, the extraction capacity and the second-order extraction rate constant were calculated using the model. Kinetics of oil extraction from sandalwood by microwave-assisted hydrodistillation proved that the extraction process was based on the second-order extraction model as the experimentally done in three different steps. The initial extraction rate, h, was 0.0232 g L^{-1} min^{-1}, the extraction capacity, C_S, was 0.6015 g L^{-1}, the second-order extraction rate constant, k, was 0.0642 L g^{-1} min^{-1} and coefficient of determination, R^2, was 0.9597.

Keywords: sandalwood oil; microwave-assisted hydrodistillation; extraction; kinetic model; second-order model

1. Introduction

Sandalwood and its oil, is one of the oldest known perfume materials and has a long history (more than 4000 years) of use as mentioned in Sanskrit manuscripts. Sandalwood is still used in religious rituals in India and is also used as a medium from which to carve deities and temples. The ancient Egyptians imported the wood and used it in medicine, embalming and ritual burning to venerate the gods. The oil has been in public use since the early 1800s [1]. Sandalwood has developed into a commercial timber crop over the past 10-15 years, with substantial sandalwood plantations established in India, China and Australia, and more modest plantings being established in Indonesia, Fiji, Vanuatu, Hawaii and Sri Lanka [2].

Sandalwood oil plays an important role as an export commodity in many countries and its widely used in the food, perfumery and pharmaceuticals industries. Sandalwood oil is used as a flavor component in many food products, including alcoholic and non-alcoholic beverages, frozen dairy desserts, candy, baked goods and gelatin and puddings at use levels generally below 0.001% (10 ppm) except in hard candy. The highest maximum use level for sandalwood oil in food products is approximately 90 ppm. Sandalwood oil is generally used as a natural flavoring substance or in conjunction with other flavor ingredients [3].

In perfumery, sandalwood oil is used extensively. It blends well with rose, violet, tuberose, clove, lavender, oakmoss and labdanum products. It is also commonly used in woody, wood-floral and Oriental-floral bases. The oil is also used as a base for co-distillation of other essential oils. It blends well with other chemicals such as ionones and methylionones. Because of the scarcity of sandalwood oil, the search for efficient synthetic substitutes has been conducted extensively to investigate the structure odor relationship of sandalwood constituents [4,5]. Several investigators have extensively studied and attempted to synthesize santalol derivatives with comparable olfactory activity [6,7,8]. Opdyke (1974) [9] reported the use of sandalwood oil in fragrances in the USA to be approximately 48,000 lb/year.

Sandalwood oil is used medicinally for common colds, bronchitis, fever, infection of the urinary tract, inflammation of the mouth and pharynx, liver and gallbladder complaints and other maladies [10]. In the Indian system of medicine (Ayurvedic), sandalwood oil is largely used as a demulcent, diuretic and mild stimulant [11]. The daily recommended dosage of sandalwood oil *per* the German Commission E review is 1–1.5 g for not more than six weeks [12]. Imdorf et al. (1999) [13] reported that sandalwood oil acts as a repellent of the pest *Varroa jacobsoni* Oud., in honey bee colonies and has been used as an acaricide. Choi et al. (2006) [14] reported modest activity against *Lycoriella mali* (the mushroom fly).

The main methods to obtain essential oils from the plant materials are hydrodistillation, steam distillation, solvent extraction, supercritical fluid extraction (SC-CO_2) and liquid CO_2 extraction have been used to obtain the volatile oil from sandalwood [15,16]. Among these method, hydrodistillation has been the most common approach to extract the essential oils from the medicine herbs and plants [17]. Alternative methods, employing microwaves, have been developed in order to shorten extraction time, improve the extraction yield, and reduce the operational costs. Microwave-assisted procedures for isolating essential oils have become attractive for use in laboratories and industry. The advantages of using microwave energy for oil extraction are more effective heating, fast energy transfer, faster response to process heating control, faster start-up, increased production, and elimination of some process steps.

However, to the best of authors' knowledge no work has been published on the extraction of essential oil from *Santalum* species using microwave ovens for heating. Therefore, the objective of this study were to investigate the potential of microwave-assisted hydrodistillation for the extraction of essential oils from sandalwood. In this study, the author also attempted to know and verify the kinetics and mechanism of microwave-assisted hydrodistillation of sandalwood based on a second-order model.

2. Materials and methods

2.1. Raw materials

The main raw material used in this study is sandalwood that comes from the Kupang, East Nusa Tenggara, Indonesia in powder form. All other chemicals and solvents used were of analytical grade.

2.2. Microwave extraction of sandalwood oil

A domestic microwave oven (EMM-2007X, Electrolux, 20 L, 800 W; variable in 200 W increments, 2.45 GHz) was modified for microwave-assisted hydrodistillation operation. The dimensions of the PTFE-coated cavity of the microwave oven were 46.1 cm x 28.0 cm x 37.3 cm. Twenty grams of sandalwood powder samples were placed in a 1 l flask containing deionized water (400 mL). The flask was setup within the microwave oven cavity and a condenser was used on the top (outside the oven) to collect the extracted essential oils (figure 1). The microwave oven was operated at 600 W power level for a period of 2.5 h. This period was sufficient to extract all the essential oils from the sample. The extraction was carried out in ten extraction cycles of 15 min to 2.5 hours. To remove water, the extracted essential oils were then dried over anhydrous sodium sulfate, weighed and

IOP Conf. Series: Materials Science and Engineering **128** (2016) 012009 doi:10.1088/1757-899X/128/1/012009

stored in amber vials at 4 °C until they were used for analysis. The yield of sandalwood oil was found by the following equation

$$y = \frac{V}{W} \text{ X } 100 \tag{1}$$

where y is the sandalwood oil yield (%, w/w), V is the weight or mass of extracted sandalwood oil (g) and W is the weight or mass of sandalwood powder (g).

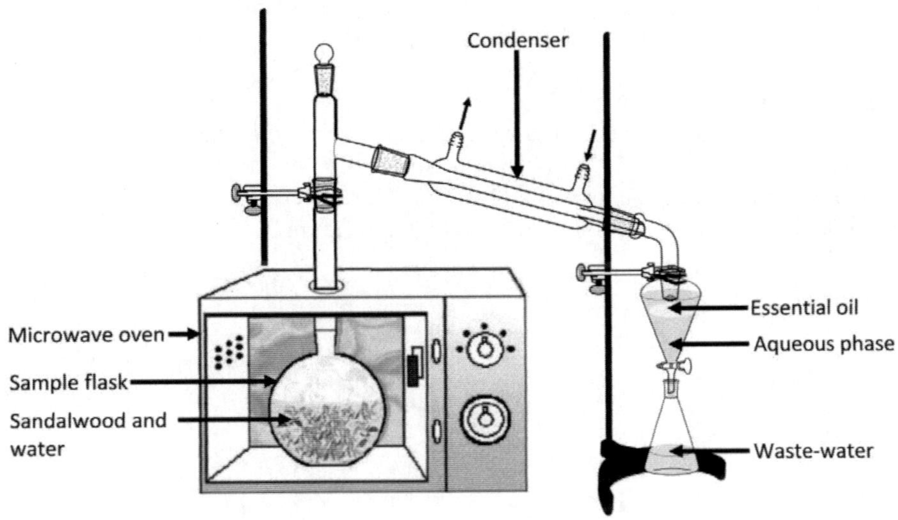

Figure 1. Schematic representation of the microwave-assisted extraction apparatus used in this study

2.3. Kinetic model

Second-order mechanism model means that the extraction occurs in two simultaneous processes. The amount of extracted oil increases rapidly with time at the beginning and then decreases slowly with the time until the end of extraction process [18,19,20,21]. The rate of dissolution for the essential oil contained in the solid to solution can be described by equation (2)

$$\frac{dC_t}{dt} = k(C_S - C_t)^2 \tag{2}$$

where k is the second-order extraction rate constant (L g^{-1} min^{-1}), C_S the extraction capacity (concentration of essential oil at saturation in g L^{-1}) and C_t is the concentration of sandalwood oil at any time t (min).

By considering the initial and boundary conditions, $t = 0$ to t and $C_t = 0$ to C_t, the integrated rate law for a second-order extraction was obtained:

$$C_t = \frac{C_S^2 kt}{1 + C_S kt} \tag{3}$$

By transforming equation (3), a linear form shown in equation (4) can be obtained and the extraction rate can be written as equation (5):

$$\frac{t}{C_t} = \frac{1}{kC_S^2} + \frac{t}{C_S} \tag{4}$$

$$\frac{C_t}{t} = \frac{1}{(1/kC_S^2) + (t/C_S)} \tag{5}$$

The initial extraction rate, h, as Ct/t when t approaches 0, can be defined as:

$$h = kC_S^2 \tag{6}$$

and, the concentration of essential oil at any time can be expressed after rearrangement as:

$$C_t = \frac{t}{(1/h) + (t/C_S)} \tag{7}$$

The initial extraction rate, h, the extraction capacity, C_S, and the second-order extraction rate constant, k, can be determined experimentally from the slope and intercept by plotting t/C_t versus t.

3. Results and discussion

As shown in figure 2, the rate of extraction was increased as the time of extraction increase until it reached plateau or constant after 120 min of extraction. 0.93% extractable oil was obtained in the 1.5 hours of extraction until it became plateau (1.02%). The experimental result was analyzed using second-order model by plotting t/C_t versus time as shown in figure 3.

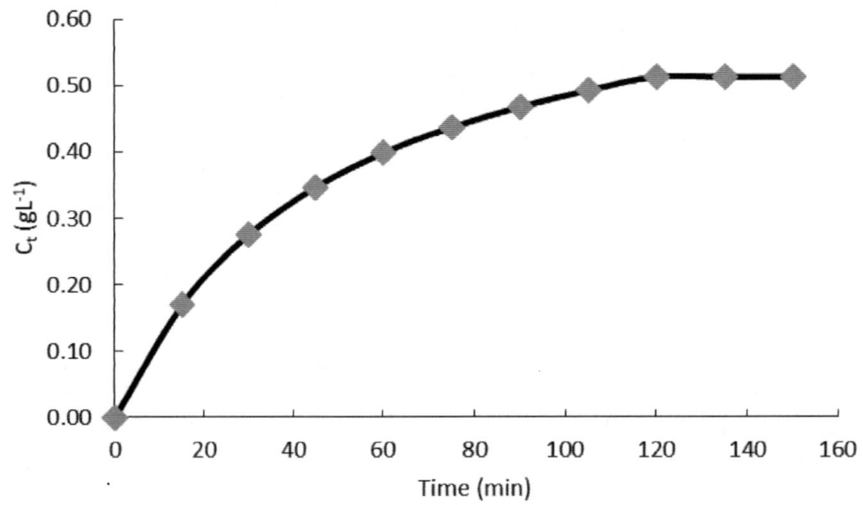

Figure 2. The concentration of sandalwood oil in the solution at any time, C_t (gL^{-1}) versus time (min)

According to figure 2, the rate of extraction was fast at the beginning and slow until the end of the extraction process. The extraction process takes place in three different steps: an equilibrium phase where the phenomena of solubilization and partition intervene, in which the substrate is removed from the outer surface of the particle at an approximately constant velocity. Then, this stage is followed by an intermediary transition phase to diffusion. The resistance to mass transfer begins to appear in the solid–liquid interface; in this period the mass transfer by convection and diffusion prevails. In the last phase, the solute must overcome the interactions that bind it to the matrix and diffuse into the extracting solvent. The extraction rate in this period is low, characterized by the removal of the extract through the diffusion mechanism. This point is an irreversible step of the extraction process; it is often regarded as the limiting step of the process [22]. Diffusion rate decreased as the time of extraction incressed due to the high solute concentration in liquid at the third stage. Even though the extraction time incressed after the maximum sandalwood oil was extracted, it did not show any changes or significant in amount of oil extracted. The trend of sandalwood oil recovery under extraction time of 60 min (77.58%), 105 min (18.46%) and 150 min (3.96%).

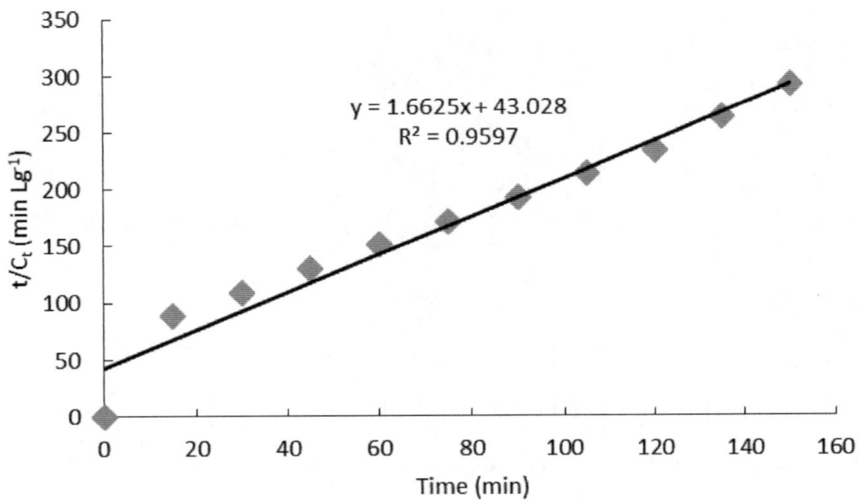

Figure 3. Second-order extraction kinetics of sandalwood

The initial extraction rate, h, the extraction capacity, C_S, the second-order extraction rate constant, k, and coefficient of determination, R^2, were calculated experimentally by referring to the linier curve in figure 3. From graph t/C_t versus time, slope is equal to $1/C_S$, and intercept is equal to $1/h$. The data showed in table 1.

Table 1. Linierization of second-order kinetic model of microwave-assisted hydrodistillation of sandalwood

C_S (g L^{-1})	k (L g^{-1} min^{-1})	h (g L^{-1} min^{-1})	R^2
0.6015	0.0642	0.0232	0.9597

For this study, the maximum yield oil extracted by microwave-assisted hydrodistillation is higher compared to conventional hydrodistillation. In the work of Hettiarachchi et al. (2010) it was observed that sandalwood oil yield extracted by conventional hydrodistillation for 9 hours is 0.43%. The fundamentals of the microwave-assisted hydrodistillation process are different from those of conventional methods because the extraction occurs as the result of changes in the cell structure caused by electromagnetic waves. In microwave-assisted hydrodistillation, the process acceleration and high extraction yield may be the result of a synergistic combination of two transport phenomena: heat and mass gradients working in the same direction [23]. On the other hand, in conventional extractions the mass transfer occurs from inside to the outside, although the heat transfer occurs from the outside to the inside of the substrate (figure 4). In addition, although in conventional extraction the heat is transferred from the heating medium to the interior of the sample, in microwave-assisted hydrodistillation the heat is dissipated volumetrically inside the irradiated medium.

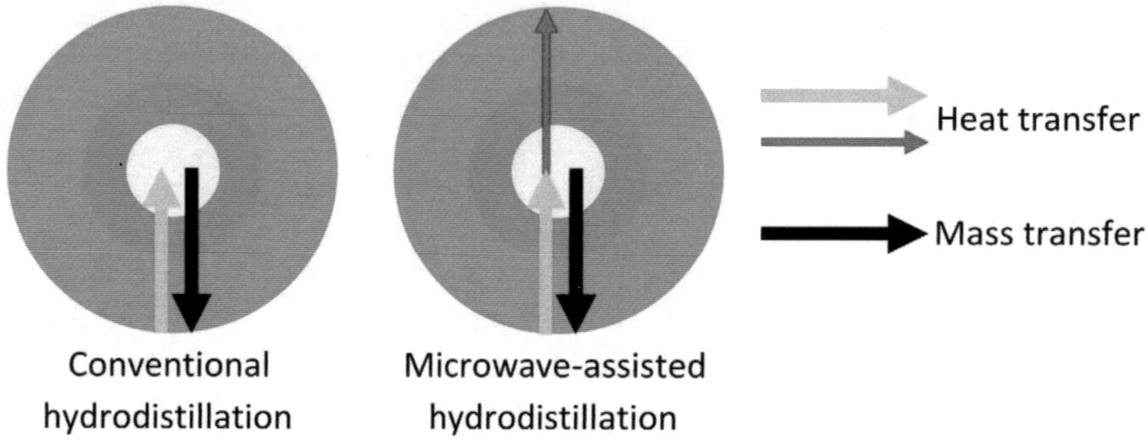

Figure 4. Heat and mass transfer mechanisms in microwave-assisted hydrodistillation and conventional hydrodistillation of sandalwood oil

4. Conclusion

Kinetics of oil extraction from sandalwood by microwave-assisted hydrodistillation proved that the extraction process was based on the second-order extraction model as the experimentally done in three different steps. The initial extraction rate, h, was 0.0232 g L^{-1} min^{-1}, the extraction capacity, C_S, was 0.6015 g L^{-1}, the second-order extraction rate constant, k, was 0.0642 L g^{-1} min^{-1} and coefficient of determination, R^2, was 0.9597.

References

[1] Arctander, S., 1960. *Sandalwood oil East India. Perfume and Flavor Materials of Natural Origin* (NJ: Elizabeth) pp 574–576
[2] Hettiarachchi D S, Gamage M and Subasinghe U 2010 Oil content analysis of sandalwood: A novel approach for core sample analysis *Sandalwood Research Newsletter* **25** 1-4
[3] Burdock G A and Carabin I G 2008 Safety assessment of sandalwood oil (*Santalum album* L.) *Food and Chemical Toxicology* **46** 421-432
[4] Shvets N M and Dimoglo A S 1998 Structure–odour relationships: results of an applied electron-topological approach *Nahrung* **42** 364–370
[5] Bajgrowicz J A and Frater G 2000 Chiral recognition of sandalwood odorants *Enantiomer* **5** 225–234
[6] Buchbauer G, Winiwarter S and Wolschann P 1992 Surface comparisons of some odour molecules: conformational calculations on sandalwood odour V *Journal of Computer-Aided Molecular Design* **6** 583–592
[7] Buchbauer G, Zechmeister-Machhart F, Weiss-Greiler P and Wolschann P 1997 Structure–activity relationships of sandalwood odorants: synthesis and odour of methyl-beta-santalol *Archives of Pharmacology (Weinheim)* **330** 112–114
[8] Buchbauer G, Sunara A, Weiss-Greiler P and Wolschann P 2001 Synthesis and olfactoric activity of side-chain modified beta-santalol analogues *European Journal of Medicinal Chemistry* **36** 673–683
[9] Opdyke D L J 1974 Reviews on fragrance raw materials. Sandalwood oil, East Indian *Food and Cosmetics Toxicology* **12** (Supp.) 989–990
[10] PDR Herbal 2004 *Sandalwood. Santalum album. PDR for Herbal Medicine*, third ed. (Montvale, NJ: Medical Economics Company) pp 702–703
[11] Pande M C 1977 Medicinal oils and their importance *Medicine and Surgery* **17** 13–16

[12] Anonymous 1998 *Sandalwood white. In: Blumenthal, M. (Ed.), The Complete German Commission E Reviews: Therapeutic Guide to Herbal Medicines* (Boston, MA: American Botanical Council, Integrative Medicine Communications) p 199

[13] Imdorf A, Bogdanov S, Ibanez O R, Calderone N W and Spivak M P 1999 Use of essential oils for the control of *Varroa jacobsoni* Oud in honey bee colonies; special issue – dynamics and control of varroa parasitism on apis *Apidologie* **30** 209–228

[14] Choi W K, Park B S, Lee Y H, Jang D Y, Yoon H Y and Lee S E 2006 Fumigant toxicities of essential oils and monoterpenes against *Lycoriella mali* adults *Crop Protection* **25** 398–401

[15] Moretta P, Ghisalbert E L, Piggott M J and Trengove R D 1998 Extraction of oil from *Santalum spicatum* by supercritical fluid extraction *ACIAR Proceedings Series* **84** 83-85

[16] Marongiu B, Piras A, Porcedda S and Tuveri E 2006 Extraction of *Santalum album* and *Boswellia carterii* Birdw. volatile oil by supercritical carbon dioxide: influence of some process parameters *Flavor and Fragrance Journal* **21(4)** 718 – 724

[17] Chinese Pharmacopoeia Committee 2010 *Chinese Pharmacopoeia*, 9th edition (Beijing, China: China Medical Science and Technology Press) Appendix 62

[18] Ho Y S, Oumarou H A H, Fauduet H and Porte C 2005 Kinetics and model building of leaching of water-soluble compounds of *Tilia* sapwood *Separation and Purification Technology* **45** 169-173

[19] Rabesiaka L R, Havet J, Porte C and Fauduet H 2007 Solid–liquid extraction of protopine from *Fumaria officinalis* L.-Analysis determination, kinetic reaction and model building *Separation and Purification Technology* **54** 253-261

[20] Uhm J T and Yoon W B 2011 Effects of high-pressure process on kinetics of leaching oil from soybean powder using hexane in batch systems *Journal of Food Science* **76** 444-449

[21] Meizane S and Kadi H 2008 Kinetics and thermodynamics of oil extraction from olive cake *Journal of the American Oil Chemist's Society* **85** 391-396

[22] Raynie D E 2000 *Extraction.* In: Wilson I D, Adlard E R, Cooke M, Poolie C F (eds) *Encyclopedia of separation science* (San Diego: Academic Press)

[23] Chemat F, Abert-Vian M and Zill-e-Huma Y J 2009 *Microwave assisted separations: green chemistry in action. In: Pearlman JT (ed) Green chemistry research trends* (New York: Nova Science Publishers) pp 33–62

A Comparative of business process modelling techniques

I R H T Tangkawarow* and J Waworuntu

Universitas Negeri Manado , North Sulawesi, Indonesia

*Coresponding author: irene_tangkawarow@yahoo.com

Abstract. In this era, there is a lot of business process modeling techniques. This article is the research about differences of business process modeling techniques. For each technique will explain about the definition and the structure. This paper presents a comparative analysis of some popular business process modelling techniques. The comparative framework is based on 2 criteria: notation and how it works when implemented in Somerleyton Animal Park. Each technique will end with the advantages and disadvantages. The final conclusion will give recommend of business process modeling techniques that easy to use and serve the basis for evaluating further modelling techniques.

1. Introduction

Nowadays, business competition causing companies to optimize existing business processes within the organization. Analysis the business process modeling is a tool to evaluate and make improvements over the business process there. Through the analysis of business they can decide which one is optimal or not optimal run and give attention to it.

Business process modeling is that activity aimed at the representation of all or some elements in order to produce a cohesive model of the behavior required to deliver a service and/or product to a customer or another part of the organization [2]. There are some techniques to model the business process. In practice it is not easy to determine which techniques are suitable and easily understood by stakeholders. Due to frequent misunderstanding between analyst system with a business owner or even between analyst system with the team (eg. Programmers). It required research on comparative of business process modeling techniques to overcome the above problems. This research is limited in four business process modeling techniques which often used the comparative analysis phase. The four business process modeling techniques are (1) Data Flow Diagram (DFD), (2) Business Process Modelling Notation (BPMN), (3) Activity Diagram, and (4) Integration Definition for Function Modelling (IDEF0).

The most interesting issues in business process modelling is the enormous availability of different techniques for the representation of organizational processes and their requirements [2]. Individual techniques can focus on different facets of process modelling. For example, while Activity Diagrams emphasis on the interaction between roles in the organization [8], Data Flow Diagrams (DFD) focus on the flow of data through a system [7]. This paper presents a comparative analysis each modeling techniques to explain the definition, notation and how it works implemented in the same process. Breeding process in the Somerleyton Animal Park is the scenario which used to show the difference each model.

As previous comparative analyses, Aldin L and de Cesare S [2] focus on comparison criteria to provide the reason overview. This paper presents the difference between four business process model

IOP Conf. Series: Materials Science and Engineering **128** (2016) 012010 doi:10.1088/1757-899X/128/1/012010

with comparison criteria, implementation using the same scenario, advantages and disadvantages each model.

2. Comparative Analysis

In the subsections that follow a breeding process in the Somerleyton animal park is used as a scenario to exemplify the different types of notation. The breeding process scenario refers to the process for animal breeding. The system admin checks into the system which animal is ready to breed and check if there is an animal in the park is available to breed for. When the animal and his partner is available, then do the breeding by animal keeper. But if not, then system admin has to contact another animal park, which having an animal as partner and arrange schedule when and where the breeding will be done.

2.1 Data Flow Diagram (DFD)

Data flow diagram used to describe how the system is processing data and information, in terms of input and output [8].

- The general principle in Data Flow Diagramming is that a system can be decomposed into subsystems, and subsystems can be decomposed into lower level subsystems, and so on.
- Each subsystem represents a process or activity in which data is processed. At the lowest level, processes can no longer be decomposed.
- Each 'process' (and from now on, by 'process' mean subsystem and activity) in a DFD has the characteristics of a system.
- Just as a system must have input and output (if it is not dead), so a process must have input and output.
- Data enters the system from the environment; data flows between processes within the system; and data are produced as output from the system

2.1.1 Notation

In this paper will use DFD notation from Yourdan and De Marco [13], which explain below

Entity

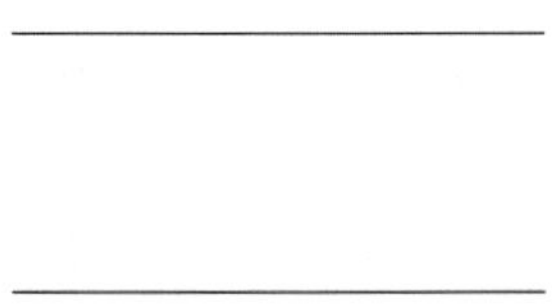

An entity is the source or destination of data. The source in a DFD represents these entities that are outside the context of the system. Entities either provide data to the system (referred to as a source) or receive data from it (referred to as a sink) [4][13].

Process

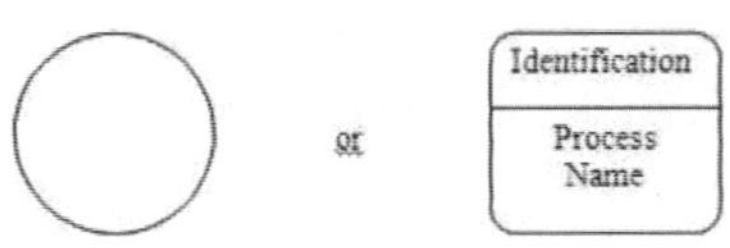

There are two notation for process in DFD from Yourdon & De Marco and Gene & Sarson. Processes can be symbolized as a circle or segmented rectangle that consist of an identifier like process number and process name [4].

Data Store

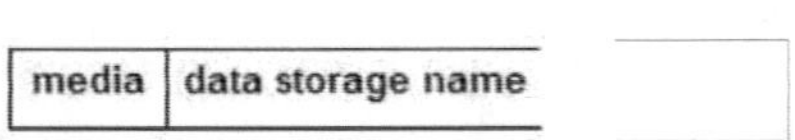

A data store is where a process stores data between processes for later retrieval by that same process or another one. The naming data store should be simple but meaningful. The data store is the place to store data in the system [13].

Data Flow

Data Flow is place to flow the information. Describe with straight lines that connected all the

component of the system. Data flow showing with arrow and given name line or data flow[13]. This data flow stream between process, data store, and external entity and showing data flow can be input for the system or output system.

2.1.2. Structure

Context Diagram

Only represent single process, show the whole activities of the process and give the boundaries of the system. The one process connects to external entities and the data flow that connect it to the process [13].

Level 0 Diagram

Showing the overall process of the system, each part of functional requirement of the system represent by one mayor process. Can use one single data store as an aggregate to represent all data that is used in this level.

Level-n Diagram

A detailed processes of a level-0 diagram. Each process is represented by its own DFD. Data flows of a higher level break into component pieces. With the components shown on lower level and each lower level shows greater and greater detail.

2.1.3 Applied to Somerleyton Animal Park

Context Diagram

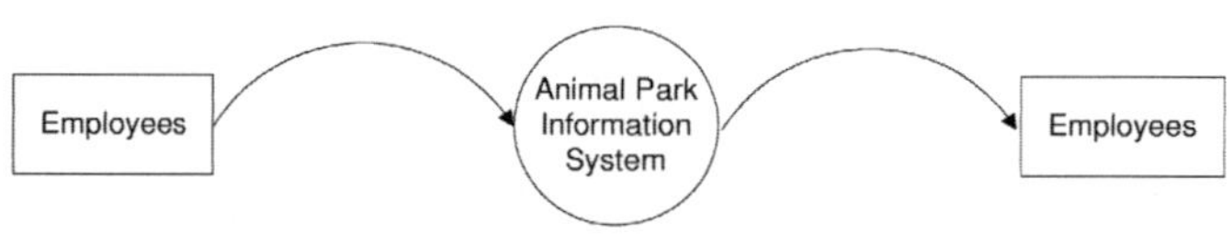

Figure 1. Context Diagram

The context diagram shown there are two employees. The first entity is employee as system administrator who works directly to Animal Park Information System. The second entity is employee as animal keeper who take care the animal.

DFD Level 0

In the DFD level 0 shows there are 3 process such as (1.0) breeding animal, (2.0) feeding the (3.0) animal and order food. Breeding the animal process is the process that we want to breakdown into DFD Level 1.

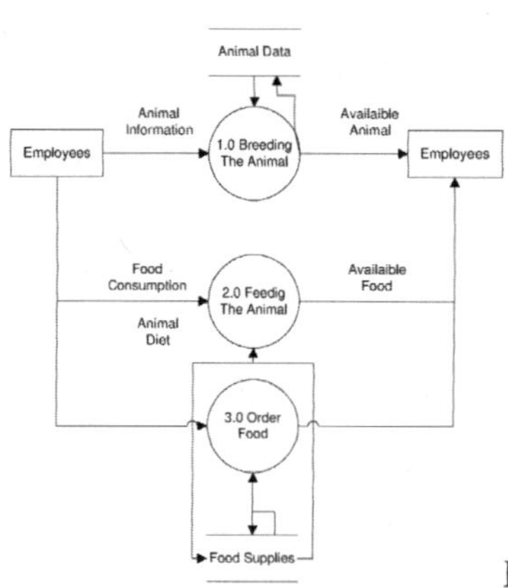

Figure 2. DFD Level 0

DFD Level 1- Breeding Animal

Breeding the animal process is breakdown, there are 4 sub process in breeding the animal show in DFD level 1 in pictures below:

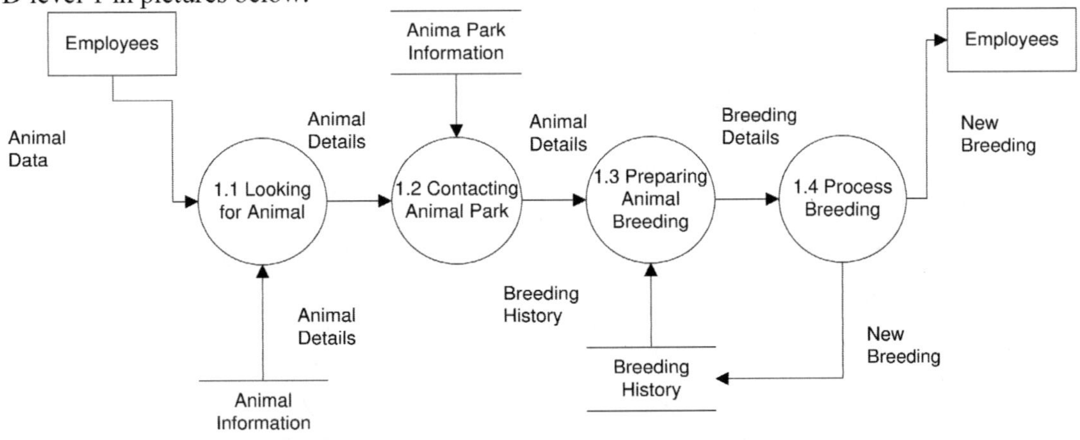

Figure 3. DFD Level 1

2.1.4 Advantages And Disadvantages

- Easy to understand for technical or non-technical persons because it uses simple symbols.
- Because it is very high level, it helps to define the boundaries of the system.
- The ability to represent the system in different levels. A global model in the first level can be worked out in more detail in the lower levels.
- Represent the information of data movements, this gives a clear view of what comes in the process and what not.
- Some symbols are being used in different kind of situations, this can be quite confusing.
- Because DFD is very high level, it isn't very useful when modeling complex processes.
- Modeling a large system will result in big models that are vague and probably will not cover the whole system.
- There is no possibility to model decisions in one model. A decision has to be modelled in more different models where each model stands for a decision-option. In a large, complex system this will result in hundreds of models with no real structure.
- Useless if without the pre-requisite details of the system.

2.2 Business Process Modeling Notation (BPMN) Version 1.1

The Business Process Modeling Notation (BPMN) is a graphical notation that depicts the steps in a business process [11]. BPMN shows the end to end flow of a business process. The notation has been specifically designed to coordinate the sequence of processes and the messages that flow between different process participants in a related set of activities [11] [12].

2.2.1 Notation

Events

An event is represented by a circle and is something that "happens" during the course of a business process [11]. These events affect the flow of the process and usually have a cause (trigger) or an impact (result). Events are circles with open centers to allow internal markers to differentiate different triggers or results. There are three types of Events, based on when they affect the flow: Start, Intermediate and

End. Events are also classified as Catching (as in, they might catch an incoming message to start the process) or Throwing (as in, they might throw a message at the end of the process)[12].

Figure 4. events [11]

Start event

Start Event indicates where a particular process will start. A start event is a circle that have to be drawn in thin line [11].

End event

Represents the result of a process; indicated by a single thick or bold border; and can only Throw, so is shown with a solid icon [11].

Intermediate event

Represents something that happens between the start and end events; is indicated by a tramline border; and can Throw or Catch (using solid or open icons as appropriate) [11] for example, a task could flow to an event that throws a message across to another pool and a subsequent event waits to catch the response before continuing.

Activity

An Activity is represented by a rounded-corner rectangle and is a generic term for work that company performs. An Activity can be atomic or nonatomic compound). The types of Activities are: Task and Sub-Process. The Sub-Process is distinguished by a small plus sign in the bottom center of the shape. [11]

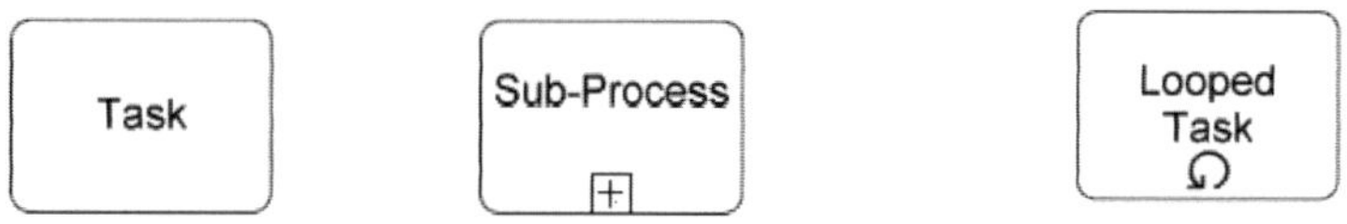

Figure 5. Activity [11]

Task

A task represents a single unit of work that is not broke down to a further level of business process detail without diagramming the steps in a procedure (not the purpose of BPMN) [11]

Sub-process

Used to hide or reveal additional levels of business process detail - when collapsed a sub-process is indicated by a plus sign against the bottom line of the rectangle; when expanded the rounded rectangle expands to show all flow objects, connecting objects, and artifacts.[11]

Has its own self-contained start and end events, and sequence flows from the parent process must not cross the boundary.

Transaction

A form of sub-process in which all contained activities must be treated as a whole, i.e., they must all be completed to meet an objective, and if any one of them fails they must all be compensated (undone). Transactions are differentiated from expanded sub-processes by being surrounded by a tramline border [11].

Gateway

A Gateway is represented by a diamond shape and will determine forking and merging of paths depending on the conditions expressed [11].

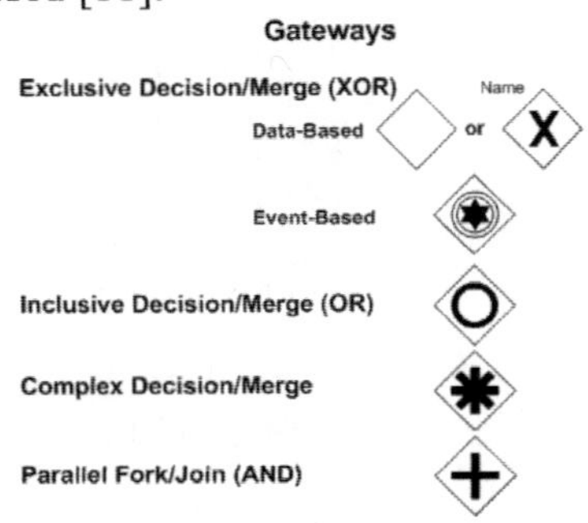

Figure 6. gateway symbol [11]

Object Connector

Flow objects are connected to each other using Connecting objects, which consist of three types (Sequences, Messages, and Associations) [11]:

Sequence Flow

A Sequence Flow is represented by a solid line with a solid arrowhead and is used to show the order (the sequence) that activities will be performed in a Process. The sequence flow may also have a symbol at its start, a small diamond indicates one of a number of conditional flows from an activity while a diagonal slash indicates the default flow from a decision or activity with conditional flows.

Message Flow

A Message Flow is represented with a dashed line, an open circle at the start, and an open arrowhead at the end. It tells us what messages flow across organizational boundaries (i.e., between pools). A message flow can never be used to connect activities or events within the same pool. Message flow used to show the flow of messages between two separate Process Participants (business entities or business rules) that send and receive them. In BPMN, two separate Pools in the Diagram will represent the two Participants.

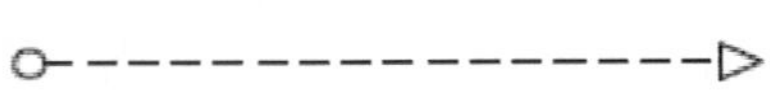

Association

An Association is represented by a dotted line. It is used to associate an Artifact or text to a Flow Object, and can indicate some directionality using an open arrowhead (toward the artifact to represent a result, from the artifact to represent an input, and both to indicate it is read and updated). No directionality would be used when the Artifact or text is associated with a sequence or message flow (as that flow already shows the direction).

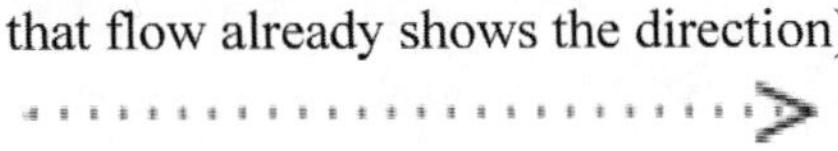

Swimlane

Many processes modeling methodologies utilize the concept of swimlanes as a mechanism to organize activities into separate visual categories in order to illustrate different functional capabilities or responsibilities. BPMN supports swimlanes with two main constructs. The two types of BPD swimlane objects are:

IOP Conf. Series: Materials Science and Engineering **128** (2016) 012010 doi:10.1088/1757-899X/128/1/012010

Pool

A Pool represents a Participant in a Process. It is also acts as a graphical container for partitioning a set of activities from other Pools, usually in the context of B2B situations.

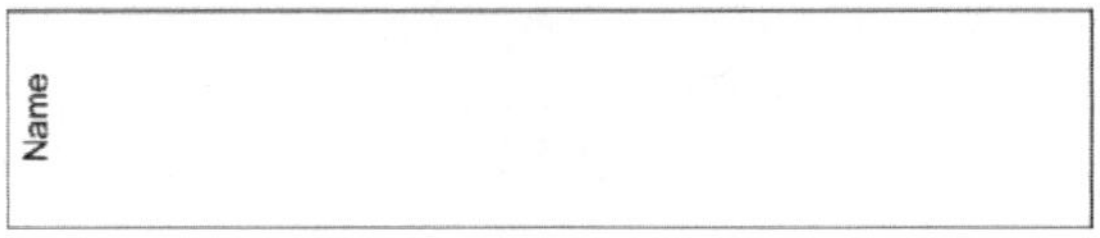

Figure 7. pool

Lane

Lane is a sub-partition within a Pool and will extend the entire length of the Pool, either vertically or horizontally. Lanes are used to organize and categorize activities.

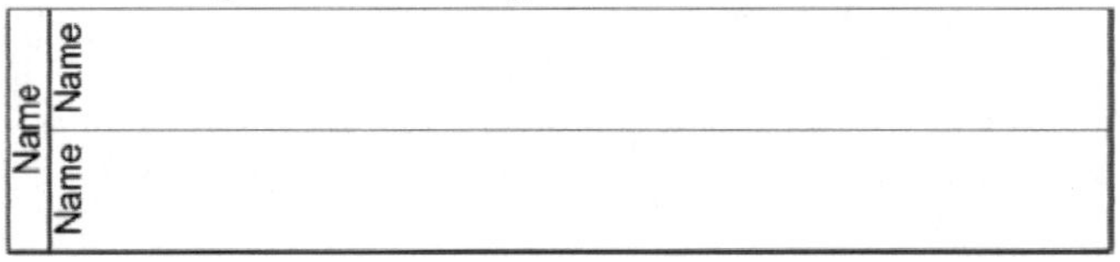

Figure 8. lane

Artifacts

BPMN was designed to allow modelers and modeling tools some flexibility in extending the basic notation and in providing the ability to additional context appropriate to a specific modeling situation, such as for a vertical market (e.g., insurance or banking). Any number of Artifacts can be added to a diagram as appropriate for the context of the business processes being modeled. The current version of the BPMN specification pre-defines only three types of BPD Artifacts, which are:

Data Object

Data Objects are a mechanism to show how data is required or produced by activities. They are connected to activities through Associations.

Figure 9. data object

Group

A Group is represented by a rounded corner rectangle drawn with a dashed line. The grouping can be used for documentation or analysis purposes, but does not affect the Sequence Flow.

Figure 10. group

IOP Conf. Series: Materials Science and Engineering **128** (2016) 012010 doi:10.1088/1757-899X/128/1/012010

Annotation

Annotations are a mechanism for a modeler to provide additional text information for the reader of a BPMN Diagram.

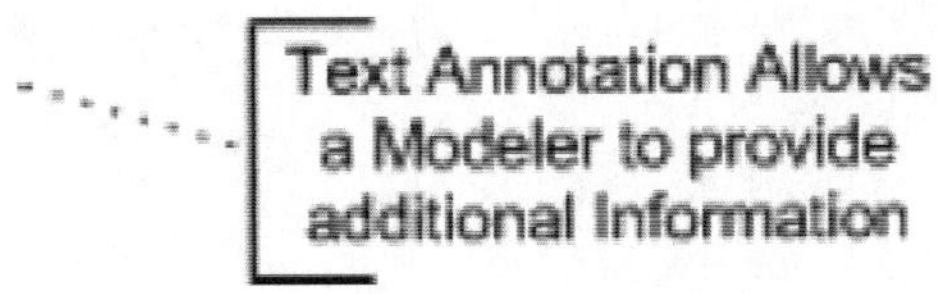

Figure 11. annotation

2.2.2 Applied to Somerleyton Animal Park

In the figure below show the pool indicated as process breeding the animal and the lane as headkeeper or animal keeper. There are 5 tasks inside the lane of animal keeper.

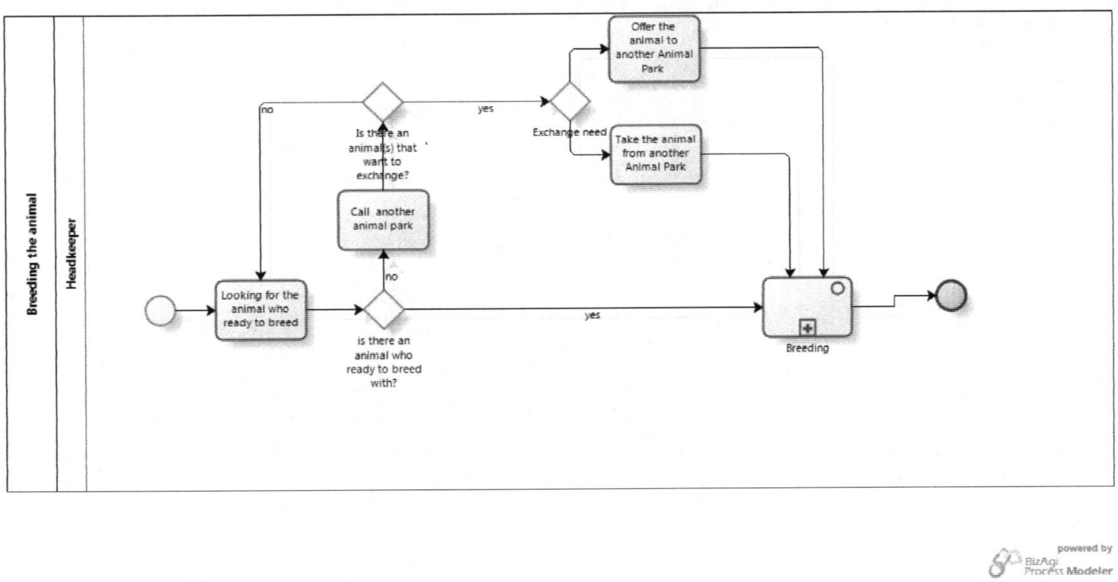

Figure 12. BPMN breeding animal

2.2.3 Advantages and Disadvantages

- Easy to understand the flow and the process
- It bridges the gap between business and technical systems personnel.
- It can be scaled for any kind of project ranging from smaller activities to complex activities.
- More likely to find the skills without having to pay for proprietary language specialists.
- It offers a process flow modeling technique that is more conducive to the way business analyst model
- Completeness notation depicting various processes.
- There are ambiguity and confusion in sharing BPMN models, for example: It is possible to model the same process with different notation variations
- There is no standardized file format for exchange BPMN models between different tools

IOP Conf. Series: Materials Science and Engineering **128** (2016) 012010 doi:10.1088/1757-899X/128/1/012010

2.3 Activity Diagrams

Activity diagrams are used to map workflows. It is a graphical representation of the steps, actions and decisions that are made in a certain process [10]. Activity diagrams are part of the modeling language UML (Unified Modeling Language), that is a set of different graphic notation techniques to create visual models.

2.3.1 Notation

Initial Activity

The initial activity is a big black circle that shows the starting point of the flow [10].

Activity

Represents an activity as a rounded rectangle [10].

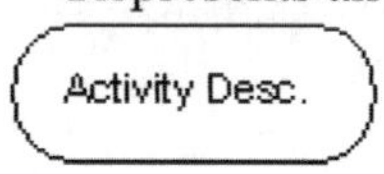

Decisions

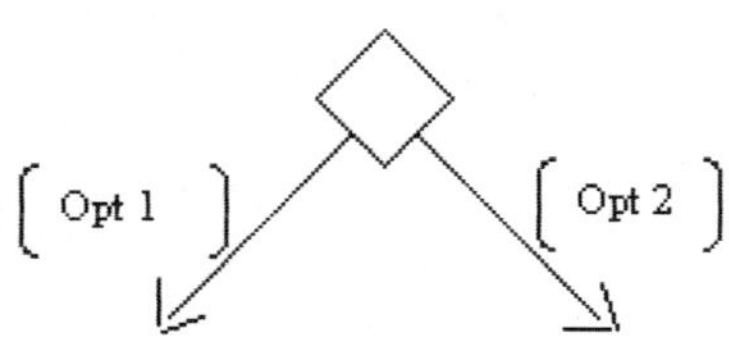

When a decision has to be made a diamond with different lines is stated [10]. The options of the decision are written next to each line. A decisive element can have as many options that are needed, but has to have at least two.

Signal

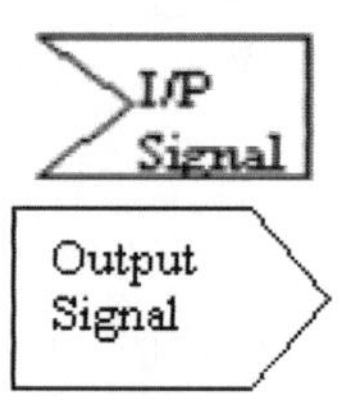

When an activity, sends or receives a message, that activity is called a signal. There are two different types of signals. An input signal and an output signal. An input signal is a message that is received by an activity and a output signal is a message that is sent by an activity [10].

Concurrent Activities

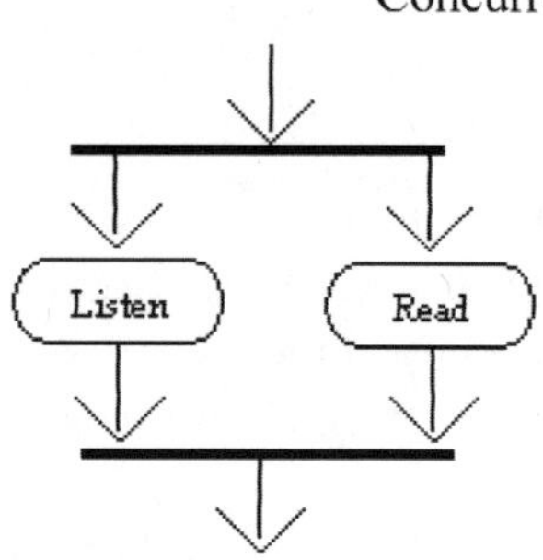

When activities take place at the same time they are called concurrent activities. For instance, it's possible to listen and read at the same time. These activities, then parallel with each other. The horizontal thick black lines say that the process waits for both parallel activities to be completed before moving to the next activity.[10]

Final activity

The end of the activity diagram is shown by a black spot within a white circle. It means that all activities in the diagram are ended. [10]

IOP Conf. Series: Materials Science and Engineering **128** (2016) 012010 doi:10.1088/1757-899X/128/1/012010

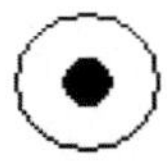

2.3.2 Applied to Somerleyton Animal Park

The figure below shows the activity diagram for breeding the animal process. The activity diagram consists of three initial activity: Looking for animal who ready to breed, breeding, and contact the animal park.

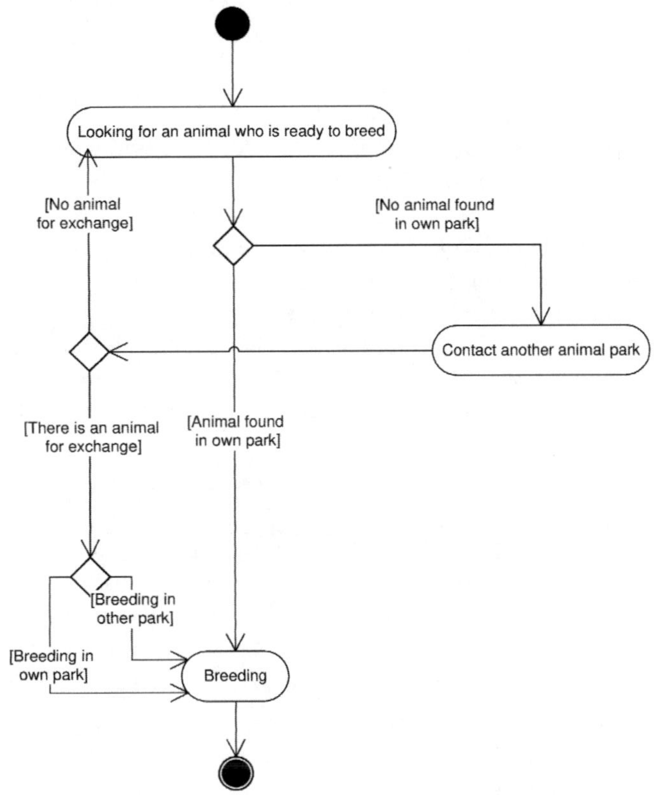

Figure 13. activity diagram breeding animal

2.3.3 Advantages and Disadvantages

- Activity diagrams are high level. This can be a big advantage when discussing business processes with the customer in an early stage of the development.
- Activity diagrams are easy to understand, also for someone with no technical background if it comes to modeling business processes.
- Activity diagrams have a flow so that it's clear which activities are performed by the outcome of the different decisions in the diagram.
- In activity diagrams it is possible the determine that activities run parallel with each other.
- There are no restrictions on what can or can't be an activity.
- Because activity diagrams are high level, it's difficult to describe complex processes.
- With activity diagrams it's not possible to describe which different objects or actors are performing which actions.
- There no distinguishes between activities, this can be confusing.

2.4. Integration Definition for Function Modelling (IDEF0)

IDEF0 is a modeling technique which can be used to display the actions, decisions and activities of an organization of system [1]. The model helps the analyst by analyzing the system and are commonly used to communicate the analysis to the customer. The activities can be described by their inputs, outputs, controls, and mechanisms.

IDEF0 can be used for new to develop systems or for analyzing existing systems. For new systems IDEF0 can be used to determine the requirements and functions for the system. In existing systems IDEF0 can be used to analyze the functions of the system and their mechanisms (Who is executing the function).

2.4.1 Notation

The IDEF0 notation consists of boxes that represents activities in the system and arrows that describes the relations between the activities [1].

- Activities (box)
    - Activities can be worked out in more detail (decomposition)
- Data and Object, that connects the activities together (arrow)
    - Inputs: transformed by Activity into outputs
    - Controls: dictate & constrain under what conditions transformations occur
    - Outputs: created as a result of activity transformations
    - Mechanisms: describe how the activity is to be carried out (resource, process, etc.)

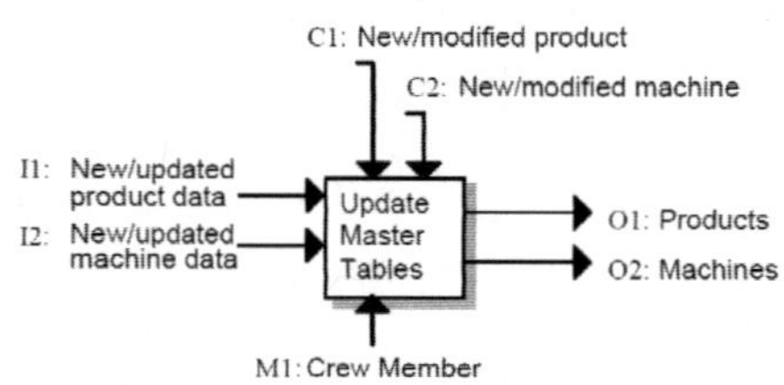

Figure 14. IDEF0 notation [1]

When we connect activities together, we have to keep in mind that only the following is allowed:

- To go from output to input;
- To go from output to control;
- To go from the output to mechanism;

2.4.2 Applied to Somerleyton Animal Park

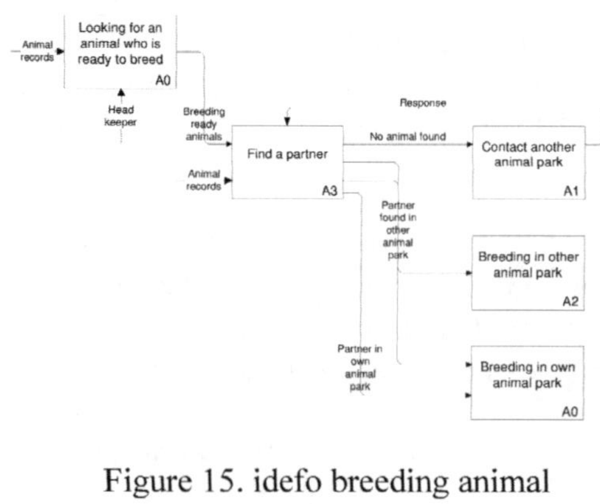

Figure 15. idefo breeding animal

2.4.3 Advantages and Disadvantages

- The distinction between input, controls, output and mechanisms will help the reader understanding the activity better.
- You can start with a high level model and work it out over time in more details
- Provides strict rules for modeling what will result in a clear and high quality model.
- Difficult to model, read and understand
- Not a very commonly used modeling technique

6. Result and Discussion

In this part we are going to compare the different techniques with each other. As an example, we use the process 'animal breeding' of the Somerlayton Animal Park. The models for this process are listed previously, but are repeated here to make things a little bit easier.

Model 1, DFD

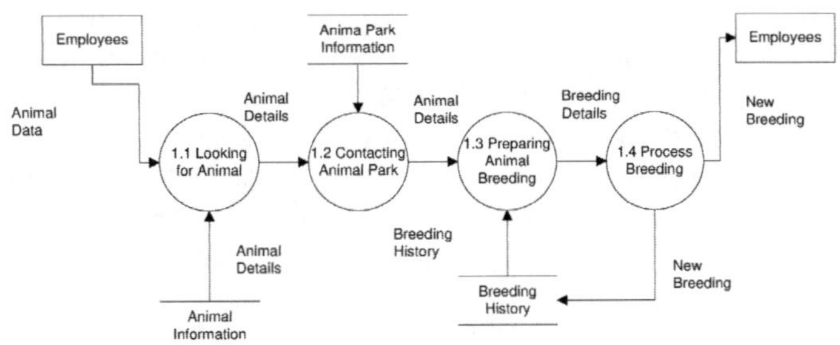

Figure 16 DFD Breeding The Animal Process

One reason for these differences are of course the developers. Not every developer thinks completely the same and can't empathize himself in the situation as good as another. The other reason is the features of each modeling technique. For instance, in

Model 1the DFD model for the process breeding an animal is listed. If you compare this model with the others you will see that in every other model some decisions are present. With DFD listing a decision in one model isn't possible, so in

Model 1 some important information is missing. In this case a user will not know that the breeding activity can be in their own zoo but also in another zoo.

Model 2, BPMN

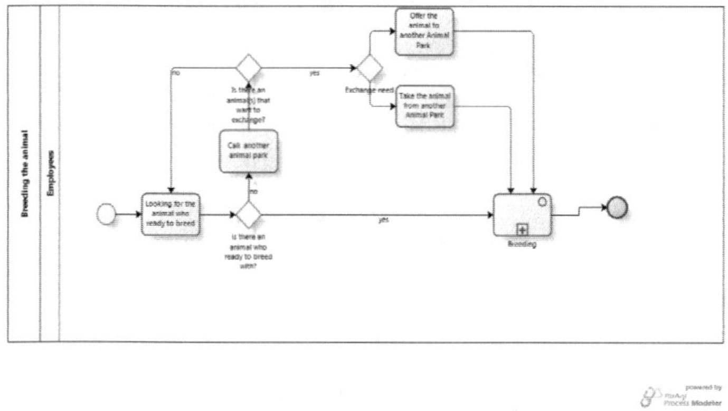

Figure 17 BPMN breeding the animal process

Everything about the process 'breeding an animal' is present in Model2. Decisions are clear and due the simplicity in symbols, the flow is easy to follow. The possibility of making decisions is also possible in activity diagrams and IDEF0 so that isn't a big plus compared to the other modeling techniques. What we do think that is a big plus are the symbols. In BPMN there are a lot of different symbols, but there are grouped in types. This keeps things simple, but also gives the analyst the choice in which detail he wants to model the process. Also the use of colors is handy, in One reason for these differences are of course the developers. Not every developer thinks completely the same and can't empathize himself in the situation as good as another. The other reason is the features of each modeling technique. For instance, in

Model 1the DFD model for the process breeding an animal is listed. If you compare this model with the others you will see that in every other model some decisions are present. With DFD listing a decision in one model isn't possible, so in

Model 1 some important information is missing. In this case a user will not know that the breeding activity can be in their own zoo but also in another zoo.

Model 2 the green round means the start of the process and the red one means the process stops. This sounds very logical, but the other techniques don't use colored symbols.

The negative side of BPMN is that despite it's easy to understand, we find it quite hard to model. In our experience modeling with BPMN takes a lot of time and arouses many discussing. Also, due all the possibilities, it's possible that different looking models are describing exactly the same process. These equivalent models can cause some confusion, especially when the analysts aren't very experienced.

Model 3, Activity Diagram

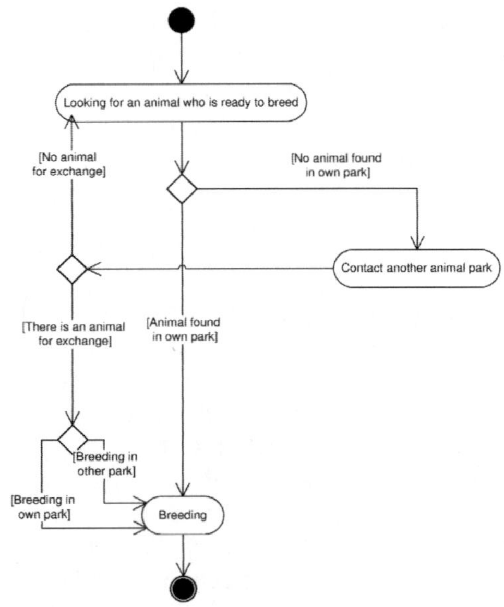

Figure 18. activity diagram breeding the animal process

Comparing Model3 with other models we can say that activity diagrams have the biggest similarity with BPMN models. It has the same rounded rectangles for the actions and also uses the diamonds for decision making. We see activity diagrams as a simple version of BPMN. The features of activity diagrams are also included in BPMN and are done almost the same way.

When comparing the different modeling techniques were wondering, does every model tell the same story? Basically, that answer is yes. But when these models are given, that are each made by using a different technique and by different developers the outcome will not be very equivalent.

Model 4, IDEF0

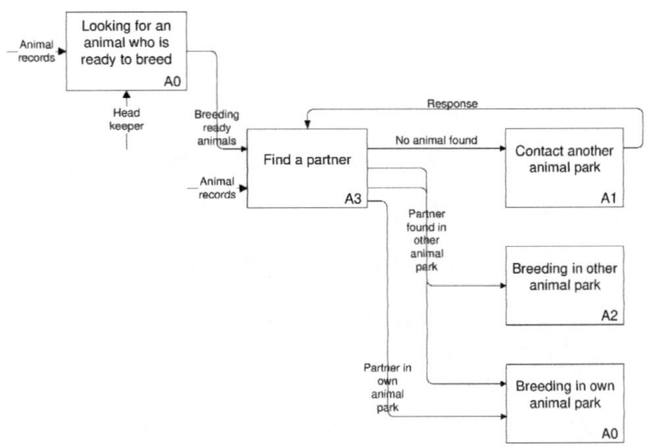

Figure 19. IDEF0 breeding the animal process

Model 4 is the toughest model to make and to understand, but provides the most information. The model of BPMN and the Activity Diagram says nothing about the input, output, mechanisms and controls. And that is what makes IDEF0 really interesting. The mechanisms and controls are sometimes hard to define, but the possibility to make clear what the input and the output of an activity area, is really a big plus of IDEF0. The feature of defining these so called ICOM's helps the developer to understand the process even better. Now he knows who and what is involved in each activity and what the result of each activity is. This advantage of IDEF0 can also be seen as a disadvantage because these kind of technical details is not interesting for example to the customer.

7. Conclusion

The conclusion is that no modeling technique is beforehand better than another. It all depends on the situation. DFD give the overview about the flow of data in the system about the source and destination of the data. DFD is not powerful to show the whole process that happened in the system. DFD does not have symbols to display branching process in the system.BPMN has a clear symbol to overview the whole process so people can easily understand the business process. Business people use BPMN to express and understanding process and technical based people use it to translate process into the system.Activity Diagram analyzes the case with describing what actions need to do and when that action should occurred. Activity diagram can describe complicated sequence of the system and modeling the parallel process in the system. IDEF0 modeling activity sequences whenever needed. Sometimes only the domain expert can understand the sequence and the activities represented by IDEF0.

8. References

[1] Ross D. T, SofTech. Inc. (1993, December 21). *IDEF Integrated DEFinition Method.* Retrieved 06 23, 2015, from IDEFØ Function Model Method: http://www.idef.com/pdf/idef0.pdf

[2] Aldin L. , de Cesare S. (2009). A Comparative Analysis of Business Process Modelling Techniques. *U.K. Academy for Information Systems (UKAIS 2009), 14th Annual* (pp. 1-17). Oxford: UKAIS 2009.

[3] Alzubidi N.H., Recker J., Bernhard E. (2011). *A Study of the Use of Business Process Modelling at SunCorp.* Brisbane: Business Process Management Research Group QUT.

[4] Braghetto K.R, Ferreira J.E, Vincent J. (2009). *Comparison of Modeling Approaches to Business.* France: Rapport de recherche n° 7065.

[5] DeMarco, J. (1979). *Structured Analysis and System Specification.* Englewood: Prentice-Hall.

[6] Fortis, A. (2006). Business Process Modelling Notation- An Overview. *Annals. Computer Science Series. 4th Tome 1st Fasc*, 41-49.

[7] H Shen, B Wall, M Zaremba, Y Chen, J Brownie. (2004). Integration of business modelling methods for enterprise information system analysis and user requirement gathering. *Computers in Industry volume 54, 54*(Computers in Industry), 307-323.

[8] Ould, M. (1995). *Business processes: modelling and analysis for re-engineering and improvement.* Wiley.

[9] Pressman, R. (2010). *Software Engineering: A Practitioner's Approach, Seventh Edition.* Mc Graw Hill Higher Education.

[10] Rumbaugh J, Jacobson I, Booch G. (1999). *The Unified Modeling Language Reference Manual.* Massachusetts: Addison-Wesley.

[11] White, S. (2006, October 16). *Introduction to BPMN.* Retrieved January 10, 2011, from IBM Software Group: http://www.bpmn.org/Documents/OMG_BPMN_Tutorial.pdf

[12] White, S. A. (2006). *Introducion to BPMN.* IBM Corporation.

[13] Yourdan E, L. L. (1979). *Structured Design: Fundamentals of a Discipline of Computer Program and System Design.* Englewood: Prentice-Hall.

IOP Conf. Series: Materials Science and Engineering **128** (2016) 012011 doi:10.1088/1757-899X/128/1/012011

Effect of addition of semi refined carrageenan on mechanical characteristics of gum arabic edible film

D Setyorini* and P R Nurcahyani

Program Studi Pendidikan Teknologi Agroindustri, Fakultas Pendidikan Teknologi dan Kejuruan, Universitas Pendidikan Indonesia.
Jalan Dr. Setiabudhi No 207 Bandung.

*Corresponding author: puji.rahmawati@upi.edu

Abstract. Currently the seaweed is processed flour and Semi Refined Carraagenan (SRC). However, total production is small, but both of these products have a high value and are used in a wide variety of products such as cosmetics, processed foods, medicines, and edible film. The aim of this study were (1) to determine the effect of SRC on mechanical characteristics of edible film, (2) to determine the best edible film which added by SRC with different concentration. The edible film added by SRC flour which divided into three concentrations of SRC. There are 1.5%; 3%; and 4.5% of SRC, then added 3% glycerol and 0.6% arabic gum. The mechanical properties of the film measured by a universal testing machine Orientec Co. Ltd., while the water vapor permeability measured by the gravimetric method dessicant modified. The experimental design used was completely randomized design with a further test of Duncan. The result show SRC concentration differences affect the elongation breaking point and tensile strength. But not significant effect on the thickness, yield strength and the modulus of elasticity. The best edible film is edible film with the addition of SRC 4.5%.

1. Introduction

Indonesia has become the largest producer of seaweed in the world, particularly *Eucheuma cottonii*. Statistical data while the UN Food and Agriculture Organization (FAO) published in March 2015 mentioned the production of Indonesian seaweed species *Eucheuma cottonii* in 2013 ranks first in the world that is as much as 8.3 million tones. As for the seaweed *Gracilaria* sp. in the same year, Indonesia ranks second after China, with a production of 975 000 tones [11]. Indonesia ministry of industry supports the cessation of exports of seaweed gradually.

Eucheuma cottonii seaweed has been used as a source of iodine to prevent goiter on the thyroid, fertilizer to the energy source of bioethanol [6]. Because of its function as a source of iodine, it was an opportunity using algae as a mixture of dodol [13]. A development of algae as an edible film carrageenan was reviewed [4]. Edible film may be made of hydrocolloids, lipids, and composites [8]. When this was developed on the basis of hydrocolloids edible film, including the use of polysaccharides [16, 17, 18]. Films made of composite materials with k-carrageenan and locust bean gum [10]. A mixture of two polysaccharides enables better mechanical properties and water vapor permeability. Gum arabic is one of main material base for edible film [9]. Edible films gum arabic can be combined with chitosan [15]. But the packaging must remain flexible so that the packaging is not easily torn. In addition, the packaging must be able to withstand the vapor permeability of water the material is packed. Therefore, it takes a lipid as a plasticiser to increase the flexibility of the package based on polysaccharides [14]. To improve the mechanical properties of the film of gum arabic can be added Semi Refined Carrageenan (SRC) and glycerol.

SRC is a product derived from seaweed *Eucheuma cottonii* that alkalized. Carrageenan alkalization process decrease a part of sulphate content. This can improve the strength of the carrageenan gel better than carrageenan. The aim of this study were (1) Knowing the mechanical characteristics of the SRC edible film, (2) Determine the best treatment of SRC edible film. SRC flour divided into three concentrations of 1.5%; 3%; and 4.5% with the addition of glycerol 3% and 0.6% arabic gum.

2. Materials and methods

SRC used in this study were made from seaweed powder. Each edible film added by SRC flour with different concentrations, and then carried out a mechanical characteristics and the water vapor permeability measurement.

2.1. Materials used.

The main material used is seaweed of *Eucheuma cottonii* which bought from traditional market in Bandung, West Java. The chemicals used are Potassium Hydroxide (Merck), Gum arabic (Bratachem), glycerol (Bratachem) and distilled water. The glasswares used are glass beaker, hot plate stirrer, thermometer, measuring cup, Erlenmeyer, aluminum foil, pan and the oven fan.

2.2. SRC flour preparation.

SRC made from *Eucheuma cottonii* extract powder with a size of 60 meshes. SRC flour preparation process is modified from [3]. Algae washed to remove impurities such as sand and gravel. Then algae dipped in 3% of Potassium Hydroxide solution for 30 minutes to remove salts and impurities. Then algae boiled in 8% of Potassium Hydroxide solution with a ratio of dry seaweed and solution of 1:8 (w/v) at 75° C ± 2 °C for 3 hours. Wash algae until the wash water has a neutral pH. Then slice algae with a size of 0.5 – 1 cm. Then algae should be dried at 37°C for 1-2 days. Dried algae grinded and filtered with a 60 mesh screen.

2.3. Edible film preparation.

Edible films made from 0.6% gum Arabic, 3% glycerol and SRC with concentration of 1,5%; 3% and 4,5% respectively. A method of preparation an edible film refers to [12] with modifications. The first step in edible film preparation is dissolution of flour SRC into 80 ml of distilled water (mixture A). Once uniformly mixed, heating the mixture A on a hot plate stirrer for 15 minutes at a temperature of 50°C. Then add the mixture B (3% glycerol and 0.6% gums Arabic was dissolved in 20 ml distilled water) and heat to 60 °C for 15 minutes. Then the air needs to reduce with a vacuum pump (degassing) for 10 minutes. Films printed on glass mold as edible film A (SRC 1.5%), edible film B (SRC 3%), and edible film C (SRC 4.5%). Each edible film put in a drying oven for 10-12 hours.

2.4. Characterization and Measurements.

2.4.1. Measurement of the mechanical characteristics of edible film

The samples that were printed then tested the straining until the film broken. This test used a universal testing machine Oriented Models Co. Ltd. UCT-5T.

2.4.2. Measurement of the speed of water vapour transmissions

The test is performed using Standard Test Methods for Water Vapor Transmissions of Materials [1] modified. SRC edible film to be tested is placed in a porcelain dish containing 10 g of silica gel. The edge of the cup covered with insulation. The cup and edible film were weighed, put into a plastic jar containing 100 ml of 40% sodium chloride solution, and then the jar is sealed. Each time the cup is weighed. This observations for 8 hours. Vapor permeability of water is expressed as the slope of the cup-weight gain (g / h) divided by the area of the test sheet (m^2). The water vapor transmission is calculated by the formula:

$$WVT = \frac{\Delta W}{t\ x\ A}$$

Where, W is weight changes edible film after 8 hours, t is time (hours), A is area of the surface area of the film (m^2).

3. Experimental design

This study used a completely randomized design (CRD), which is arranged with one factor, namely the concentration of the SRC. SRC concentration composed of three levels, namely 1.5%; 3%; and 4.5%. On mechanical examination five repetitions while the transmission permeability of water vapor carried with two repetitions long test times for eight hours. Data were initially evaluated by analysis of variance (ANOVA), and significant results were further analyzed using Duncan's multiple range tests ($p < 0.05$) to compare the mean values of films' properties.

4. Result and Discussion

4.1. The thickness of edible film

Based on the average results can be described an increase of the thickness of the added edible film SRC. The higher concentration of SRC added to the edible film, the thickness of the edible film will be even greater. The average result of edible film thickness has shown on Figure 1.

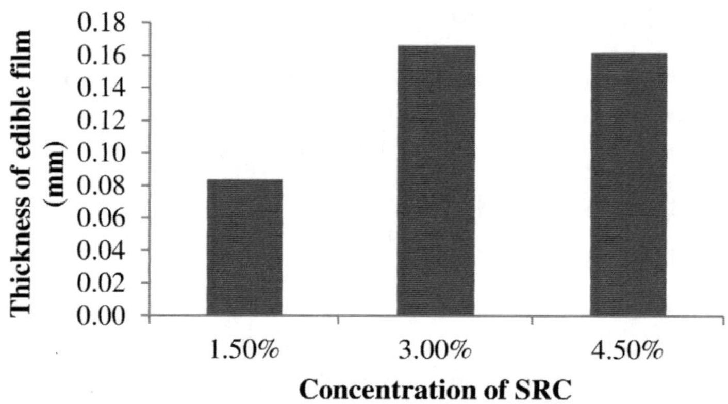

Figure 1. The average value of edible film thickness which SRC added.

Solution of edible film A molded more easily than edible film B and C. It shows the concentration of SRC affects the texture of the solution. The more liquid of solution make easier to print uniformly, but if the solution more viscous and have jendal properties, it should use a tool to flatten the surface of the film. Jendal properties in edible film solution are influenced by the concentration of the gum arabic and SRC. Gum arabic absorbs the free water contained in the solution of the edible film SRC. The fluid which trapped in the gum solution is then formed a gel [5].

Gum arabic contains a protein consisting of amino acid and hydroxyl groups which are hydrophilic. This is the hydrophilic group which can form hydrogen bonds with one or more molecules of water, so that it can absorb water and keep it in a molecule. Then, a liquid form a viscous colloidal with gel structure. The addition of less concentration of gum and SRC could form a viscous solution. The polysaccharide gum group and its derivatives (including carrageenan) are characterized by its ability to produce highly viscous solutions at low concentrations [7]. The result of thickness test obtained there is not significantly different between edible film B and C. But the best thickness is edible film A, its about 0,082 mm.

4.2. Tensile strength of the edible Film
The higher concentration of SRC increased the tensile strength of the film. The results of variance calculations show that the F-results > F-table (28.52) > F (3.89) in 5% and 1% level (6.93). There is a significantly different treatment. Edible film A and B have SRC concentrations lower than edible film C, so the film has a tensile strength value less than C. The less solids content in the solution is lower tensile strength values. The average of SRC tensile value shown in Figure 2.

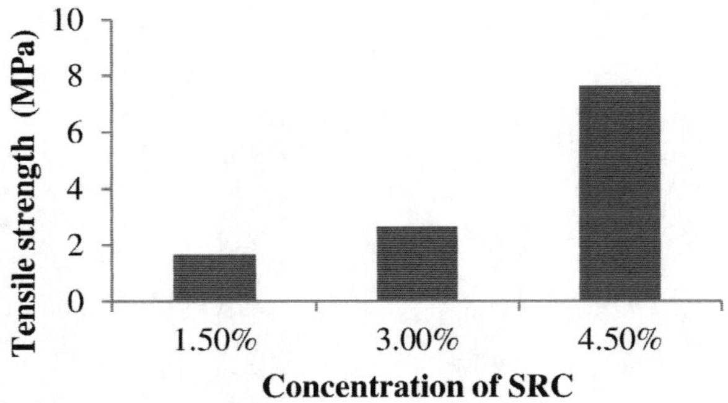

Figure 2. The average tensile strength value of SRC edible film

4.3. Elongation of the edible film
Elongation of an edible film is provided by the number of force which makes an edible film broken. Then measured a number of edible film lengthening before separating, then we calculate the average of the results. The higher concentration of SRC added make the edible film texture becomes more rigid. The higher concentration of SRC, elongation or lengthening becomes lower. The addition of the gum arabic in edible film compositions also affect to the texture of the edible film. The average elongation value of edible film shown in Figure 3.

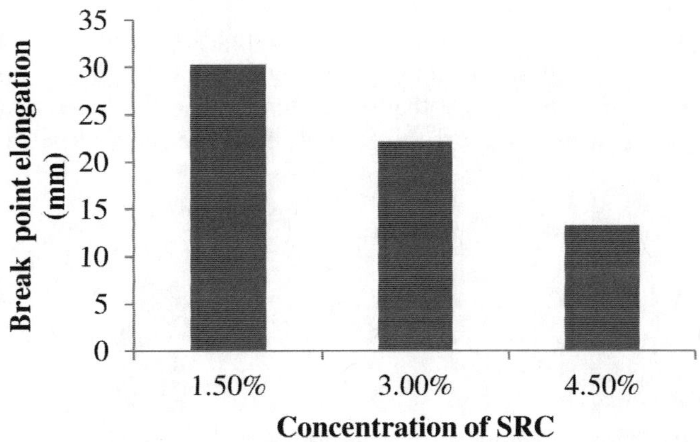

Figure 3. The average elongation value of SRC edible film.

Indeed, both SRC and gum arabic is a substance which can bind the water such that the higher concentration of the SRC edible film solution is more viscous. Once dry, edible film with the higher concentration of SRC made the edible film become harder. The best results edible film elongation values are edible film C. The solid film that is formed, it is increasingly difficult for the elongated film so that it reduces the percentage of the value of the extension. Powerful edible film should pack the food.

4.4. Yield stress of SRC edible film.

Yield is the edible film indicated value before changes when pressurized and forced. In the case of linear polymers, the yield point is often defined as the expected maximum stress. Indeed, the measurement of the yield expected maximum elasticity, very sensitive to how the load applied during the tensile test [2].

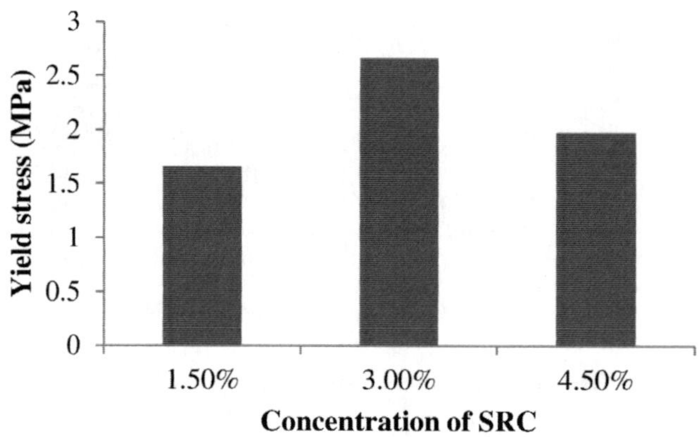

Figure 4. The average yield stress value of SRC edible film.

The results of tests shown the addition of SRC on different concentrations of edible film does not provide a real effect on the yield of edible film. It is assumed lack of accuracy when testing the yield strength during the test. The yield stress value must be measured accurately before edible film changes due to the force on the edible film. The average yield stress value of SRC edible film shown in Figure 4.

4.5. Modulus of elasticity of edible film.

Modulus of elasticity is a linear elasticity which uses a comparison between the strain and the stress. The test results showed that higher addition of SRC concentration causes higher the modulus of elasticity. The elasticity differences suggest that the differences in the concentration of edible SRC to provide a different modulus of elasticity of the film. The average modulus elasticity value of edible film shown in Figure 5.

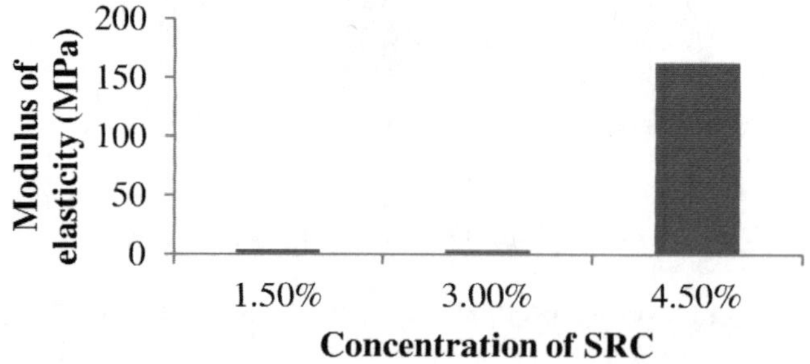

Figure 5. The average modulus elasticity value of SRC edible film.

4.6. Water vapor permeability of edible film.

The results of the vapor permeability of water from the edible film for 8 hours showed that the higher concentration of the added SRC causes decreasing of vapor permeability of water. The results of analyzing of variance show that there was no significant difference, either at the 5% and 1%. However, a good edible film is edible film with the lowest vapor permeability of water. Test results showed that the edible film to the value of the lower water vapor permeability is an edible film of C, with an average of 0.975 g-water / m^2.8h. Water vapor permeability of SRC edible film shown in Figure 6.

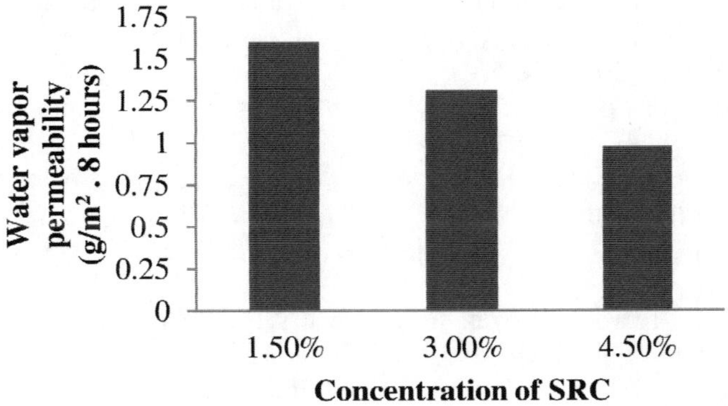

Figure 6. Water vapor transmission of SRC edible film.

5. Analysis of variance the mechanical characteristics of the edible film variance

Table 1 shows the variance analysis results of the mechanical characteristics of edible film SRC. Other tests have shown that the addition of SRC Duncan 3% and 4.5% on the same influence edible film thickness statistically. With regard to the modulus of elasticity characteristics known to increase by 4.5% SRC differ significantly from 1.5% to 3%. Meanwhile, further SRC 1.5% and 4.5% did not differ significantly, but significantly over the addition of 3% SRC. But adding CBC with different concentrations had a significant effect on the elongation point of the parameters at break.

Table 1. The variance analysis results and duncan test results.

Mechanical chracteristics	Concentration of SRC		
	1,5% (A)	3% (B)	4,5% (C)
Thickness (mm)	0,0838^{a}	0,1660^{b}	0,1620^{b}
Tensile strength (MPa)	1,6615^{a}	2,6564^{b}	7,6352^{c}
Break point elongation (mm)	30,2750^{a}	22,1342^{b}	13,2522^{c}
Yield stress (Mpa)	1,6586^{a}	2,6634^{b}	1,9796^{a}
Modulus of elasticity (MPa)	3,9907^{a}	3,5460^{a}	162,8544^{b}

Conclusion

The result show SRC concentration differences affect the elongation breaking point and tensile strength. But not significant effect on the thickness, yield strength and the modulus of elasticity. The best edible film is edible film with the addition of SRC 4.5%.

Acknowledgment

The authors are grateful for the support of the civitas academica of Program Studi Pendidikan Teknologi Agroindustri, Fakultas Pendidikan Teknologi dan Kejuruan, Universitas Pendidikan Indonesia.

References

[1] American Society for Testing of Material. Standart Test Methods for Water Vapor Transission of Materials. ASTM.e96.1995.

[2] ASM International. Tensile Testing, Second Edition, Introduction to Tensile Testing. 2004.

[3] E. N. Dewi, Y. S. Darmanto and Ambariyanto. Characterization and Quality of Semi Refined Carrageenan Products from Different Coastal Waters Based on Fourier Transform Infrared Technique. Journal of Coastal Develpopment 16 (2012) 25 – 31.

[4] E. T-Kafrani, H. Shekarchizadeh, M. M-Behabadi Development of edible films and coatings from alginates and carrageenans. Carbohydrate Polymers 137 (2016) 360–374.

[5] F. G. Winarno. Kimia Pangan dan Gizi. Gramedia Pustaka Utama, Jakarta. 2002.

[6] J. F. N. Abowei and E. N. Ezekiel. The potentials and utilization of Seaweeds. PSCI Publications, Sci. Agri. 4 (2013) 58-66.

[7] J. M. deMan,. Principles of Food Chemistry. Aspen Publishers, Inc. Gaithersburg, Maryland. 1999.

[8] I.G. Donhowe and O.R. Fennema. Edible films and coatings: characteristics, formation, definitions, and testing methods. in: J.M. Krochta, E.A. Baldwin, M. Nisperos-Carriedo (Eds.) Edible Coatings and Films to Improve Food Quality. Technomic Publ. Co., Inc, Lancaster, PA; 1994:1.

[9] J. Milani, & G. Maleki. Hydrocolloids in food industry. In B. Valdez (Ed.), Food Industrial Processes – Methods and Equipment. 2012 : 17–38. InTech.

[10] J. T. Martins, M. A. Cerqueira, A. I. Bourbon, A. C. Pinheiro, B. W. S. Souza, A. A. Vicente. Synergistic effects between k-carrageenan and locust bean gum on physicochemical properties of edible films made thereof. Food Hydrocolloids 29 (2012) 280-289.

[11] Kkp.go.id. Indonesia produsen rumput laut terbesar dunia. 2015.

[12] Layuk, Payung; Djagal W.M. dan Haryadi. Jurnal Teknologi dan Industri Pangan: Karakterisasi Komposit Film Edible Pektin Dading Buah Pala (*Myristica fragrans Houtt*) dan Tapioka. 13 (2002).

[13] M. Astawan , S. Koswara and F. Herdiani. Pemanfaatan rumput laut (*Eucheuma cottonii*) untuk meningkatkan kadar iodium dan kadar serat pangan pada selai dan dodol. Jurnal.Teknol. dan Industri Pangan, 25 (2004) 61-69.

[14] M. Jouki, N. Khazaei, M. Ghasemlou and M. HadiNezhad. Effect of glycerol concentration on edible film production from cress seed carbohydrate gum. Carbohydrate Polymers 96 (2013) 39– 46.

[15] M. Maqbool, A. Ali, P. G. Alderson, N. Zahid, and Y. Siddiqui. Effect of a Novel Edible Composite Coating Based on Gum Arabic and Chitosan on Biochemical and Physiological Responses of Banana Fruits during Cold Storage. *J. Agric. Food Chem.*, 59 (2011) 5474–5482.

[16] N. Khazaei, M. Esmaiili, Z. E. Djomeh, M. Ghasemlou and M. Jouki. Characterization of new biodegradable edible film made from basilseed (*Ocimum basilicum L.*) gum. Carbohydrate Polymers 102 (2014) 199– 206.

[17] Q. Ma, D. Hu, H. Wang and L. Wang. Tara gum edible film incorporated with oleic acid. Food Hydrocolloids, 56 (2016) 127–133.

[18] S. Galus and J. Kadzinska. Food applications of emulsion-based edible films and coatings. Trends in Food Science & Technology 45 (2015) 273-283.

Analysis of temperature difference on the total of energy expenditure during static bicycle exercise

Sugiono*

Industrial Engineering Department, Univeristas Brawijaya (UB), Indonesia

*Corresponding author: 1sugiono_ub@ub.ac.id

Abstract. How to manage energy expenditure for cyclist is very crucial part to achieve a good performance. As the tropical situation, the differences of temperature level might be contributed in energy expenditure and durability. The aim of the paper is to estimate and to analysis the configuration of energy expenditure for static cycling activity based on heart rate value in room with air conditioning (AC)/no AC treatment. The research is started with study literatures of climate factors, temperature impact on human body, and definition of energy expenditure. The next step is design the experiment for 5 participants in 2 difference models for 26.80C – 74 % relative humidity (room no AC) and 23,80C – 54.8 % relative humidity (room with AC). The participants' heart rate and blood pressure are measured in rest condition and in cycling condition to know the impact of difference temperature in energy expenditure profile. According to the experiment results, the reducing of the temperature has significantly impact on the decreasing of energy expenditure at average 0.3 Kcal/minute for all 5 performers. Finally, the research shows that climate condition (temperature and relative humi dity) are very important factors to manage and to reach a higher performance of cycling sport.

Keywords: energy expenditure, climate factors, static bicycle, hearth rate

1. Introduction

Bicycling is growing in popularity both as a form of transportation and for recreation in Indonesia. Many people start to realize the importance of sport and leisure in keeping and increasing human healthy. It is widely acknowledged that cycling is one of the cheapest and the best ways for people to achieve good health and fitness. People who cycle regularly could be live longer and stronger than those who do not exercise by bicycle. Moreover, the cycling facilities, participants and industry can be a significant driver of and contributor to Indonesian local economies. Based on the research before, bicycling gave tangible positive outcomes for economic, social, climate change and healthy lives [1]. These researches typically include categories such as industry (which includes manufacturing, sales and repair) and tourism. The some advantages of bicycling sport are: build strength muscle, increasing stamina, improves cardio-vascular fitness, eats up calories, improves heart health, improves human body coordination, and reducing stress [2].

According to explanation above, it can be concluded that bicycling is very popular for Indonesian community for doing sport and for recreation. Hence, it is reasonable that the activity should be conducted to physical workload and physiological condition. Level of heart rate (LHR) will be used to evaluate and to monitor the cyclist during their activity in order to keep in highest performance. The normal people generally has rest hearth rate at averages 60 to 80 beats per minute and for highly trained endurance athletes has as low as 30 bpm [3]. Bedny Z. Gregory and friends in their paper of "A heart rate evaluation approach to determine cost-effectives an ergonomics intervention" noted that the total amount of heart rate is not exceed than 100 bpm and 4.17 kcal/min [4]. The average of LHR more than 100 bpm requires additional break time. When studying dynamic physical work, heart rate is the most informative measure for evaluating physical workload and environmental stress [4].

The importance of the hearth rate information can be used for managing the energy expenditure during bicycling activity. Ergonomics intervention is suitable for solving the problem. Knowledge of the metabolic rate is required to assess the working practices, energy cost of specific jobs or sport activities, and the total energy cost of an activity. As long as cyclist can control their heart rate or energy spend are not exceed than 100bpm or 4,17 kcal/ min, the top performance and the durability will be achieved [5]. Controlling and monitoring the energy expenditure are not easy for scientist. There are many factors influence in the problems e.g. body posture, human healthy, age, weight, climate (temperature, relative humidity), track, etc. The research will be focused on the impact of temperature in bicycling activity correlated in managing the energy expenditure. Static bicycle with difference workload will be applied in the experiment. In sort, the purpose of the paper is to investigate the difference impact of energy expenditure for cyclist in load variably. As a result, top performance and durability for the activity can be managed easily.

2. Experimental Methods

2.1. Research Materials

Cycling is one of human activity that correlated with muscular effort, work physiology and anthropometry. It is called a physical ergonomics which affected the capabilities of worker (human) to create an efficient job. A lot of thinks to study the physical ergonomics, such as: metabolism, heart rate, blood pressure, respiratory system (VO2), energy expenditure etc. To study the worker's performance based on the physical ergonomics is very complex which influenced by human posture, mental, habit and food consumption.

A lot of research used energy expenditure to measure the human performance in many goals as sports, health and recovery. The British Nutrition Foundation (BNF) defined an energy expenditure as the total of the basal metabolic rate (the amount of energy expended while at complete rest), the thermal effect of food (TEF, the energy required to digest and absorb food) and the energy expended in physical activity. Energy expenditure is proportional to hearth rate and VO2 in doing activities [6]. There are two kinds of measurement the energy expenditure for direct and indirect calorimetric. Direct calorimetry (metabolic chamber) is determined by calculating the production of heat by the body. On the other hand, indirect calorimetry is determined by employed VO2 during rest and activity. Oxygen consumption is considered as the standard for measuring the physiological intensity of exercise. The maximal oxygen uptake (VO2max) is important in that it is generally considered the main factor of the aerobic fitness of a cyclist. VO2 and energy expenditure are calculated for athletics, workers hospitals to determine the capability of the body to perform work, to measure the efficiency of performing various activities. In this case, the efficiency is defined as work done divide energy used.

Recent studies have shown that heart rate can successfully be used as a proxy for the measurement of rate of oxygen consumption or metabolic [7]. The heart rate investigation such as measurement of oxygen consumption (VO2max) needs special technical knowledge, skill and sophisticated instrumental setup. It becomes difficult to determine the oxygen consumption of a human with limited resources. As consequence, many researches exploit the relationship among heart rate, blood pressure and O2 consumption. According to the journal noted that the VO2max can be calculated as formula 1 bellow [8]:

$$VO2max = 15\frac{mliter}{Kg.minute}x\frac{HRmax}{HRrest}$$

Where: VO_{2max} = Maximum oxygen consumption (mliter)

HR_{max} = maximum heart rate (bpm)

HR_{rest} = Heart rate in rest condition (bpm)

2.2. Experiment Design

According to the aim of the research as mentioned before, the research will investigate the correlation between differences of the environment temperatures and the energy expenditure in cycling activity. The research used a static bicycle to test the correlation for five male cyclists. Table 1 show personal details for five participants included age, weight and height. The data is used to determine the total of energy expenditure during activity based on heart rate and VO2.

Table 1:List data of five personal cyclists

No	Participant	Age	Weight	Height
1	Participant 1	22 years	175 cm	78 Kg
2	Participant 2	21 years	153 cm	44 Kg
3	Participant 3	22 years	176 cm	70 Kg
4	Participant 4	21 years	171 cm	71 Kg
5	Participant 5	20 years	161 cm	65 Kg

Following is the steps for doing the research to know the impact of difference temperature in cycling activity. The difference temperature is created by using air conditioning (AC) and without air conditioning (no-AC). To do the test, two days experiment have been done for the first day for test in normal environment condition with T = and RH = (no AC) and the second day for test in design environment condition with T = 23.8^{0}C, RH = 54.8 % (AC). The steps of the research for the first day are:

- Measuring heart rate and blood pressure for all participants in the rest condition without air conditioning (no-AC). The participants have the same treatment before do the test which having breakfast 4 hours before cycling with the same menu.
- Warming up using static bicycle with the same speed and workload in 4 minute.
- Measuring heart rate and blood pressure for all participants each minute for 5 minute cycling in no AC condition.

The steps of the research for the second day (using AC) are:

- Designing the environment condition for AC until steady condition which having same condition of air quality, temperature and relative humidity.
- Measuring heart rate and blood pressure for all participants in rest condition with air conditioning. The participants have the same treatment before do the test which having breakfast 4 hours before cycling with the same menu.
- Warming up using static bicycle with the same speed and workload in 4 minute.
- Measuring heart rate and blood pressure for all participants each minute for 5 minute cycling in no AC condition.

3. Results and Discussion

Measuring heart rate and blood pressure are used to analysis the demand of energy expenditure correlated with difference of environment temperature. Figure 1 explains the comparison of heart rate for five participants in no AC and environment with AC condition. It can be concluded that by using AC, it will increase the average of heart rate for participant from 86.6 bpm to 90.26 bpm or 4.2 %.

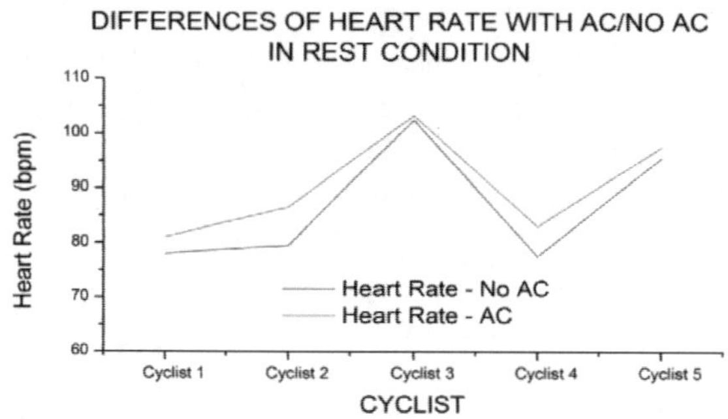

Fig. 1. Heart rate comparison for participants in rest condition with AC and without AC

Figure 2 explains the comparison of blood pressure systolic/diastolic for five participants in no AC and environment with AC condition. It can be concluded that AC increased the average of blood pressure systolic for participant from 123.6 mmHg to 137.4 mmHg or 11.2 % and from 71.8 mmHg to 81.4 mmHg or 13.4 % for diastolic. It can be declared that AC in rest condition will increase the blood pressure for participants (cyclist) with average age 21.2 years old, weight 65.6 Kg and height 167.2 cm. Subjectively, participants in tropical condition will not enjoy in the low environment temperature as indicated by increasing both of blood pressure and heart rate.

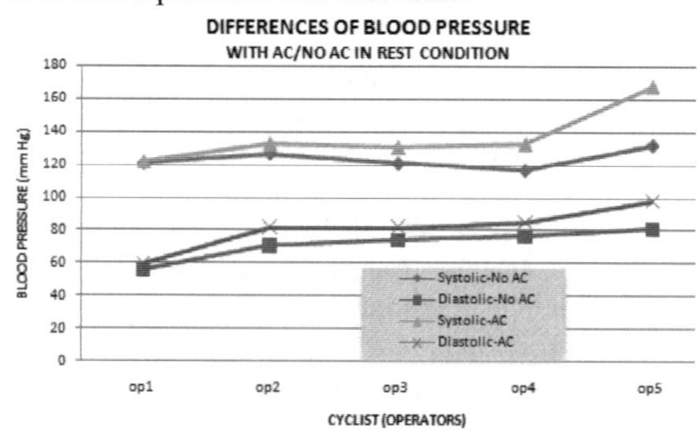

Fig.2. Blood pressure comparison for participants in rest condition with AC and without AC

Figure 3 explains the comparison of heart rate for five participants in cycling condition with AC/no AC condition. It can be concluded that AC decreased the average of heart rate for participant from 111.2 bpm to 107.7 bpm or 3.1 %. It can be declared that generally AC in working (cycling) condition will decrease the heart rate for participants (cyclist). Except, the participant 5 was still has higher heart rate in AC condition than in no AC with delta heart rate = 2.7 bpm.

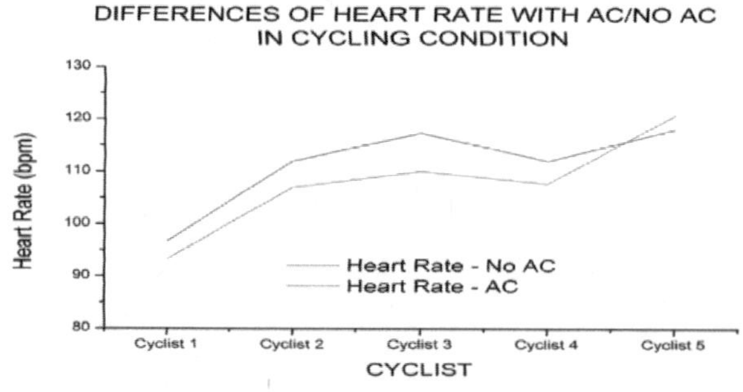

Fig.3. Heart rate comparison for participants in cycling condition with AC and without AC

Figure 4 explains the comparison of blood pressure systolic/diastolic for five participants in cycling activity with AC/ no AC condition. It can be concluded that AC decreased the average of systolic-blood pressure for participant from 134.2 mmHg to 128.9 mmHg or 4.0 % and from 79.1 mmHg to 73.6 mmHg or 7.0 % for diastolic. It can be declared that AC in working (cycling) condition will decrease the blood pressure for participants (cyclist).

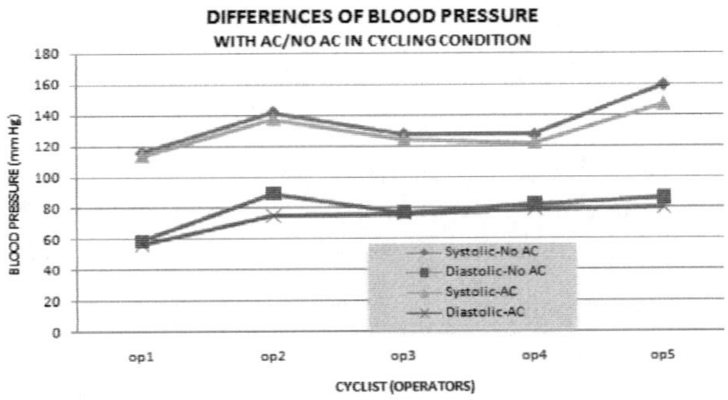

Figure 4. Blood pressure comparison for participants in cycling condition with AC and without AC

The total of physical energy expenditure is calculated by multiply the amount of O2 intake (VO2) with 4.8 Kcal/minute [5]. The VO2 can be calculated by equation 1, as example the participant 3 with average heart rate for working (6 minute in cycling) = 117.3 bpm, heart rate for rest = 102.5 bpm age = 21 years old, weight = 70 Kg, height = 171 cm cycling in no AC condition will need VO2:

$$VO2 = \frac{15\, mliter}{Kg\ x\ minute}\, x\, \frac{117.3}{102.5}$$

$$= \frac{15\ \ x\ 70\ x\ 6\, mliter}{1}\, x\, \frac{117.3}{102.5}$$

$$= 1.20\ \text{Liter/minute}$$

As a result, the participant 3 spend the energy expenditure = 1.20 liter/minute x 4.8 kcal/liter or 5.76 kcal per minute for static cycling test. Figure 5 shows the differences of energy expenditure for all cyclists in AC and no AC condition. From the graph, it can be inferred that energy expenditure for cyclists in AC condition needs less energy than cyclists in non AC condition.

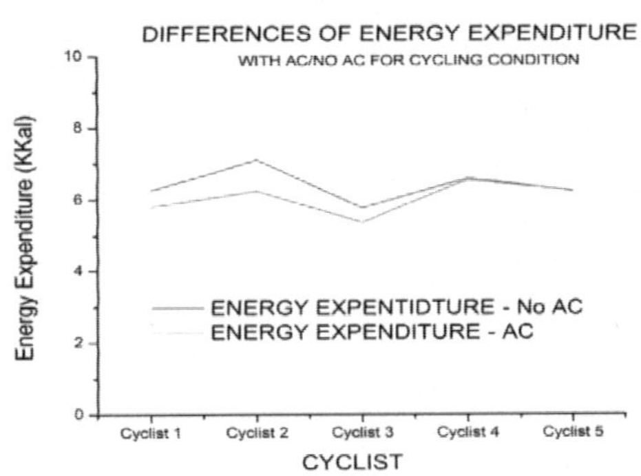

Fig.5. Difference of the total of energy expenditure for cyclists in AC and no AC

4. Conclusions

The research has successfully presented the correlation of heart rate/ blood pressure, human activity, energy expenditure and environment condition (temperature and relative humidity) in tropical condition. According to 5 participants, low temperature (using AC) will generally increase the heart rate/ blood pressure in rest condition. Whereas, the AC condition (cooling room) gave less heart rate than no AC condition in static cycling test. As consequence, the total amount of energy expenditure for cycling test in AC condition is less than in no AC condition. In sort, lower temperature is a benefit for cyclists and it can help them to manage and to keep more energy in their body.

Acknowledgment

Thanks to the Ministry of National Education of the Republic of Indonesia for supporting this paper. The authors are also grateful to the laboratory of work design and ergonomics, Industrial Engineering Department, the University of Brawijaya (UB), Malang Indonesia for their extraordinary courage.

References

[1] Cavill N, Davis A., "Cycling and Health: what's the evidence?". Cycling England, 2007.

[2] Rojas-Rueda D, de Nazelle A, Tainio M, Nieuwenhuijsen MJ, 2011. "The health risks and benefits of cycling in urban environments compared with car use: health impact assessment study", British Medical Journal BMJ 2011; 343:d4521

[3] de Hartog JJ, Boogaard H, Nijland H, Hoek G, 2010. Do the health benefits of cycling outweigh the risks?. Environmental Health Perspectives

[4] Bedny Z.Gregory, Karwowski Waldemar, Seglin H. Mark, "A heart rate evaluation approach to determine cost-effectives an ergonomics intervention", International Journal Of Occupational Safety And Ergonomics 2001, Vol. 7, No. 2, 121–133.

[5] Mikell P. Groover, "Work Systems and the Methods, Measurement, and Management of Work", Pearson Education, Inc., 2007.

[6] Akihiko Tajima, "Assessment of energy expenditure: an association between heart rate and oxygen uptake in daily physical activity" International Journal of Analitycal Bio-Science, Vo.2, No.4 (2014)

[7] Sanchita Ghosha, Rauf Iqbalb, Amitabha Dec and Debamalya Banerjeed, "Relationship of Heart Rate with Oxygen Consumption of adult male workers from Service and Manufacturing Sectors", International Journal of Physical Education, Fitness and Sports Vol.3.No.3 | September'2014.

[8] Uth, Niels; Henrik Sørensen; Kristian Overgaard; Preben K. Pedersen (January 2004). "Estimation of VO2max from the ratio between HRmax and HRrest--the Heart Rate Ratio Method". *Eur J Appl Physiol.* 91 (1): 111–5

Analyzing traffic source impact on returning visitors ratio in information provider website

A Prasetio* [1,2], **P K Sari** [2], **O O Sharif** [2] **and E Sofyan** [2]

[1)] Universitas Pendidikan Indonesia, Jl. Setiabudi 229 Bandung, Indonesia;
[2)] Universitas Telkom, Jl. Telekomunikasi No. 1 Bandung 40257, Indonesia;

*Corresponding author: adhipras@gmail.com

Abstract. Web site performance, especially returning visitor is an important metric for an information provider web site. Since high returning visitor is a good indication of a web site's visitor loyalty, it is important to find a way to improve this metric. This research investigated if there is any difference on returning visitor metric among three web traffic sources namely direct, referral and search. Monthly returning visitor and total visitor from each source is retrieved from Google Analytics tools and then calculated to measure returning visitor ratio. The period of data observation is from July 2012 to June 2015 resulting in a total of 108 samples. These data then analysed using One-Way Analysis of Variance (ANOVA) to address our research question. The results showed that different traffic source has significantly different returning visitor ratio especially between referral traffic source and the other two traffic sources. On the other hand, this research did not find any significant difference between returning visitor ratio from direct and search traffic sources. The owner of the web site can focus to multiply referral links from other relevant sites.

1. Introduction

The Internet has grown enormously during the last decade with an estimated around 3 Billion users around the globe in 2014 [1] and brings various services along with it. The most popular service that runs through The Internet infrastructure is World Wide Web or web [2]. Due to its popularity, web site becomes an important factor in e-commerce. Although company can use many channels to get in touch with their customers, like e-mail and social media, but in the end it is important for any company to point their campaign to their web site. Web site also becomes a place where transaction usually happens.

Since high returning visitor is a good indication of a web site's visitor loyalty, it is important to find a way to improve this metric. Visitors of a web site come from different sources and can be categorized into three major source called Direct, Referral and Search [3][4][5]. Direct visitors are visitor who type web site address directly in a browser's address bar. Referral visitors come from other web site where provide a link to destination web site. Search visitors are those who find out the web site from search engine search result based on specific keyword.

Previous study showed that visitors come from different sources have different characteristic due to the nature of how they come to a web site. Different traffic sources could leads to different Page views [4] or duration [5] characteristic. This research investigated if there is any difference on returning visitor metric among three web traffic sources namely direct, referral and search. Even though [3] has done the similar research, but this study will use different approach by using new metric called returning visitors ratio instead of returning visitor amount and use ANOVA instead of linear regression. Unlike previous study by [3], the metric and analysis method in this research compare each traffic source and will reveal if there is any returning visitor's ratio characteristic difference between them.

An information provider web site named prothelon.com will be used as data source in this study. This web site provide tutorial about web site development and rely on affiliation as its revenue source. Its monthly visitors are ranged between 30.000 to 40.000 visitors and dominated by search engine traffic source.

This study aimed to answer the following research questions:
1. How is returning visitor ratio characteristics based on traffic source?
2. Are there differences in returning visitor ratio based on traffic source?

The results of this study will allow site owners to determine which source of visitor's traffic that will improve its returning visitor. This information can be used to decide strategies to increase that traffic source.

2. Literature Review

2.1 Returning Visitors

E-commerce is the use of the internet specifically the web, for business transactions [2]. E-commerce transaction accomplished through digital transactions between buyer and seller. Web site is important in e-commerce system [2]. Web site has many goals based on organization's business model. Some of popular goals are encourage visitors to buy something, read information , fill out registration form, etc.

To measure web site effectiveness there are some online marketing metrics that can be used such as Page views, Stickiness, Unique visitors, etc [2]. Due to the nature of internet users who wants relevant information immediately, they tend to leave a web site as soon as they thought the information does not match what they are looking for. Some of the visitors will stay for a while in a web site if they think it's worth their attention. Some of these visitors will come back to the same web site if they need more information. These returning visitors are important because it indicate visitors loyalty to a web site. The more loyal visitors usually mean more opportunity that they will act according to a web site goals.

A web site owner can do a lot of things to improve web site performance. Some previous research already done to find out visitors characteristic based on various online marketing metrics. The result showed that different visitors groups lead to different browsing behaviour characteristic. Visitors of a web site could come from various traffic source group. Usually they are grouped into three different sources namely direct traffic, referral traffic and search traffic. [3][4][5] showed that some traffic source groups has similar characteristic, but other groups could have significant differences on their browsing behaviour.

[3] has found that different traffic source will affect differently to the amount of returning visitors. The study utilize linear regression to address its research questions. This study will try to use different

approach by introducing new metric called returning visitor ratio. This metric derived from two metrics namely the amount of returning visitors and total visitors by using this equation:

$$R = r / t \tag{1}$$

Where R is returning visitor ratio, r is the amount of returning visitor and t is the amount of total visitor

2.2 Site Management Tools
In order to improve a web site performance, we need to measure the degree of success of a website. Data to measure the level of success and performance of a website is generally monitored and retrieved from a software tool known as site menagement tools. [2] stated that the site management tools is very important to make web site work, and to determine how well a web site works. Furthermore [2] also mentioned that site management tools could become handy when we need to understand visitor behavior on a web site, observing the customer purchases more effectively, observe marketing campaigns, information hits and visits. Site management tools can also be implemented using web analytics technology to obtain more complete information. With web analytics, we can optimise web sites in order to accomplish business goals and/or to improve customer satisfaction and loyalty [6].

One of popular web analytics is Google Analytics provided by Google. Google Analytics is the most widely used tool to measure the performance of a website. Statistical data from [7] showed that in September 2015, Google Analytics is used by 59% of the entire internet that use site analytics in their web page. [6] and [8] mentioned that data provided by Google Analytics metrics have adequate types and can be used as a source of research data related to e-commerce. Some previous studies also used Google Analytics as their source of data [3][4][5][9].

3. Methodology

This study used comparative quantitative research. This type of research tries to find out the cause for the difference behaviour or status within group [10]. Subject of this study were the returning visitor ratio of prothelon.com web site. Monthly returning visitor and total visitor from three web traffic sources (direct, referral, and search) are retrieved from Google Analytics tools and then calculated using equation (1) to measure returning visitor ratio. The period of data observation is from July 2012 to June 2015 resulting in a total of 108 samples.

This study compared prothelon.com web site's returning visitor ratio based on three sources of traffic: direct traffic, referrals, and search. One-way ANOVA (Analysis of Variance) test two variance under the null hypothesis (H0) wheter the two variances are equal [11]. The first variance is the variance between groups (among groups) and the second variance is the variance in each group (within groups). According to [12] it is called one-way ANOVA since only one independent variable involved (traffic sources). Minimum data must be at least in interval scale, so the ratio data in this study, which level is higher than the interval, meet the requirement.

Initial hypothesis for ANOVA is expressed as follows:

H0: $\mu_1 = \mu_2 = \mu_3$, there is no difference in the average returning visitor ratio that come from direct, referral, or search.

H1: not all μ_j are the same, there is a difference in the average returning visitor ratio of the traffic source that come from direct traffic, referral, or search.

where μ_1 is the average returning visitor ratio of the source of direct traffic, μ_2 is the average visit duration of the source of referral traffic, μ_3 is the average visit duration of search traffic source, and $j = 1, 2, 3$.

Although the hypothesis initially see the difference average between groups, but the one that tested is the difference of variance between groups (among groups) and the variance within each group [11]. Variance between groups known as Sum of Square Among Groups (SSA) and the variance in each group is the Sum of Square Within Groups (SSW). When SSA value greater than SSW value, the initial hypothesis is not accepted, and consequently the alternative hypothesis is proven. In this case, the conclusion is that there are differences in the average returning visitor ratio based on direct traffic sources, referrals, or search.

4. Results and Discussion

4.1 Descriptive Analysis

In average, 17% of total visitor who come from search traffic source, 19% of total visitors from direct traffic source, and 25% of total visitors from referral visitors are returning visitors. The returning visitors ratio from referral traffic source has the biggest average of 25%.

4.2 ANOVA

The requirement to run ANOVA test is that data should be normal. All returning visitor ratio data from each group (direct, referral, and search) has passed this requirement.

Table 1. ANOVA Result-Returning_Visitor

	Sum of Squares	df	Mean Square	F	Sig.
Between Groups	.127	2	.064	27.811	.000
Within Groups	.240	105	.002		
Total	.367	107			

As shown in Table 1, since p-value = $0.000 < \alpha = 0.05$, we can conclude that there are significant differences among average returning visitor ratio coming from direct, search, and referral traffic source.

Table 2. Multiple Comparison Result

(I) Traffice_Source	(J) Traffice_Source	Mean Difference (I-J)	Std. Error	Sig.	95% Confidence Interval	
					Lower Bound	Upper Bound
Direct	Search	.015192305	.011263893	.372	-.01158653	.04197114
	Referral	-.063956101*	.011263893	.000	-.09073494	-.03717726
Search	Direct	-.015192305	.011263893	.372	-.04197114	.01158653
	Referral	-.079148406*	.011263893	.000	-.10592724	-.05236957
Referral	Direct	.063956101*	.011263893	.000	.03717726	.09073494
	Search	.079148406*	.011263893	.000	.05236957	.10592724

*. The mean difference is significant at the 0.05 level.

Tukey analysis in Table 2 showed that there is no significant difference between average returning visitor ratio from direct and search traffic sources. On the other hand, the average returning visitor ratio from referral traffic source proven to have significant difference either with direct or search traffic sources.

4.3 Discussion

The result of this study showed that different traffic sources could lead to different average returning visitor ratio characteristic. This result consistent with previous result that showed significantly different visitor behaviour based on their traffic sources [3][4][5][9]. Returning visitor usually lead to higher stickiness. This happens because the returning visitor already consider that the web site contains useful and relevant information for them. This result support previous study [13] which found out that referral visitors lead to significantly higher stickiness measured by their visit duration.

5. Conclusions and Implications

5.1. Conclusions

The average returning visitor ratio from referral traffic source has the highest percentage of 25% compared to 17% from search traffic source and 19% from direct traffic source. The results showed that different traffic sources have significantly different returning visitor ratio especially between referral traffic source and the other two traffic sources. On the other hand, this research did not find any significant difference between returning visitor ratio from direct and search traffic sources.

5.2.Implications

The results showed that referral visitors had the highest average returning visitor ratio. The owner of the web site can focus to multiply referral links from other relevant sites.

References

[1] Internet World Stats 2015. *World Internet Users and 2015 Population Stats.* http://www.internetworldstats.com/stats.htm. Accesed 20 September 2015
[2] Laudon K C and Traver C G 2014 *E-Commerce Business Technology Society 10E* Pearson
[3] Plaza B 2009 Monitoring web traffic source effectiveness with Google Analytics *Aslib Proceedings: New Information Perspectives Vol. 61 No. 5, 2009 pp. 474-482*
[4] Omidvar M A, Mirabi V R and Shokry N 2011 Analyzing the Impact of Visitors on Page Views with Google Analytics *International Journal of Web & Semantic Technology, Vol. 2 No. 1.*
[5] Prasetio A, Sharif O O, Alamanda D T and Partono A 2013 The Impact of Traffic Source on Page Views *Proceedings: The 7th International Conference on Information & Communication Technology and Systems pp. 15-16*
[6] Hasan L, Morris A and Probets S 2009 Using Google Analytics to evaluate the usability of e-commerce sites *IN: Kurosu, M. (ed.). Human Centered Design, HCII 2009, Lecture Notes in Computer Science 5619, pp. 697-706.*
[7] BuiltWith 2015 *Statistics for websites using Analytics technologies* http://trends.builtwith.com/analytics accesed 20 September 2015
[8] Prasetio A 2011 Studi Penggunaan Google Analytics dalam Penelitian E-Commerce *Jurnal Manajemen Indonesia, Vol. 11, No. 2, Mei 2011, 69-74*
[9] Plaza B 2011 Google Analytics for measuring website performance *Tourism Management 32 (2011) 477-481*
[10] Sekaran U 2011 *Research Method for Business: A Skill Building Approach* John Wiley & Sons
[11] Levine D M, Krehbiel T C and Berenson M L 2006 *Busines Statistics: A First Course (4th Ed)* New Jersey: Prentice Halls
[12] Zikmund W G, Babin J B, Carr J C and Griffin M 2009 *Business Research Methods 8th Ed* Cincinnati, Ohio: South-Western College Pub.
[13] Prasetio A, Sharif O O, Perdana I and Alamanda D T 2015 Analyzing the Impact of Traffic Source on Visit Duration *Proceedings: The 7th Indonesia International Conference on Innovation, Entrepreneurship, and Small Business (IICIES 2015)*

IOP Conf. Series: Materials Science and Engineering **128** (2016) 012014 doi:10.1088/1757-899X/128/1/012014

The effect of fly ash and coconut fibre ash as cement replacement materials on cement paste strength

R Bayuaji*, R W Kurniawan, A K Yasin, H AT Fatoni and F M A Lutfi

Departement of Civil Engineering, Institut Teknologi Sepuluh Nopember, Surabaya, Indonesia 60111

*Corresponding author: bayuaji@ce.its.ac.id

Abstract. Concrete is the backbone material in the construction field. The main concept of the concrete material is composed of a binder and filler. Cement, concrete main binder highlighted by environmentalists as one of the industry are not environmentally friendly because of the burning of cement raw materials in the kiln requires energy up to a temperature of 1450° C and the output air waste CO_2. On the other hand, the compound content of cement that can be utilized in innovation is Calcium Hydroxide (CaOH), this compound will react with pozzolan material and produces additional strength and durability of concrete, Calcium Silicate Hydrates (CSH). The objective of this research is to explore coconut fibers ash and fly ash. This material was used as cement replacement materials on cement paste. Experimental method was used in this study. SNI-03-1974-1990 is standard used to clarify the compressive strength of cement paste at the age of 7 days. The result of this study that the optimum composition of coconut fiber ash and fly ash to substitute 30% of cement with 25% and 5% for coconut fibers ash and fly ash with similar strength if to be compared normal cement paste.

1. Introduction

Concrete material is the fundamental material in the construction area. In the other side, Indonesia is very famous with local materials but they have not been fully utilized in concrete technology primarily for binder and filler material. Cement, concrete main binder highlighted by environmentalists as one of the industry are not environmentally friendly because of the burning of cement clinker raw materials requires energy with temperatures up to 1450° C and the output air waste CO2. A fact which needs serious attention to environmental issues that in the production of every ton of cement produces one ton of CO2 emissions released into the air [1]. On the other hand, Indonesia is one of the country's largest producer of foodstuffs, such as rice, maize, sugar cane and coconut. Indonesia has the largest oil fields in the world to the region approaching the 3.74 ha [2]. Coconut is one of the human needs that the value of commodity consumption is high, various preparations can be made from coconuts.

Coconut Fibre Ash (CFA) performed Center for Environmental Health Engineering, obtained by the composition of the compound Silica (SiO2) amounting to 67.55%. This composition proves that CFA has the potential to collaborate with fly ash (FA) as a cement replacement materials (CRMs). Previous studies that the characteristics of local pozzolan concrete example: rice husk ash [3-5], Mud Sidoarjo's ash [6], Bromo Volcanic's ash [7] and other materials. These innovations reduce the impact of CO2 emissions in the cement manufacturing process. Positive effect on concrete technology covers still meet the standards of strength, economical and environmentally friendly This research will discuss the effect of the use of composite CFA and FA to subtitute cement materials.

2. Fundamental Theory

Pozzolanic materials are defined as cementitious materials that commonly compose silica or alumina and silica [8]. The pozzolan material as cement material replacement was clasified by their chemical and mineralogical composition and particle characteristics [9]. The one of significant parameter that have effectiveness to react chemically with Calcium hydroxide, Ca(OH)2, and to form Calcium Silicate Hidrate is the fineness of pozzolan material particle. In addition, this parameter affects the strength and durability of concrete [10]. Pozzolanic reaction is defined that composing a pozzolan (S) and calcium hydroxide (CH) in the presence of water (H). The simple equation shown in Equation (1) [11].

Pozzolan + CH + H ==> C-S-H (1)

The pozzolan materials have ability to convert calcium hydroxide to calcium silicate hydrate (C-S-H). This concept give opportunity for concrete technology, to reduce cement as binder. Furthermore, the capillary voids are either eliminated or reduced in size. This in turn increases cement-concrete material such as strength and durability of the hydrated paste.

This concept makes the standpoint of alternative materials to reduce the use of cement and automatically reduce energy and resource that to produce cement.

3. Methodology

To achieve the purposes of this research, an experimental laboratory study in the Laboratory of Civil Engineering Diploma ITS was developed using the following materials:

The Ordinary Portland Cement (OPC) Type 1 was used throughout this research. The physical and chemical properties as listed in Table 1. Type 1 was chosen because of the observation on mortar properties could be done in normal hydration process without any addition of admixture to the concrete, hence the advantages of CFA and FA usage in cement paste can be maximally observed. The cement density and the surface area were found to be 3.15 and 359 m2/kg respectively. The composition of the cement (in oxides) is presented in Table 1.

TABLE 1 Cement Properties [12]

Parameter	SNI 15-2049-2004	Test Result
Chemical Composition		
Al_2O_3		6.03
SiO_2		20.65
Fe_2O_3		3.44
CaO		65.96
MgO	Max 6.00	2.05
SO_3	Max 3.50	2.22
Loss on Ignition	Max 5.00	3.96
Free Lime		1.09
Insoluble residue		2.37
Alkali (Na_2O+0,658K_20)		0.38
X-Ray Difraction Properties		
Tricalcium Silicate (C3S)		56.68
Dicalcium Silicate (C2S)		11.00
Tricalcium Aluminate (C3A)		7.72

Parameter	SNI 15-2049-2004	Test Result
Tetracalcium Aluminate Ferrite (C4AF)		8.91
Physical Properties		
Fineness		
-Blaine specific surface (m^2/kg)	Min 280	359
Time of setting, Vicat Test:		
- Initial set (minutes)	Min 45	125
- final set (minutes)	Max 375	240
Autoclave test:		
-Expansion (%)	Max 0.8	0.10
Compressive strength:		
- 3 days (kg/cm^2)	Min 125	255
- 7 days (kg/cm^2)	Min 200	316
- 28 days (kg/cm^2)	Min 280	411
False set (%)	Min 50	77,36

One of the CRMs materials of cement paste is obtained from burning FA Paiton with the physical and chemical test results. Physical and chemical test data shown in Table 2

Table 2 Physical and chemical test results of Fly Ash

Chemical test results	
Silica (SiO_2)%	54,5
Alumina (Al_2O_3)%	23,0
Oksida Besi (Fe_2O_3)%	5,3
Kapur (CaO)%	9,5
Oksida Magnesium (MgO)%	2,0
Sulphuric Anyhydrate (SO_3)%	0,5
Others %	5,2
Physical test results	
Fineness (m^2/kg)	287
Specific gravity ($gram/cm^3$)	2,49

Coconut husk used in this study is a waste of coconut on Pasar Mangga Dua, Wonokromo, Surabaya. This study used coconut husk which has been dried and then performed some process that consists of combustion, grinding and sieving to obtain available chemical and mineralogical composition of coconut fibre ash CFA as cement replacement materials.

The combustion process carried out in the laboratory of Concrete and Building Material Department of Civil Engineering ITS Sukolilo with a total time of burning 6 hours and a maximum temperature of 600° C.

To obtain the chemical and mineralogical composition of coconut fibre ash CFA as cement replacement materials. Because of that, the temperature was raised gradually as long as combustion process. There were three steps temparature 100°C, 300°C and 600°C. It was shown completely in Figure 1.

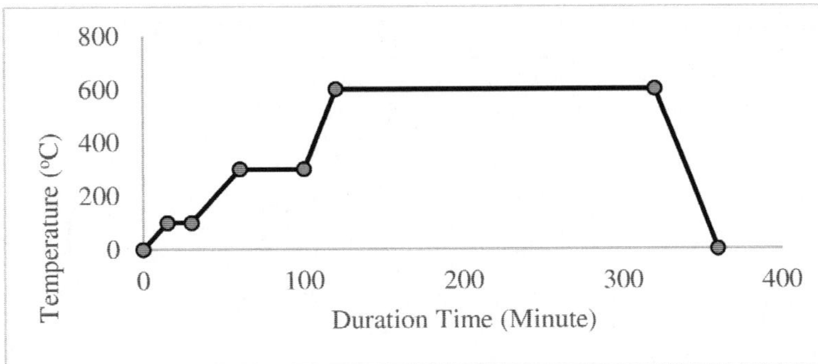

Fig. 1: The heat profile for coconut husk burning.

To accomplish the optimum mix proportion of cement paste that utilizing CFA and FA are used a identification test: compressive strength testing standard, SNI-03-1974-1990[13]. Pasta sample with a cylinder, diameter and height 2.5 cm and height 5 cm respictively. Comparison of water cementitious ratio (w/c) used fixed at a value of 0.3 as a control variable is OPC cement 100%. The hardened samples were then cured with soaked in fresh water then to conduct compressive strength test at 7 days using Universal Testing Machine UTM. The constituent cement paste using CFA and FA is described in Table 3.

Table 3. Mix proportion of cement paste using CFA and FA

No	Percentage by weight (%)			Weight of each material (gram)			
	Cement	**FA**	**CFA**	**Cement**	**FA**	**CFA**	**Cement**
1	100,00	0,00	0,00	14,73	108,97	0,00	0,00
2	90,00	0,00	10,00	14,73	98,07	0,00	4,37
3	80,00	0,00	20,00	14,73	87,18	0,00	8,73
4	70,00	0,00	30,00	14,73	76,28	0,00	13,10
5	90,00	5,00	5,00	14,73	98,07	4,23	2,18
6	80,00	5,00	15,00	14,73	87,18	4,23	6,55
7	80,00	10,00	10,00	14,73	87,18	8,46	4,37
8	80,00	15,00	5,00	14,73	87,18	12,68	2,18
9	70,00	5,00	25,00	14,73	76,28	4,23	10,91
10	70,00	10,00	20,00	14,73	76,28	8,46	8,73
11	70,00	15,00	15,00	14,73	76,28	12,68	6,55
12	70,00	20,00	10,00	14,73	76,28	16,91	4,37
13	70,00	25,00	5,00	14,73	76,28	21,14	2,18

4. Results and Discussion

Compressive strength test was used to identify the effectiveness of CFA and FA as cement replacement materials on cement paste strength. In this session to elucidate the result of strength and simple mechanism of CFA and FA on cement paste. The result of paste cement compressive strength that contain OPC, CFA and FA on 7 days was shown on Table 4.

Table 4. The compressive strength of cement paste using CFA and FA at 7 days

No of samples	Compressive Strength (MPa)
	7 days
1	66,18
2	15,27
3	10,18
4	15,27
5	81,45
6	78,77
7	76,05
8	67,91
9	66,2
10	61,08
11	45,82
12	15,27
13	5,09

The compressive strength of number of sample 9 was composed CFA and FA 30% by weight of cement. This result had similar result with control (number of sample 1). This is a proof that CFA and FA be able to convert calcium hydroxide on cement paste to calcium silicate hydrate (C-S-H) as strength product. SNI 15-2049-2004 determine the minimum strength of cement paste on 7 days is 20 MPa.

5. Conclusion

The pozzolan materials of CFA and FA have ability to convert calcium hydroxide on cement paste to calcium silicate hydrate (C-S-H) as strength product. Furthermore, the capillary voids are either eliminated or reduced in size. This in turn increases cement-concrete material such as strength and durability of the hydrated paste. To sum up the study of CFA and FA characteristic that these materials can be utilized as cement replacement material on cement paste. They can substitute cement 30% by weight on cement paste with similar strength if to be compared normal cement paste.

Acknowledgment

The authors would like to thank DIKTI for provided financial assistance to complete this laboratory research grants. In addition, the support of Laboratory of Testing Materials of Diploma Civil Engineering Program ITS is greatly appreciated.

References

[1] H.G. Van Oss, A.C. Padovani, Cement manufacture and the environment-Part I: Chemistry and technology, Journal of Industrial Ecology, 6 (2002) 89-106.

[2] Kementan, Prospek dan Arah Pengembangan Agribisnis Kelapa, in, 2005.

[3] R. Bayuaji, M.F. Nuruddin, Influence of Microwave Incinerated Rice Husk Ash on Hydration of Foamed Concrete, Advances in Civil Engineering, (2014).

[4] R. Bayuaji, The Influence Of Microwave Incinerated Rice Husk Ash On Foamed Concrete Workability And Compressive Strength Using Taguchi Method, Jurnal Teknologi, 75:1 (2015) 265–274.

[5] R. Bayuaji, M.F. Nuruddin, S. Francis, J. Ekaputri, S. Junaedi, H. Fansuri, Mechanical Properties of MIRHA-Fly Ash Geopolymer Concrete, in: Materials Science Forum, Trans Tech Publ, 2015, pp. 49-57.

[6] M.F. Nuruddin, R. Bayuaji, Sidoarjo mud: A potential cement replacement material, Civil Engineering Dimension, 12.1 (2010) 18-22.

[7] R. Bayuaji, M.S. Darmawan, I.P. Rahardjo, N.A. Husin, Characterization of Pozzolanicity Bromo's Volcanic Ash in: International Seminar on Applied Technology, Science, and Arts (3rd APTECS), Surabaya, 2011.

[8] A. Thevarasah, V. Perampalam, M. Selvaratnam, Some Studies on Pozzolanic Cement, J Natn. Sci. Coun. Srilanka 7 (1) (1979) 57-63.

[9] V.S. Ramachandran (ed), Concrete Admixtures Handbook 2nd ed, USA, 1995.

[10] A.M.Neville, Properties of Concrete, Longman, London, , 2006.

[11] P.K. Mehta, P. Kumar Mehta, P.J.M. Monteiro, Concrete: microstructure, properties, and materials., 2006.

[12] P.T. Semen-Gresik, Test Result of Portland Cement Type I PT Semen Gresik (Persero) Tbk January 2011, in: N. 001.C/SGPCI/3010/02.11 (Ed.), PT Semen Gresik (PERSERO) Tbk, Gresik, 2011.

[13] SNI-03-1974-1990, Metode Pengujian Kuat Tekan Beton, Departemen PU.

Comparison of performance of partial prestressed beam-column subassemblages made of reactive powder concrete and normal concrete materials using finite element models

S A Nurjannah*, B Budiono, I Imran and S Sugiri

Graduate Program of Civil Engineering, Faculty of Civil and Environmental Engineering, Institut Teknologi Bandung
Jl. Ganesha No. 10, Bandung, Indonesia

*Corresponding author: sanurjannah@gmail.com

Abstract. Research on concrete material continues in several countries and had produced a concrete type of Ultra High Performance Concrete (UHPC) which has a better compressive strength, tensile strength, flexural strength, modulus of elasticity, and durability than normal concrete (NC) namely Reactive Powder Concrete (RPC). Researches on structures using RPC material showed that the RPC structures had a better performance than the NC structures in resisting gravity and lateral cyclic loads. In this study, an experiment was conducted to apply combination of constant axial and lateral cyclic loads to a prototype of RPC interior partial prestressed beam-column subassemblage (prototype of BCS-RPC) with a value of Partial Prestressed Ratio (PPR) of 31.72% on the beam. The test results were compared with finite element model of beam-column subassemblage made of RPC by PPR of 31.72% (BCS-RPC-31.72). Furthermore, there was BCS-RPC modeling with PPR of 21.39% (BCS-RPC-21.39) and beam-column subassemblages made of NC materials modeling with a value of PPR at 21.09% (BCS-NC-21.09) and 32.02% (BCS-NC-32.02). The purpose of this study was to determine the performance of the BCS-RPC models compared to the performance of the BCS-NC models with PPR values below and above 25%, which is the maximum limit of permitted PPR. The results showed that all models of BCS-RPC had a better performance than all models of BCS-NC and the BCS-RPC model with PPR above 25% still behaved ductile and was able to dissipate energy well.

Introduction

Research on Ultra High Performance Concrete (UHPC) has produced a high-performance concrete in compressive strength, tensile strength, flexural strength, modulus of elasticity, and durability namely Reactive Powder Concrete (RPC). RPC consists of micro-sized material composed of silica sand, silica flour, silica fume, and superplasticizer, that leads it has a compact behavior better than Normal Concrete (NC). To increase the compressive strength since early age and accelerate the hydration process, RPC was treated in a steamed room or placed in hot water under temperature of 90° Celsius for three days. In an ideal condition of laboratory, the compressive strength of 120-230 MPa was achieved by RPC cylinders, while the RPC cylinders restrained by steel sleeves achieved the compressive strength of 490-680 MPa [1].

Literature Review

The use of open frame structures reinforced by partial prestressed strands are commonly used to reduce the dimensions of the structural elements and expand the space. The partial prestressed structures are more ductile and able to dissipate earthquake energy better than fully prestressed structures [2]. Prestressed strands in the plastic hinges are placed unbondedly and the maximum limit of Partial Prestressed Ratio (PPR) is 25% to ensure the structure remains ductile under earthquake loads [3].

The results of research on RPC material that had been done by Richard and Cheyrezy were developed by Gowripalan, et. al. [4] and Graybeal [5] by changing the composition of the material forming the RPC. The results showed that the RPC composed by different material composition had a higher performance than NC and had a compressive strength that was relatively similar to RPC had been created by Richard and Cheyrezy. Menefy [6] did research of bending load on the RPC beams used RPC mixture composition that developed by Gowripalan and produced cylinder compressive strength of 125-194 MPa. The RPC beams showed better performance than NC beams in the experiments. In this study, the composition of the RPC refered to a composition that had been developed by Naibaho, et al. [7]. There was an additional volume fraction of polypropylene fiber up to 0.08%.

The beam-column subassemblage structures were analyzed using discrete element models of three-dimensional finite element. Finite element modeling of reinforced concrete was done by setting concrete, mild steels and prestressed strands as separate discrete elements. Finite element analysis performed by ANSYS program. Concrete elements were modeled as SOLID65 elements which are three dimensional elements and has eight nodals. Each node has three translational degrees of freedom in the direction of X, Y, and Z axes as shown in figure 1. SOLID65 was modeled as element that fractured by tensile stress, crushing due to compressive stress, plastic deformation, and creep. Modeling of longitudinal mild steels, transversal mild steels, and longitudinal prestressed strands elements used LINK8 elements. The forces acting on the elements were axial forces on the ends of the elements as shown in figure 2.

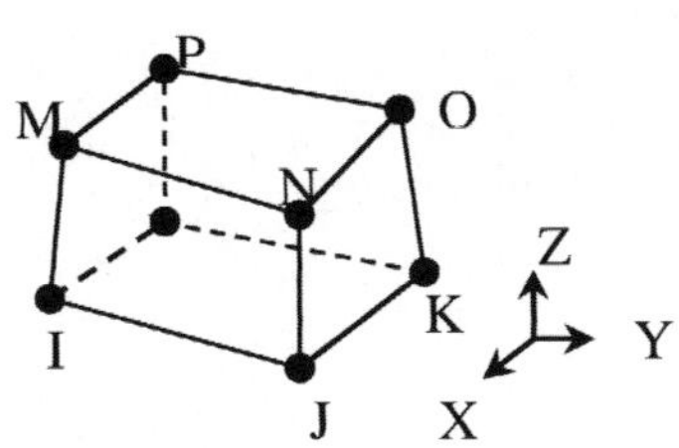

Figure 1. Test Set Up of SBC-RPC Prototype.

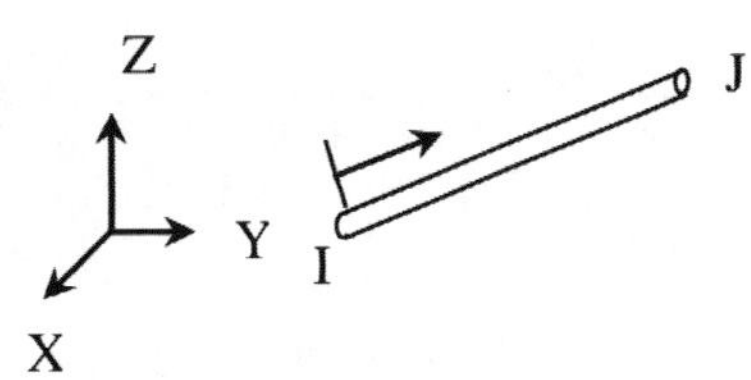

Figure 2. Cyclic Lateral Loading History Based on Deflection Control.

In the structure field, the equilibrium equation for the linear system is expressed as:

$$[K]\{u\} = \{F^a\} \quad (1)$$

where:

$[K]$: structure stiffness matrix
$\{u\}$: degrees of freedom vector
$\{F^a\}$: load vector

In the nonlinear case, equation (1) could not be directly used. The iteration process was required to obtain a solution of the equation. In the ANSYS program, there were several methods that can be used to obtain convergent solutions. One of them was the Newton-Raphson method. This method is an iterative process in solving nonlinear equations. In this study, the method chosen for the ANSYS program was Full Newton-Raphson method, where the stiffness matrix was updated in each iteration as expressed in equation (2) and equation (3).

$$[K_i^T]\{\Delta u_i\} = \{F^a\} - \{F_i^{nr}\} \quad (2)$$

$$\{u_{i+1}\} = \{u_i\} + \{\Delta u_i\} \quad (3)$$

where:

$[K_i^T]$: structure stiffness matrix
$\{u_i\}$: degrees of freedom vector
$\{F_i^{nr}\}$: load vector

Prototype and Experiment Methods

In experimental studies, there was a prototype of interior Beam-Column Subassemblage made of RPC (BCS-RPC) with Partial Prestressed Ratio (PPR) of 31.72% on the beam. This prototype were part of five prototypes that used for experiments on the influence of PPR variations to the beam-column subassemblages hysteretic behavior. Details of dimension and reinforcement of interior SBC-RPC prototype with PPR of 31.72% is shown in figure 3. The test set up is shown in figure 4. The prototype resisted a combination of a constant axial load of $\boldsymbol{0.1 f_c' A_g}$ at the top end of the column and lateral cyclic load at a distance of 350 mm from the upper end of the column based on deflection control [8] as shown in figure 5. The experiment was conducted in the Laboratory of Structure and Building Construction, Research and Development of Housing and Settlement Center, Ministry of Public Works, Indonesia.

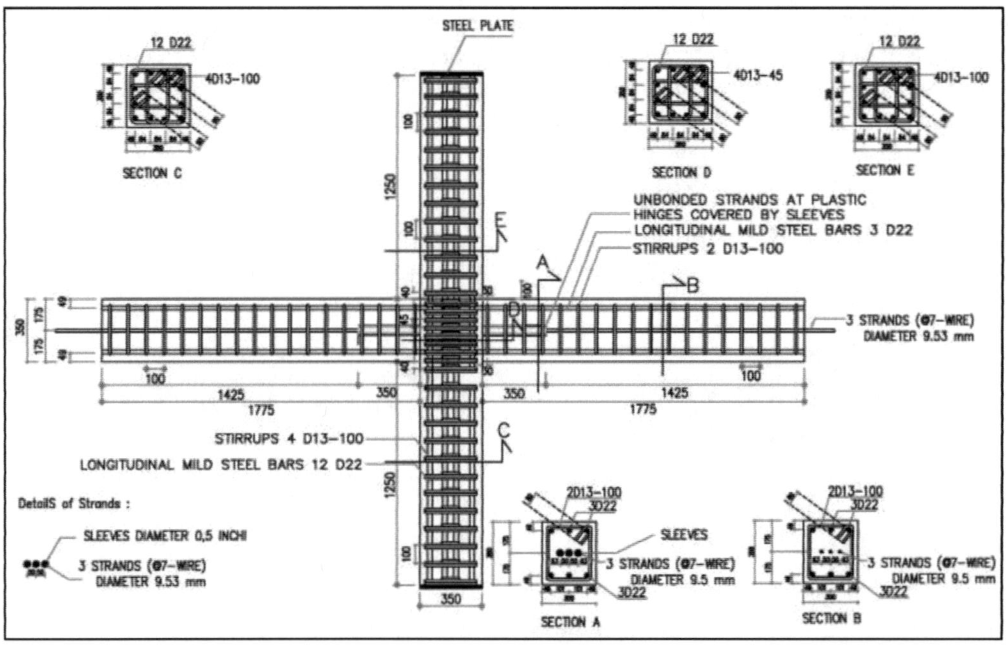

Figure 3. Details of Dimensions and Reinforcements of SBC-RPC Prototype.

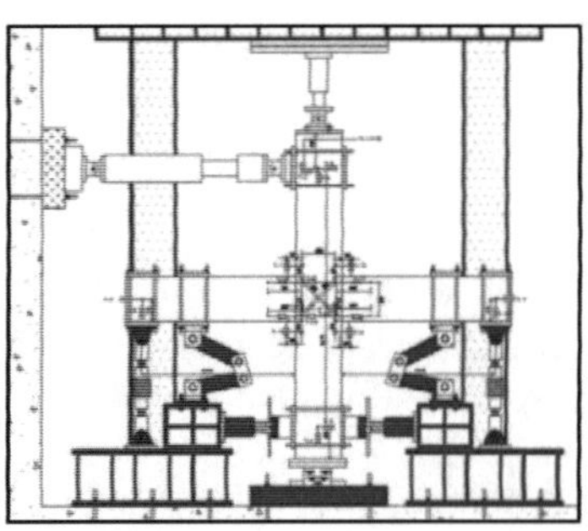

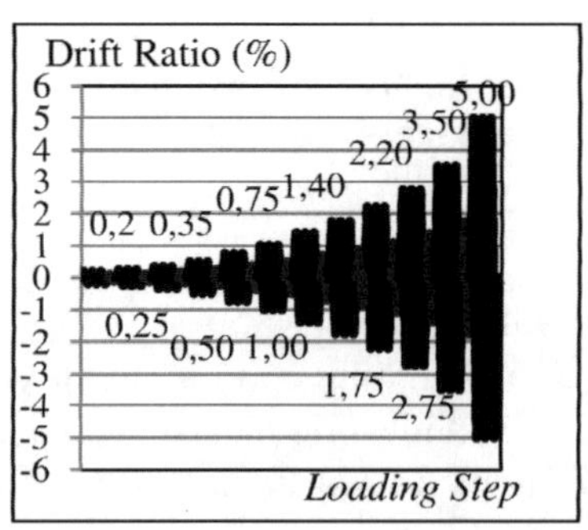

Figure 4. Experiment Set Up of SBC-RPC Prototype.

Figure 5. Cyclic Lateral Loading History Based on Deflection Control.

Modeling of Beam Column Subassemblages by Finite Element Methods

The beam-column subassemblage models **were** created by **a** non**linear** finite element program with varied values of **PPR** below and above 25**%,** while the concrete material structure was RPC and NC. PPR by 25**%** is the maximum value allowed as it had been presented in section 2. In the area of plastic hinges, **the prestressed strands were placed** unbondedly. **The** program input were the dimensions and details of the structure, as well as the stress-**strain** curves of each material that were parts of the structural models. The average compressive strength of RPC at 28 days **for** the program input was based on material experiments which will be elaborated in section 5. The NC compressive strength **was** set of 45 MPa. With the value of the **NC** compressive strength**, the NC** tensile strength and modulus of elasticity can be estimated using the following equations:

$$f_t = 0,62\sqrt{f'_c}\ \text{(T Class)} \quad (4)$$

$$E_c = 4700\sqrt{f'_c} \quad (5)$$

where:

f'_c **: compressive strength of concrete (MPa)**
f_t **: tensile strength of concrete (MPa)**
E_c **: modulus of elasticity of concrete (MPa)**

Thus, **the values of** tensile strength and modulus of elasticity **of** NC **was** 4.16 MPa and 31528 MPa, **respectively.** In the modeling, there were PPR and beam nominal moment variation as shown in table **1.**

Tabel 1. Partial Prestressed Ratio and Nominal Moment of BCS-RPC and BCS-NC Models.

Strand	**Strand Area**	**Concrete Type**	**Partial Prestressed Ratio**	**Model Code**	**Beam Nominal Moment**
A_p	A_p		*PPR*		$M_{n\ Beam}$
(mm²)	**(mm²)**		**(%)**		**(kN.m)**
1D12.7	98.71	RPC	21.39	BCS-RPC-21.39	140.19
3D9.5	164.52	RPC	31.72	BCS-RPC-31.72	154.87
2D9.5	109.68	NC	21.09	BCS-NC-21.09	152.39
2D12.7	197.42	NC	32.02	BCS-NC-32.02	171.22

Results and Analysis

In this study**, there were** material experiments, structure experiment **of** the BCS-**RPC** prototype with PPR of 31.72**% (prototype** BCS-**RPC-31.72),** and **structure** modeling used finite element method program**.** The material experiments results were used as input of BCS-**RPC-31.72** modeling**. Then, the hysteretic curve of the modeling** result was compared with **one of the structure** experiment result of BCS-**RPC-31.72** prototype. The **hysteretic curve** of model that behaved **relatively** close to the result of the experiment became a reference **of** BCS-**RPC** model with PPR of 21.39**%** (**BCS-RPC-21.39)** and BCS-**NC models** with PPR of 21.09**% (BCS-NC-21.09)** and 32.02**%** (BCS-**NC-32.02**).

1.1. Testing Results of Reactive Powder Concrete Mixture Composition

The compressive tests on cylindrical RPC samples were conducted at the age of 4, 7, 14 and 28 days. The first test was conducted at the age of four days because the samples were steamed for three days before. At the age of 42 days, the compressive strength tests were conducted because there was a lateral cyclic loading experiment of the prototype BCS-RPC-31.72. The test results are displayed in the form of curves in figure 6. The figure shows that the compressive strength (f_c) of concrete was relatively the same since the age (t) of 4 days up to 42 days. The RPC average compressive strength at the age of 28 and 42 days were 120.65 MPa and 117.84 MPa, respectively.

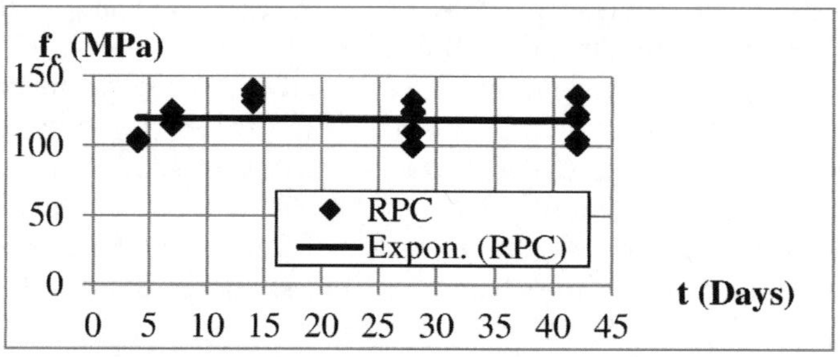

Figure 6. Relationship of Age with Compressive Strength of RPC.

1.2. Comparison of BCS-RPC Prototype and Finite Element Model Hysteretic Curves

The results of the BCS-RPC-31.72 prototype experiment were compared with the BCS-RPC-31.72 model. Figure 7 shows that the hysteretic curve of BCS-RPC-31.72 model was relatively close to the hysteretic curve of BCS-RPC-31.72 prototype with almost the same deflections (δ) and loads (**F**). Figure 8 shows that the gradient of the backbone curve of the BCS-RPC-31.72 model was close to one of the BCS-RPC-31.72 prototype.

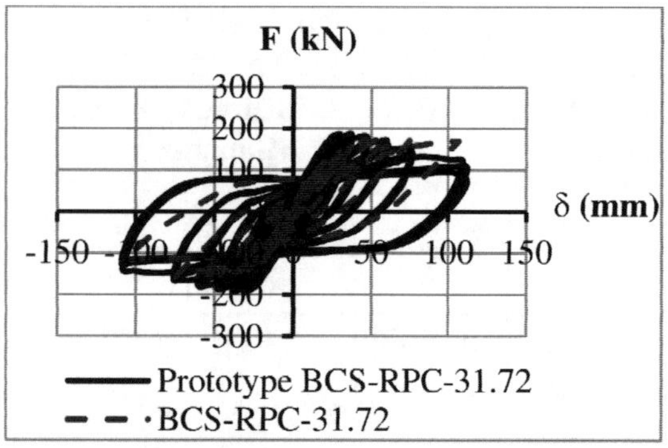

Figure 7. Hysteretic Curves of Prototype and Model of BCS-RPC-31.72.

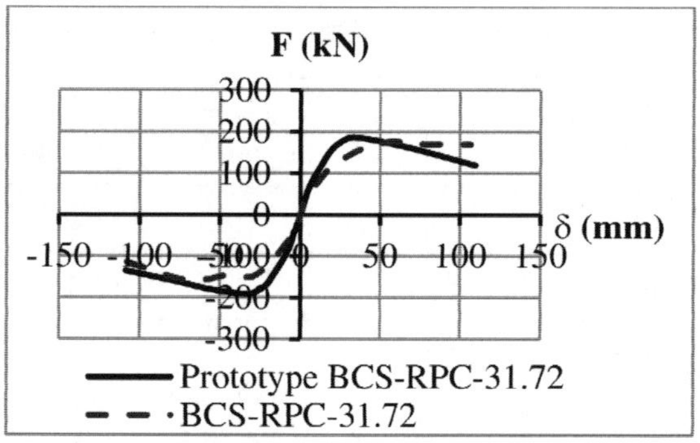

Figure 8. Backbone Curves of Prototype and Model of BCS-RPC-31.72.

1.3. Behavior of BCS-RPC and BCS-NC Models

Modeling of BCS-RPC and BCS-NC generated hysteretic curves (load-deflection relation) as shown in figure 9. The BCS-RPC-21.39 model achieved drift ratio of 5.00% under push loading (positive) and 3.50% under pull loading (negative). The BCS-RPC-31.72 model achieved drift ratio of 5.00% both under push and pull loadings. Under push loading, the BCS-NC-21.09 and BCS-NC-32.02 models achieved drift ratio of 3.50% and 5.00%, respectively. While under pull loading, the BCS-NC-21.09 and BCS-NC-32.02 models achieved the same drift ratio of 3.50%. All BCS-RPC models had higher initial stiffnesses than all BCS-NC models as shown in figure 10. This was because the RPC has a higher compressive strength than NC at the same strain. The values of maximum drift ratio, maximum deformation, and yield deformation are shown in table 2.

IOP Conf. Series: Materials Science and Engineering **128** (2016) 012015 doi:10.1088/1757-899X/128/1/012015

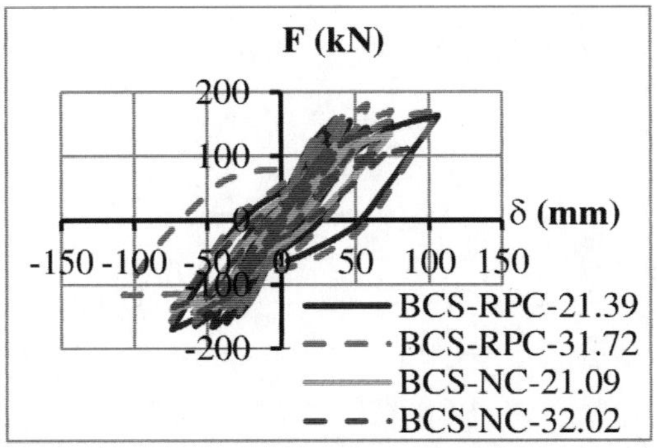

Figure 9. Hysteretic Curves of BCS-RPC-and BCS-NC Models

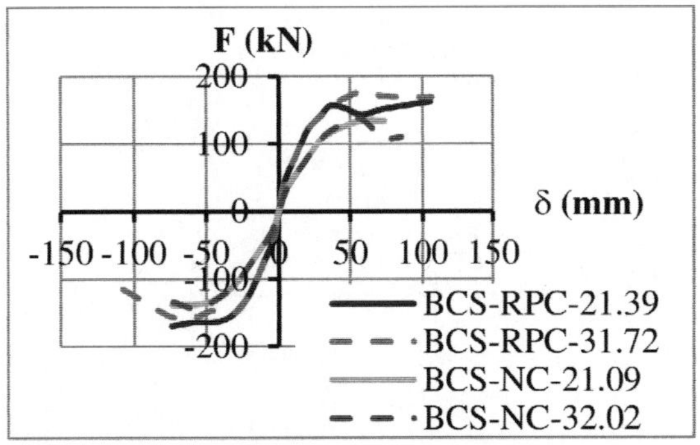

Figure 10. Backbone Curves of BCS-RPC-and BCS-RPC Models.

Table 2. Achieved Drift Ratio, Maximum Deformation, and Yield Deformation.

Model	**Achieved *Drift Ratio***		**Maximum Deformation**		**Yield Deformation**	
	(+)	(-)	$\delta_{max\,(+)}$	$\delta_{max\,(-)}$	$\delta_{y\,(+)}$	$\delta_{y\,(-)}$
	(+)	(-)	**(+)**	**(-)**	**(+)**	**(-)**
	(%)	(%)	**(mm)**	**(mm)**	**(mm)**	**(mm)**
BCS-RPC-21.39	5.00	3.50	105.91	-74.02	15.80	-15.81
BCS-RPC-31.72	5.00	5.00	106.91	-107.48	15.80	-15.81
BCS-NC-21.09	3.50	3.50	73.67	-73.70	15.76	-15.77
BCS-NC-32.02	5.00	3.50	85.25	-73.08	15.76	-15.77

The stiffness degradation (***K/K$_y$***) of all BCS-RPC and BCS-NC models were relatively similar since the drift ratio of 0.2% until the maximum achieved drift ratio of each model both under push and pull loading as shown in figure 11 and figure 12, respectively. All BCS-RPC models showed lower strength degradation (***F/F$_y$***) than all BCS-NC models under push and pull loading as shown in figure 13 and figure 14, respectively. It indicates that all BCS-RPC models had more stable strength than all BCS-NC models. The strength degradation of BCS-RPC-21.39 model decreased at drift ratio of 2.75% and increased at drift ratio of 3.50%. The BCS-RPC-31.72 and BCS-RPC-21.39 models showed relatively equal strength degradation up to drift ratio of 1.75%. The BCS-NC-21.09 and BCS-NC-32.02 models had relatively same strength degradation under push and pull loading up to the drift ratio of 2.75%. The BCS-NC-32.02 model had decreased strength degradation at the drift ratio of 3.50% under push loading.

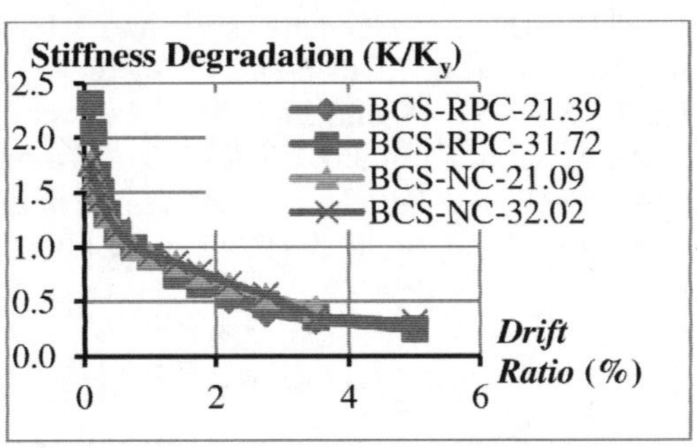

Figure 11. Stiffness Degradation under Push Loading.

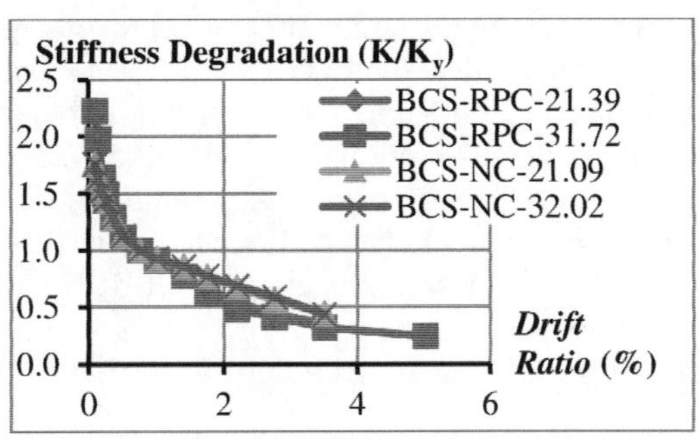

Figure 12. Stiffness Degradation under Pull Loading.

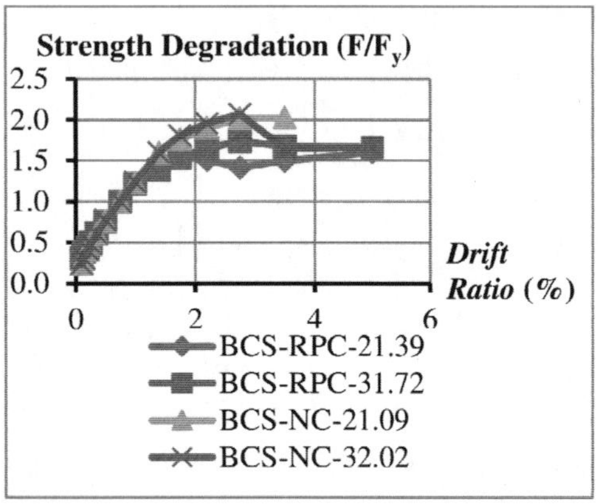

Figure 13. Strength Degradation under Push Loading.

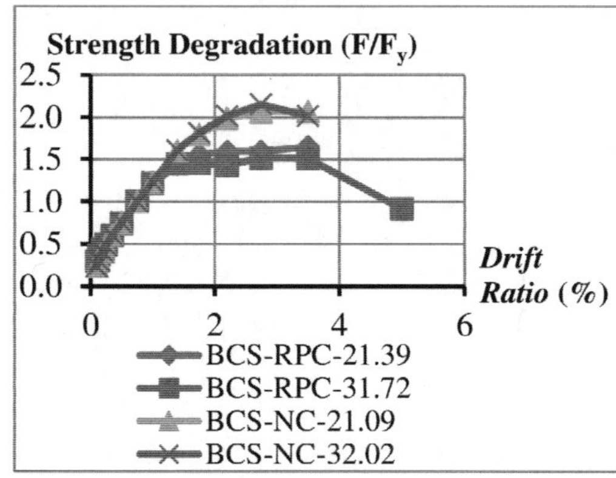

Figure 14. Strength Degradation under Pull Loading.

Table 3 shows the values of ductility (μ) and energy dissipation (E_d) of all models. The achieved maximum drift ratio affected the ductility and energy dissipation of each model. Despite BCS-RPC-31.72 model had PPR over 25%, it developed the highest ductility than other models because it was able to achieve the drift ratio up to 5.00% under push and pull loadings. It indicates that the PPR of 31.72% still allows the model to behave ductile. Under push loading, the values of the next highest ductility achieved by the BCS-RPC-21.39, BCS-NC-32.02, and BCS-NC-21.09 models, respectively. While under pull loading, the values of the next highest ductility achieved by the BCS-RPC-21.39, BCS-NC-21.09, and BCS-NC-32.02 models, respectively.

Table 3. Ductility of BCS-RPC dan BCS-NC Models.

Model	Ductility		Ductility/Maximum Ductility		Energy Dissipation	
	$\mu_{(+)}$	$\mu_{(-)}$	$\mu/\mu_{max\,(+)}$ (%)	$\mu/\mu_{max\,(-)}$ (%)	E_d (kN.m)	$E_d/E_{d\,max}$ (%)
BCS-RPC-21.39	6.70	4.68	1.00	0.69	23.57	76.91
BCS-RPC-31.72	6.77	6.80	1.01	1.00	30.64	100.00
BCS-NC-21.09	4.68	4.67	0.70	0.69	11.10	36.24
BCS-NC-32.02	5.41	4.63	0.81	0.68	9.98	32.58

The highest energy dissipation was achieved by the BCS-RPC-31.72 model because it had a higher nominal moment than the BCS-RPC-21.39 model, that allowed it to resist loads better and achieved the drift ratio up to 5.00% under the push and pull loadings. The BCS-NC-21.09 and BCS-NC-32.02 models only had energy dissipation by 36.24% and 32.58% of the energy dissipation of BCS-RPC-31.72 model, despite had nominal moment nearly equal to or greater than all BCS-RPC models. This was due in high drift ratio loadings, the NC ability was lower than RPC's in resisting lateral loads, that caused all BCS-NC models had a lower strength than all BCS-RPC models.

Conclusion

From the finite element method modeling, some conclusions can be written as follows:

- The modeling of partial prestressed beam-column subassemblages made of Reactive Powder Concrete (RPC) material was able to provide information of performance of prestressed beam-column subassemblage prototypes as proven by the hysteretic curve shape of the model was relatively close to the hysteretic curve of prototype test results. Then the model can be used as a platform for modeling the beam-column subassemblages made of Normal Concrete (NC).

- Analysis of all of beam-columns subassemblage models made of RPC and NC showed that all BCS-RPC models had better performance in terms of strength, stiffness, ductility, and energy dissipation than all BCS-NC models.
- The BCS-RPC-31.72 model that had PPR over 25% exhibited ductile behavior up to drift ratio of 5.00% under push and pull loadings and dissipate energy better than BCS-RPC-21.39 model which had PPR less that 25%.

Acknowledgment
The primary financial support for the research program was provided by P.T. Wijaya Karya Beton, Indonesia under joint research with the Faculty of Civil and Environment Engineering, Bandung Institute of Technology on contract number KU.09.09/OA.WB.191/2014 and 167/I1.C09/DN/2014.

References

[1] Richard P and Cheyrezy M, Composition of Reactive Powder Concrete, *Cement and Concrete Research*, Vol. 25 No.7 (1995) 1501 – 1511.
[2] Thompson K J dan Park R, Ductility of Prestressed and Partially Prestressed Concrete Beam Sections, *Precast/Prestressed Concrete Institute Journal*. (March-April 1980) 46-70.
[3] National Standardization Board, SNI 03-2847-2013 Procedure of Calculation of Concrete Structures for Buildings, Jakarta, Indonesia. (2013)
[4] Gowripalan N, Watters R, Gilbert R I, and Cavill B, Reactive Powder Concrete for Precast Structural Concrete-Research and Development in Australia, *21st Biennial Conference of The Concrete Institut of Australia*, Brisbane, Australia (2003) 99-108.
[5] Graybeal B A, Compressive Behavior of Ultra-High-Performance Fibre-Reinforced Concrete, *ACI Materials Journal*, Vol. 104, No. 2. (March-April 2007)
[6] Menefy L, Investigation of Reactive Powder Concrete and Its Damping Characteristics When Utilized in Beam Eelements. *Thesis of Doctor Philosophy*, Griffith School of Engineering, Griffith University, Gold Coast Campus, Australia. (2007)
[7] Naibaho P R, Budiono B, Surono A, Pane I, Experimental Study of Reinforcement Addition on Exterior Beam-Column Connections Using Reactive Powder Under Cyclic Load, *Proceedings of the National Conference in Graduate Program of Civil Engineering*, Bandung Institute of Technology. (2014)
[8] ACI Comittee, Acceptance Criteria for Moment Frames Based on Structural Testing and Commentary, ACI 374.1-05, Farmington Hills, USA. (2005)

The values of culture and architecture typology of Mandailing traditional house

P L A Luthan*

Faculty of Engineering, UniversitasNegeri Medan

*Corresponding author: putri.lynna@gmail.com

Abstract. Mandailing is one of bataknesse ethnic in North Sumatera which belong to strong to hold the customs and traditions Mandailing culture which reflected in the parts of the traditional houses. However rapidly the construction modern house make possibility the traditional house which enriched with culture value will be extinct.This studyto identify the value of culture and architecture typology of traditional house in Mandailing Natal North Sumatera.A case study using phenomenologyc method that is recording and serving the phenomenon which found in field as evidence that include existing condition from building. The finding show that 1)religion system or reliance in the front of location election, ornamentation orientation, construction system and room arrangement, 2) kinship system in the front of room organization, room amount and the completeness, 3) live philosophy found in shape and room arrangement which can accomodate all activity, 4) leadership system in the fornt of ornamentation shape and the meaning, pillar shape, house size vertically, and 5) social system in the front of roof shape, the used of decoration style and room arrangement. Architecture typology of Mandailing house dug from the shape and the construction, room arrangement, and roof shape.

1. Introduction

Mandailing is the one of bataknesse ethnic in Mandailing Natal district of North Sumatera province, Mandailing Natal district is on west beach territory North Sumatera which border on West Sumatera. Some years recently in that area the housing construction grow rapidly implemented in a row with population growth. House live which build inclined adopting modern house shape likes in big cities generally. The condition likes that would causing the extinction Mandailing traditional live house which has value of culture that was handed down from one generation to the next.

Traditional building as the one of archeological architecture traditional which have the important meaning to reflecting idea and attitude a supporter of society. Traditional building which created based on culture values need understanding in developed the culture values in physical form. Nowadays, buildings with traditional architecture not too much found in Mandailing, mostly had been extinct by the time and the rest had been long standing. Facing with that condition, we know that awareness to protect or preserve the traditional culture values not grow yet in that Mandailing society. Instead today generation mostly does not care and respect with their own culture. This thing happened may be caused by social-culture changing which had been made symbols on that traditional building and not relevant morewith nowadays condition. More understanding about the various symbolic meaning which found in traditional buildings, so that worried the local wisdom values time to time will be lost caused by culture development. This thing marked by emerge the phenomenon to rebuild traditional building by some society who have ability in financial side as the effort to hold high the identity and true self their culture (Nasution, 2006). But, the new shape so different with the traditional form. Any desire from the ethnic society to build house live based on ethnic in form and way which more modern is the factor who become main priority in directness traditional culture preservation. To make a live tradition or preserve the tradition in modern live is not an easy thing because on and another be in contradiction. But awareness that tradition is the one which appear in a long process, agreed on group, have history value, spiritual, morality, art, myth and local wisdom, and others.

In order to not happened the wild construction and worrying future prospect traditional architecture, so needed the identification of shape and structure that traditional building. The identification is done to get the information about architecture typology Mandailing traditional house. Typology is the study about types to make classification which based on the similarity of object character (Galih, 2010), in research traditional architecture, typology used as the tools to analyzing the object. Any two aspects which analyzed in study about traditional house that are shape and room typology (Aufa, N, 2010). Classification the shape of traditional house distinguished based on building element that consist of roof, wall and pillar (Ryeung, S, 2012). Meanwhile, room typology consists of the pattern and room function. Survey conducted in two location which relative still many be found traditional house likely in Mandailing Godang and Mandailing Julu.

Ryeung, S (2012) states that to analyze the typology of traditional houseneeded investigation to three element of house building, there are; roof, wall, and pillar. Hosein, A. (2012), in research about traditional house states that privacy concept and room optimum, simplicity, multi-functionality, family propinquity, room arrangement which opened are social-culture value which potential significantly implemented in contemporary house design. Based on both of the opinion above can be concluded that in designing a Mandailing traditional house must be considering the kinship with the room arrangement which is fair and flexible and also the house form that focused on the roofs, walls and pillar.

The main problem in this research is to see the originality of Mandailing house typology which become reference for society to rebuild traditional house live based on Mandailing ethnic. Another section of your paper

2. Research and Methodology

This research using phenomenology method which is recording and serving the phenomenon that found in field as fact which including existing condition in traditional building. Phenomenology according to Husserl (Smith,2007;Crowell, 2001) is an attempt to understand the consciousness as experienced from the first -person perspective. The essence of Husserl's phenomenology according to Smith (2007) is the awareness that is directed or intentionality (consciousness directivity).

Interview technique which opened and detailed conducted on respondent, this research also seeing the relevancy information that given with surrounding environment condition, the existence of traditional house , and the house history (Satori and Komariah, 2009). Information that given by informant validated with another informant that met in research location, likes society figure, the house owner, or house tenant. Collecting data of traditional house physically conducted by do measuring and recording, meanwhile secounder data conducted by journal and similar research that also discuss about typology some traditional house The election of house which will observed with sampling purposive method, with criteria election likes the originality of shape/house architecture and house which still occupied.

Data analysis conducted through descriptive-qualitative approachment. Finding condition in field analyzed based on the interpretation result researcher on the condition of existing house which found in field.

3. Discussion

3.1. The culture values of Mandailing traditional house

Luthan, PLA., 2013. Generally traditional society live and attitude based on social-culture values which hold high. That values contained in a system which consist of live philosophy, social system, leadership system, and family system or religion. Obviously, that systems formed in attitude and materiil culture (that is artefak; traditional building)

Live philosophy

Mandailing society have live philosophy holong and domu, that is love and affection between fellow society member (Lubis, R, 2006). Love and affection not only in context people to people, but people with another creature, also people with God.

Mandailing society is the society group who has behave to erect the justice and peacefullness, togetherness value which strong, also hold high unity values. Togetherness values very strong effecting room arrangement shape in Mandailing traditional house (especially King-Raja Bogas house), be found some rooms that can accomodating all togetherness activity on that society.

Social value

Implementation and custom law in daily live or in custom ceremony Mandailing society done based on structure and social system which called markoummarsisolkot (family custom in group), or often called as daliannatolu. Dalian natolumeans pillar which three, containing the meaning that Mandailing society trusting on social system which joined together in one structure arrangement that consist of kahanggi ,mora and boru (Lubis, R, 2006).

In discussion activity daliannatolu component there any roled as highest decision taker. This highest decision taker based on tradition ceremony types which implemented, people who become family head, also who become mora, kahanggior anakboru when the ceremony went on. This thing make possibility any distribution of special rooms for the third component. Dalian natoluin king house or bagasgodang custom ceremony. Also make possibility any distribution of sit position for daliannatolu component based on each role ofmora, kahanggi, or anakboru. Moreover, live which based on discussion result make possibility for Mandailing society to have a building that specially intended for that discussion activity.

In social system Mandailing society not known any social stratification. Social level just used to distinguished between role and postion of king with king family member, king assistant, and common people. This level distinguish size and completeness of the traditional house. Common people house do not allowed using roof shape that same with king, also do not allowed using decoration style. This social system also emerge when organizing room inside the king building house.

Leadership value

When implementing society government onMandailing customs law, government institution and the customs sit some traditional leader figure which called as namoranatorasand lead by someone who has status as raja panusunanor raja panusuk. Raja panusunanis head of government in hutainduk(mother vllage), meanwhile raja panusukis head of government in hutaanakor developed from hutainduk. But, each hutarunning the government autonomously. King is primus interparis(leader) who lead democraticly appropriate with discusiion result between the head of village or also called as namoranatoras.

Politic lives in Mandailing society describing any law values and customs which hold high in every lives principle. King is someone who considered important and respected the existence, guarding his citizen fairly and full of affection. So that, king is the center of social-culture activity Mandailingsociety., it menas the place where people tell all of their live problems and asking the moril helped even material.

Family value

In Mandailing society any three family group (kingrous), there are mora, kahanggi (people who have sama sure name or have family relationship brother-sister) and anakboru. The third group used by Mandailing society as pillar component for their social system which called daliannatolu (the three pillar). In architecture context, family structure or kinship structure explains about pattern and manner to have relation which can forming something interaction that can known as occupied patterns. Family structure in connection with the relation between parents-children that can reflecting in room arrangement pattern that can forming privacy hierarcy. Meanwhile, family structure explains about the relation of individual role in big family.

IOP Conf. Series: Materials Science and Engineering **128** (2016) 012016 doi:10.1088/1757-899X/128/1/012016

Religion value and reliance

Religion values that trusted by Mandailing society can seen clearly on determining of building orentation and building appearance vertically. Building orientation determined by following the sunrises direction and do not allowed back to back it. This thing believed by society that sun and the come direction of light is considered as the live source. So the society trust that house obtaining the live source from the sun (the main source from God).

Moreover, building appearance which is house on stilts and consists of three parts among others, tarumanibagas (space underneath of house), bagas (house to live), parapi(roof) believed as cosmos creation or world. Space underneath of house believed as under world symbol which is death world and evil souls as the creation of people relation with another creature. The middle part of house believed as middle world symbol which is life, the place for people to do their daily activity, where happened relationship between people to people. And the top part of house is holy and sacred world as the creation of relationship between people with God. The culture values that contained in live of Mandailing traditional society hold high as guidance to live in form of group can seen in table 1.

Table 1. The culture values of Mandailing traditional building
Source: analysis, 2014

Culture value identification	Traditional house
Religion system or reliance	Election of location, Orientation, Ornamentation, Construction system, Room arrangement, Mass shape
Family system	Room organization, The amount of room and the completeness (custom building, rice barn, yard)
Live philosophy	Arrangement room shape, The house completeness as symbol of individu and social activity
Leadership system	The ornamentation shape and the meaning, Pillar shape, House size vertically
Social system	Roof shape, Used of various decoration, Room organization

3.2. The shape of traditional house in Mandailing

The house shape in Mandailing different among king's house and common people's. The difference seen clearly from building mass, room size, room arrangement pattern, decoration style application, roof shape, and dimention. Generally, the king's house were in an area which is completed with yard (allamanbolak), rice barn (sopoeme), and custom building (sopogodang). Meanwhile, common people do not have that certainty. The difference of building shape can seen clearly in Figure 1-4

King house in Habincaran village

Common people house in Habincaran village

Figure 1 : The difference between king's house and common people's in Mandailing
Source : Field Survey, 2013

IOP Conf. Series: Materials Science and Engineering **128** (2016) 012016 doi:10.1088/1757-899X/128/1/012016

Bagasgodang	*Sopogodang*

AlamanbolakdanSopoeme	

Figure 2 : Traditional house area in Hutagodang village – Mandailing.
Source : Field Survey, 2013

*Bagasgodang*inPanyabungan Tonga	*Bagasgodang*in PidoliDolok

Figure 3: BagasGodang inMandailingGodang
Source : Field survey, 2013

Bagasgodang inHutagodang village	BagasgodangSingenguvilag e	*Bagasgodang* in Habincaran village

*Bagasgodang*in	*Bagasgodang*AlahanKae	*Bagasgodang* in Tolang

KayuLaut village	village	village

Figure4 :Bagas Godang in Mandailing Julu
Source : Field Survey, 2013

3.3. The structure and construction of traditional house in Mandailing

From structure and construction side, houses in Mandailing using house on stilts construction with wooden structure and natural stone as the ground. Vertically, the bouse divided into three parts that are parapi(upper world)as banuaparginjang, bagas (middle word) as banuatonga, and tarumanibagas (bottom world) as banuapartoru. Mandailing people had a belief that this universe is divided into three parts or called Banua, they are :BanuaParginjang (upper world), this is a place where God, the human master that called Datu Natumompa Tano Nagumorga Langit as the creator and the owner of this sky and earth; Banua Tonga (middle world), this is a place where human does daily life activities. This place is symbolized by red color; BanuaPartoru (bottom world), this is a place for dead people or called as spirit world. This place is symbolized by black color (Nuraini, 2011).The third parts centered in pantartonga(center room) is the room that considered holy, in this part made center pillar or tonga pillar. The construction shape of traditional house can seen in Figure 5

Figure5 : Mandailing people cosmology house
Source : Field survey, 2013

In Figure 6 can seen each form of building component that found on king house in Mandailing.

Ground construction of natural stone.	Pillar construction and beam of wood.	Room condition (*pantartonga*) in*bagasgodang*	Decoration style in ari cover side. (*bindu*)

Figure 6 : The element of bagas Godang building in Mandailing
Source : Field survey, 2013

3.4. Construction model of ground connection and coloumn.

Whole form and ground system of traditional house in Mandailing, even king house and common people, the pillar arrangement have the form of octagonal which put over pebble flat without loop or connection like house on stilts that many made by this time. Place or surface to put the stone usually

IOP Conf. Series: Materials Science and Engineering **128** (2016) 012016 doi:10.1088/1757-899X/128/1/012016

above the land. But, in some traditional house found the surface that had been permanent with make massive surface and had been made harden or cemented.

Keterangan : a) Panyabungan Tonga, b) Hutagodang, c) PidoliDolok, d) Singengu, d) common people's house foundation

Figure 7 : Foundation shape of bagasGodang and Mandailing society house
Source : Field survey, 2013

3.5. The room arrangement of Mandailing traditional house

The room management in Mandailing traditional house not regardless from the occupant activity everyday.Bagasgodangarrangement (king's house) and common people's house certainty different. If in king's house always held the activities that involve about customs and society, so in common people's house only have occupying activity likes sleep, cook and together with family member, and socialization with neighbour. Activities which held inside house give effect to arrangement pattern and room size that needed to accomodate the activity. In example, in king's house found room which the size big enough in middle room, that called as pantartonga, and this room used as gathering room for king's family (kahanggi, mora, anakboru) and common people to do custom ceremony activity. Moreover, in front side of king's house any room that used to socializicing with neighbour and family member which called parangin-anginan,or another term called as terrace. Besides custom activity, in king's house also live family member with daily activity likes sleep, eat, cook, take a bath, and so on.

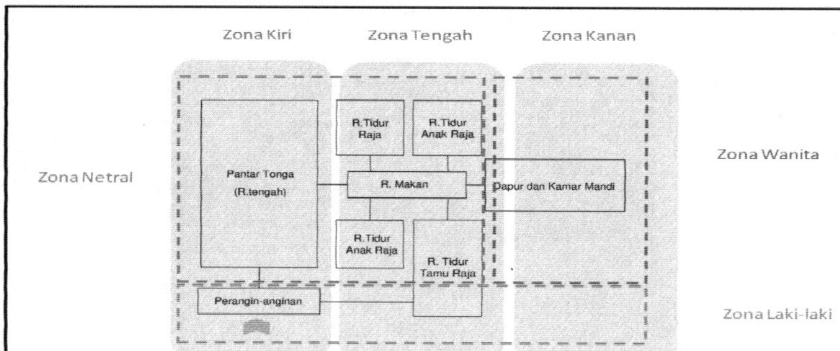

Figure 8 : The room arrangement pattern of BagasGodang
Source : Redrawing fromNasution (2006)

3.6. The roof shape of Mandailing traditional house

Roof shape has ridge line which warping or called as silingkungdolokpancucuran roof. The shape of silingkungdolokpancucuran roof means as modesty that do not distinguish someone level nor position, innocent attitude to accept the changing and emendation., also helpful behavior to help another people who need. Special for king's house has thin cover (triangle shape in roof) in each sides, any 4 thin

IOP Conf. Series: Materials Science and Engineering **128** (2016) 012016 doi:10.1088/1757-899X/128/1/012016

cover which completed with ornament and added one piece in stairs roof side. These thin cover made from wavy zinc and in front side of roof found eaves trough the rain water from roof to the disposal channel. Roof in this king's house consist of main roof, kitchen roof (addition), and stairs room, that can marked from the ridge line.

Roof shape *silingkungdolokpancucuran*applicated in King's house	Roof shape*sarotole*applicated in King's house and his family.	Roof shape *sarocino*applicated in common people's house

Figure 9 : Roof shape of Mandailing traditional house
Source : Field Survey, 2013

4. Conclusion

Based on the typology identification found concept implementing banuaseen clearly in shape of Mandailing traditional house which divided into 3 (three) groups that are Banuaparginjang (upper world), Banuatonga (middle world), Banuapartoru (bottom world).

The values that contained inside 1) religion system or reliance marked by election of location, ornamentation location, construction system, and room arrangement, 2) family system marked by room organization, the amount of room and completeness, 3) live philosophy found in shape and arrangement of room that can accomodate all activity, 4) leadership system marked by ornament style and the meaning, pillar shape and house size vertically, and 5) social system marked by roof shape, the using of decoration style and room organization.

References

[1] A Aufa, N. 2010.TipologiRuangdanWujudArsitektur Masjid Tradisional Kalimantan Selatan.Journal of Islamic Architecture Volume 1 Issue 2.

[2] Crowell S Galt (2001) Husserl, Heidegger and the Space of Meaning : Paths toward Transcendental Phenomenology, Norhwestern University Press, USA.

[3] Galih, W. 2012.Tipologi Nusantara Green Architecture DalamRangkaKonservasidanPengembanganArsitektur Nusantara BagiPerbaikanKualitasLingkunganBinaan.JurnalRuas, Volume 10 No. 2.

[4] Hoseini, AHG. 2012. The Essence of Malay Vernacular Houses : Towards Understanding the Socio-Cultural and Environmental Values. Journal of the International Society for the Study of Vernacular Settlements.

[5] Lubis,L, 2006, PartuturonDalamMasyarakatAngkola, JurnalIlmiahBahasadanSastra, Volume II, April 2006.

[6] Luthan, PLA, et all, 2013.Model Development Building Construction in North Sumatra Traditional Design House Live For Free Ethnic-Based North Sumatera, Proceeding Big Star Bandung.

[7] Luthan, PLA, et all, 2014, "Pengembangan Model Konstruksi Bangunan Tradisional Sumatera Utara Sebagai Panduan Perancangan Rumah TinggalBerbasis Etnis Sumatera Utara", Jurnal Saintika, Volume 16, March 2014.

[8] Luthan, PLA, et all, 2014, Pengembangan Model Konstruksi Bangunan Tradisional Sumatera

Utara Sebagai Panduan Perancangan Rumah Tinggal Berbasis Etnis Sumatera Utara" Laporan Penelitian Hibah Bersaing, Lembaga Penelitian Unimed, Medan.

[9] Luthan, PLA, et all, 2014, "Struktur Bangunan Tradisonal Mandailing" Jurnal Saintika, Volume 17, September 2014.

[10] Nasution, IN. 2006, "Analisis Penyebab Perbedaan Bentuk Bagas Godang di Wilayah Mandailing Godang dan Mandailing Julu". Thesis Institut Teknologi Bandung, Bandung.

[11] Nuraini, Cut. 2011, The Concept of Banua in Mandailing Architecture," Proceeding CIB International Conference – Local Wisdom in Global Era", at UKDW Yogyakarta, 21-22 Januari 2011.

[12] Ryeung, S., et all. 2012. Modernization of the Vernacular Malay House in Kampong Bharu, Kuala Lumpur. Journal of Asian Architectture and Building Engineering-Vol. 11 No. 2 May Page 95-102.

[13] Smith, W David (2007) Husserl London and New York : Routledge,

Disaster-friendly sundanese traditional building construction

J Maknun*, T Busono and Nuryanto

Department of Architectural Engineering Education
Faculty of Technology and Vocational Education
Universitas Pendidikan Indonesia

*Corresponding author: joharmaknun@upi.edu

Abstract. Indonesia is one country that is highly prone to earthquakes because it is located at the juncture of four tectonic plates, namely the Asian continental shelf, the continental shelf of Australia, the Indian Ocean plate and the Pacific plate. In the southern and eastern parts of Indonesia there is a volcanic belt that extends from the island of Sumatra, Java and Nusa south eastern Sulawesi, the sides in the form of old volcanic mountains and the lowlands are mostly dominated by swamps. The condition of potential and proneness to disasters such as volcanic eruptions, earthquakes, tsunamis, floods, and landslides. Indonesian society have a local knowledge for facing disasters. Local communities in particular areas have local knowledge for facing disasters. Such knowledge is commonly kept by members of the communities and applied to their environments, including houses. This research aims to describe disaster-friendly Sundanese traditional building construction. It employs the evaluation method, by comparing Sundanese traditional house construction to the standards of disaster-friendly construction. The results indicate that the Sundanese traditional building constructions have been qualified as disaster-friendly buildings.

1. Introduction

The phenomenon of an earthquake is one of the natural disasters that we can not predict around the world, including in Indonesia. Naturally these natural phenomena can not be avoided. This is because the plates that are in the earth's crust move actively. Effect of movements and collisions between the plates is what produces the tremor. One form of disasters that often occurs in Indonesia is earthquakes.

Problems faced by the countries in the world that is prone to earthquakes is how to make friendly home construction models earthquake to be safe and comfortable for humans which is connected with the local wisdom communities in Indonesia, especially Sundanese communities. Some examples of earthquakes in Indonesia, especially in West Java which many took many lives and property were earthquakes with a strength of 6.8 magnitude that a potential tsunami. Most of the fatalities were caused by falling debris died houses or buildings that do not withstand earthquake vibrations. This was due to a house or building was not designed to withstand earthquake loads. The concept of earthquake resistant buildings is essentially an effort to make the building into a coherent whole, which can not be separated by the earthquake. The application of this concept is one of them make a good and strong bond between building elements, the selection of appropriate construction materials, and the implementation of good and appropriate standards. In addition, local knowledge in the form of Sundanese traditional architecture should be one of the main reference in designing the model homes that are friendly to earthquakes.

In a static state, a building only bears the gravity, which refers to its own weight and live load (if any). When the ground shakes, the building receives the vibration which is transmitted to the top part through the foundation. If the building is rigid/stiff, it will follow the movement of the ground. The surface acceleration of a local earthquake directly affects building construction; this results in the maximum horizontal force on the building. The size of the force is equal to the mass of the building multiplied by the acceleration of the ground (Canadian Wood Council, 2003).

To minimize the impacts of earthquakes, the Indonesian Government has established SNI-176-2002, which is the planning standard of earthquake resistance for building structures. Some limitations of earthquake resistant building construction planning are:

- Building plans should be simple, symmetrical and not too long, 6 meters;
- Roof construction should use a material that is lightweight and simple;
- The foundation land should be dry, dense, and has uniform hardness (distributed hardness), and the basic foundation should be deeper than 45cm;
- The foundations need to be tied to one another by means of foundation beams.

The community awareness of the importance of *Pengurangan Resiko Bencana* (PRB) or Disaster Risk Reduction is getting better; this is indicated by many emergency simulations which are conducted in affected areas. The supporting facilities provided by the Government through the National Disaster Management Agency (BNPB), such as emergency shelters and early warning systems, are adequate. However, the community awareness is not accompanied by the understanding of how to create building structures and constructions that can reduce the risks of disasters, which can be learned from local architectures. Local knowledge (wisdom) is the knowledge developed in a society, which is obtained through a process of trials and errors to the physical environment, such as against earthquakes, floods, and many others. Such knowledge is mostly kept by local communities; it is commonly applied to their environments, such as buildings (houses).

2. Method

This research aims to analyse the Sundanese traditional building construction in withstanding earthquakes. It was conducted by using the descriptive-evaluative method. The research process was conducted by doing inventory analysis of building construction elements and comparing them with the construction standards of simple earthquake-resistant buildings from SNI-176-2002 Standard Planning Earthquake Resistance for Building Structure Building Institute in Indonesia.

3. Results and Discussion

The most typical Sundanese traditional building is imah panggung, which is a house with a pit under the floor, approximately 40-60cm. Panggung derived from the word pang and gung, which mean being put on the highest place (Nuryanto, 2006). Panggung is the most important form for Sundanese people, with long suhunan and jure. Long suhunan is the roof with the shape of a long ridge from front to back (like a saddle), while jure is a roof with a ridge forms a shorter or a so called triangular roof shield. The panggung shape dominating building/construction system in Tatar Sunda (another nickname for West Java Province in Indonesia) has engineering/technical and symbolic functions. Technically, imah panggung has three functions: not interfering water absorption field, the pit functioning as a conditioning space with air flow crossing for warmth and coolness, and the pit for storing firewood supply (Adimihardja, 2008).

The symbolic function of imah panggung is based on the belief of Sundanese people that the world is divided into three elements: handap (under world), ambu luhur (top world), and tengah (middle world). Tengah is the centre of the universe and man put himself at this centre. Thus, human habitation should be located in the middle, not at ambu handap (underworld/the earth) or ambu luhur (the world top/sky). Houses must be supported by poles which serve to divide the whole house with the lower and upper world. Furthermore, the poles should not be located directly on the ground; it should be given pedestals that separate the poles/house from the land. The pedestals are rocks called umpak (Adimihardja, 2008).

The structure and construction of the Sundanese imah panggung are light and simple, because most of the materials used are from the local environment and self-made (Figure 1). This can be seen from the foundation of the houses, which are stones taken from rivers, hills, or mountains. The walls are made of woven bamboos or wooden planks; the floor is made of talupuh or palupuh, which are shredded bamboos or woods. The roof frame is made of bamboos and woods, with hateup kiray and injuk as the roof cover. While the construction seems light and simple, it is strong and sturdy. It is proven by the durability/endurance of the construction during and after earthquakes. This condition can be seen in Kampung Baduy, Naga, Kasepuhan Ciptagelar, and Dukuh. The houses are sturdy; none of which collapsed.

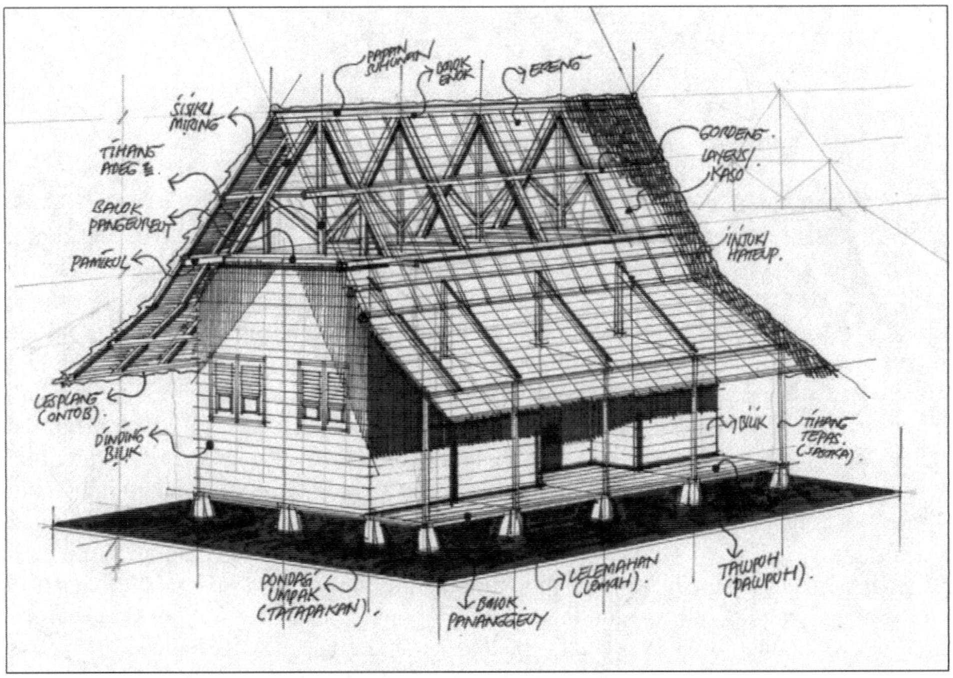

Figure 1. Structure and construction of Sundanese traditional *imah pangggung* (Nuryanto, 2015)

In terms of building materials, traditional Sundanese buildings have met one of the requirements of earthquake resistant buildings, which is made of lightweight materials, such as bamboos and woods. The advantages of wood as a building material have been revealed by Brostow et al. (2010); wood consists of two parts: the center which can resist compression, and the outer part which can resist tension (Figure 2). Wood, having low water content, has a better ability to resist compression. Furthermore, wood cells can continue the compression/pressure. Wood is a material which can support the development of green architecture.

Another material which is commonly used for Sundanese traditional buildings is bamboo. According to Sharma et al. (2015), bamboo has many advantages as a construction material; it is renewable and has wood-like mechanical properties. Bamboo fiber has many variations; it can be used for interior and exterior building materials. Tomas (2014) asserts that bamboo is a very flexible material and widely available; therefore, it should be used as an engineering material for houses and other buildings.

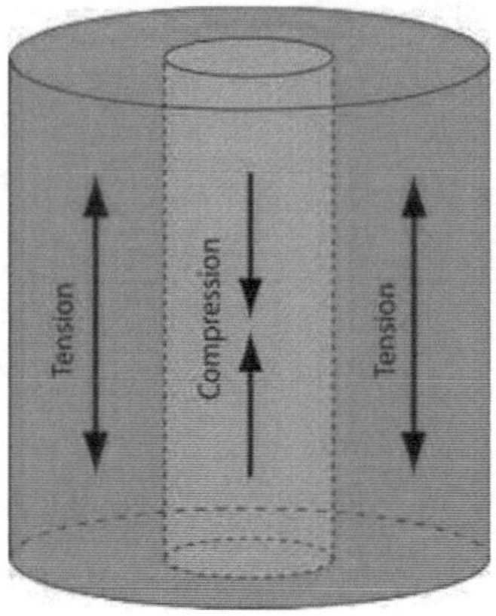

Figure 2. Tree trunk regions in compression and tension (Brostow et al., 2010)

In general, the strength system of Sundanese traditional imah panggung uses bonding, pupurus connection, and paseuk (pegs). On the frames of the floor, walls, and foundations, beams are installed and connected, both vertically and horizontally using the pupurus (pen and holes); the connection uses ropes or rattan and wooden pegs. No nails, nuts, and bolts are used because they are prohibited by the custom and contrary to the rules of their ancestors (taboo). According to Felix (1999), connection pegs have an efficiency level of 60%, which is better than bolt connections that have an efficiency level of 30%, or nail connections that have an efficiency level of 50%. Structures and construction have close links; if one of them does not exist, the building cannot be constructed. "Euweuh rarangka can teu ngarangka, euweuh ngarangka wangunan can TEU ngadeg" – this means that there is no frame of the house that can be constructed. The arrangement of structure and construction of houses of the Sundanese community are based on the panggung form; they are divided into two types: handap and luhur. Handap is a structure located under the floor of the house, consisting of lelemahan (ground) and umpak/tatapakan (foundation). On the other hand, luhur is a structure located on the floor, such as pangadeg/adeg (wall), lalangit/palapon (ceiling), and rarangka (frame). The structure is a separable part of the construction, because its functions support the strength of the building.

Umpak is constructed above the ground level, which is the foundation made of stone with a defined construction technique (Figure 3). Three types of Umpak are known: buleud, lisung, and balok. Buleud is a type of umpak which is round; lisung is trapezoidal and balok resembles a cube. Generally, people use the round umpak. According to local residents, umpak can be mounted in two ways: dina luhur taneuh, which is above the ground, and dina jero taneuh, which is partially buried in the ground. Commonly, they mount the umpak by burying it in the ground. Based in the analysis of the requirement for earthquake resistant buildings, the foundation used in Sundanese traditional buildings has met the requirement. They use foundations which have been bound to one another.

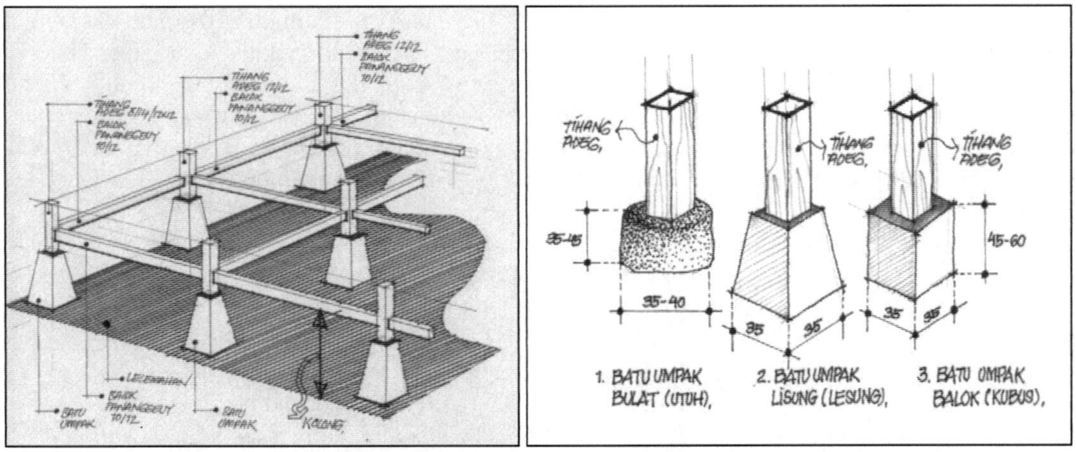

Image 3. *Handap* structure consisting of *lelemahan* and *umpak* foundation (Nuryanto, 2006).

Luhur structure is differentiated into three parts: pangadeg, lalangit, and rarangka. Pangadeg structure is a frame collated based on two components: wall and floor. The wall structure is arranged based on three components: tihang, pangadeg/tihang adeg, pananggeuy and tihang nu ngabagi. It is made of bamboo woven with a braid system, and the board wall with sirih stacking system (see Figure 4). The pananggeuy and tihang nu ngabagi construction uses pupurus connection technique, and oblique hooked lip reinforced with nails, pegs, and ropes; it is also applied to sunduk awi.

On the floor structure, Sundanese community has three types of flooring: talupuh, papan (board) and bilik (chamber). Talupuh or palupuh is floor made of bamboo which has been stoned to a certain size as needed. The bamboo used is usually the Gombong type or Wulung type, which has a diameter of ± 15-20 cm and thickness of ± 12-15 mm. When it is cleaved and stoned, the width can reach ± 30 cm. The Sundanese community believes that the tapuluh floor has some advantages, such as easy to make and fit, inexpensive, as well as more beautiful look.

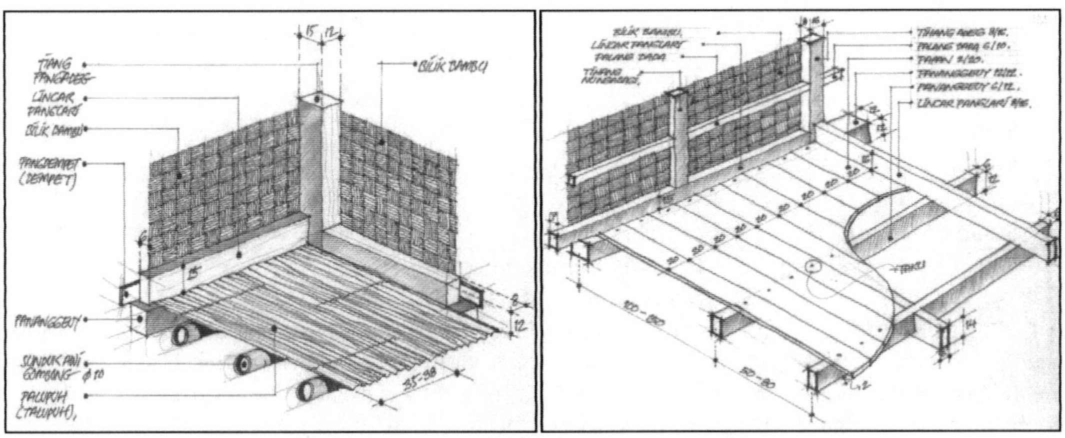

Image 4. *Talupuh* floor and floor boards in *imah panggung* (Nuryanto, 2006).

The lalangit and rarangka are suhunan structures, which are roof truss arranged based on two components: frames and ceilings (see Figure 5). The frames consist of two parts: nu mikul and nu dipikul. Nu mikul is the main frame, while nu dipikul is the supporting frame. The whole suhunan structure uses wooden and bamboo frames formed triangle frames. For the nu mikul structure, makelar adeg (10x15cm) is placed perpendicular on top of the pamikul. Pamikul (8x15cm) is mounted horizontally, connecting the makelar adeg. On the right and left of the makelar adeg, jure suhunan is mounted with a slope of ± 30-45. Jure suhunan (8x15cm) connects the pamikul block, which lies beneath the makelar adeg at its upper end. On top of the makelar adeg, some members of the Sundanese community mounted a chock (5x10cm) horizontally. To strengthen the jure suhunan position, a chock (8x12cm) is mounted on the left and right of the makelar adeg.

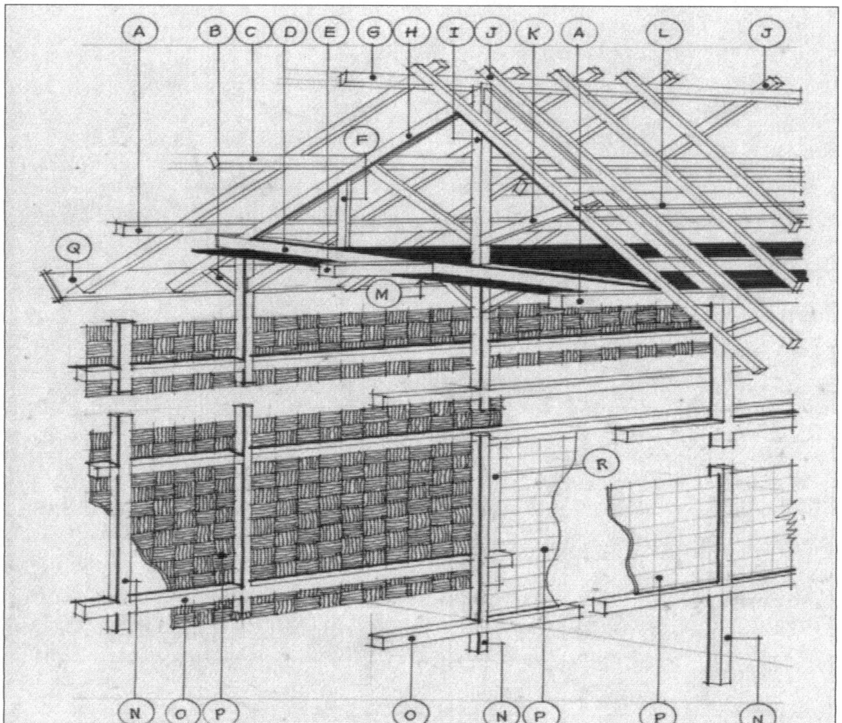

	Beams *pangeureut*	G.	Beams *nok suhunan*	M.	*Ceiling*
	Beam *sokong*	H.	Beams *jure suhunan*	N.	*Tihang pangadeg*
	Beams *gordeng*	I.	Beams *realtor adeg*	O	Beams *pananggeuy*
	Beams *pamikul*	J.	*Layeus* / kaso-kaso	P.	Bamboo booth
	Beams *panglari*	K.	Beam *sokong*	Q.	*Lesplang* board
	Beams *nunjang*	L	*Ereng/reng*	R.	*Tihang nu ngabagi*

Figure 5. Friendly home construction earthquakes
(Nuryanto, 2006).

In the structure of nu dipikul, there is gordeng or gording (6x12cm), which is placed on the jure suhunan horizontally based on the span of the frames. Some part of the roof does not use gordeng. Above the gordeng, people install layeus or rafters (5x7cm) vertically in the direction of the jure suhunan. The layeus connects the pamikul with the chock. Above the layeus, ereng or sloping (2x3cm) is placed parallel to the jure suhunan beam, starting from the bottom to the top. In general, bamboo layeus and sloping were used. To strengthen the position of jure suhunan, side brackets (5x5cm) were also mounted underneath with a certain slope. Some people do not use gapit, lesplang, pangereut, panglari, nunjang, wind ties and locking beams, due to material efficiency. Once the frame is complete, the roof cover can be fitted. The installation of hateup roof cover is easier than injuk, because the construction is not too complicated. In hateup construction, ereng is not needed because the roof cover has been adjusted according to the layeus distance ± 45-50 cm. Generally, the wood connection techniques used are hooked-straight lip, hooked-oblique, and pen-holes, reinforced with rope and rattan. The connection and bonding techniques require thoughts and attention, because mistakes will cause serious damages.

On the suhunan structure or the roof of imah panggung, Sundanese traditional community, such as in Baduy Kajeroan, Kasepuhan Ciptagelar, Kampung Naga, and Dukuh Pantang, uses roof tiles (made of clay). Because it is prohibited by their ancestors, they call it teu wasa or teu wani. In their cosmology, using roof tiles (made of clay) means burying themselves alive, because only dead people deserve to be in tombs. "Jelema hirup keneh kunaon kudu di ruang", meaning that living people should not be in the grave. Furthermore, using roof made of clay from the ground also means committing adultery with ‘mother’, as in their belief the land represents the earth which has the meaning of ka indung. “Manusa hirup tina saripatina taneuh” means that humans live from the essence/core of land. Taneuh or soil also has the meaning of death. Those who dare to use the roof will be kabendon (received anger) by the ancestors, such as pain, bad luck, and hard life. This also applies to other materials, such as asbestos, zinc roof, and other materials made of clay/land core. The materials are prohibited to use as it is contrary to the ancestors’ rules. The roof structure of Sundanese traditional buildings uses a lightweight material, which is in line with the requirements of earthquake resistant building construction. For example, a traditional house on Cikondang - located at RT 03, RW 03, Cikondang, Lamajang Vilage, Pangalengan, Bandung Regency – still stands although the area has been affected by an earthquake measuring at 7.3 magnitude occurring Wednesday, September 2, 2009. The local knowledge is reflected by the house, answering the question whether local buildings can withstand an earthquake (Pikiran Rakyat, September 7, 2009).

Unlike hundreds of houses which were cracked up or collapsed, the 12 x 8m house stands strong. Almost nothing has changed. The roof is made of fiber and the wall is made of woven bamboo. There was no damage, except factors related to the age of the building and lack of maintenance of the building, which can be seen from the windows of the house and the door, as well as the 9 sereseg on each window. The advantage of this house is it is earthquake-resistant. It is due to the belief that it is taboo to make houses from stones, because it is prone to earthquakes. The use of roof which is made of clay is also taboo, as it is believed by the Sundanese community that living people should not live under the land (represented by the clay).

Figure 6. The various typologies of vernacular houses in the southern part of West Java (Triyadi & Harapan, 2008)

The results of research conducted by Triyadi and Harapan (2008) showed that the vernacular buildings in the southern part of West Java (Figure 6) are examples of houses that have withstood earthquakes striking the area. The buildings are able to survive, while many other buildings (non-vernacular) collapsed. This is evidence showing the application of indigenous knowledge of Sundanese (East Java) community to buildings. Such knowledge is one of potentials which should be explored through research/fieldworks, so that data on the indigenous knowledge can be used by communities, especially local communities.

4. Conclusions

The strength of Sundanese traditional imah panggung uses bonding, pupurus connection, and paseuk (pegs). On the frame of the floor, walls and foundation, the beams installed use pupurus (pens and holes); they are also tied with fiber ropes or rattan and wooden pegs. No nails, nuts and bolts are used, because they are prohibited by customs and contrary to the rules of their ancestors (taboo).

In Sundanese traditional imah panggung, iron nails or modern construction/buildings parts are rarely found. For connecting the poles, bamboo paseuk or rope made of palm/coconut fibre are used. For the roof, the houses use fibers, palm leaves or thatch leaves. It is very rare that Sundanese traditional houses use roofs made from clay. Another interesting aspect is the materials used in the houses. It is impossible that booths with thin walls and wooden floors are used by a community which has barbaric civilization. Houses for Sundanese communities are not for protection from enemies, but from nature, such as rains, wind, sunshine, and animals.

References

[1] Adimihardja, K. (2008). The dynamics of Local Culture. Bandung: Study Center LBPB.

[2] Brostow, W., Datashvili, T., and Miller, H. (2010). W ood and W ood D erived M aterials. Journal of Materials Education Vol. 32 (3-4): 125-138.

[3] Canadian Wood Council. (2003), wood frame construction - meeting the challenges of Earthquakes, building performance series 5. Canada; CWC.

[4] Dep. PU. (2002). Earthquake Resilience Planning Standards For Building Structure: ISO-1726-2002. Bandung: Puslitbangkim PU.

[5] Felix. (1999). Construction Wood. Bandung: CV Trimitra Mandiri.

[6] Nuryanto (2006): Continuity and Change Patterns Village and House Live from Kasepuhan Ciptarasa to Ciptagelar, in the district. Sukabumi (south), West Java. Thesis Research Master of Architecture, Architectural Studies Program SAPPK-ITB, Bandung (not for publication);

[7] Nuryanto (2015): Architecture archipelago, Sunda Traditional Architecture Series: Traditional Architecture in Frame Architecture Sunda Archipelago: Introduction to Architecture and House Village Stage.Textbook Archipelago Architecture Architectural Engineering Program, Department of Architectural Engineering Education FPTK Indonesia University of Education, Jakarta: PT. Rajawali Grafindo Persada.

[8] Harian Umum Pikiran Rakyat. 7 September 2009.

[9] Sharma, B., Gatoo, A., Bock, M., and Remage, M. (2015). Engineered bamboo for structural applications. Construction and Building Materials 81 (2015) 66-73

[10] Thomas and Ganiron. (2014). Investigation on the Physical Properties and Use of Lumampao Bamboo Species as Wood Construction Material. International Journal of Advanced Science and Technology Vol.72 (2014), pp.49-62

[11] Triyadi, S and Harapan, A. (2008). Local Wisdom House Vernacular In West Java, South Part in Responding to Earthquake. Journal of Science and Technology EMAS, Vol. 18, No. 2, May 2008.

Effects of using passive filter for reduce electrical load harmonics

T Sucita*

Department of Electrical Engineering Education, Indonesia University of Education
Jalan Setiabudhi, 229 Bandung 40154 Indonesia

*Corresponding author: tasmasucita@upi.edu

Abstract. Due to the use of electrical current load that uses a lot of electronic components (passive non-linear electrical loads), so the impact will cause harmonics in the electrical network system. These harmonics can unwittingly cause a relatively large loss in electrical energy consumption and can lower the power factor of an electrical installation. Limits how much the harmonic distortion that is installed on the load adjusted to the IEEE 519-1992 standard. The study was conducted by taking data on a network of electrical installation of a building using measuring devices Fluke 43B Power Quality Analyser. The data is then processed and consulted with the standard IEEE 519-1992. Once the data has a discrepancy with the standard, further made the filter design using linear passive components. The design is then installed on the network installation by means of simulated order harmonic losses can be overcome so that the circuit meets the IEEE standard installation by changing the parameters of the linear load L and C. The results of this study indicate that THDi value decreased after the installation of filters for phase R fell by 9.39%, the S phase decreased by 7.54% and for the T phase decreased by 16.88%. So that meets the IEEE standard by 15%.

1. Introduction

Harmonics is a phenomenon arising from the operation of the non-linear electrical load, which is a source of wave formation at high frequencies that are multiples of its fundamental frequency as 100 Hz, 150 Hz, 200 Hz, 300 Hz, and so on. It can interfere with the electrical system at the fundamental frequency is 50 Hz, so that the current and voltage waveforms are purely sinusoidal ideally will become disabled as a result of harmonic distortion that occurs. The main effect of voltage and current harmonics in the power system [2] are (1) Addition of harmonics due to the level of the series and parallel resonant relationship, (2) The decrease in the efficiency of power generators, transmission and utilization, (3) Interference with telephone circuits (telecommunications) and transmitter as the zero sequence harmonic currents, (4) errors in the meter-gauge rotating disc of energy meter. Harmonics need special attention because they can unwittingly lead to relatively large losses, especially on large energy users such as office buildings and industrial sectors. To dampen harmonics on non-linear load can be used filters. The use of filters in a power system with a high content of harmonics will be very helpful in dealing with disruptions caused by the effect of the harmonics [3]. Filter is a device that has a function to skip a certain frequency. With the addition of a filter on a power system containing the sources of harmonics, harmonic currents then spread to the entire network can be reduced as small as possible. In addition, the frequency of the fundamental harmonic filter can compensate the reactive power and also used to improve the power factor of the electricity grid system [4]. Based on the above

IOP Conf. Series: Materials Science and Engineering **128** (2016) 012018 doi:10.1088/1757-899X/128/1/012018

the writer is interested in conducting research as outlined in the title: "Planning Passive Linear Filters To Reduce Electric Charging of Non Linear Harmonics". The problem in this study is: "How do I cope with Harmonics on a network of electrical installations caused by the use of non-linear electrical loads?" In order to widen the issues discussed are not too large, then the scope of the discussion is limited to the analysis of harmonic voltage (THDV) and harmonics current (THDi) as well as the drafting of the filter circuit PSIM simulation program to simulate the measurement results [9].

2. Methodology

Step-by-step outline of research described in the flowchart below.

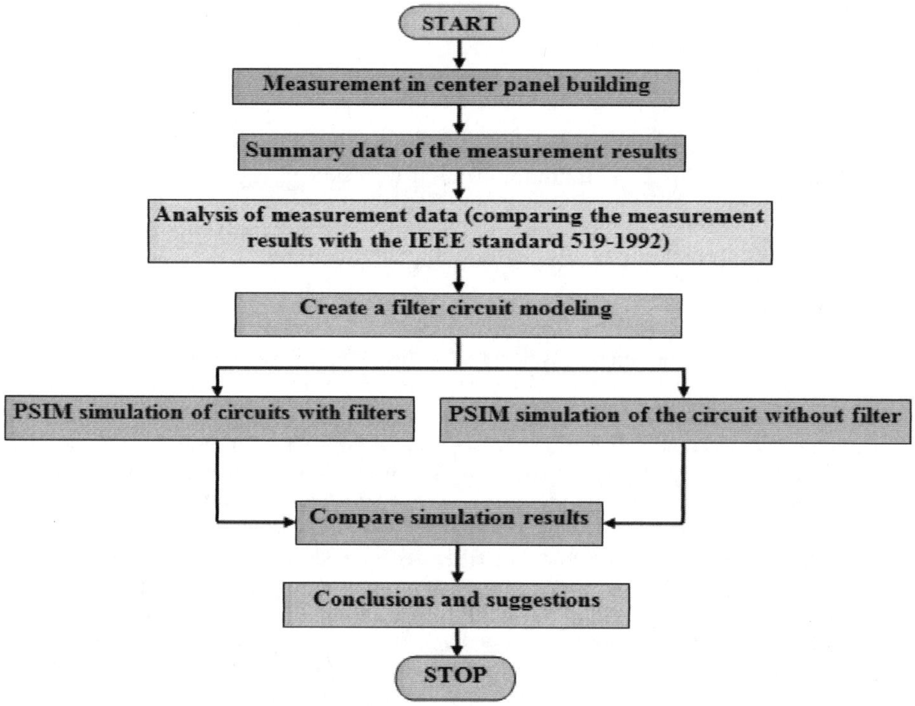

Figure 1. Flowchart Research

Data collection was conducted in this study is to perform a direct measurement of the field at the side of the panel supply electrical installations at the time of the operation state. Measuring instruments used for data collection in this study is the Fluke 43B Power Quality Analyzer. This measure can be used to measure current, voltage, power, power factor, harmonics (THD) and others.

Measurement data is limited to the data analyzed and % THDV %THDi. Data THDV measurement results compared to standard IEEE No. 519 -1992 on voltage. Value % is permitted THDV IEEE No. 519-1992 for voltages below 68 kV is <5%, if the measurement value exceeds 5% of the declared voltage harmonic distortion has occurred on the system. While for the data % THDi compared to standard IEEE No. 519 -1992 of the current [5]. Before comparing with the standard, it must first be sought value ISC/IL. IL is the current value of the measurement results while the value of ISC is the short circuit current value, which can be found using the following formula [4]:

$$I_{SC} = \frac{Full\ Load\ Current}{Impedance} = \frac{I_{FL}}{Z} \quad \text{and} \quad I_{FL} = \frac{S}{380 \times \sqrt{3}} \times 1000$$

Weight used in the modeling circuit includes: a resistive load resistor, inductor as the load inductor and bridge diode as a source of harmonics. Resistor and inductor values can be determined from the measured data that is of value Vrms, Irms, frequency and cos φ [7].

In this modeling series filter consists of a input filters and output filters. Installation of the filter on the input side aims to reduce the current harmonics that occur at the source, while mounting the filter on the output side aims to reduce harmonics that occur at the rectifier output voltage or the load [4].

3. Results

Data measurement results can be seen in the table below.

Table 1 Recapitulation of Data Measurement Results

Parameters	Phase R	Phase S	Phase T
Voltage rms (V)	213,8	217,8	219,7
Current rms (A)	45,8	81,1	62,5
THD_V (%)	5,3	5,7	5,3
THD_I (%)	23	7,4	15,3
Frequency (Hz)	50	49,9	50,1
P (kW)	9,2	16,9	13,4
S (kVA)	9,7	17,0	13,7
Q (kVAR)	3,0	2,7	2,9
PF	0,95	0,99	0,98
DPF	0,98	1	1

Graphic measurement results of each phase can be explained in the following picture:

Phase R

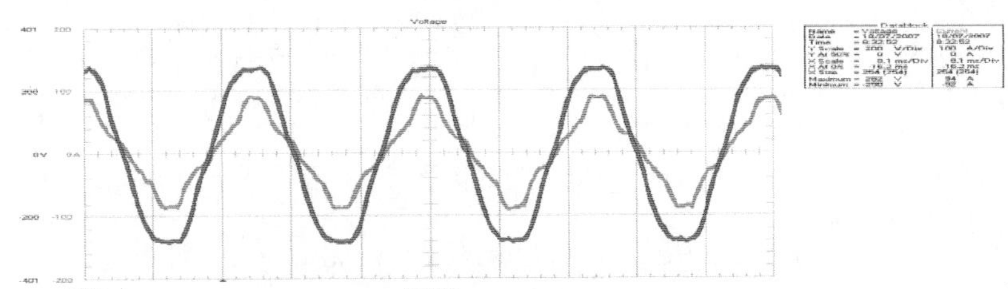

Figure 2. GraphVoltage and Current

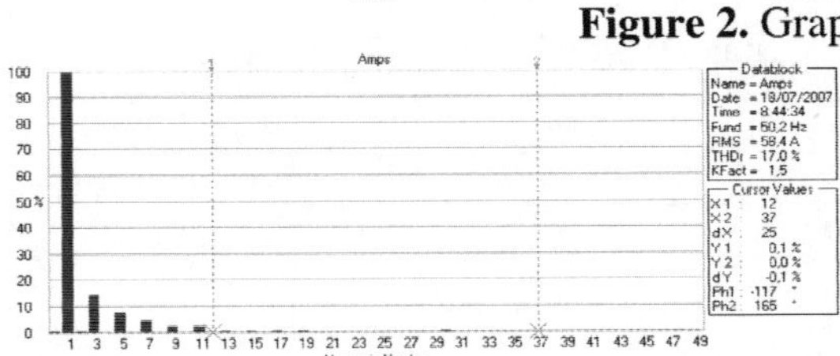

Figure 3. Graph Harmonic of Current

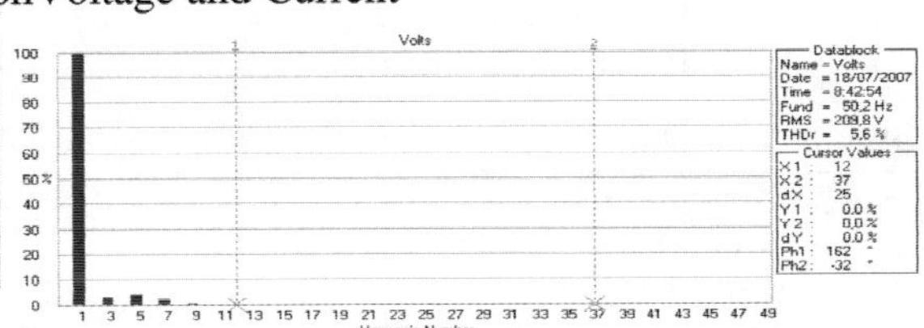

Figure 4. Graph Harmonic of Voltage

Phase S

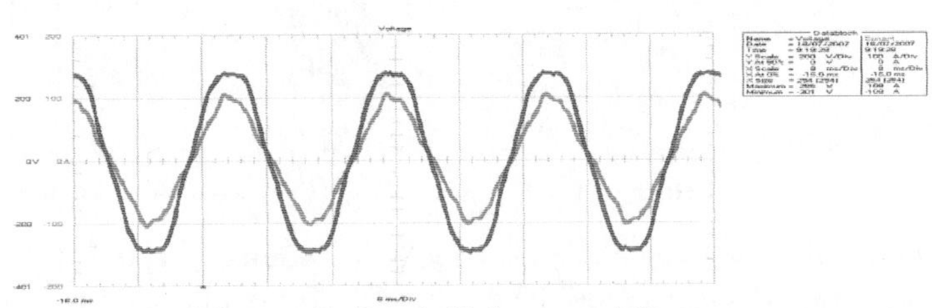

Figure 5. Graph Voltage and Current

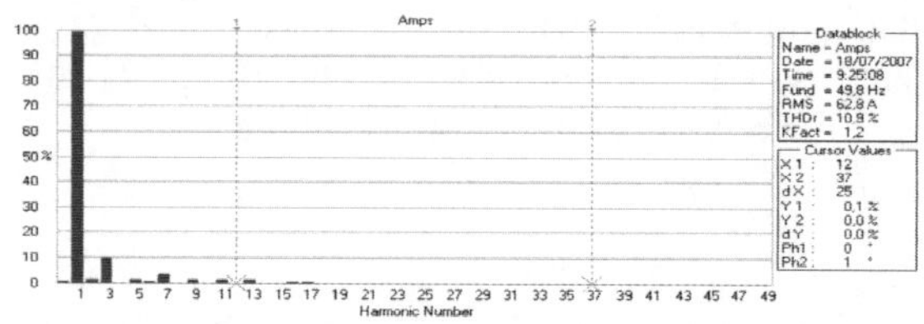

Figure 6. Graph Harmonic of Current

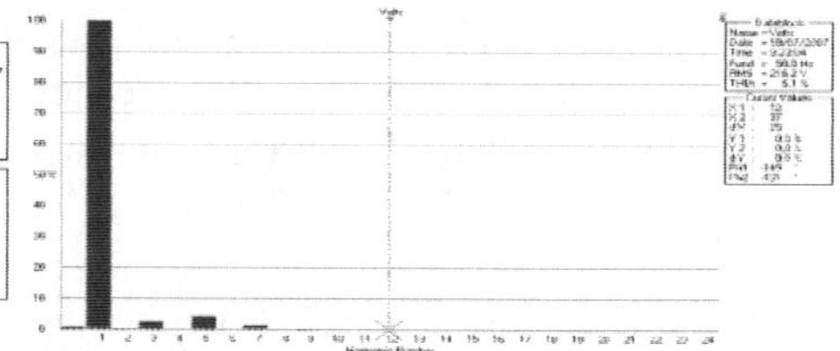

Figure 7. Graph Harmonic of Voltage

Phase T

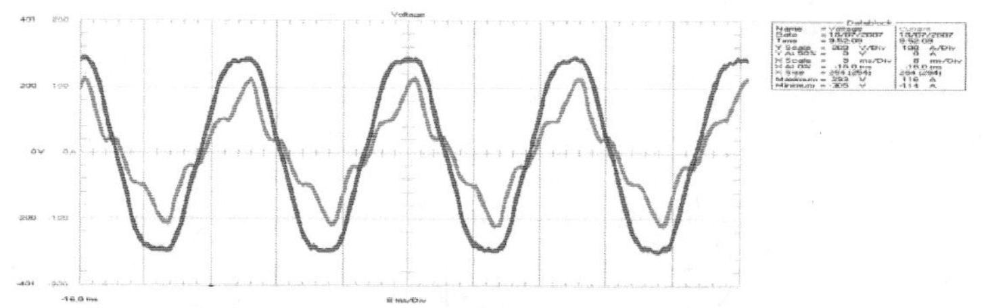

Figure 8. GraphVoltage and Current

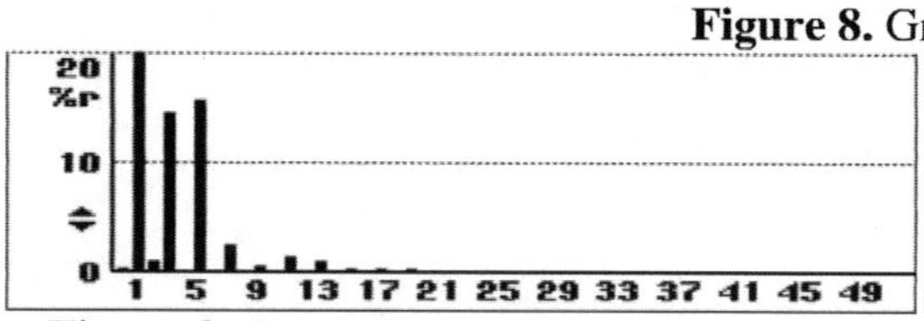

Figure 9. Graph Current Harmonic

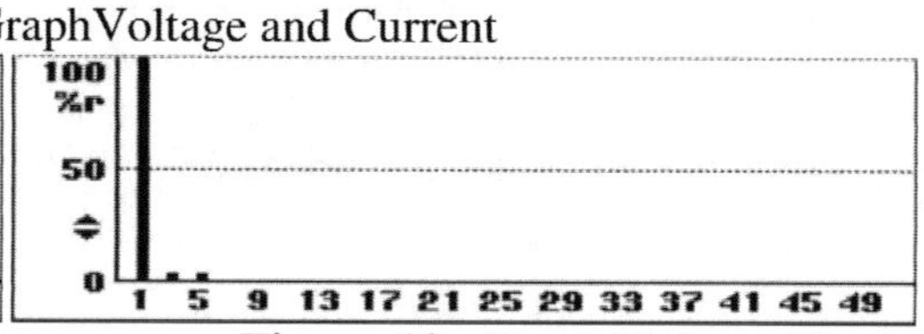

Figure 10. Graph Voltage Harmonics

Standard %THDv allowed by the IEEE Standards No. 519-1992 for voltages below 68 kV was < 5%. The measurement data are:

Table 2 Results Analysis of Measurement THDv

	Time	Measurement THDv (%)	Standard THDv (%)	Analysis
Phase R	13.11	5,3	5	Not suitable
Phase S	13.31	5,7	5	Not suitable
Phase T	13.57	5,3	5	Not suitable

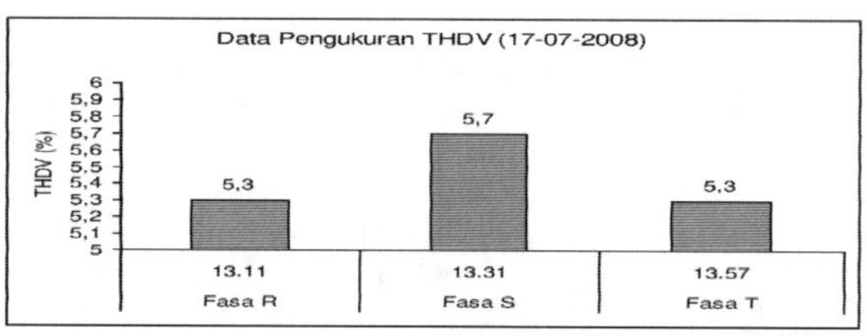

Figure 11. Graph Results Measurement THDv

Based on the above data, the value of the average measurement THDV not in accordance with the IEEE 519-1992 standard which exceeds the value of 5%. To find current standard % THD can be determined by calculating the ISC / IL, where ISC is the short circuit current and the maximum current IL. ISC value can be determined by the calculation below and to the value derived from the value of IL. IRMS measurement results using 43B Power Quality Analyzer. The results of the calculation can determine the value of % THDi by looking at Table 2 current standard IEEE 519-1992.

Following calculation to determine the value of ISC: Note, Power Transformer = 400 kVA; Working voltage of 20,000 V / 380 V / 220 V; Impedance (Z) = 4%. Calculation of Short-circuit Current:

$$I_{SC} = \frac{Full\ Load\ Current}{Impedance} = \frac{I_{FL}}{Z}\ ;\ I_{FL} = \frac{\text{DayaTrafo}}{380\times\sqrt{3}}\times 1000\ ;\ I_{FL} = \frac{400\ \text{kVA}}{380\sqrt{3}}\times 1000 = 607{,}737\ \text{A}$$

$$I_{SC} = \frac{I_{FL}}{Z}\ ;\ I_{SC} = \frac{607{,}737}{0.04} = 15.193{,}425\ \text{A}$$. So the value of I_{SC} is 15193.425 A

Table 3 Results Analysis of Measurement THDi

	At	I_{SC} (A)	I_L (A)	I_{SC}/I_L	Measurement THDI (%)	Standard THDI (%)	Analysis
Phase R	13.11	15.193,425	45,8	331,73	23	15	notsuitable
Phase S	13.31	15.193,425	81,1	187,34	7,4	15	appropriate
Phase T	13.57	15.193,425	62,5	243,09	15,3	15	notsuitable

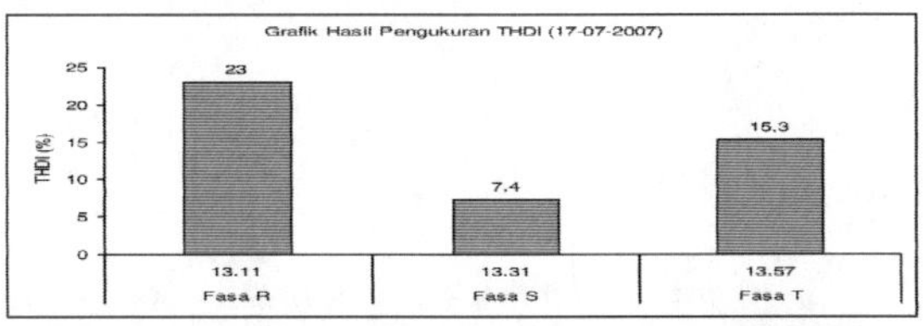

Figure 12. Graph Results Measurement THDi

Based on the above data, it can be concluded that most of the measured values THDi not in accordance with the IEEE standard 519-1992 which exceeds the value of 15%. For the R phase, the measurement data THDi measured value at 13:11 hours is equal to 23%. For the S phase, the measurement data THDi value measured at 13:31 hours that is equal to 7.4% of this value does not exceed the standard. For phase T, the measurement data THDi measured value at 13:57 hours that is 15.3%.

Measurement data is then made modeling using PSIM 4.1 program. In the modeling will be made a model with a large load current load current approaches the measurement results. Load modeling is made using 1-phase full-wave rectifier with load R and L are used to describe the harmonic currents. Here will be described a series of modeling, waveform, shape FFT spectrum analysis results of PSIM program and the current value of each order harmonics of each measurement before and after using the filter using a filter.

To create a network test on PSIM required parameters on components used in network modeling. Component calculation load on each phase, taking the measurement data can be explained as follows. Phase R. When the voltage of 217.1 volts, current is 55.4 amperes, frequency of 50.1 Hz, and cos φ of 0.97 lagging, then raised the resistance of 3.8 ohms and inductance of 3.02 mH. Similarly to the S phase, when the voltage of 223.8 volts, current is 44.3 amperes, and setting frequency of 49.9 Hz, and cos φ of 0.96 lagging, then raised the resistance of 4.85 ohms and inductance of 4, 5 mH. As for Phase T, when voltage of 218 volts, current at 56.5 ampere, and setting frequency 49.9 Hz, and cos φ of 0.98 lagging, then raised the resistance of 3.78 ohms and inductance of 2.46 mH.

3.1. Calculation filter L and C

3.1.1. filter Input

Filter on the input side aims to reduce harmonics that occur in the source stream. To determine the value of the filter L and C, can be calculated using the formula : $f_0 = \frac{1}{2\pi\sqrt{LC}}$

The result of the calculation gives a chance filter with resonance frequency of 50 Hz, and 20 m H inductance value, the value of C of 352 uF.

3.1.2. filter Output

Installation of filters on the output side aiming to reduce the amount of harmonics that occur in the rectifier output voltage or the load. The formula for determining the value of L and C in the same filter

as the output filter insert. Resonance frequency of the output filter to be set to 100 Hz and L = 5 mH, the obtained C of 507 uF.
In a series of simulations were made to approach the ideal measurement values, then there are some components that are adjusted value is the value of L in S phase and phase T.

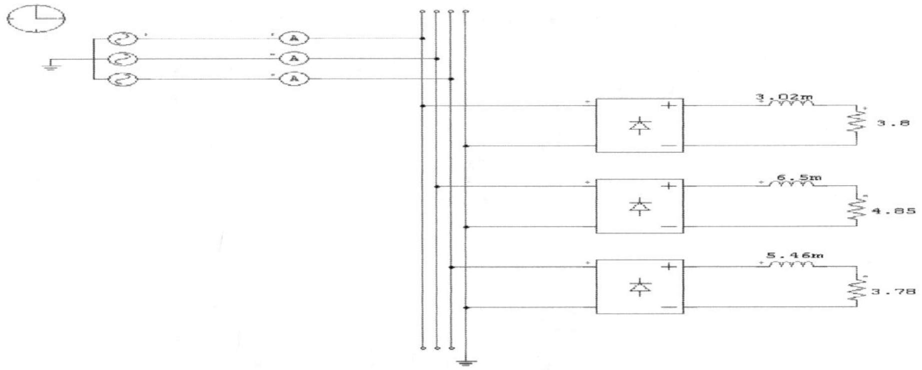

Figure13. Network Modeling Before Using Filter

Modeling the circuit in the image above drawing load on the measurement, where harmonic load on each phase containing the third and fifth harmonics as well as the seventh. In a series of surrogate modeling made on 4 pulse rectifier load with a load R, L which can result in the third and fifth harmonics. Data obtained from the simulation results that we are setting the value of the measurement approach. The simulation results above produces the following waveform before using filters.

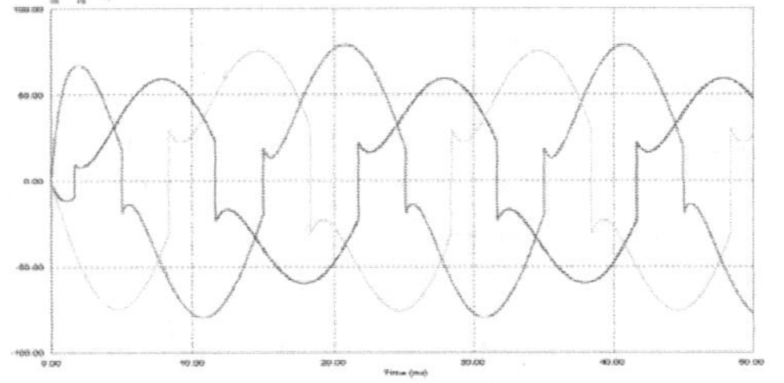

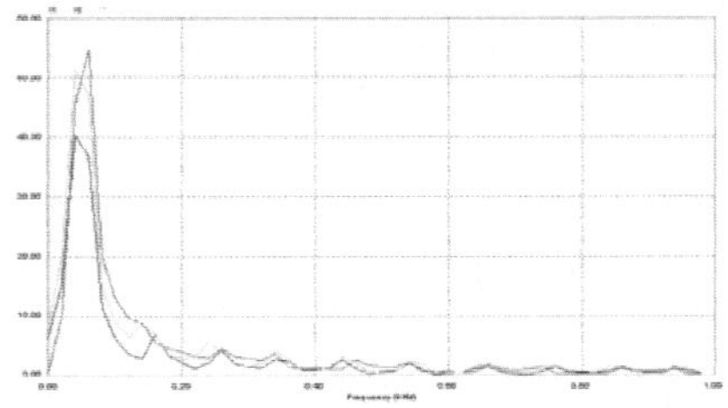

Figure 14 Form of Current Wave Figure 15 Shape Spectrum Current Harmonic

From Figure 14, it appears that wave form is characterized by containing ripple waveform is not smooth, it shows that the system of harmonic disturbances. With FFT analysis will be more visible harmonics that occur, as it is known-order harmonics where the greatest. As shown in Figure 15, the shape of the current spectrum is not flat on the odd-order harmonics multiples.

From the above simulation results obtained percentage value THDi flowing in each phase is the phase of 18:38% R, S phase of 19:19% and 22:38% for the T phase, where the value exceeds the specified standard (15%).

3.2. After a series of Simulation Filter Installation

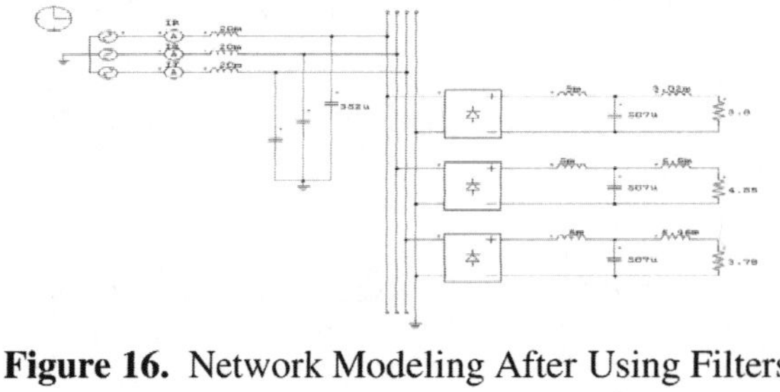

Figure 16. Network Modeling After Using Filters

IOP Conf. Series: Materials Science and Engineering **128** (2016) 012018 doi:10.1088/1757-899X/128/1/012018

Network modeling on the network in picture 16 have been added to the input filter on the source and output load filter consisting of components A and C, the shape of the waveform and spectral flow after installation filters as below.

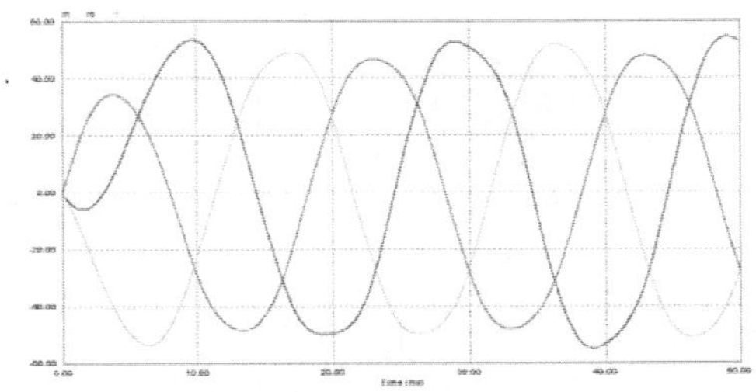

Figure 17. Shape of Current Wave

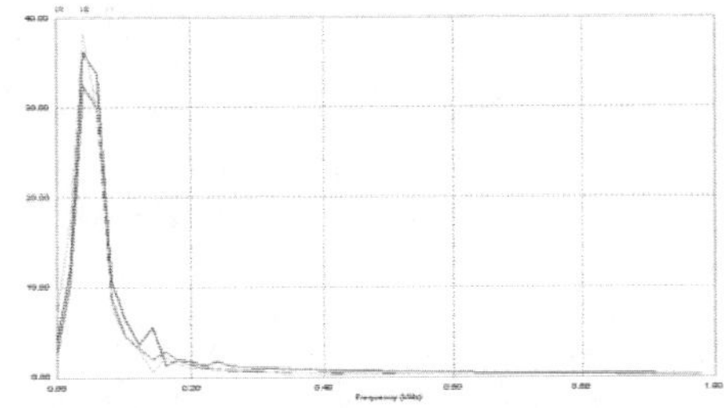

Figure 18. Form of Spectrum Current Harmonics

From the above simulation results obtained percentage value THDi flowing in each phase after the phase R filter mounted at 8.68%, the S phase by 12.4% and amounted to 8.73% T phase, so that the value meets the specified standards (15%). From the simulation results it is known that the value of harmonic currents that occur at each order harmonics decreased significantly..

To find large changes in the value of harmonic currents before and after the installation of filters at each phase, can be seen in the following table.

Table 4 Comparison of Current Harmonic Value Phase R

Order Harmonics	Measurement results [ampere]	Before Filter Simulation Results [ampere]	Simulation Results After Filter [ampere]
1	52,2	50,13	31,18
3	8,4	7,25	2,46
5	3,3	3,8	0,9
7	2,3	2,61	0,57
9	1,6	1,9	0,42
11	1,1	1,57	0,33
13	0,1	1,2	0,28
15	0,6	1,1	0,24
17	0	0,9	0,21
19	0,1	0,8	0,19
THDI	**18,26%**	**18,28%**	**8,89%**

Table 4 shows that the total current harmonic (THDi) for phase R after the installation of the filter is equal to 8.89%. Whereas before the installation of the total current harmonic filters (THDi) is approximately 18:28% where there has been a decline of 9:39%. So the value THDi meet IEEE 519-1992 standard that is equal to 15%.

Table 5 Comparison of Current Harmonic Value Phase S

Order Harmonics	Measurement results [ampere]	Before Filter Simulation Results [ampere]	Simulation Results After Filter [ampere]
1	41,2	38,54	34,97
3	7,6	4,98	3,48
5	3,6	3,3	1,5
7	2,1	2,67	0,96
9	1,21	2,07	0,75
11	1,03	1,67	0,62

13	0,76	1,4	0,51
15	0,23	1,18	0,45
17	0,34	1,09	0,39
19	0,19	1,01	0,35
THDI	**21,50%**	**19,32%**	**11,78%**

Table 5 shows that the total current harmonic (THDi) in the S phase after the installation of the filter is equal to 11.78%. Whereas before the installation of the total current harmonic filters (THDi) is approximately 19.32% where there has been a decline of 7.54%. So the value of THDi meet IEEE 519-1992 standard that is equal to 15%

Table 6 Comparison of Harmonic Phase Value Phase T

Orde Harmonics	Measurement results [ampere]	Before Filter Simulation Results [ampere]	Simulation Results After Filter [ampere]
1	52,8	49,17	33,76
3	8,4	7,91	1,41
5	7,7	4,95	0,9
7	1,2	3,5	0,53
9	0,5	2,7	0,42
11	0,7	2,2	0,33
13	0,8	1,9	0,27
15	0,2	1,7	0,23
17	0,1	1,5	0,2
19	0,1	1,37	0,18
THDI	21,82%	22,47%	5,59%

Table 6 shows that the total current harmonic (THDi) on phase T after the installation of the filter is equal to 5.59%. Whereas before the installation of the total current harmonic filters (THDi) is approximately 22:47% where there has been a decline of 16.88%. So the THDi meet IEEE 519-1992 standard that is equal to 15%.

4. Conclusion

Based on the data of measurement and data analysis can then be concluded as follows:

1. Data from the study showed that the total harmonic current distortion (THDi) measured at the sample panel are out of bounds (above) specified by the IEEE 519-1992 standard (the standard limit of 15%), that for the phase R THDi is the largest of 26.4%; phase S of 21.1%; and to phase T of 21.6%.
2. Data from the study showed that the total harmonic voltage distortion (THDV) are also exceeds the IEEE 519-1992 standard (the standard limit of 5%), which is to phase R of 6.1%; phase S of 5.2%; and to phase T of 5.4%.
3. Based on data resulting from the simulation has been done to show that the decline THDi done after the installation of filters. As the figures decline can be explained as follows: for the phase R down of 9,39% (from 18:28% to 8.89%), to phase S drop by 7.54% (from 19:32% to 11.78%), and phase T down of 16.88% (from 22.47% to 5.59%). As the simulation results show that the third phase of the THDi to meet IEEE 519-1992 standard that is below the 15%.

In connection with these results, the researchers recommended that the building be carried out installation of filters to improve the efficiency of electricity consumption associated with to reduce harmonic interference has occurred.

References

[1] Collombet C., Lupin, J.M., Schonek, J. (1999, Desember). *Harmonic disturbances in networks, and their treatment*. Cahier technique no. 152. Tersedia: http://www.schneider-electric.com.tr/schneider _en/pdf/publications_ect/ ECT152.pdf. [29 April 2008].

[2] De La Rosa, Francisco C. (2006). *Harmonics and Power Systems*. New York: Taylor & Francis, LLC.

[3] Hadibowo, E. (2007). *Perencanaan Filter Pasif untuk Mesin PP1 Winder Extruder 1 Di PT Yanaprima Hastapersada Sidoarjo*. Tugas Akhir Sarjana pada FTI UKP Surabaya: tidak diterbitkan.

[4] Hibbard, John F., Lowenstein, Michael Z. *Meeting IEEE 519-1992 Harmonic Limits (Using Harmonic Guard Passive Filters)*.[online]. Tersedia: http://www.transcoil.com/meetieee.pdf. [09 Februari 2008].

[5] Masri, S. (2004). *Analisis Kualitas Daya Sistem Distribusi Tenaga Listrik Perumahan Moderen*. Jurnal Rekayasa Elektrika, Volume 3 No. 2. Tersedia: http://www.ft.elektro.usk.ac.id/rekayasa/ 2004/321_2004.pdf.[06 Juni 2008].

[6] Muljono, S. (2006). *Pengaruh Harmonisa pada Energy Quality untuk Catuan Perangkat Infokom*. Artikel [online]. Tersedia: http://www.ristishop.com/portal/portalarticledetail.php?id=354& lang. [20 Maret 2007].

[7] Powersim Inc. (2003). *PSIM User's Guide Version 6.0*.

[8] Sankaran, C. (2002). *Power Quality*. Florida: CRC Press LLC.

[9] Sucita, T. (2013). *Analisis Pengujian Harmonisa Pada Beban Listrik Berdasarkan Standar IEEE 519-1992*. Forum Pendidikan Tinggi Teknik Elektro Indonesia (FORTEI). Padang. [Internet : http://fortei2013.com]

[10] Sutanto, J., Buhron, H. *Implikasi Harmonisa dalam Sistem Tenaga Listrik dan Alternatif Solusinya*.[online].Tersedia:http://plnkc.or.id/library/download/attach_harmonisa%20yusher pln%20jabar.doc [02 Oktober 2007].

[11] The Institute of Electrical and Electronics Engineers, Inc. (2004). *IEEE Recommended Practices and Requirements for Harmonic Control in Electrical Power Systems*. [online]. Tersedia: http://seniordesign.engr. uidaho.edu/2005_2006/hanfilco/docs/IEEE519-1992.pdf. [18 April 2008].

[12] Tribuana, N., Wanhar. (1999, April). *Pengaruh Harmonik pada Transformator Distribusi*. Elektro Indonesia, Nomor 25, Tahun V.[online]. Tersedia: http://www.elektroindonesia.com/elektro/ener25. html. [14 April 2008].

Sentiment analysis enhancement with target variable in Kumar's Algorithm

A A Arman*, A B Kawi and R Hurriyati

Sekolah Teknik Elektro dan Informatika
Institut Teknologi Bandung
Jl. Ganesha 10 Bandung, Indonesia

*Corresponding author: arry.arman@yahoo.com

Abstract. Sentiment analysis (also known as opinion mining) refers to the use of text analysis and computational linguistics to identify and extract subjective information in source materials. Sentiment analysis is widely applied to reviews discussion that is being talked in social media for many purposes, ranging from marketing, customer service, or public opinion of public policy. One of the popular algorithm for Sentiment Analysis implementation is Kumar algorithm that developed by Kumar and Sebastian. Kumar algorithm can identify the sentiment score of the statement, sentence or tweet, but cannot determine the relationship of the object or target related to the sentiment being analysed. This research proposed solution for that challenge by adding additional component that represent object or target to the existing algorithm (Kumar algorithm). The result of this research is a modified algorithm that can give sentiment score based on a given object or target.

1. Introduction

Information has a very important role at this time. Information used primarily as a support in the decision making or simply just to add insight. At the beginning of Data Warehouse and Business Intelligence era, the data sources for analytical mostly come from internal. The data sources usually come from operational databases that spread in many databases in many business units across the organization. By this system, manager can decide something by doing cross analytical between databases from different business unit or different application, i.e. analysing relationship between selling and amount of marketing budget.

In other side, the concept of competitive intelligence start to implement widely. Organization start to collect or capture strategic information from the outside and combine with internal data sources to decide strategic decision. At the beginning, external information can be collected by conventional way. Today, when more and more business going online, when more and more people talk through social media, collecting information must conducted by different way.

Sentiment analysis (also known as opinion mining) refers to the use of text analysis and computational linguistics to identify and extract subjective information in source materials. Sentiment analysis is widely applied to reviews discussion that being talked in social media for many purposes, ranging from marketing, customer service, or public opinion of public policy. The result of sentiment analysis can be used as a source from external side.

There is a lot of sentiment research has been done. Sentiment analysis research used many different approaches such as corpora based approach or statistical-based approaches (Naive Bayes, EM, SVM).

Many of these studies were based on foreign language (mostly for English) that require adaptation in other languages. One of the popular algorithm for Sentiment Analysis implementation is Kumar algorithm that developed by Kumar and Sebastian [1]. Kumar algorithm can identify the sentiment score of the statement, sentence or tweet, but cannot determine the relationship of the object or target related to the sentiment being analysed. This research will try to propose solution for that challenge by adding additional component that represent object or target to the existing algorithm (Kumar algorithm). By this modification, modified algorithm can give new score that represent sentiment that already corrected by the factor of relationship to the object or target being analyzed.

2. Related Work

Research conducted by Go and friend [2] uses emoticons for classifying sentiment in English which previously inspired by research conducted by the Read [3]. Their research using data taken from the Twitter API as training data for their systems. The weakness of this work is they only divide sentiment polarities into 2 types positive and negative without neutral sentiment. Later research conducted by Mr. and Paroubek [4] who classify sentiment to the neutral class by using tweets coming from the official account of mass media as training data for neutral sentiment.

Different studies also conducted by Xue Chen [5] adapting algorithm that has long been used to topic modeling. Sentiment classification is done using DLDA algorithm. This algorithm is inspired by the first LDA algorithm that intended to classify documents into certain topics [6]. They use this algorithm because the underlying assumption is topic is a collection of words (bag of words).

Other research conducted by Chenliang Li [7] propose an algorithm for segmentation that called HybridSeg. Tweet segmented into sections that have meaning or context information and then it can be easily used by other applications. Li Chenliang get good accuracy results when used in NER (Named Entity Recognition).

The research that use data from Indonesian language is by Wicaksono,et al [8]. It used lexicon approach in building their corpora. The study showed an increase in performance with the use of lexicon. Research conducted by Kumar and Sebastian [1] using syntax approach applying to the "bag of words". The research conducted by Jiang [9] who classify with target approach show improved performance compared to the classification solely on features of independent targets.

The majority of existing research is not considering object as the main feature and only focusing in the value of the sentiment. For example, "*happy happy happy*" making it as a positive sentiment but we have no idea to whom the text is intended.

Refer to sentiment definition from *cambridge.org*, sentiment is "a thought, opinion or idea that is based on the situation or way of thinking from something". Objects become important and influential because of it is the purpose of sentiment. Target recognition in the text will determine the subjectivity of tweet. Definition of sentiment by Liu [10]. Liu said that the target is part of the sentiment that can be seen in this relation (*ei,aij,sijkl,hk,mp*) or *entity, aspect, sentiment, holder, time*.

Based on these, this research use the object as one of its features in sentiment classification. This Research will design, develop, and implement a modified algorithm to calculate sentiment related to particular object. The algorithm is an enhancement from an algorithm developed by Kumar and Sebastian. We use this algorithm based on the advantage from its simple computational level that can quickly deliver results, and good enough to rely on with the vocabulary provided. All these advantages are not complete without the ability of the algorithm to handle the targets of sentiment. Target of sentiment is very important when the algorithm is applied to the real world or used in the specific environment.

3. Understanding Kumar Algorithm [1]

Kumar algorithm is a sentiment classification algorithm developed by Akshay Kumar and Teeja Mary Sebastian with the main objective is to get the semantic orientation of opinion words in tweets [1]. Kumar algorithm distribute sentence or tweets into positive, neutral, or negative categories. This algorithm is using a hybrid approach to utilize the advantages from both corpus and dictionary. This algorithm also uses the Machine Learning and Natural Language Processing techniques.

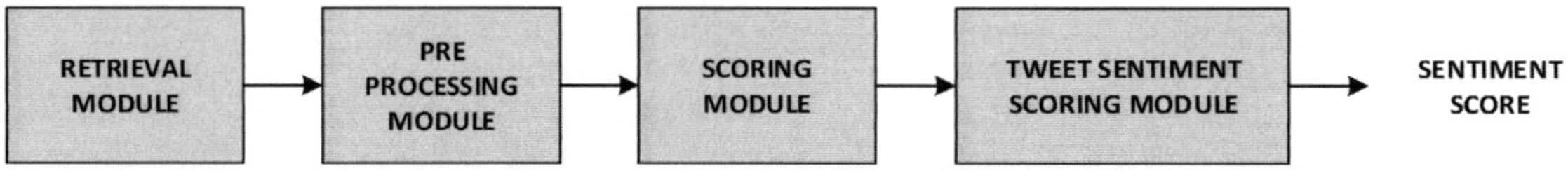

Figure 1. System Architecture of Sentiment Analysis System (modified from [1])

Classification is conducted by taking into account the emergence of words and characters that are considered to have roles in determining sentiment of tweets such as adjective, verb, and adverb. The calculations were carried out to map tweets into the range of -1 to 1. The advantages of these algorithms is the simplicity of its calculation as shown in equation 1. The weaknesses of this algorithm is that its performance depends on the completeness of the vocabulary.

$$S(T) = \frac{1 + \frac{Pc + \log(Ns) + \log(Nx)}{3}}{|OI(R)|} * \sum_{i=1}^{|OI(R)|} S(AG_i) + S(VG_i) + N_{ei} * S(E_i)) \qquad \text{Eq (1)}$$

Where,

$\lvert OI(R) \rvert$	denotes the size of the set of opinion groups and emoticons extracted from the tweet,
Pc	denotes fraction of tweet in caps,
Ns	denotes the count of repeated letters,
Nx	denotes the count of exclamation marks,
S (AGi)	denotes score of the i-th adjective group,
S (VGi)	denotes the score of the i-th verb group,
S (Ei)	denotes the score of the i-th emoticon
Nei	denotes the count of the i-th emoticon

The classification process is conducted by looking at the pieces of words contained in the tweet after pre-processing. In scoring module, sentence process to determine the value of the adjective group and adverb group. Value of adjective group is a combination of the occurrence of the adjective word with the adverb word. Its value is calculated by the following equation.

$$AG = nAdj * nAdv \qquad \text{Eq (2)}$$

Where:

AG denotes value of Adjective Group.
nAdj denotes Adjective value.
nAdv denotes Adverb value.

The calculation of the value of the verb group which is the combination of the occurrence of the verb word and adverb is done by the following equation.

$$VG = nV\,b * nAdv \qquad \text{Eq (3)}$$

Where:

VG denotes Value of Verb Group.
nVb denotes Verb value.
nAdv denotes Adverb value.

IOP Conf. Series: Materials Science and Engineering **128** (2016) 012019 doi:10.1088/1757-899X/128/1/012019

Table 1. Table of Verbs and Adverbs

Verb	Score	Adverb	Score
cinta	1	sempurna, tuntas, sembuh, sehat	1
kagum	0.9	paling, amat	0.9
suka	0.8	sangat	0.8
nikmat	0.7	selalu	0.7
senyum	0.6	sekali	0.6
terkesan	0.5	banyak	0.4
tertarik	0.4	cukup	0.3
senang	0.3	lebih	0.2
santai	0.2	beberapa	-0.2
bosan, tolak	-0.2	agak	-0.3
jijik	-0.3	sedikit	-0.4
derita	-0.4	kurang	-0.6
tidak suka	-0.6	jarang	-0.8
sebal	-0.9	tidak	-0.9
benci	-1	tidak pernah	-1

Table 2. Table of Emoticons

Emoticon	Meaning	Score
:D	Laugh out loud	1
BD	Laugh out loud with glasses	1
XD, :)), ;), =D>	Laugh	1
\m/	Hai5	1
:), =), :-), :3	Smile	0,5
:*	Kiss	0,5
:|	Poker face	0
:\	Not decided	0
:(	Sad	-0,5
>/3	Broken heart	-0,5
B(	Sad with glasses	-0,5
:(	Crying	-1
X-(	Angry	-1

Table 3. Table of Adjectives

Adjective	**Score**
bahagia, ceria, gemar, gembira	0,5
babil, awawarna, awamineral, azal, baki, bangkas, batangan	0
agresif, akut, alergis, ambekan, aneh, amburadul, ceroboh	-0,5

The second calculation of the above equation is conducted by keeping in mind that if the pair not found, its value will be multiplied by 0.5 and the value of each word is taken based on research conducted by Kumar and Sebastian. If not found it will look for synonyms of words in Indonesian. The value of emoticons are also taken from Kumar and Sebastian's research with additions and adjustments as needed. The value of adjective's word is done by looking at the meaning of the word. Adjective words that have positive meanings will be given a value of 0.5, -0.5 for negative, and 0 for neutral.

Research conducted by Go and friend [2] uses emoticons for classifying sentiment in English which previously inspired by research conducted by the Read [3].

4. Proposed Algorithm

The proposed modification of Kumar algorithm only concern in the Tweet Sentiment Scoring Module. Again, the original algorithm has no parameter related to the object or target being analyzed. In order to add object parameter, we need to insert additional parameter that represent object, say it called *RS(Obj)*. After addition, the modified equation will be transform into new form as can be seen in Equation 4 and Equation 5.

$$SM(T, Obj) = S(T) * RS(Obj) \quad \text{Eq (4)}$$

$$SM(T, Obj) = \frac{1 + \frac{Pc + \log(Ns) + \log(Nx)}{3}}{|OI(R)|} * \sum_{i=1}^{|OI(R)|} S(AG_i) + S(VG_i) + N_{ei} * S(E_i)) * RS(Obj) \quad \text{Eq (5)}$$

Where

SM(T,Obj)	denotes the modified sentiment score of modified algorithm for Text *T* and Object *Obj*
S(T)	denotes the sentiment score of Kumar algorithm
RS(Obj)	denotes relationship score of *Obj* in the sentence *RS* will be zero if *Obj* not exist in the sentence *RS* will have positive value and less than 1

There is several requirements to define the *S(Obj)*. First requirement, *S(Obj)* value should be always positive to maintain the polarity of previous sentiment result. Second requirement, the value should be zero when the object is not found. Its mean that no related sentiment to that object. In this case, *S(T)* can have some value, but the sentiment value is not related to the object.

$$RS(Obj) = \begin{cases} 0 \; if \; Obj \; not \; exist \; in \; T \\ 1 - \frac{(i-1)}{N} \; ; \; where \; i = word \; position \; of \; Obj \; in \; T; N = number \; of \; word \; in \; T \end{cases}$$

The second equation of RS will generate value correspond to the position of *Obj Word* in sentence *T*. Assume that position around the beginning is more important than the position around the end. First extreme condition, the *Obj Word* exist at the first position, i=1, RS(Obj)=1. Second extreme condition, the *Obj Word* exist at the last position (i.e N=8), i=7, RS(Obj) = 1-7/8=1/8. Value of RS will be maximum (RS=1) if *Obj Word* exist in the first position. The value will decrease if the position of *Obj Word* moving to the next position up to the end. At the end, the value become close to zero, but never reach zero. The value zero will be assign by the first equation (or rule) if *Obj Word* not exist in the sentence *T*. *SM(T,Obj)* score will be same with the *S(T)* for the case that *Obj Word* exist in the fisrt position.

To compute the relationship score *(RS),* the system must scan the sentence T to find the position of Obj word. If it is found, it will be the value of i, and RS can be compute by the second equation of RS above. If Obj word not found, RS will be assigned as zero.

Obj Word can be given manually when user want to know sentiment score for any **Object** (represent by *Obj Word*). The other alternative, additional process can be apply in pre-processing module. There is mechanism to identify all *Obj Word* in each sentence T. Finally, in Tweet Sentiment Scoring Module the computation of *S(T)* and *SM(T,Obj)* will be repeated according to the number of *Obj Word*. The result will be *SM(T,Obj)* value for each *Obj Word*.

5. Evaluation

System evaluation will be conducted in the follows steps.

1) Collect sample of data to be classified from Twitter (1751 tweet samples).
2) Conduct "Manual Sentiment Analysis" from collected samples to categorize each sample into neutral, positive, or negative sentiment (will be used as a reference).
3) Preparing *Obj Word*
   a) Identify sentence that have no object, and assign Obj Word as any NOUN, i.e. "people". Remember, this type of sentence have no NOUN.
   b) Identify all sentence that have object, choose one object (if have more than one object), and assign *Obj Word* with a correct object that available in that sentence.
   c) Change several *Obj Word* in step (b) to the other Noun that not available in that sentence. The aim of this step to verify if the modified algorithm can identify *Obj Word* that not exist in the sentence.
4) Compute *S(T)* and *SM(T,Obj)* for each sentence *T* and *Obj*.

Table 4. Experiment Result

Analytical Methods	Neutral (0)	Positive (+)	Negative (-)	Change (+) to (0)	Change (-) to (0)	Change (+) to (-) or (-) to (+)
Manual	1026	549	176	-	-	-
Kumar Algorithm (*S(T)*)	865	676	210	-	-	-
Modified Algorithm (*RS(T,Obj)*)	897	653	201	23	9	0

Table 5 show the result of the experiment. Manual Sentiment Analysis of 1751 sentences identify that 1026 sentences is neutral, 549 sentences is positive, and 176 is negative. Identification of sentiment category by Kumar algorithm show that 865 sentence is neutral, 676 is positive, and 210 is negative. These result show that Kumar algorithm have accuracy less than 100%. But, the percentage is not important, because this research not to measure Kumar algorithm accuracy.

Put attention to the result of Modified Algorithm that compute *RS(T,Obj)*. The number of neutral sentiment is increased, and the number of positive sentiment and negative sentiment is decreased. Its happen because some sentence that previously in the category of positive (23) and negative (9) category was changed. In our experiment several sentence have no NOUN, so it will be change to neutral when it compute by modified algorithm, because any NOUN asn an object will not found in the sentence. For other case, several sentence give the wrong object, so the Obj Word will be not found and give the result as neutral. In other hand, there is no sentence that change the polarity, because the algorithm guarantee that it will not happen. Changing from neutral to any polar also will not happen, because the RS(Obj) maximum value is 1, and always positive. So the original value from Kumar algorithm will never increase.

6. Conclusion

Based on the evaluation result that already discuss in part 5, it can be concluded that in this paper already propose the modification of Kumar Algorithm. The modified algorithm add additional component to the original algorithm. This additional component, *RS(Obj)*, represent the relationship of the object (Obj) being analysed. The list characteristics of the modified algorithm describe in the following lists.

1) The value range of *RS* is 0 to 1.
2) The value of *RS* will be zero if *Obj Word* that represent Object (or target) was not found in the sentence being analysed
3) The value of *RS* will be maximum (value=1) if the *Obj Word* found in the first position in the sentence.
4) The value of *RS* will be decreased if the position of *Obj Word* moving to the end of the sentence. At the end position, the value will close to zero, but never zero.
5) The final value of new algorithm (*SM(T,Obj)*) never bigger than the value from original Kumar algorithm (*S(T)*).

References

[1] A. Kumar and T. M. Sebastian, "Sentiment analysis on twitter," *IJCSI International Journal of Computer Science Issues*, 2012.

[2] A. Go, R. Bhayani, and L. Huang, "Twitter Sentiment Classification using Distant Supervision," *Processing*, vol. 150, pp. 1–6, 2009. [Online]. Available: http://www.stanford.edu/~alecmgo/papers/TwitterDistantSupervision09.pdf

[3] J. Read, "Using Emoticons to reduce Dependency in Machine Learning Techniques for Sentiment Classification," *ACL Student Research workshop*, no. June, pp. 43–48, 2005.

[4] A. Pak and P. Paroubek, "Twitter as a Corpus for Sentiment Analysis and Opinion Mining." *Lrec*, pp. 1320–1326, 2010. [Online]. Available: http://incc-tps.googlecode.com/svn/trunk/TPFinal/bibliografia/PakandParoubek(2010).TwitterasaCorpusforSentimentAnalysisandOpinionMining.pdf

[5] X. Chen, W. Tang, H. Xu, and X. Hu, "Double lda: A sentiment analysis model based on topic model," in *Semantics, Knowledge and Grids (SKG), 2014 10th International Conference on*, Aug 2014, pp. 49–56.

[6] D. M. Blei, A. Y. Ng, and M. I. Jordan, "Latent Dirichlet Allocation," *Journal of Machine Learning Research*, vol. 3, pp. 993–1022, 2012.

[7] C. Li, A. Sun, J. Weng, and Q. He, "Tweet Segmentation and its Application to Named Entity Recognition," *IEEE Transactions on Knowledge and Data Engineering*, vol. 4347, no. 2, pp. 1–1, 2014. [Online]. Available: http://ieeexplore.ieee.org/lpdocs/epic03/wrapper.htm?arnumber=6823714

[8] A. F. Wicaksono, C. Vania, B. D. T, and M. Adriani, "Automatically Building a Corpus for Sentiment Analysis on Indonesian Tweets," 2014, pp. 185–194.

[9] L. Jiang, M. Yu, M. Zhou, X. Liu, and T. Zhao, "Target-dependent Twitter Sentiment Classification," *Computational Linguistics*, pp. 151–160, 2011. [Online]. Available: http://www.aclweb.org/anthology/ P11-1016

[10] B. Liu, *Sentiment Analysis and Opinion Mining*, 2012, vol. 5, no. May.

Extract transformation loading from OLTP to OLAP data using pentaho data integration

R J Salaki*, J Waworuntu and I R H T Tangkawarow

Pendidikan Teknologi Informasi dan Komunikasi, UNIMA, Manado, Indonesia

*Corresponding author: reynaldojsalaki@unima.ac.id

Abstract. The design of the data warehouse in this case is expected to solve the problem of evaluation of learning results as well as the relevance of the information received to support decision-making by the leader. Data warehouse design is very important, which is designed to utilize the existing resources of information. GPA (Grade Point Average) data warehouse can be used for the process of evaluation, decision making and even further planning of the study program of PTIK. The diversity of data sources in the course PTIK make decision-making and evaluation process becomes not easier. Pentaho Data Integration is used to integrate data in PTIK easy. CPI data warehouse design with multidimensional database modeling approach using the dimension tables and fact tables.

1. Introduction

The use of information technology that is integrated with the process of work in an institution or corporation has become an absolute necessity today. This is caused by the need of the institution to optimize the ability to analyze the problems encountered which will affect the decision-making process. The availability of complete and accurate data is a measurement of the viability of the institution.

One indication of the success of a program of study is the absence of data GPA of students that can be accessed quickly and accurately. Problems that occur at this time generally lies in the variety of input data, causing delays in data processing. This study is expected to overcome the diversity of data sources by using Pentaho Data Integration-Kettle. This application can integrate the data that is ready to be processed in the data warehouse so it can later be presented with accurate and timely. The integration process is well known in the Business Intelligence as a process Extract Transformation Loading (ETL). ETL process will change the data On-Line Transactional Processing (OLTP) into data On-Line Analytical Processing (OLAP).

This research built data warehouse to display the GPA of students, especially in PTIK Study Program. The Manado State University established in 1955, which was originally the Teachers' Training College Manado. The PTIK study program was officially open in 2010.

The data were collected from the student data and students' GPA. This research is expected to provide precise and accurate data, especially for evaluation, decision making and planning for the development of the study program.

2. Literature review

2.1. 2.1 Business Intelligence

Business Intelligence describes a concept and method of how North to improve the quality of decision business decisions based system based data. BI often be treated equate as briefing books, report and query tools, and information systems executive. BI is a support system based decision making data (Imelda, 2015).

Research Herring, 1999, confirms that the companies without intelligence needs facing performance bad and frustrated at their CI departments. On the other hand company with a form of intelligence requires the identification process benefit from their successful CI programs. But why do not have many companies have a CI department? Reason for this might be difficult to track down intelligence needs the right to a certain strategy (Budi Harijanto, 2013).

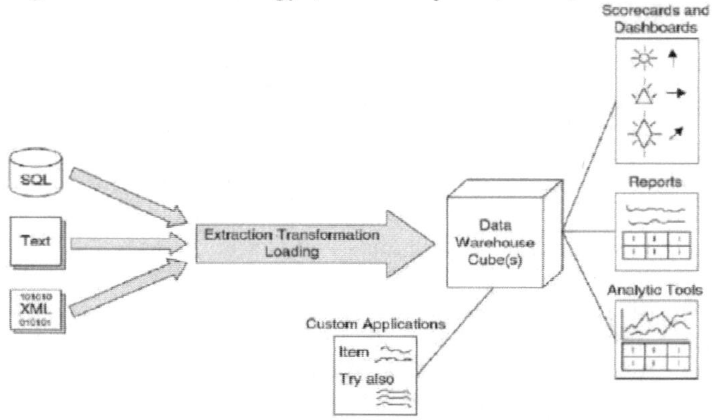

Figure 1. BI Process Steps (Source : Prastuti Sulistyorini, 2010)

2.2 Data Warehouse

The process of data warehousing carried out in three main steps , also known as extraction, transformation and loading (ETL).Extraction program retrieves data from various database Heterogeneous operations based on specific models. Metadata describes the model and the definition of the data source element (Budi Harijanto, 2013).

According to pioneer the concept and term of data warehouse, William Inmon, the definition of the data warehouse is a collection of data subject-oriented, integrated, non-volatile, and time-variant in order to support decisions of management (Stephanie Pamela Adithama, 2013).

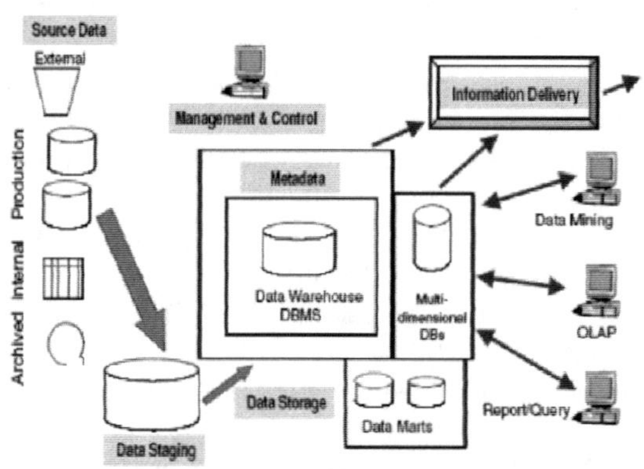

Figure 2. Component of Data Warehouse
(Source : Stephanie Pamela Adithama, 2013)

3. Research Methodology

Design method multidimensional data modelling according Kimball used includes 4 stages (Kimball & Ross, 2002). The fourth stage these are:

1. Select the Business Process
   The first step in the design is to decide what business process(es) to model by combining an understanding of the business requirements with an understanding of the available data.
2. Declare the Grain
   Once the business process has been identified, the data warehouse team faces a serious decision about the granularity. What level of data detail should be made available in the dimensional model? This brings us to an important design tip.

IOP Conf. Series: Materials Science and Engineering **128** (2016) 012020 doi:10.1088/1757-899X/128/1/012020

3. Choose the Dimensions
   Once the grain of the fact table has been chosen, the date, product, and store dimensions fall out immediately.
4. Identify the Facts
   The fourth and final step in the design is to make a careful determination of which facts will appear in the fact table. Again, the grain declaration helps anchor our thinking.

4. Design Data Warehouse

4.1. Data Warehouse

Data Warehouse is a special database that is used as a "data warehouse" or data which has been consolidated from various data sources of existing information systems in an organization/company. According to Kimball, there are some requirements for the data warehouse, some of which are :

- Data warehouse must make information from a company/institution/organization easily accessible.
- Data warehouse must display information about the company/institution constantly.
- Data warehouse must present data that will be used as a basis or guidelines for decision making.

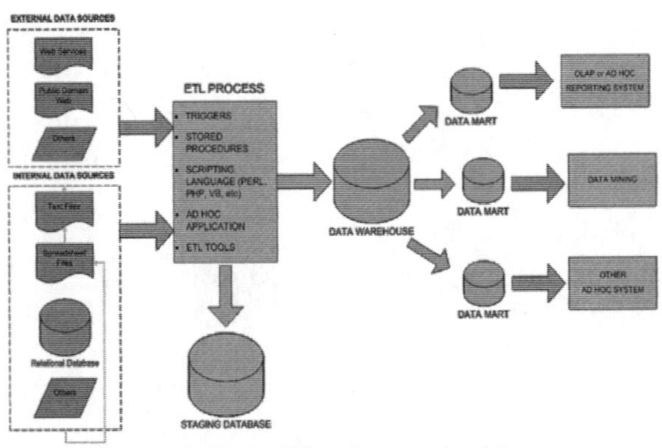

Figure 3. Data Warehouse Architecture
(Source : Kimball, 2002)

4.2. Architectural Design Data Warehouse PTIK.

Source of data to be processed is taken from academic databases PTIK Study Program. Such data contain all the academic data of the students in the study program PTIK that would normally be in upload for each semester to be displayed on the Academic Information System Manado State University. Sorting data is done to sort out what data is to be used without interfering with operational data while in use. The staging process is also done to facilitate the ETL process later, because the data are taken really only the data required for the Data Warehouse GPA student. Here is the data warehouse architecture design PTIK Study Program :

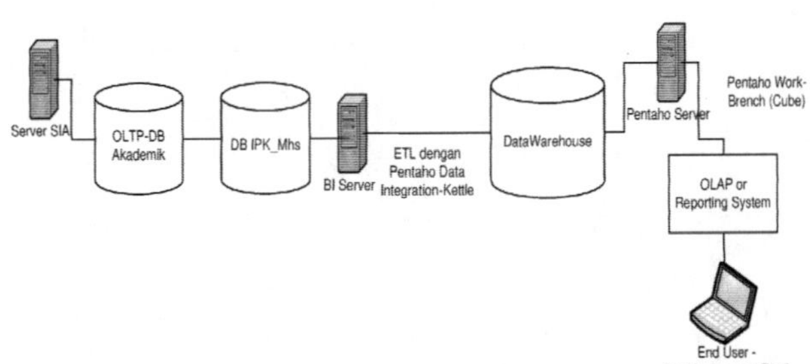

Figure 4. Physical Data Warehouse Architecture PTIK Study Program

4.3. Dimensional Data Modeling

Modeling multidimensional database will consist of fact tables (fact tables) and dimension tables (dimension tables) are interlinked. A fact table contains various value aggregations on which the measurement (measure) as well as some of the key that corresponds to the dimension tables that will be the viewpoint of the measure.

The structure of the fact table and the dimension table has a design scheme that can improve performance and ease of translation.

OLAP systems schema is the basis for doing data warehousing. Two schema most commonly used by the various OLAP engine are the star schema (Star Schema) and scheme snowflakes (Snowflake Schema). In this research multidimensional depiction of data using the Star Schema.

Dimension table is a table that contains the data that show the results of a review of various points of view. Table dimensions will construct cube. Table dimensions are available :

Table dim_mhs

The dimension tables contain data on students who are limited only students' register number (nim), the name of the place and date of birth, gender and year of admission.

Table 1. Students Dimension Table

dim_mhs	
PK	**sk_mhs**
	nim nama_mhs tempat_lahir tanggal_lahir jk thn_masuk kd_prodi

Table dim_dosen

The dimension tables contain data lecturers only limited to the name of the lecturer, and lecturer code. For lecturers code is NIDN.

Table 2. Lecturer Dimension Table

dim_dosen	
PK	**sk_dosen**
	kd_dosen nip nama_dosen Gelar_dosen TempatLahir TanggalLahir JK No_KTP Kd_Prodi Kd_MK Nama_MK

Table dim_mk

The dimension tables contain data subjects, the weight of credits from the courses and in the semester how mk (subjects) exists.

Table 3. Subjects Dimension Table

dim_mk	
PK	**sk_mk**
	kd_mk nama_mk sks semester GanjilGenap Kd_Dosen

Table dim_waktu

The dimension tables contain data such as time of day, quarter, semester, month, year and date.

Table 4. Time Dimension Table

dim_waktu	
PK	sk_waktu
	hari kuartal semester bulan tahun tanggal

table dim_nilai

The dimension tables contain data on the value of each course students each academic year and the first semester and the second semester.

Table 5. Value Dimension Table

dim_nilai	
PK	sk_nilai
	nim tahun_akademik ganjilgenap kd_mk kelas nilai_akhir (0-4) grade(A-E)

Fact table is a table containing the facts of business, generally a table the details of transactions that have occurred (Mulyana, 2015) . Fact Table designed in a data warehouse is taken relating to the evaluation of data from PTIK Study Program. Fact_ipk only one fact table is actually what is enough GPA display data from year to year. But considered necessary also for information about the value of each student.

Table fact_nilai

This table contains all the data values PTIK students in each course that is already in the contract.

Table 6. Value Fact Table

fact_nilai	
FK1 FK2 FK3	sk_mhs sk_dosen sk_waktu

Here is a picture of a star schema GPA study programs and student grades.

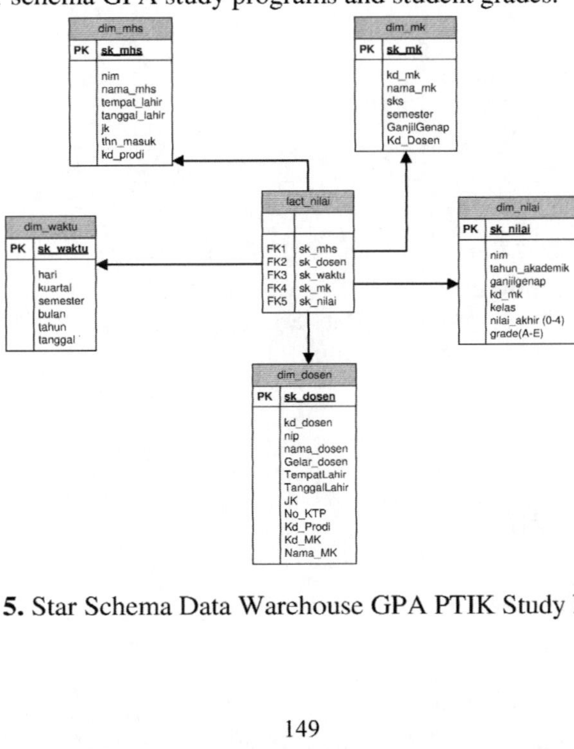

Figure 5. Star Schema Data Warehouse GPA PTIK Study Program

5. Experiment Results

5.1. Pentaho Data Integration

Pentaho Data Integration (PDI) or Kettle ETL utility is open source under Pentaho Corp.Amerika. This project was originally an initiative of Matt Casters, a programmer and consultant Business Intelligence (BI) from Belgium who has managed projects for enterprise BI big.

Currently Kettle ETL is a utility that is very popular and one of the best on the market. Some advantages are as follows :

- Have a collection of data processing modules that quite a lot. More than 100 modules or step.
- Have a module that facilitates the design of the data warehouse model as Slowly Changing Dimensions Dimension and Junk.
- Performance and scalability are well-proven.
- Can be developed with a variety of additional plugins.
- Utility Kettle to be used in the integration of this data using the Spoon.

5.2. Data Integration for Dimension Table

By using PDI Kettle, diversity of data available on PTIK study program can be integrated into the database with the database platform used is MySQL. Tables of this dimension that will form the Data Warehouse GPA for PTIK Study Program.

Dimension table Students (Dim_Mhs)

Data source of dim_mhs table is derived from a list of names of students who enrolled in the first semester of academic year 2014 (1). Where the total registered students is 1293. However, this data is only as names and NIM are stored in Excel files. While the data warehouse requires a complete student data, such as place of birth, date of birth, gender, year in, etc. Therefore, data collaboration with master data that is used in PDPT (Database PT) needs to be done. The PDPT the data is the data Ms.Access. Merging the data source from Excel files and data on Ms.Acces can be done using this PDI. Here is an overview transformation design to use PDI-Kettle.

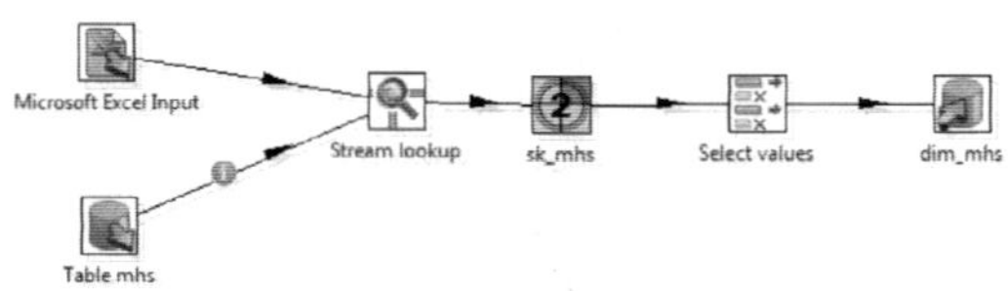

Figure 6. The transformation to a table dim_mhs

Once the transformation is done resulting student dimension tables as follows :

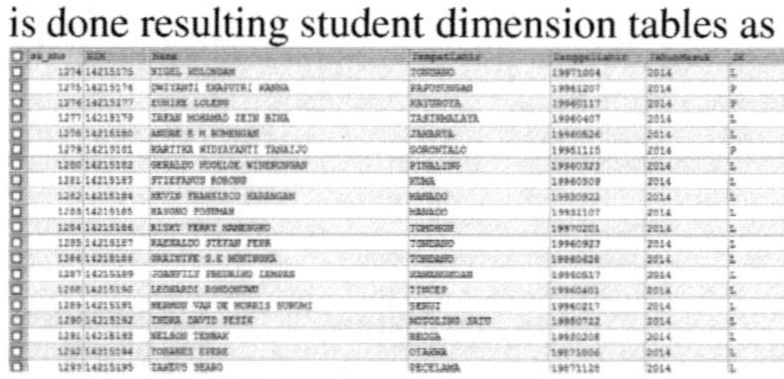

Figure 7. Student Dimension Table

Dimension table Lecturer (dim_dosen)

Data lecturer there in PTIK Study Program (Data Source) is the data stored in Excel files. Merging data is done to obtain data on the complete faculty.

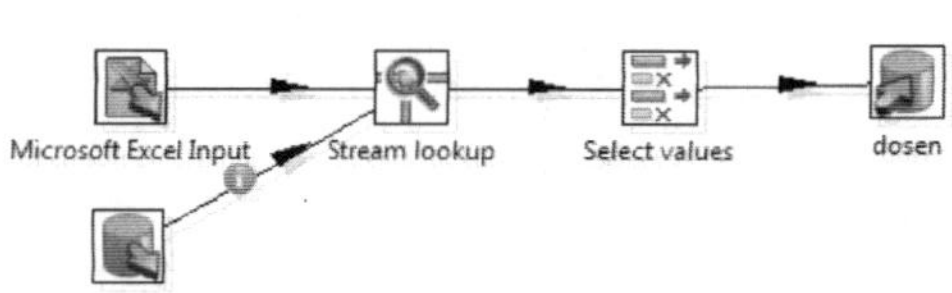

Figure 8. Transformation for table dim_dosen

Once the transformation is done resulting lecturer dimension tables as follows :

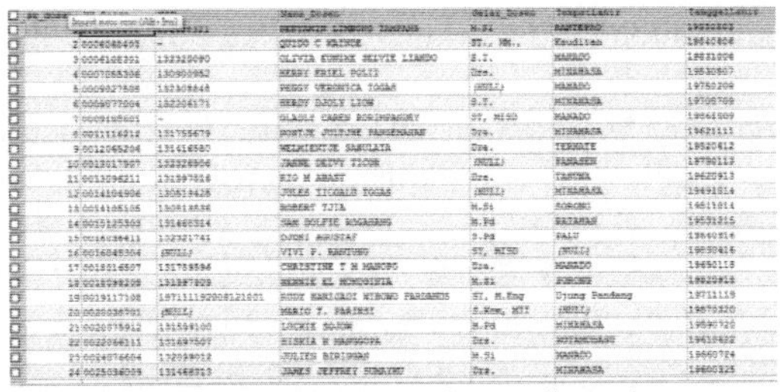

Figure 9. Table data from dim_dosen

Table Fact Value (fact_nilai)

Fact tables are created in the transformation by combining the dimension tables that dim_dosen, dim_mk, dim_mhs, dim_nilai, dim_waktu who had previously designed.

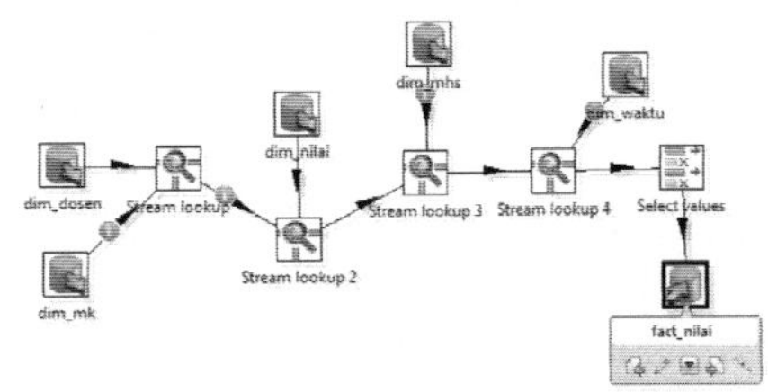

Figure 10. Transformation ETL for table fact_nilai

Table fact_nilai after transformation is executed is as follows :

sk_mhs	sk_mk	sk_dosen	sk_waktu	sk_nilai
21	65	18	20160102	1
21	67	22	20160102	2
21	66	32	20160102	3
21	(NULL)	(NULL)	20160102	4
21	(NULL)	(NULL)	20160102	5
21	(NULL)	(NULL)	20160102	6
(NULL)	66	32	20160102	7
(NULL)	(NULL)	(NULL)	20160102	8
(NULL)	67	22	20160102	9
(NULL)	66	32	20160102	10
203	44	29	20160102	11
203	78	7	20160102	12
203	(NULL)	(NULL)	20160102	13
203	54	17	20160102	14
203	64	4	20160102	15
203	56	3	20160102	16
203	80	20	20160102	17
203	(NULL)	(NULL)	20160102	18
203	81	19	20160102	19
203	66	32	20160102	20
203	43	24	20160102	21

Figure 11. Table Data fact_nilai

6. Conclusion

The diversity of data into an existing data source in Prodi PTIK can be overcome by using Pentaho Data Integration (PDI)-Kettle. Data warehouse design results generated GPA makes the data can be viewed in tabular form a more orderly so easily processed. This is the data that is ready to be processed into the Pentaho schema workbench and then the dashboard can be presented using Pentaho Business Intelligence (BI) Server.

References

[1] Budi Harijanto, Gunawan Budiprasetyo. Perancangan Aplikasi Business Intelligence Hasil Proses Belajar Mengajar (Studi Kasus Program Studi Manajemen Informatika). Jurnal ELTEK, Vol.11 No.1, April 2013.

[2] "CIBIA Courseware", 2015, Multimatics, Jakarta.

[3] Imelda. Business Intelligence. Majalah Ilmiah UNIKOM Vol.11 No.1 2015.

[4] JRP, Mulyana., 2014, "Pentaho: Solusi Open Source untuk Membangun Data Warehouse", ANDI, Yogyakarta.

[5] Kimball, Ralph., Ross, Margy., 2002, "The Data Warehouse Toolkit, Second Edition", John Wiley & Sons, Inc, Canada.

[6] Kimball, Ralph., Caserta, Joe., 2004, "The Data Warehouse ETL Toolkit", John Wiley & Sons, Inc, Canada.[online].

[7] Meta Suzanam, Jemakmun, Suyanto. Analisis dan Perancangan Data Warehouse Rumah Sakit Umum di Daerah Palembang Bari. Jurnal Ilmiah dan Teknik Informatika Universitas Binadarma Vol1.No.1 November, 2013.

[8] Prastuti Sulistyorini. Business Intelligence dan Manfaatnya Bagi Organisasi. Majalah Ilmiah IC Tech Vol.5 No.2 Mei,2010.

[9] Stephanie Pamela Adithama, Irya Wisnubhadra, Benyamin L. Sinaga. Analisis dan Desain Real-Time Business Intelligence untuk subjek Kegiatan Akademik pada Universitas menggunakan Change Data Capture. SENTIKA Yogyakarta, 9 Maret 2013.

[10] Thia Feris, "Pentaho Knowledge Based", [online], (http://pentaho-en.phi-integration.com)

Growth mechanism of GaAs1-xSbx ternary alloy thin film on MOCVD reactor using TMGa, TDMAAs and TDMASb

A Suhandi* [1], Y R Tayubi[1] and P Arifin[2]

1) Departemen of Physics Education FPMIPA UPI, Jl. Dr. Setiabudhi 229 Bandung
2) Departement of Physics FMIPA ITB, Jl. Ganeca 10 Bandung

Corresponding author: a_bakrie@yahoo.com

Abstract. . Metal Organic Chemical Vapor Deposition (MOCVD) is a method for growing a solid material (in the form of thin films, especially for semiconductor materials) using vapor phase metal organic sources. Studies on the growth mechanism of GaAs1-xSbx ternary alloy thin solid film in the range of miscibility-gap using metal organic sources trimethylgallium (TMGa), trisdimethylaminoarsenic (TDMAAs), and trisdimethylaminoantimony (TDMASb) on MOCVD reactor has been done to understand the physical and chemical processes involved. Knowledge of the processes that occur during alloy formation is very important to determine the couple of growth condition and growth parameters are appropriate for yield high quality GaAs1-xSbx alloy. The mechanism has been studied include decomposition of metal organic sources and chemical reactions that may occur, the incorporation of the alloy elements forming and the contaminants element that are formed in the gown thin film. In this paper presented the results of experimental data on the growth of GaAs1-xSbx alloy using Vertical-MOCVD reactor to demonstrate its potential in growing GaAs1-xSbx alloy in the range of its miscibility gap.

1. Introduction

Recently the ternary alloy GaAsSb has been studied intensively as the candidates for long-wavelength infrared laser diode. This material has a range of values of the energy gap (Eg) between 0.72 to 1.42 eV depends on the content of Sb elements. This energy gap range corresponding with wavelength range required in the optical fiber communication systems [1,2,3,4].

In growing GaAsSb epitaxial layers, there are several major challenges to obtain high-quality and high-performance epitaxially grown layers. One of all is the wide alloy composition range of the GaAsSb miscibility gap when mixing GaAs and GaSb. If two binary semiconductor materials that have large difference in lattice parameters, for example GaAs and GaSb, are mixed together to form an alloy, it is hard to avoid the phase separation of two binary materials below the critical temperature (Tc). Thermodynamics can be used to predict the occurrence of spinodal decomposition or phase separation. By using a straightforward regular-solution model can be calculated the excess-free energy curves as a function of Sb composition x at various temperatures for GaAs1-xSbx as shown in Figure 1 [5]. The large positive mixing enthalpy with a large difference in lattice constant can overwhelm the negative mixing entropy, resulting in mixing Gibb's free energy with an upward curve in the center of the composition region, below the critical temperature (in Figure 1, Tc is around 760°C). This means that at equilibrium, the mixing composition within the upward curved region is not stable and the alloy decomposes into a mixture of two phases. Therefore, it is theoretically impossible to grow an alloy within the miscibility gap by any equilibrium growth technique e.g., liquid phase epitaxy (LPE), at temperatures below Tc. However, under certain conditions, the alloys in the miscibility gap can be successfully grown even at lower temperatures than Tc by non-equilibrium growth techniques such as MBE or MOCVD. In the MBE and MOCVD growth processes, there are two components of physical processes that occur they are thermodynamics and kinetics. The thermodynamic determine the driving force for overall process of the film growth, while the kinetics describe the rate at which various processes occur. Controlling the kinetic parameters can disturb the equilibrium thermodynamics, so that fluctuations in the composition due to the phase separation can occur at temperatures around Tc, consequently the composition of the mixture can be in a metastable state [6]. Therefore, with using both modern growth techniques it is possible to grow a ternary alloy GaAs1-xSbx in the range composition of its miscibility gap.

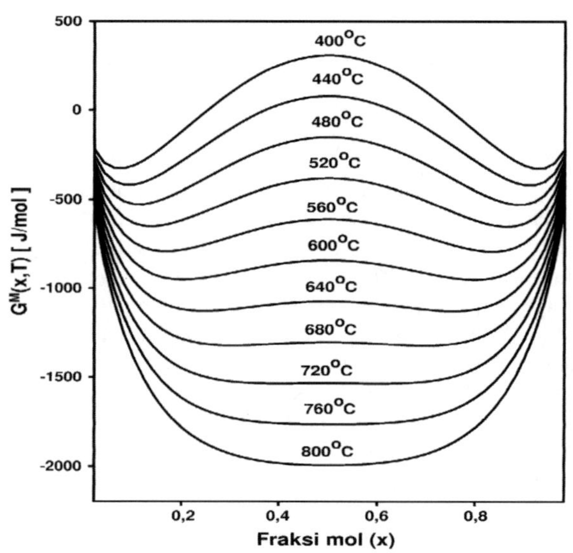

Figure 1. Calculated excess-free energy as a function of *x* at different temperatures for $GaAs_{1-x}Sb_x$.

This paper describes the results of a study on the mechanisms of physical and chemical in the growing process of ternary alloy GaAs1-xSbx by MOCVD technique using metalorganic sources TMGa, TDMAAs, and TDMASb, and the characterization of GaAs1-xSbx thin film grown by MOCVD technique.

2. Methods

To probe the feasibility of growing a $GaAs_{1-x}Sb_x$ thin film by MOCVD technique, a series of experiments about growth of GaAsSb thin films has been conducted. The MOCVD reactor system used in this experiment is vertical type-MOCVD reactor which available in Laboratory for Electronic Material Physics, Prodi Fisika FMIPA ITB. Semi-insulating GaAs crystal (100) was used as the substrate. Schematically, MOCVD reactor system that is used, is shown in Figure 2. The conditions and parameters of growth of $GaAs_{1-x}Sb_x$ films are shown in Table 1.

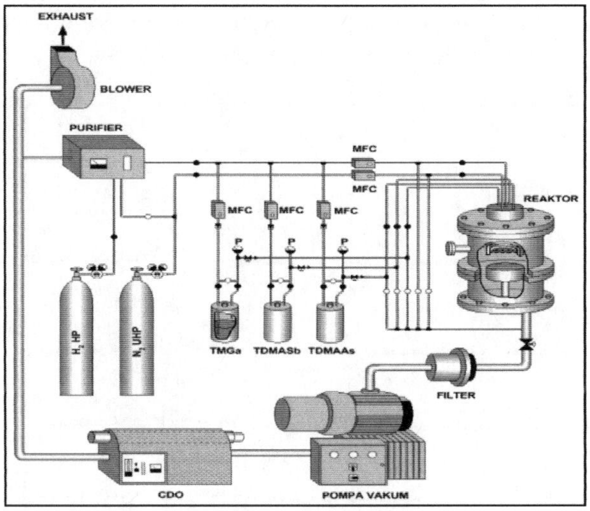

Figure 2. Schematic of Vertical Type MOCVD reactor

Table 1. The conditions and parameters for growing $GaAs_{1-x}Sb_x$ thin films

No	Parameter penumbuhan	Kondisi
1	Tekanan *chamber* reaktor	~ 50 Torr
2	Laju aliran gas Nitrogen pengencer	300 Sccm
3	Laju aliran gas Hidrogen pengencer	300 Sccm
4	Temperatur *bubbler* TMGa	-10°C
5	Tekanan *bubbler* TMGa	10 psi
6	Temperatur *bubbler* TDMAAs	24°C
7	Tekanan *bubbler* TDMAAs	0,5 bar
8	Temperatur *bubbler* TDMASb	26°C
9	Tekanan *bubbler* TDMASb	0,5 bar
10	Rasio V/III	1,2 – 4,8
11	Temperatur penumbuhan	520 – 600°C
12	Fraksi mol masukan uap sumber TDMASb (x_v)	0 - 1

Structural properties of the films were characterized by X-ray diffraction (XRD), the element composition of the films were examined using energy dispersive spectroscopy (EDS), and the electrical properties of the $GaAs_{1-x}Sb_x$ films were investigated using the room-temperature semi-automated Van der Pauw Hall-effect measurement to determine the type of semiconductors, carrier mobility, and carrier concentration.

3. Results and Discussion

Figure 3 shows the X-ray diffraction pattern of a several $GaAs_{1-x}Sb_x$ films grown at growth temperature of about 580°C and the V/III ratio of about 4.8 with the inputs composition of the Sb vapor source (xv) is varied. The orientation of the crystal plane were observed represented by (400) plane, because the whole grown films have a single crystal orientation.

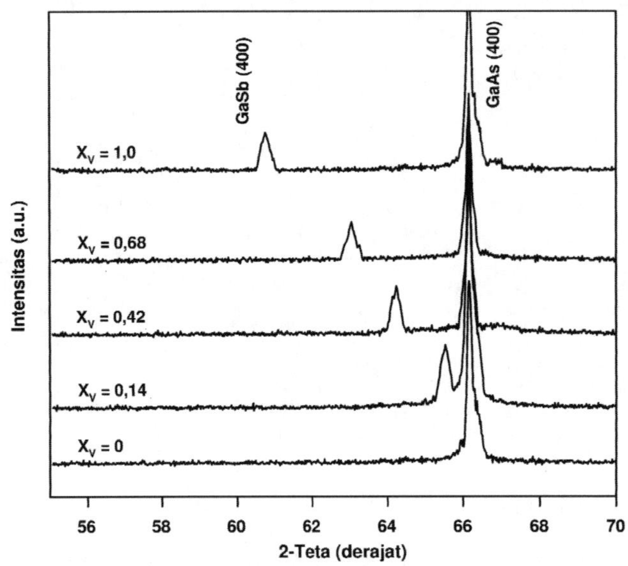

Figure 3. XRD pattern for (400) plane from several $GaAs_{1-x}Sb_x$ films samples growth at growth temperature of about 580°C and V/III ratio of about 4,8 with varied x_v

It appears that a shift in the peak angle of the (400) plane towards smaller 2θ when the composition of the Sb vapor source inputs increase. This shift indicates changes in the value of the lattice constant of the material towards a greater, that is from the lattice constant of GaAs to the lattice constant of GaSb. According to the Vegard equation, a shift toward smaller diffraction angle indicating incorporation Sb atoms into binary GaAs form a ternary $GaAs_{1-x}Sb_x$.

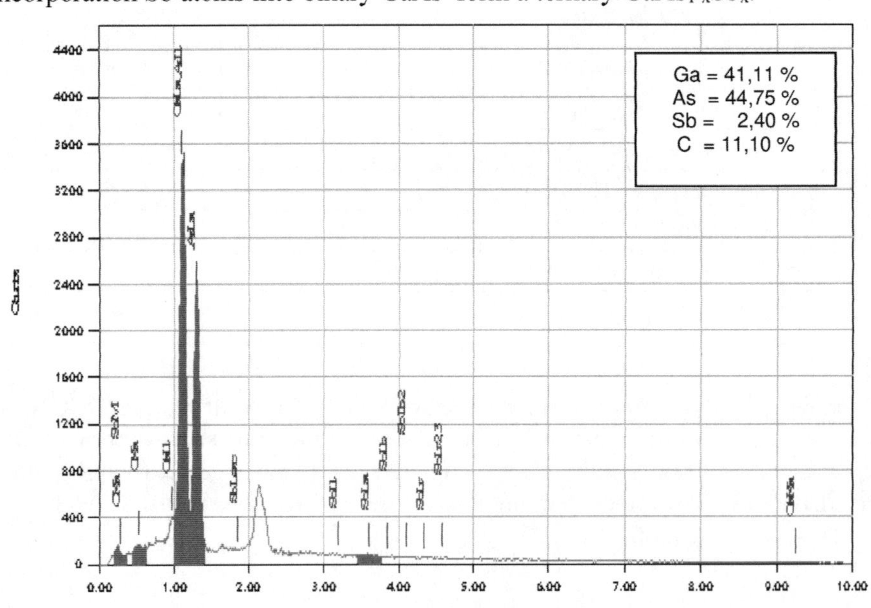

Figure 4. Results of EDS characterization of GaAsSb film

Figure 4 shows the percent composition of the elements attached to the $GaAs_{1-x}Sb_x$ thin film sample grown at a temperature of 580°C and the V/III ratio of about 4.8 with mole fraction of Sb source vapour input (xv) of about 0.14, resulting from EDS (energy dispersive spectroscopy) measurement. The EDS data showed that the three elements of the alloy that is Ga, As, and Sb has been incorporated in the film. In addition there are a number of the carbon (C) elements in the film. The results of this characterization has shown that the $GaAs_{1-x}Sb_x$ thin films have been successfully grown by MOCVD on the range of its miscibility-gap.

Growing process of $GaAs_{1-x}Sb_x$ films by MOCVD method can be described as shown in Figure 5 [7]. Because of heat gradient vertically upwards due to heating at the bottom, then the metalorganic precursors are introduce into the reaction chamber will undergo pyrolytic decomposition at the top. Reactants of decomposition will move across the boundary layer to the substrate surface. Getting closer to the substrate, the decomposition will be complete, and when it reaches the substrate will occur heterogeneous vapor /solid phase reaction form a thin film of solid material. There are several parameters that control the film formation reaction in the MOCVD reactor, such as growth temperature, reaction pressure, and flow rate V/III ratio [8]. So to obtain good quality films, it is necessary to optimization of the three parameters of this growth. By using concepts of physics and chemistry, MOCVD growth mechanism can be studied. Good knowledge of this mechanism is very useful in determining the fit growth conditions and parameters for growing process.

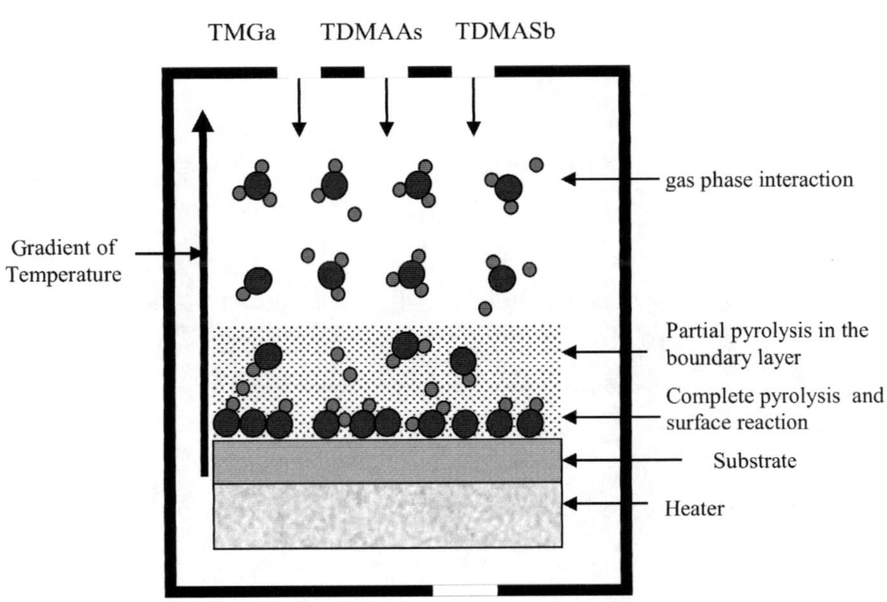

IOP Conf. Series: Materials Science and Engineering **128** (2016) 012021 doi:10.1088/1757-899X/128/1/012021

Figure 5. Schematic process of growing $GaAs_{1-x}Sb_x$ films in the Vertical-MOCVD reactor

Pyrolytic decomposition of TMGa will produce Ga atoms and methyl radicals [6], while the pyrolytic decomposition of TDMAAs (or TDMASb) will produce As (or Sb) atoms, reactive aminos such as $N(CH_3)_2$ and aziridin ($HN(CH_2)_2$), and H atoms [9]. Decomposition mechanism of each metalorganic sources are shown in Figure 6a to 6c.

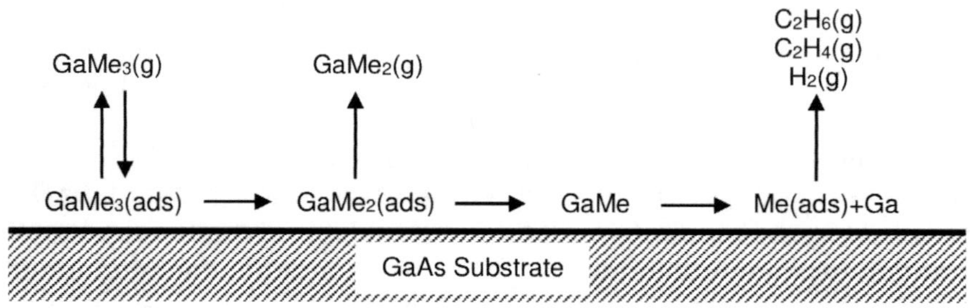

(a)

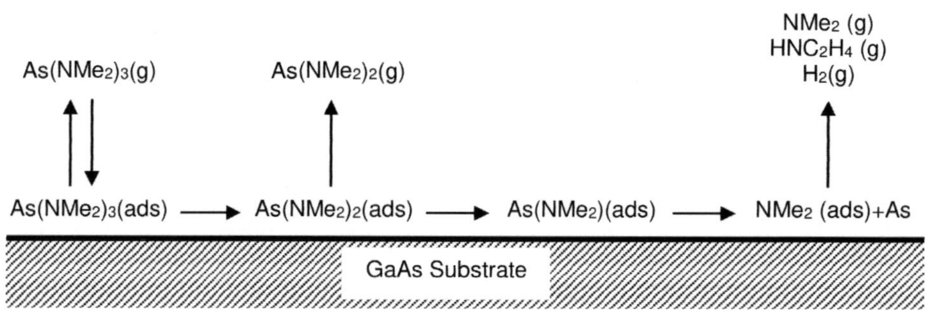

(b)

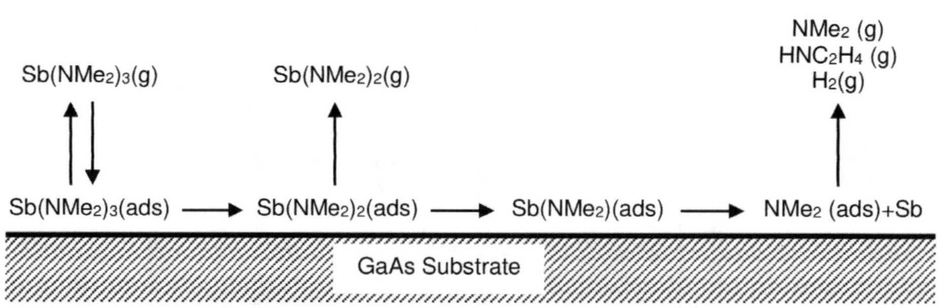

(c)

Figure 6. Schematic pyrolitic decomposition for each precursor on the GaAs substrate; (a) TMGa, (b) TDMAAs, (c) TDMASb

Based on the decomposition mechanism, the chemical reactions that occur in the growth of of GaAs, GaSb, and $GaAs_{1-x}Sb_x$ thin film by using TMGa, TDMAAs, and TDMASb on the the MOCVD reactor is predicted as follows:

$$Ga(CH_3)_{3(V)} + xSb[N(CH_3)_2]_{3(V)} + (1-x)As[N(CH_3)_2]_{3(V)} \Rightarrow$$
$$GaAs_{(1-x)}Sb_{(x)(S)} + 2N(CH_3)_{3(V)} + HN(CH_2)_{2(V)} + CH_{4(V)} \quad (1)$$

Based on the results of measurements of the charge carrier concentration using Hall Van der Pauw method showed that the charge carrier concentration of the order of 10^{17} cm^{-3}, with the majority carrier is hole. The formation of holes as the dominant type of charge carrier in the $GaAs_{1-x}Sb_x$ is due to the contamination of carbon (C) derived from metalorganic sources used. In TMGa chemical structure, there is a direct bond between Ga atoms and carbon atoms (C). Due to the dehydrogenation the species surface [Ga = CH_3], then it is possible to form gallium carbene species [Ga = CH_2] who has a strong bond with the substrate surface. This species [Ga = CH_2] can be react with free Ga atoms, which results in displaced of As atoms and its position occupied by carbon atoms [8]. Carbon is a group IV atom that has four valence electrons. When this atom is bonded with Ga atoms and occupies a position of As or Sb atoms which is pentavalent atoms, there will be generate a hole in the film due to vacancy of electrons in this bond. Hole is a positive charge carrier, so from this process will form a p-type semiconductor. The mechanism of incorpotion of carbon atoms into the grown films are shown in Figure 7.

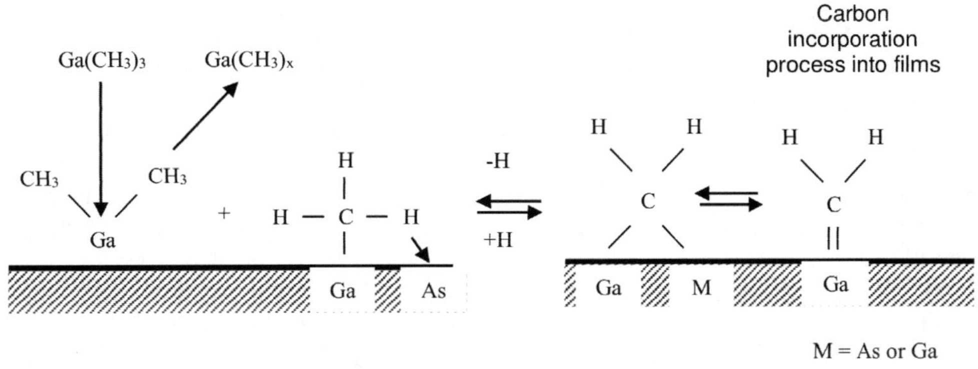

Figure 7. The mechanism of carbon incorporation into the film $GaAs_{1-x}Sb_x$

Order of the charge carrier concentration (hole) is lower than that obtained by Fink, et. al [10]. They get a charge carrier concentration (hole) of the order of 10^{18} to 10^{19} cm^{-3} when using metalorganic sources TMGa, TBAs (Tetrabuthyl Arsine), TMSb (Trimethyl Antimony), and AsH_3 (Arsine) to grow $GaAs_{1-x}Sb_x$ films. This shows that the use of new metalorganic sources TDMAAs and TDMASb can reduce contamination of carbon (C) in the $GaAs_{1-x}Sb_x$ Film. On the metalorganic sources TDMAAs {As [$N(CH_3)_2]_3$} and TDMASb{Sb [$N(CH_3)_2]_3$} there are no direct bonding between As or Sb atoms with C atoms, but is buffered by N atom. So the probability of incorporation of carbon atoms that are in both these sources into grown $GaAs_{1-x}Sb_x$ thin films is relativelly small [9].

4. Conclusion

Metalorganic Chemical Vapor Deposition (MOCVD) technique as a semiconductor thin film deposition techniques that its process conditions are not in thermodynamic equilibrium, have a good potential to be used as a growth technique for $GaAs_{1-x}Sb_x$ ternary alloy thin film. The existence of this potential demonstrated by the success of grow $GaAs_{1-x}Sb_x$ films by MOCVD technique in the range of its miscibility gap. The XRD and EDS characterization results showed that the grown thin films was in the form of $GaAs_{1-x}Sb_x$ ternary alloy. Use TDMAAs and TDMASb combine with TMGa as a metalorganic sources produce carbon (C) contamination in the $GaAs_{1-x}Sb_x$ film fairly low, of the order of 10^{17} cm^{-3}, one to two order lower than that resulting from the use of conventional metalorganic sources in form methyl compounds such as TMSb, TMAS, and TBAs or Arsine.

References

[1] Sun, X., Wang, S., Hsu, J. S., Sidhu, R., Zheng, X. G., Li, X., Campbell, J. C., Holmes, A. L. Jr., (2002) GaAsSb : a novel material for near infrared photodetectors on GaAs substrate, IEEE J. Sel. Top. Quantum Electron. **8**, 817.

[2] Yeh, J. Y., Mawst, L. J., Khandekar, A. A., Kuech, J. F., Vurgaftman, I., Mayer, J. R., Tnsu, N., (2006) Long wavelength emission of InGaAsN/GaAsSb Type II 'W' quantum wells, Applied Physics Letters **88**, 051115.

[3] Sadovyev, Y. G., Samal, N., Photoluminiscence and band alignment of strained GaAsSb/GaAs QW structures grown by MBE on GaAs (2010), Materials, 3, 1497-1508.

[4] Kawamura, Y., Sahashi, T., (2014) 2.86 mm room-temperature light emission of InGaAsN/GaAsSb type II quantum well diode grown on InP substrates, Japanese Journal of Applied Physics, 53, 2.

[5] Noh, M., S., (2003), Material Growth and Characterization of GaAsSb on GaAs Grown by MOCVD for Long Wavelength Laser Applications, Dissertation, The University of Texas at Austin, USA.

[6] Razegi, M., (1989) The MOCVD Challenge, Vol. 1, p. 18, Adam Higler, Philadelphia.

[7] Stringfellow, G. B., (1989), Organometalic Vapor Phase Epitaxy : Theori and Practice, Academic press Inc., p.32.

[8] Jones, A. C., O'Brien, P., (1997) CVD of Compound Semiconductors : Precursor Synthesis and Applications, VCH, Weinheim, Germany.

[9] Marx, D., Asahi, H., Liu, X. F., Higashiwaki, M., Villaflor, A. B., Miki, K., Yamamoto, K., Gonda, S., Shimomura, S., Hiyamizu, S., (1995), Low temperature etching of GaAs substrate and improved morphology of GaAs grown by metalorganic molecular beam epitaxy using trisdimethylaminoarsenic and triethylgallium, Journal of crystal growth **150**, 551-556.Ikossi-Anastasiou, K., GaAsSb for heterojunction bipolar transistors, *IEEE Trans. Electron. Devices* **40**(5), (1993). Matthews, J. W., dan Blakeslee, A. E. Defect in epitaxial multilayers: I. Misfit dislocation, *J. Cryst. Growth* **27**, (1974).

[10] Fink, V., Chevalier, E., Pitts, O., Drovak, M. W., Kavanagh, K. L., Bolognesi, C. R., Watkins, S. P., Humel, S., Moll, N.,(2001) Anisotropic resistivity correlated with atomic ordering in p-type GaAsSb, Appl. Phys. Lett., 79, 15.

Implementation of cloud computing in higher education

Asniar* and R Budiawan
School of Applied Science, Universitas Telkom

*Corresponding author: asniar@telkomuniversity.ac.id

Abstract. Cloud computing research is a new trend in distributed computing, where people have developed service and SOA (Service Oriented Architecture) based application. This technology is very useful to be implemented, especially for higher education. This research is studied the need and feasibility for the suitability of cloud computing in higher education then propose the model of cloud computing service in higher education in Indonesia that can be implemented in order to support academic activities. Literature study is used as the research methodology to get a proposed model of cloud computing in higher education. Finally, SaaS and IaaS are cloud computing service that proposed to be implemented in higher education in Indonesia and cloud hybrid is the service model that can be recommended.

1. Introduction

Cloud computing is a new trend in distributed computing, where people can develop SOA-based service in Internet network. There are numerous companies that calculate and evaluate the advantages of cloud service implementation. Cost effectiveness is the main thing that is calculated in order to implement cloud computing. Although there are other factors that must be thought in order to determine the implementation of new technology.

Higher education in Indonesia is institutions that have potential to use cloud computing service. But, operational management is not aware of the importance cloud computing to support academic and operational activities. There are many reason of this condition. But, usually the thing that can be a reason behind this condition is the difficultness to choose an implementation model that support academic and operational activities.

Based on the problem above, this research tries to conduct the needs and the suitability of the cloud computing implementation in universities in Indonesia and then propose cloud computing models as the first step to determine a cloud models that support academic and operational activities.

2. Methodology

Literature study in order to propose service cloud computing model in higher education is used as the research methodology.

3. Cloud Computing

Cloud computing is an internet network-based computation style where resource dynamically changed as the service. Cloud refers to the hardware and software for data centre that support the needs of users such as storing data and remote-host application. In the implementation of cloud computing, infrastructures can be cut to reduce cost. An example for this thing is using outsourcing data storage, so hardware device can be neglected to provide in physique.

Cloud computing is a style of computing where dynamically scalable and virtualized resources are provided as a service over the Internet. The cloud refers to the data centre hardware and software that supports a client's needs, often in the form of data stores and remotely hosted applications. These infrastructures enable companies to cut costs by eliminating the need for physical hardware, allowing companies to outsource data and computations on demand [4].

There are several clouds computing service that can be used in a higher education such as [2]:

- Infrastructure as a Service (IaaS): can be used to satisfy the infrastructure needs of the students, faculties or researcher globally or locally with some specific hardware configuration for a specific task.
- Platform as a Service (PaaS): certain providers are opening up application platforms to permit customers to build their own application without the cost and complexity of buying and managing the underlying hardware and software layers.
- Software as a Service (SaaS):the application service provider is hosting the application which runs and interacts through web browser, hosted desktop or remote client. It eliminates the need to install and run the application on customer own computer and simplifying maintenance and support.
- Computing as a Service (CaaS):providers offer access to raw computing power on virtual server such as Amazons, EC2 service.

There are the deployment models of cloud computing [4]:

- Private Cloud: The cloud infrastructure is operated solely for an organization. It may be managed by the organization or a third party, and may exist on premise or off premise.
- Community Cloud: The cloud infrastructure is shared by several organizations and supports a specific community that has shared concerns (e.g., mission, security requirements, and policy and compliance considerations). It may be managed by the organizations or a third party, and may exist on premise or off premise.
- Public Cloud: The cloud infrastructure is made available to the general public or to a large industry group, and is owned by an organization selling cloud services.
- Hybrid Cloud: The cloud infrastructure is a composition of two or more clouds (private, community or public) that remain unique entities but are bound together by standardized or proprietary technology that enables data and application portability (e.g., cloud bursting for load balancing between clouds).

4. Cloud Computing In Higher Education

In many technology arenas, higher education exhibits two behaviors. As regards networking and high-performance computing, higher education enjoys a reputation as an innovator [3]. The use of Cloud Computing in higher education must be analyzed both from the benefits point of view, as well as from that of the risks and limitations presented in Table 1 [1].

Table 1. Main Benefit And Limitation Of Cloud Computing In Higher Education

Benefit	Limitation
Access to applications from anywhere	Offline usage with further synchronization opportunities
Support for teaching and learning Risks related to data protection and security and accounts management	Risks related to data protection and security and accounts management
Software free or pay per use	Organizational support
24 hours to access infrastructures and contents	Dissemination politics, intellectual property

After the analysis, one or more models of Cloud Computing may be chosen to be used. The decision must take into account the real needs and be aligned with the higher education strategy.

5. Proposed Model of Cloud Computing Service in Higher Education

The main users of Information Technology services in higher education can be grouped into 4 classifications: students, faculty (researchers), administrative staff, and IT staff (developers). Based on this condition, there is a model of cloud computing that depend on user. This model is shown by Table 2 below.

Table 2. Proposed Model of Cloud Computing Services in Higher Education

Users	Service		
	SaaS	PaaS	IaaS
Students	√	-	√
Lecturer/Researcher	√	-	√
Administration Staff	√	-	√
Staff IT (developer)	-	√	√

Table 2 shows that SaaS (Software as a Service) and IaaS (Infrastructure as a Service) are services can be proposed for students, lecturer/researcher, and administrative staff. Meanwhile, PaaS (Platform as a Service) and IaaS (Infrastructure as a Service) are services can be proposed to IT staff. For distributed cloud computing service model, hybrid cloud is the suitable one to be implemented. Hybrid cloud is a combination of public cloud and private cloud. Public cloud is owned and managed by the cloud provider. This means the higher education does not have full control over it, and have limited access to the private cloud as a temporary subscriber. Access can be made by user that shows based on cloud computing service.

6. Conclusion

Cloud computing service is an important development in distributed computing where people can develop and create many application and service in Internet network. A higher education is a potential institution to use cloud computing service. In order to use cloud computing service in higher education, SaaS and IaaS are proposed to be implemented. Both of them are very useful for student, researcher, administration staff. Cloud hybrid is the service model that can be recommended because its flexibility and blended characteristic between public cloud and private cloud. Public cloud is managed by cloud provider with limitation access as a subscriber, meanwhile private cloud means the service is owned by the higher education with access limited only for students, researchers, administration staff, and IT staff (developer).

IOP Conf. Series: Materials Science and Engineering **128** (2016) 012022 doi:10.1088/1757-899X/128/1/012022

Based on the explanation above, cloud computing can be a new things for higher education to be implemented. Cloud computing can support tridarma activities for a higher education, especially in Indonesia. Besides, it has many potential opportunities to make revolution in education area.

References

[1] Marinela Mircea and Anca Ioana Andreescu, "Using Cloud Computing in Higher Education: A Strategy to Improve Agility in the Current Financial Crisis", Communications of the IBIMA," http://www.ibimapublishing.com/journals/CIBIMA/cibima.html Vol. 2011 (2011), Article ID 875547, 15 pages DOI: 10.5171/2011.87554, IBIMA Publishing

[2] Mathew, Saju, "Implementation of Cloud Computing in Education - A Revolution," International Journal of Computer Theory and Engineering, Vol. 4, No. 3, June 2012

[3] Richard Katz, Phil Goldstein, dan Ron Yanosky , "Cloud Computing in Higher Education"

[4] Samah A. Massadeh and Muhammad A. Mesleh, "Cloud Computing in Higher Education in Jordan," World of Computer Science and Information Technology Journal (WCSIT) ISSN: 2221-0741 Vol. 3, No. 2, 38-43, 2013

IOP Conf. Series: Materials Science and Engineering **128** (2016) 012023 doi:10.1088/1757-899X/128/1/012023

Novel binary PSO algorithm based optimization of transmission expansion planning considering power losses

Astuty* and T Haryono

Electrical and Information Technology Departement
Universitas Gadjah Mada, Yogyakarta

*Corresponding author: astuty.stl13@mail.ugm.ac.id

Abstract. Transmission expansion planning (TEP) is one of the issue that have to be faced caused by addition of large scale power generation into the existing power system. Optimization need to be conducted to get optimal solution technically and economically. Several mathematic methods have been applied to provide optimal allocation of new transmission line such us genetic algorithm, particle swarm optimization and tabu search. This paper proposed novel binary particle swarm optimization (NBPSO) to determine which transmission line should be added to the existing power system. There are two scenerios in this simulation. First, considering transmission power losses and the second is regardless transmission power losses. NBPSO method successfully obtain optimal solution in short computation time. Compare to the first scenario, the number of new line in second scenario which regardless power losses is less but produces high power losses that cause the cost becoming extremely expensive.

1. Introduction

The additional of high scale power generating capacity increases over the year to fulfill the electricity demand. Most of generating plants are located far from the load center. Transmission network is an important infrastructure that must be considered in generation expansion planning. To avoid line overloading, net transmission capacity (NTC) has to meet reliability criteria while delivering electric power to the load. The construction of transmission line which needs a high budged becoming an issue in transmission expansion [1]. Transmission expansion planning (TEP) aims to minimize the line construction which directly relates to investment cost. It determines which line must be added to existing power system by considering technical, economic and reliability aspect.

There are a lot of mathematical methods that have been used to do TEP optimization. Genetic algorithm has been used in several researches in binary and decimal codification model [2]. Discrete PSO with decimal codification also has been applied in STEP optimization. The precision and convergence speed of discrete PSO are better and faster than genetic algorithm. In addition to discrete PSO, traditional PSO with real number for continuous model also have been used to find the optimum transmission line considering line loading [4]. The other methods such as discrete evolutionary particle swarm optimization, simulated annealing algorithm, tabu search, and hybrid algorithm successfully provide optimum solution [5]-[8]. Ordinal optimization and mixed integer linear programming also have been applied to solve TEP problem [9]-[11].

Most of previous researches conducted TEP study to obtain optimal cost of transmission line investment regardless power losses in power system. In this research, the optimization of TEP considers investment cost, power losses and line loading transmission. Novel binary PSO algorithm is applied to solve TEP problem in this research.

2. Literature Review

2.1. Transmission Expansion Planning

In 1970, Garver proposed a research about transmission expansion planning. The research was conducted by using linear programming method. Ever since, study of TEP have been done by many researchers. Heuristic method is a developing method to solve such problem.
Based on stage of process, TEP is classified into two types, which are:

- Static transmission expansion planning (STEP)
- Dynamic transmission expansion planning (DTEP)

STEP determines which line should be added and how many lines should be added in electric power system in a certain period. DTEP determines when the lines should be added in addition to determine the location and the number of line should be added.

2.2. Particle Swarm Optimization

In 1995, Eberhart and Kennedy introduced particle swarm optimization as a population based search algorithm based on the simulation of social behavior of birds, bees or a school of fish [3]. The original PSO was developed for continuous values spaces. But in 1997, discrete PSO was introduced for discrete optimization problem.

TEP problem consist of discrete time parameter. Binary particle PSO (BPSO) and Discrete PSO (DPSO) are two methods in PSO technique that can be used to solve TEP problem. BPSO is applied in this paper to determine which line should be added in power system. In addition to TEP problem, BPSO has been applied in power generation scheduling.

In BPSO, each particle represents its position in binary values which are 0 or 1. Each particle's value can then be changed from one to zero or vice versa. In BPSO, the velocity of particle defined as probability that a particle might change its state to one. Then, a novel BPSO is introduced to improve the weakness in original BPSO. In novel BPSO, the velocity of a particle is its probability to change its state from its previous state to complement value, rather than the probability of change to one.

2.3. Novel Binary Particle Swarm Optimization

The P_{ibest} and P_{gbest} of the swarm is updated as in continuous or binary version. The mayor difference between NBPSO and original PSO is the interpretation of velocity. Here, as in continuous version of PSO, velocity of the particle is the rate at which the particle changes its bit's value. Two vectors for each particle are introduced as $\vec{V}_i^0$and $\vec{V}_i^1$. $\vec{V}_i^0$is the probability of that bits of the particle change to zero while $\vec{V}_i^1$ is the probability of that bits of the particle to change to one. Since in update equation of these velocities, the inertia term is used, these velocities are not complement. So the probability of change in *j*-th bit of *i*-th particle is simply defined as follows:

$$V_{ij}^{c} = \begin{cases} V_{ij}^{1}, if\ x_{ij} = 0 \\ V_{ij}^{0}, if\ x_{ij} = 1 \end{cases} \tag{1}$$

In this way, the velocity of particle is simply calculated. Also the update $\vec{V}_i^0$ and $\vec{V}_i^1$ is as follows: consider the best position visited so far for a particle is P_{ibes} and the global best position for the particle is P_{gbest}. Also consider that the *j*-th of *i*-th best particle is one. So to guide the bit *j*-th of *i*-th particle to its best position, the velocity of change to $1(\vec{V}_i^1)$ for that particle increases and the velocity of change to zero ($\vec{V}_i^0$) is decreases. Using this concept following rules can be extracted:

$$if\ P_{ibest}^{j} = 1\ then\ d_{ij,1}^{1} = c_1 r_1\ and\ d_{ij,1}^{0} = -c_1 r_1$$

$$if\ P_{ibest}^{j} = 0\ then\ d_{ij,1}^{0} = c_1 r_1\ and\ d_{ij,1}^{1} = -c_1 r_1$$

$$if\ P_{ibest}^{j} = 1\ then\ d_{ij,1}^{1} = c_1 r_1\ and\ d_{ij,1}^{0} = -c_1 r_1$$

$$if\ P^{j}_{gbest} = 1\ then\ d^{1}_{ij,2} = c_2 r_2\ and\ \ d^{0}_{ij,2} = -c_2 r_2$$

$$if\ P^{j}_{gbest} = 0\ then\ d^{0}_{ij,2} = c_2 r_2\ and\ \ d^{1}_{ij,2} = -c_2 r_2$$

Where $d^{1}_{ij,1}$, $d^{0}_{ij,1}$ are two temporary values, while $r_1$and $r_2$are two random variables in the range of (0,1) which are updated each iteration. Also $c_1$and $c_2$are two fixed variables which are determined by user, then:

$$V^{1}_{ij} = wV^{1}_{ij} + d^{1}_{ij,1} + d^{1}_{ij,2}$$

$$V^{0}_{ij} = wV^{0}_{ij} + d^{0}_{ij,1} + d^{0}_{ij,2} \tag{2}$$

In fact, this algorithm if the j-th bit in the global best variable is zero or if the i-th bit in the corresponding personal best variable is zero, the velocity (V^{0}_{ij}) increases. And the probability of changing to one also decreases with the same rate. In addition, if the j-th bit in global best variable is one,V^{1}_{ij} increases and V^{0}_{ij} decreases. After updating velocity of particles, $\vec{V}^{1}_{i}$and $\vec{V}^{0}_{i}$, the velocity of change is obtained as in (1).

In NBPSO, A normalization process is done by using sigmoid function as:

$$v_{ij}(t) = sig\left(v_{ij}(t)\right) = \frac{1}{1 + e^{-v_{ij(t)}}}$$

and the next particles state is computed as follows:

$$x_{ij}(t+1) = \begin{cases} \overline{x}_{ij}(t) & ,if\ r_{ij} < V_{ij} \\ \overline{x}_{ij}(t) & ,if\ r_{ij} < V_{ij} \end{cases} \tag{3}$$

That is, if $x_{ij} = 0$ then $\overline{x}_{ij} = 1$ and if $x_{ij} = 1$ then $\overline{x}_{ij} = 0$. Meanwhile r_{ij} is a uniform random number between 0 and 1.

3. Problem Formulation

The optimization in this paper will be conducted by applying NBPSO algorithm considering power losses in transmission line to get the optimal solution.

3.1. Objective function

The objective function in this TEP is to get the optimal total cost of the investment of line construction and the cost of power losses in the network [2]. The objective function of conventional TEP can be written as follows:

$$C_T = \sum_{i=1}^{NB} \sum_{j=1}^{NB} CL_{ij} n_{ij}$$

To minimize power losses in line ($P_{loss\ i}$), TEP can be written as follows:

$$C_T = \sum_{i=1}^{NB} \sum_{j=1}^{NB} CL_{ij} n_{ij} + 8760 \times C_{MWh} \times \sum_{t=1}^{NYE} P_{loss\ t} \tag{4}$$

$$P_{loss_t} = (LGC)^{t-1} \times P_{loss}$$

Where *NB* shows the number of bus. C_{MWh} represents cost per MWh in \$US. CL_{ij} for construction cost of line transmission in \$US. *LGC* is load growth coefficient. *NB* and *NYE* are number of branch and life time prediction of line expansion (year). While *i* and *j* represent branch to and from.

IOP Conf. Series: Materials Science and Engineering **128** (2016) 012023 doi:10.1088/1757-899X/128/1/012023

3.2. Constraints

Optimization in TEP has several constraint to be fulfilled.

$$P(V,\theta,n) - P_G + P_D = 0$$

$$Q(V,\theta,n) - Q_G + Q_D = 0$$

$$P_g^{min} \le P_g \le P_g^{max}$$

$$V^{min} \le V \le V^{max}$$

$$(n_{ij} + n_{ij}^0)S^{from} \le (n_{ij} + n_{ij}^0)S^{max}$$

$$(n_{ij} + n_{ij}^0)S^{to} \le (n_{ij} + n_{ij}^0)S^{max}$$

$$0 \le n_{ij} \le n_{ij}^{max}$$

$$Line_{loading} \le LL_{max}$$

Variable P_G is real power generated by electric generator, while P_D represents the real power in the load side. V, θ, n are magnitude of voltage, angle, and bus respectively. S is apparent power, the superscript from and to refer to apparent power from bus or to bus, while superscribt min and max represent the upper and lower boundary. A constraint that has to be considered is line loading in transmission must be less or equals to maximum line loading (LL_{max}).

3.3. Fitness function

The result of fitness function marks how optimal the solution we get. in this function, the total cost is added by penalty function. Penalty function itself is a function accumulating all of constraints violation [2].

$$fitness = \frac{1}{C_T + penalty\ function} \tag{5}$$

while penalty function (*pf*) is written as follows

$$pf = \alpha \sum_{i=1}^{NB} f(P_{gi}) + \beta \sum_{i=1}^{NB} f(V_i) + \gamma \sum_{j=1}^{\Omega} f(line\ loading)$$

α,β and γ are penalty coefficient. The penalty is set up as follow,

$$f(x) = \begin{cases} 0 & if\ x_{min} \le x \le x_{max} \\ x - x_{max} & if\ x > x_{max} \\ x_{min} - x & if\ x < x_{min} \end{cases}$$

4. Proposed Methodology

4.1. IEEE Garver's 6-Bus System

The NBPSO algorithm will be applied to Garver's system [2]. The system consists of five busses. To fulfill the additional demand, new power generation will be added to the existing power system.

Table 1. Garver's Bus Data

Bus Number	Max. generation (MW)	Generation (MW)	Demand (MW)
1	160	100	80
2	-	-	240
3	370	260	150
4	-	-	180
5	-	-	240
6	610	550	0

The single line diagram of Garver 6-bus system is shown in figure 1. There are 6 existing lines and 11 planning transmission network which marked by dashed line. New power plant will be placed in bus 6.

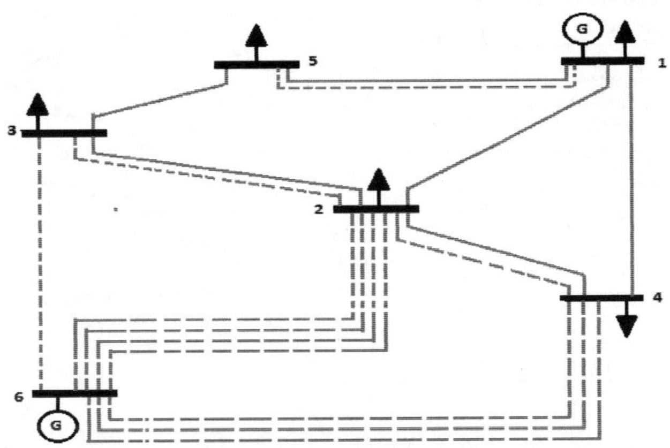

Figure 1. Garver's 6-Bus

The additional of new power generation in bus 6 will influence the performance of transmission line in carrying power to the load. To support generation expansion, transmission line should also be planned to expand. Table 2 shows the number of new transmission lines that planned to be added.

Table 2. Garver's line data

Line	Resistance (p.u)	Reactance (p.u)	Capacity (MVA)	Length (km)
1 – 2	0.0070	0.0248	750	200
1 – 4	0.0105	0.0372	750	300
1 – 5	0.0035	0.0124	750	100
2 – 3	0.0035	0.0124	750	100
2 – 4	0.0035	0.0124	750	200
3 – 5	0.0035	0.0124	750	100

4.2. Simulation

Simulation will be conducted to find the line being loaded more than 50% of the MVA rating. *LLmax* in this simulation is 750 MV. The test will also run within two scenario. First, consider power losses in TEP optimization. Second scenario is optimizing TEP without considering power losses in the network. Table 3 shows parameters that use in NBPSO algorithm.

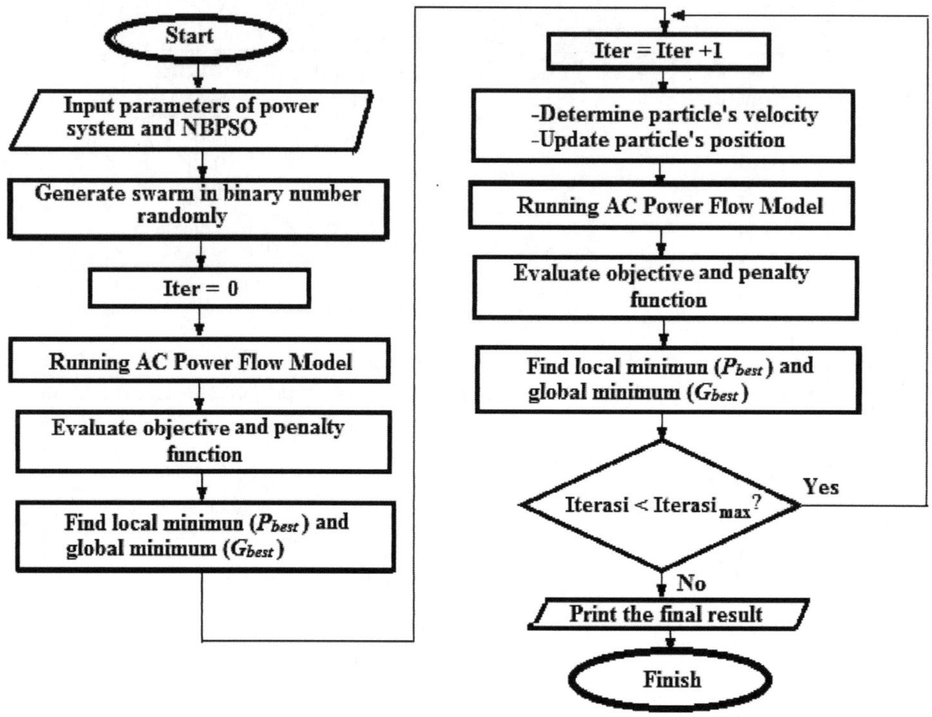

Figure 2. Flowchart of TEP Using NBPSO

Figure 2 shows the flowchart of this research. The algorithm of TEP optimizitaion by using NBPSO is initialized by generating swarm in binary value randomly. Binary value represents the statue of the line whether active or not. Evaluate the performance of each particle by running power flow. Fitness function is calculated by applied equation (5). Next step is finding the initial local best P_{best} and global best G_{best}. Changing the velocity of the particle $\vec{V}_i^1$ and $\vec{V}_i^0$ according to (2) to determine the velocity of the particle, $\vec{V}_i^c$ as in (1). After that generate the random variable r_{ij} in the range (0,1). Updating the position to a new position according to the equation (3). Running power flow to update the new P_{best} and global best G_{best} and repeat until maximal iteration is afforded

Table 3. NBPSO's Parameter Control and Object Solution

Parameter	Value
Number of population	30
Number of iteration	100
C_1 , C_2	2
v_{max}	4
NYE	20
C_{MWh}	36.1
LGC	1.08

5. Result and Discussion

The simulation was run using 2 scenarios. First scenario is optimizing TEP considering power losses in the network and the other was optimizing TEP regardless power losses. The convergence curve obtained from test is shown if Figure 3 and Figure 4.

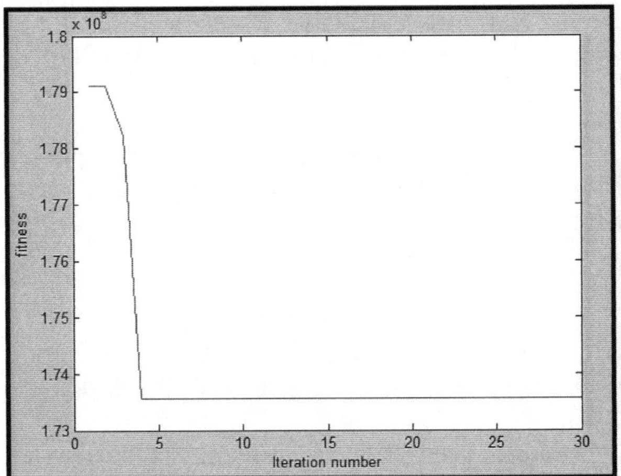

Figure 3. Convergence Curve of Scenario 1

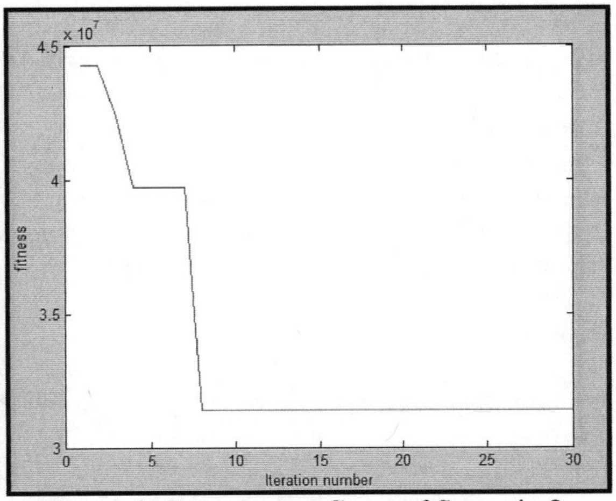

Figure 4. Convergence Curve of Scenario 2

The test was conducted in ten times for each scenario. Figure 3 shows the convergence is reached before fifth iteration in 6 seconds. The convergence time of scenario 2 is longer than scenario 1. the fitness value of scenario 2 is changing up to eighth iteration.

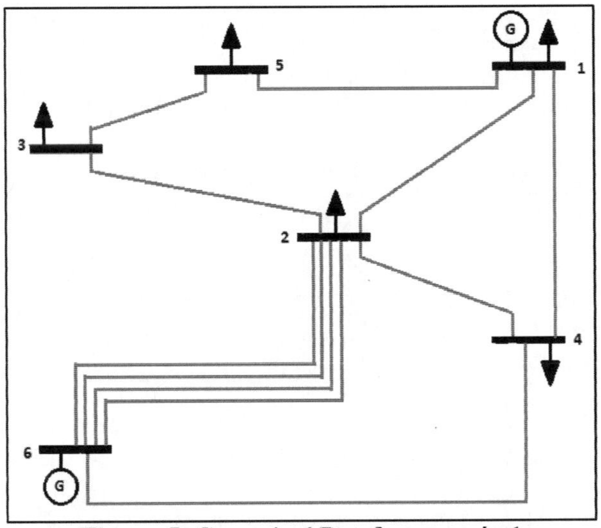

Figure 5. Garver's 6 Bus for scenario 1

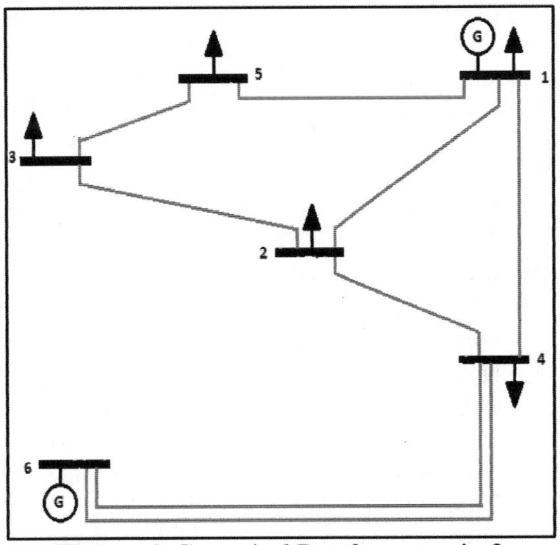

Figure 6. Garver's 6 Bus for scenario 2

The configuration for the scenario 1 is shown by Figure 5. By considering power losses in the network, there are four new transmission lines from bus 2 to bus 6 and one transmission line from bus 4 to bus 6. The optimization regardless power losses only require 2 lines from bus 4 to bus 6 as depicted in Figure 6.

Table 4. Cost of investment and losses

Type of cost	Amount of cost (U$)	
	Scenario 1	Scenario 2
Transmission	81.202.000	31.366.000
Losses	92.355.000	341.610.000

Compared to scenario 1, scenario 2 that does not consider power losses in the network only requires two new lines to support power delivery. Although it needs less additional line, the losses of power in transmission produce the cost of loss becoming extremely expensive. Based on Table 4, the investment cost for scenario 1 is more expensive, but its operational cost is lower than scenario 2.

6. Conclusion

This paper shows that novel binary particle swarm optimization method successfully finds optimal solution for transmission expansion problem. NBPSO's method has less time consume and stable convergence characteristic. The solution obtained by proposed method provide expectation for effective implementation to further research.

References

[1] H. Shayeghi, M. Mahdavi, and A. Bagheri 2010 Discrete PSO algorithm based optimization of transmission lines loading in TNEP problem *Energy Conversion and mangement* 2010, *p.* 112-121.

[2] Ikrima Alfi 2014 Transmission expansion planning using genetic algorithm considering transmission loading *Gadjah Mada University Repository* 292091

[3] Mojtaba Ahmedieh Khanesar, Mohammad Teshnehlab, and Mahdi Aliyari Shoorehdeli 2007 A novel Binary Particle Swarm Optimization *Proceeding of the 15th Mediterranean Conference on Control & Automation* T **33**001July27-29, 2007, Athens. Greece

[4] D. Khavita, and K. Shanti Swarup 2010 Transmission Expansion Planning Using LP-Based Particle Swarm Optimization *IEEE*

[5] Manuel Costeira da Rocha, João Tomè Saraiva 2011 Multiyear Transmission Expansion Planning using Discrete Evolutionary Particle Swarm Optimization*8th International Conference on the European Energy Market (EEM)*, 2011. Zagreb, Croatia.

[6] A. S. D. Braga, J. T. Saraiva 2005 A Multiyear Dynamic Approach for Transmission Expansion Planning and Long-Term Marginal Cost Computation *IEEE Trans. On Power System* **20** 1631-1639

[7] M. V. Pereira, L. M. V. G. Pinto, S. H. Cunha, G. C. Oliviera 2001 A Decomposition Approach to Automated Generation/Transmission Expansion Planning *IEEE Trans. On Power System* **16** 247-253, May 2001.

[8] Mohammad Sadegh Sepasian, Hossein Seifi, Asghar Akbari Foroud, and A. R. Hatami 2009 A Multiyear Security Constrained Hybrid Generation-Transmission Expansion Planning Algorithm Including Fuel Supply Cost *IEEE Transaction on Power System* **24**

[9] Min Xie, Jin Zhong, and Felix F Wu 2007 Multiyear Transmission Expansion Planning Using Ordinal Optimization **22** 1420-1428.

[10] Jamshid Aghaei, Nima Amjady, Amir Baharvandi, and Mohammad Amin Akbari 2014 Generation and Transmission Expansion Planning: MILP-Based Probabilistic Model *IEEE Transaction On Power System* **29** 1592-1601.

[11] H M D R H Samarakoon, R M Shrestha 2001 A Mixed Integer Linear Programming Model For Transmission Expansion Planning With Generation Location Selection **23** 285-293

Ply ThicknessFiberGlass on Windmill Drive Salt Water Pump

Agus Sifa[1], Badruzzaman[2] and Dedi Suwandi[3]
Politeknik Negeri Indramayu
Jl. Raya Lohbener No.08 Lohbener- Indramayu 45252

E-mail : agus.sifa86@gmail.com

Abstract. Factors management of salt-making processes need to be considered selection of the location and the season is very important to support the efforts of salting. Windmills owned by the farmers are still using wood materials are made each year it is not effectively done and the shape of windmills made not in accordance with the requirements without considering the wind speed and the pumping speed control influenced by the weight and size of windmill, it affects the productivity of salt. to optimize the function of windmills on pumping salt water by change the material blade on the wheel by using a material composite, composite or fiberglass are used for blades on windmills made of a material a mixture of Epoxy-Resin and Matrix E-Glass. The mechanical characteristics of the power of his blade one of determining the materials used and the thickness of the blade, which needed a strong and lightweight. The calculation result thick fiberglass with a composition of 60% fiber and 40% epoxy, at a wind speedof area salt fields9 m / s, the drag force that occurs at 11,56 kg, then the calculation result by 0,19 mm thick with a layer of 10, the total thickness of 1,9 mm, with a density of 1760 kg / m3, mechanical character of elongated elastic modulus of 46200 MPa, modulus of transverse elasticity of 10309,6 MPa, shear modulus of 3719 MPa and Poisson ratio of 0,31, then the calculation using the finite element ABAQUS obtained critical point at the confluence of the blade to the value of Von Mises tension was happening 1,158e9 MPa maximum and minimum 2,123e5 MPa, for a maximum value of displacement occurred condition at the tip of the blade. The performance test results windmills at a wind speed of 5,5 m / s wind power shows that occur 402,42 watts and power turbines produced 44,21 watt, and TSR 0,095 and the value Cp of 0,1, test results windmill in salt fields in the beginning rotation windmill lighter, able to move above wind speed of 5.5 m/s.

Keywords: Fiberglass, Weigth of Blade,Ply Thickness

1. Introduction

Indonesia is an agricultural country rich in marine resources, the development of the current era requires community to be more advanced and productive, especially in the field of food sourced from the sea, one of the food sources that currently have a high need is salt not only society but the need for food industry too high, needs high salt impact on the availability of salt was not able to be met due to the non optimal management of salt fields and farmers have not been able to implement technology that is currently being developed [1].

Factors management of salt-making processes need to be considered selection of the location and the season is very important to support the efforts of salting. Windmills owned by the farmers are still using wood materials are made each year it is not effectively done and the shape of windmills made not in accordance with the requirements without considering the wind speed and the pumping speed control influenced by the weight and size of windmill, it affects the productivity of salt. to optimize the function of windmills on pumping salt water by change the material blade on the wheel by using a material composite, composite or fiberglass are used for blades on windmills made of a material a mixture of Epoxy-Resin and Matrix E-Glass. Based on this, there should be a study to optimize the function of windmills on pumping salt water by changing the raw material of blade on the wheel by using a material composite, composite or fiberglass are used for blades on windmills made of a material a mixture of Epoxy-Resin and Matrix E-Glass. The advantages the research by reducing the weight of the blade turbines aims to windmills can be driven at low wind speeds , and is capable of moving the salt water pump to the optimum of power generated.

2. Methodology

Some steps has been done in the research as below;

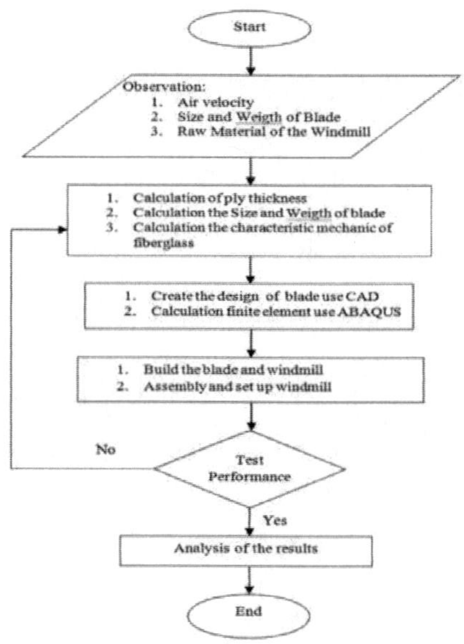

Figure 2.1 Flow chart of research steps

Table 2.1 Property Materials [3]

Material	Density $\rho\left(\frac{kg}{m^3}\right)$	Modulus Elasticity E (MPa)	Shear Modulus G (MPa)	PoissonRatio	Tensile strength $\sigma\ (MPa)$
E-Glass	2600	74000	30000	0.25	2500
Epoxy	1200	4500	1600	0.4	130

3. Result and Discussion

3.1. Calculation Thickness

The thickness of fiberglass done before Finite Element calculation as an initial condition for determining the thickness of the blade, the composition of the volume for the manufacture of blade 60% fiber volume of E-glass and 40% volume Epoxy as reinforcement, using the following equation;

$$h = m_{of}\left[\frac{1}{\rho_f} + \frac{1}{\rho_f}\left(\frac{1-M_f}{M_f}\right)\right] \quad (1)$$

$m_{of} = 300$

$Mf = Mass\ of\ fraction = \frac{Mass\ of\ fiber}{Mass\ of\ total} = \frac{1560}{2040} = 0{,}76$

$\rho_f = mass\ density\ of\ epoxy = 1200\ kg/m^3$

Fiberglass blade thickness is obtained;

$$h = 300\left[\frac{1}{1200} + \frac{1}{1200}\left(\frac{1-0{,}76}{0{,}76}\right)\right] \quad (2)$$

$h = 0{,}19\ mm$

IOP Conf. Series: Materials Science and Engineering **128** (2016) 012024 doi:10.1088/1757-899X/128/1/012024

Then, the thickness of each layer of fiberglass at 0.19 mm with a 10 ply, then the thickness of blade windmill 1.9 mm.

3.2. Mechanical characteristics of of fiberglass

3.2.1. Modulus Elasticity

- Longitudinal

$$E_l = E_f V_f + E_m V_m \tag{3}$$

E_f = Modulus Elasticity of fiber =74000 MPa
V_f = Volume fiber 60%
E_m= Modulus Elasticity of matrice epoxy = 4500 MPa
V_m = Volume matrice epoxy 40%

- Transverse

$$E_t = E_m \left[\frac{1}{(1-V_f)+\frac{E_m}{E_{ft}}V_f}\right] \tag{4}$$

3.2.2. Shear Modulus

$$G_{lt} = G_m \left[\frac{1}{(1-V_f)+\frac{G_m}{G_{flt}}V_f}\right] \tag{5}$$

3.2.3. Poisson Ratio

$$v_{lt} = v_f V_f + v_m V_m$$

v_f = poisson ratio of fiber = 0,25
v_m = poisson ratio of matrice epoxy = 0,40

The calculation results percentage composition of epoxy and E-glass of material characteristics values obtained the material a mixture of in the following table;

Table 3.1 Mechanical Caracteristics

Modulus Elasticity Longitudinal E_l (MPa)	Modulus Elasticity Tranverse E_t (MPa)	Shear Modulus G_{lt} (MPa)	PoissonRatio (v_{lt})
46200	10309,6	3719	0,31

3.2.4. Modulus Any Direction

$$E_x = \frac{1}{\frac{c^4}{E_l}+\frac{s^4}{E_t}+2c^2s^2\left(\frac{1}{2G_{lt}}-\frac{v_{lt}}{E_t}\right)} \tag{6}$$

c= cos θ
s= sin θ

IOP Conf. Series: Materials Science and Engineering **128** (2016) 012024 doi:10.1088/1757-899X/128/1/012024

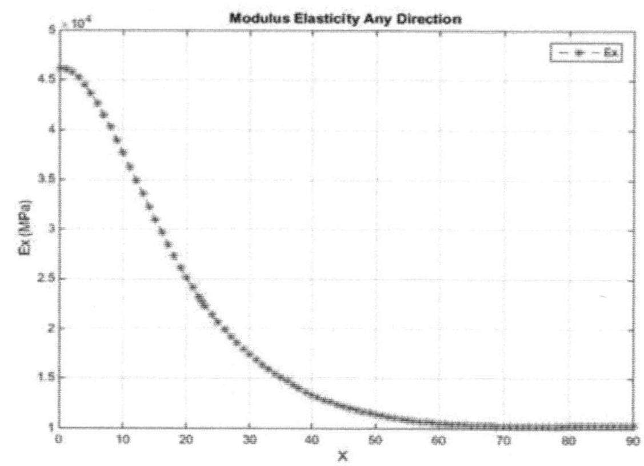

Figure 3.1. Modulus Elasticity any direction

In the process of finite element calculations need to be done with the input initial conditions, a thickness of 0.19 mm on each layer, the angle between the longitudinal and transverse layers of 0 ° and 90 ° with the character input material as in Table 1, it can be seen that the calculation results of the styles caused by wind speed, with the observation of the fields salt that occurred maximum wind speed of 9 m / s.

3.3. *Mass Density of a Ply*

The mass density of a ply can be calculated as;

$\rho_f = mass\ density\ of\ epoxy = 1200\ kg/m^3$

$\rho_m = mass\ density\ of\ E - glass = 2600\ kg/m^3$

$$\rho = \rho_f V_f + \rho_m V_m \quad (7)$$

$\rho = 1200.0{,}6 + 2600.0{,}4$

$\rho = 1760 kg/m^3$

$$m = v.\rho \quad (8)$$

$m = 130257{,}73\text{mm}^2. 1{,}9mm. 1760 kg/m^3$

$m = 0{,}0002475 m^3. 1760 kg/m^3$

$m = 0{,}4356 kg$

The calculation result windmill weight of 0.4356 kg with a thickness of 1.9 mm, while the results have been made windmill has weight with a thickness of 2 mm, has a weight of 0.510 kg.

3.4. *Finite Element*

3.4.1. *Design*

In the process of finite element calculations need to be done with the input initial conditions, a thickness of 0.19 mm on each layer, the angle between the longitudinal and transverse layers of 0 ° and 90 ° with meninput character of material as shown in Table 1, it can be seen that the calculation results of the styles caused by wind speed, with the observation of the fields salt that occurred maximum wind speed of 9 m / s.

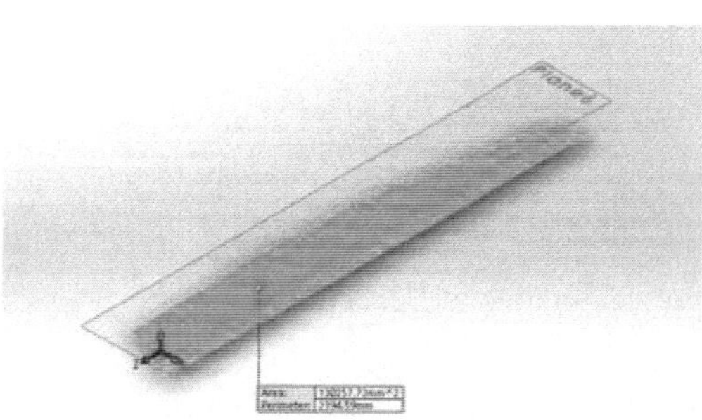

Figure 3.2Design Blade of Windmill

The design of the blade that was created with CAD-type planar surface, with a length of 1m and a the blade that will be created, with an area (A) 130,766.64 mm^2 .

IOP Conf. Series: Materials Science and Engineering **128** (2016) 012024 doi:10.1088/1757-899X/128/1/012024

To know the great value of the force caused by wind speed and on the shape of the blade has a coeficient of drag (CD) 2.3 and the air density of 1.2 kg / m3, then can be searched large drag force that occurred on the blade;

$$F_{drag} = \frac{1}{2} CD.\rho.A.v^2 \qquad (9)$$

$$F_{drag} = \frac{1}{2} 2{,}3.1{,}2kg/m^3.\, 130257{,}73\text{mm}^2.\, 9^2$$

$$F_{drag} = 11{,}56\ kg \approx 12\ kg$$

Then the pressure on the blade of a windmill;

$$P = \frac{F}{A} \qquad (10)$$

$$P = \frac{12\ kg}{130257{,}73\ \text{mm}^2}$$

P =92,125 $kg/m^2 \approx 921{,}25\ N/m^2$

3.4.2. Initial Condition

Figure3.3 Initial Condition and Meshing

3.4.3. Result

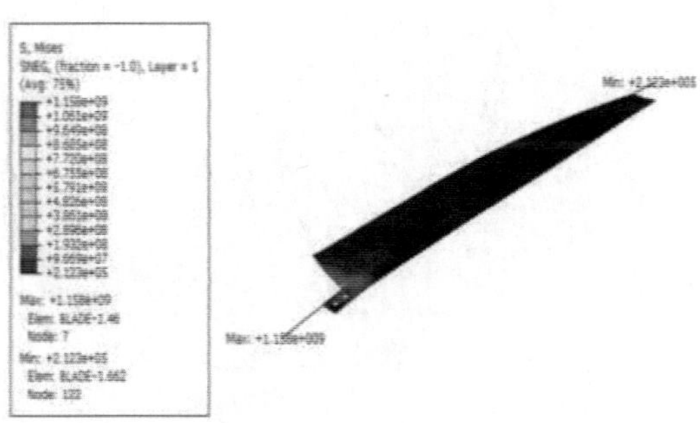

Figure 3.4 Result Stress Von Mises

Results of the finite element calculations using ABAQUS, the maximum voltage happened as a result of pressure loading occurs at the point of retaining a large hole with a maximum value 1,158e9 MPa, and minimal 2,123e5 MPa occur at the end of the the blade.

Figure 3.5 Result deplacement

In the visual image above shows the results of deplacement maximum value occurs at the end of the windmill the blade.

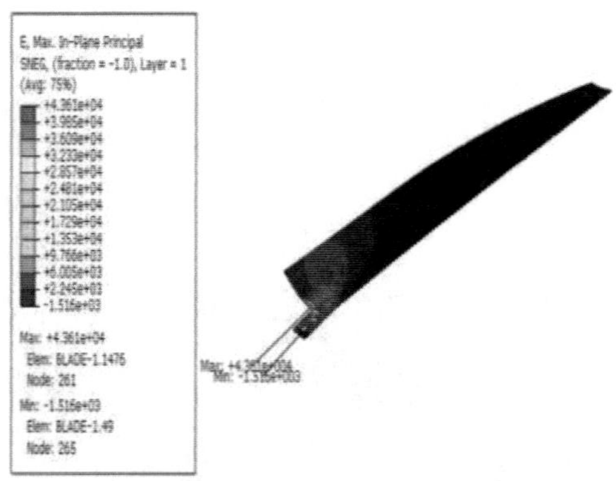

Figure 3.6 Result Modulus of Elasticity

Finite element calculation results of Figure 3.6, shows the maximum value in the area close to the pedestal, with a maximum value and minimal -1,516e3 4,362e4 MPa MPa, compared with the value that occurs in finite element calculation of the maximum values happened less than the value of the calculation of elasticity at an angle 0° of 46200 MPa.

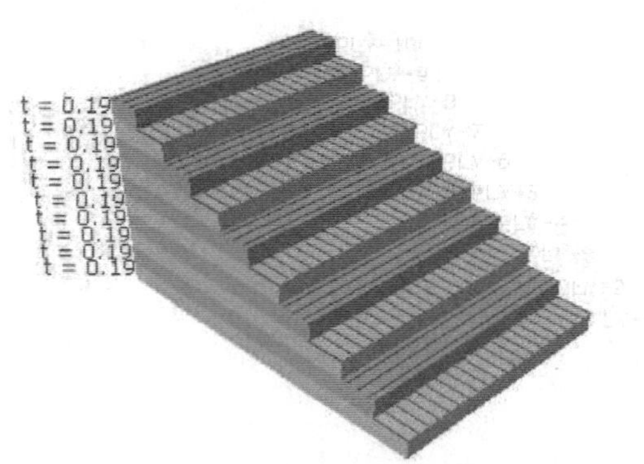

Gambar 3.7 The ply thickness of the fiberglass

In the figure above shows the direction of angle E-glass on the blade and the many layers of the calculation results are shown in figure 0.19.

3.5. Build Windmill

The process of manufacture and assembly of blades on a windmill, then conducted laboratory testing and installation of windmill in the salt fields, test results windmill in salt fields in the beginning rotation windmill lighter able to move above wind speed of 5.5 m / s.

Figure 3.6 Build and Assembly windmill

3.6. Test Performance

3.6.1. Tip Speed Ratio(TSR)

A comparison between the speed at the tip of the blade to the wind speed through it. Can be in the know by using the equation;

$$\lambda = \frac{\Omega R}{V} \quad (11)$$

$$\Omega = \frac{\pi}{60} = 0{,}523$$

$$\lambda = \frac{0{,}523\ .\ 1}{6{,}28} = 0{,}095$$

The calculation of the value of the Tip Speed Ratio is 0.095 of the value 0.095<1 refers to the Betz law, then the more experienced the blade thrust.

3.7. Air Power

$$P_{angin} = \frac{1}{2}.\rho.A.v^3 \quad (12)$$

Dimana: P= Air Power (*watt*)
ρ= Density of Air(kg/m^3)
A= Areawindmill (m^2)
v= Air Velocity (m/s)

to look for wide sweep of the turbine rotor the blade can use the following formula;

$$A = \frac{\pi}{4}.d^2 \quad (13)$$

Maka : $A = \frac{3{,}14}{4}.2{,}22^2$

$= 3{,}87\ m^2$

Obtained a sweep area of the rotor with a diameter of 2,22 meters which is 3.87 m^2.

$P_{angin} = \frac{1}{2}.1{,}25\ kg/m^3.3{,}87m^2.(5{,}5)^3 m/s$

= 402,42 watt

From the above calculation, the wind power generated at wind speed of 5,5 m / s is at 402,42 watts.

3.8. Windmill Power

Calculation the windmill power can use the equation;

$$P_{kincir} = T.\omega \quad (14)$$

Dimana : T = *torque* (Nm)

ω = blade speed (rad/s)

then torque of the test results that occur when wind speed of 5.5 m / s of 3.84 Nm. As for the seek values blade velocity (ω) can use the following equation;

$$\omega = \frac{2.\pi.n}{60} \quad (15)$$

where: n = shaft rotation (rpm)

test results kecapatan windmill with 5.5 m / s is obtained value of rotation 110 rpm.

$$\omega = \frac{2.3,14.110}{60} = 11,51\ rad/s$$

From the above calculation, the power generated by the windmill shaft rotation speed of 110 rpm is at 11,51 rad/s.

the values of that have been is obtained over the windmill power can be calculated as follows;

$$.P_{kincir} = T.\omega$$

$$P_{kincir} = 3,84\ Nm\,.\,11,51\ rad/s = 44,21\ watt$$

From the above calculation, the power generated by the windmill torque of 3,84 Nm is at 44,21 watts.

3.8.1. Calculation Coeficient Power (Cp)

To calculate the value of Cp can be calculated using the equation;

$$Cp = \frac{P_{windmill}}{P_{air}} \quad (16)$$

$$Cp = \frac{44,21\ watt}{402,42\ watt}$$

Cp= 0,1

Test operation of the windmill can be said TSR value of 0,095, and the value Cp of 0,1

4. Conclusion

The calculation result thick fiberglass with a composition of 60% fiber and 40% epoxy, at a wind speed 9 m / s,the drag force that occurs by 112 kg, then the calculation result by 0,19 mm thick with a layer of 10, the total thickness of 1,9 mm, with a density of 1760 kg / m^3, with a mechanical character of the modulus of elasticity of 46200 MPa longitudinal, transverse elasticity modulus of 10309,6 MPa, shear modulus of 3719 MPa and Poisson ratio of 0,31, then calculate by using ABAQUS finite element obtained at the critical point pedestal portion of the blade to the value of Von Mises tension was happening 1,158e9 MPa maximum and minimum 2,123e5 MPa, for a maximum value of displacement occurred condition at the tip of the blade. The elasticity happens calculation results have a greater value than the simulation results by using ABAQUS finite element, the performance test results windmills at a wind speed of 5,5 m / s wind power shows that occur 402,42 watts and power turbines produced 44,21 watt, and TSR 0,095 and the value Cp of 0,1, test results windmill in salt fields in the beginning rotation windmill lighter, able to move above wind speed of 5.5 m/s.

Recommendations

At the of fiberglass blade the critical point happens to be able to distinguish the foundation thickness at that point thicker or with a better composition in order to reduce the the critical point happened and some other variables such as the blade angle and sweep an area that affects windmills windmill

Acknowledgements

We would like to thanks to the Ministry of Research, Technology and Higher Education which has provided grant support for this research, and the salt farmers Losarang village, Indramayu.

5. References

[1] Petunjuk Teknis produksi garam menggunakan teknologi geomembran di kabupatem Indramayu http://diskanla.indramayukab.go.id/ accessed on 10 April 2014.

[2] Peter Grand dan Carl. Q Roussean.(2000). *Composite Structure* (Theory and Practice, ASTM International,USA

[3] Daniel Gay, dkk (2003). *Composite Materials Design and Applications,Editions Hermes*, California,USA

[4] Burton,Tony dkk, (2001).*Wind Energy Handbook*, John Wiley and Son.Ltd. England

[5] Ragheb,Magdi.(2011).*Fudamental and Advanced Topics in Wind Power*, ISBN:978-953-307-508-2.Intech Publisher. accessed on 15 September 2015 www.intechopen.com

[6] Hirshberg,Gary (1982).*The New Alchemy Water Pumping Windmill Book*. USA

[7] Supeni. *Design of Smart Structures for Wind Turbine Blades*. 2nd Malaysian Postgraduate Conference (MPC2012). Bond University. Queensland,Australia. PP-20-36, Juli 2012.

[8] Nicoar,Dumitru, *On the Numerical Analysis Laminate Composite*. Fascicle of Management and Technological Engineering, Annals of the Oradea University,PP.241-244,Mei 2013, dikases pada tanggal 15 september 2015 http://www.imtuoradea.ro/auo.finite/

[9] Toledo V,Miguel. *Design and Experimentation of a 1 MW Horizontal axis Wind Trubine*, Journal of Power and Nenergy Engineering, Vol.2 PP.9-16. Januari 2014. accessed on 13 September 2015 http://www.scirp.org/journal/jpee

[10] The Drag Coeficient, Glenn Research Center, National Aeronautics and Space Administration, accessed on 30 Agustus 2015 www.grc.nasa.gov

[11] Aerodynamics for students, Departement of Engineering, University of Cambridge, http://www-mdp.eng.cam.ac.uk/web/library/enginfo/aerothermal_dvd_only/aero/fprops/introvisc/node11.htmlaccessed on 1 September 2015

[12] Lemaitre,Jean and Friends, (2009) Mecanique Des Materiaux Solides 3e Edition,Paris,France- Dunond

[13] Callister,William(1994). Materials Science and Engineering and Introduction, Departement of Materials Science and Engineering The University of Utah. John Wiley and Son,Inc New York.

[14] Mercyline,Machini and Julius Mokaya,Okero,(2014), Design of A Windmill for Pumping Water, Departement of Mechanical and Manufacturing Engineering, University of Nairobi. Accesed on 30 Augtus 2015 http://mechanical.uonbi.ac.ke/

IOP Conf. Series: Materials Science and Engineering **128** (2016) 012025 doi:10.1088/1757-899X/128/1/012025

Separation of glycerolysis product using hexane

S Mujdalipah*[1], A H Sasmita[2], I K Amalia[3] and A Suryani[3]

[1] Program Studi Pendidikan Teknologi Agroindustri, Universitas Pendidikan Indonesia, Jl. Setiabudi No. 207, Bandung, Indonesia

[2] Jurusan Pendidikan Teknik Mesin, Universitas Pendidikan Indonesia, Jl. Setiabudi No. 207, Bandung, Indonesia

[3] Departemen Teknologi Industri Pertanian, Institut Pertanian Bogor, Kampus IPB Darmaga, Bogor, Indonesia

*Corresponding author: siti.mujdalipah@upi.edu

Abstract. Hexane is non-polar solvent and it can dissolve triglyceride and free fatty acid. The aim of this research is to get the hexane ratio to the glycerolysis product that can produce product that has the best characteristics. The study was conducted in five stages. In the first stage, the physicochemical properties of palm oil were analysed. In the second stage, palm oil glycerolysis was conducted. In the third stage, glycerolysis product was extracted by using hexane in various concentrations. In the following stage, physicochemical properties of extracted products were analysed. In the last stage, data were subjected to an analysis of variance. Results showed that ratio between glycerolysis product to hexane in this work did not significantly (α=0.05) affect the yield, the free glycerol content, and the ability of extracted products to reduce the surface tension of water. It was also found that the best treatment had yield of 28.5%. It was able to lower water surface tension from 72.00 to 31.80 dynes/cm and had free glycerol content of 2.40%. Base on this work, we found that further research need to be done.

1. Introduction

Monoglyceride (MAG) and diglyceride (DAG) is the emulsifier that can be gotten from palm oil. They are many used as emulsifier in the various food products, such as margarine, bread, ice cream, and chewing gum. There are 20 kinds of food emulsifiers today and 70% of them are MAG and DAG [1]. Bakery is the most food product that uses these two kinds of emulsifier.

Product glycerolysis contains many impurities such as unreacted glycerol, unreacted triglyceride, and catalyst. This process produces a lot of excess glycerol. Hui [2] described that the finished product generally contains 35 -50% mono-glycerides and the rest are mostly di-glycerides, some unreacted triglycerides (10%), residual glycerol (3-4%), and free fatty acid (1-3%). Purification process to get a product that contains a high purity of mono-glycerides in the food industry is needed. Mono-glycerides have better emulsifying properties than a mixture of different acylglycerols. Some methods to remove excess glycerol and impurities in glycerolysis product were found by the writers.

Some of them are a vacuum or molecular distillation [3 – 7], column chromatographic purification [8], decantation [4], and thin layer chromatographic purification [3]. Distillation process is an energy intensive process [5] and purification process by decanting it takes a time. Some kinds of organic solvent can dissolve oil or fat. Some of them are benzene, methanol, ethanol, isopropanol, diethyl ether, carbon tetra chloride, and trichloroethylene. Hexane is non-polar solvent. It can dissolve triglyceride and free fatty acid. Hexane has been used for decades to extract oil from oilseeds with massive construction and operational costs [9]. However, information of the use of hexane to separate excess glycerol from glycerolysis product is rarely found. The information about impact of the use of hexane information on the characteristics of the glycerolysis product is also rarely. The aim of this research is to get the hexane ratio to the glycerolysis product that produce product that has the best characteristics as an emulsifier.

2. Methodology

Main materials used in this study were palm oil, glycerol, and solvent. Chemicals used for analysis were chloroform, iodine, glacial acetic acid, potassium iodide, $Na_2S_2O_3$, amylum, KOH, HCl, ethanol, phenolphthalein, $K_2Cr_2O_7$, NaOH, alcohol 95%, and ethanol. Equipment including four-neck flask, condenser, rotary evaporator, vacuum pump, hotplate stirrer, Du Nuoy tensiometer, analytical balance, separatory flask, heater bath, desiccators, beaker glass, Buchner funnel, Erlenmeyer flask, burette, separator funnel, and other glass equipment were used. The study was done in five stages, namely 1) analysis of palm oil physicochemical properties, 2) glycerolysis of palm oil, 3) extraction of monoglyceride and dyglyceride by using hexane in various concentrations, 4) analysis of physicochemical properties of extracted products, and 5) data analysis.

2.1 Analysis of palm oil physicochemical properties

Palm oil was analyzed for its physicochemical properties including acid value and water content. The acid value was analyzed volumetrically by 0.1 N NaOH as titrant and phenolphthalein as an indicator. The water content was analysed using oven method at 105 °C for 1 hour until a constant weight.

2.2 Glycerolysis of palm oil

Glycerolysis process was conducted for two hours by using KOH as catalyst. The glycerolysis was done by using solvent with a glycerol/palm oil ratio of 3.5 in 80°C.

2.3 Extraction of monoglyceride and dyglyceride by using hexane in various concentrations

Extraction process was done by using hexane with hexane/glycerolysis product ratios of 1, 2, and 3. The process was conducted at room temperature and was continued by separating extracted product from residue of glycerol by using centrifugation.

2.4 Analysis of physicochemical properties of extracted products

Products resulted from glycerolysis were tested for their ability to lower the surface tension of water by using DuNouy tensiometer and free glycerol content [10].

2.5 Data analysis

Data were subjected to an analysis of variance by using Microsoft Excel application software.

3. Results and Discussion

3.1 Acidity and water content of Palm oil

Ma and Hanna [11] noted that free fatty acid and water can inactivate the alkali catalyst. They will consume the catalyst and reduce catalyst efficiency by causing soap formation. Atadashi *et al.* [12] noted the same thing that water from feedstock or produced during course of reaction may lead to deactivation of the alkali catalyst by hydration. Presence of water also speeds up hydrolysis of triglycerides and increases FFAs content in vegetable oils [12]. Kombe *et al.* [13] and Ding *et al.* [14] also noted those free fatty acids are not easily converted by homogeneous base trans esterification, because of the concurrent soap formation of the free fatty acids with the catalyst. The free fatty acids of 0.05% or 0.11 mg KOH/g sample in oil can inactivate catalyst as much as 0.1 kg/ton oil. Meanwhile, 0.01% of water in oil will inactivate 0.3 kg catalyst/ton oil. The acid value of palm oil was 1.31 mg KOH/g sample. Meanwhile, the water content of palm oil was 0.1%. The palm oil had acid value and water content which will bother activity of KOH as catalyst. High acid value of palm oil can lead the formation of soap as a result of the reaction between free fatty acids and KOH which will cause emulsion. The emulsion makes the product hard to separate. It was the reason why the search method for separating excess glycerol from the product needs to be done.

3.2 Effects of Concentration of Hexane on the Yield of Extracted Products

Oil or fat are soluble in organic solvent. Extraction using solvent is the standard technique. It is widely used technique because it is simple and easy to run. In this study, the yields of extracted products were from 32% to 28.25% (Figure 1). The highest yield was obtained at glycerolysis to solvent ratio of 1:3. We can see from the figure that the product yield increases with increasing concentration of solvent. Ratio of solvent to solid is one of important parameter for extraction process [15]. Dasari and Goud [15] said that extraction depends on the nature of the solvent and oil, contact time of sample with solvent, extraction temperature, particle size and solvent ratio. Increasing solvent to glycerol ratio will increase the yield. It is due to the concentration gradient becomes greater which favors good mass transfer [16].

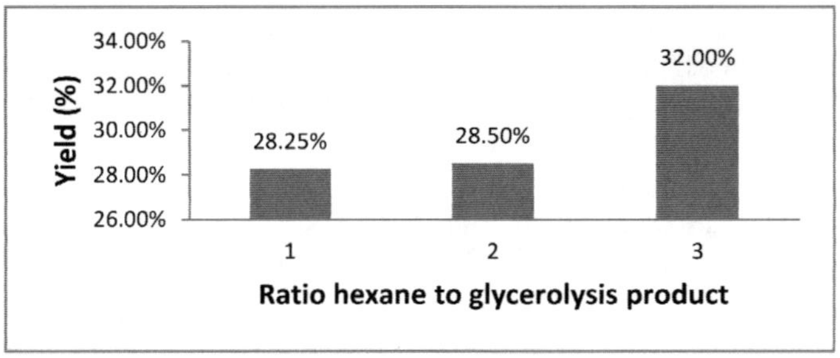

Figure 1. Relationship between ratio hexane to glycerolysis product on the yield of extracted products

Results showed that the concentration in this research did not significantly (confidence level 95%) affect the yield of extracted products. The extracted products which were extracted by using hexane with hexane/glycerolysis product ratio of 3 produced the highest yield.

3.3 Effects of Concentration of Hexane on the Ability of Extracted Products to Lower Water Surface Tension

Glycerolysis process was done as the research result of Kumoro [17]. In this study, the glycerolysis was conducted by using KOH as catalyst and isopropyl alcohol solvent at 80°C for two hours. The products were then extracted by using hexane with hexane/glycerolysis product ratios of 1, 2, and 3. In this study, the extracted products could lower water surface tension from 72 to 31.8-33.75 dynes/cm or 53.125%-55.764% (Figure 2). This proved that the products had surface active properties. Active groups contained in the products were able to lower water surface tension.

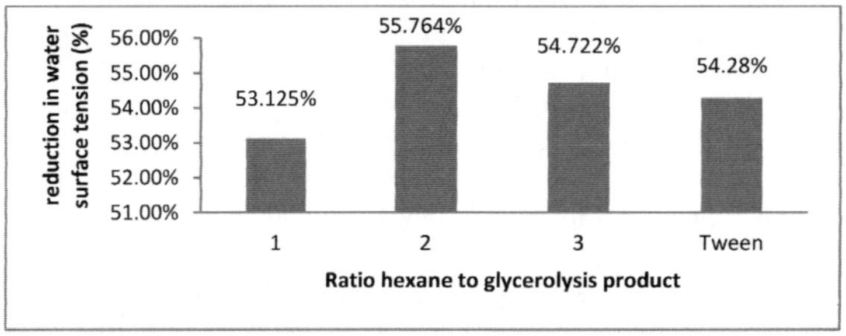

Figure 2. Relationship between ratio hexane to glycerolysis product on the ability of extracted products to lower water surface tension

Results showed that the concentration did not significantly (confidence level 95%) affect the ability of extracted products to lower water surface tension. The extracted products which were extracted by using hexane with hexane/glycerolysis product ratios of 2 and 3 had better ability to lower water surface tension than commercial emulsifier (tween).

3.4 Effects of Concentration of Hexane on the Free Glycerol Content of Extracted Products

Allowed free glycerol content in commercial glycerolysis products is no more than 7% [4]. Meanwhile, Hui [2] described that the finished product from glycerolysis process generally contains 3 – 4 % residual glycerol. Chetpattananondh and Tongurai [6] found that the impurities in monoglycerides product will reduce its emulsifying properties. Extracted products in this study contained free glycerol at a level of 2.40-3.05%. This indicated that the amount of glycerol that does not bind to fatty acid or residual glycerol which is still contained in the product was fairly low.

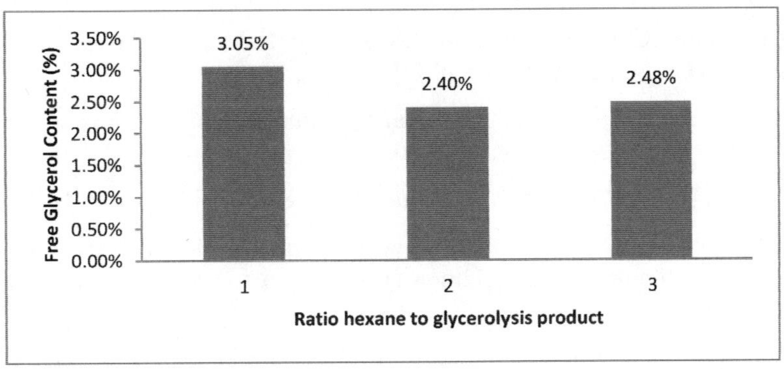

Figure 3. Relationship between ratio hexane to glycerolysis product to the free glycerol content of extracted products

Results showed that the concentration of hexane did not significantly (confidence level 95%) affect free glycerol content of extracted products. However, we can see from the figure that the free glycerol content of the product decreases with increasing concentration of hexane. Increasing solvent concentration will increase mass transfer [18]. It is also will increase contact area between products and solvent.

4. Conclusion

Hexane can extract monoglycerida and diglyceride from glycerolysis product. Concentration of hexane did not significantly (confidence level 95%) affect the yield and the free glycerol content of extracted products. It also did not affect the ability of extracted products to lower water surface tension. The extracted products was extracted with hexane/glycerolysis ratio of 2 had the best characteristics.

5. Reference

[1] Galúcio C S, Souza R A, Stahl M A, Sbaite P, Benites C I and Maciel M R W 2011 Physicochemical characterization of monoacylglycerols from sunflower oil *Procedia Food Science* **1** 1459 – 1464

[2] Hui Y H 1996 *Bailey industrial oil and fat products* vol 4 (New York, USA : John Wiley & Sons, Inc) p 569-585

[3] Kaewthong W, Sirisansaneeyakul S, Prasertsan P, and Kittikun A H 2005 Continuous Production Of Monoacylglycerols By Glycerolysis Of Palm Olein With Immobilized Lipase *Process Biochemistry* **40** 1525-1530

[4] Prakoso T and Sakanti M M 2007 Producing Monoglyceride *Journal of Teknik Kimia Indonesia* **6** 689 – 697

[5] Chew Y H, Chua L S, Cheng K K, Sarmidi M R, Aziz R A, and Lee C T 2008 Kinetic Study On The Hydrolysis Of Palm Olein Using Immobilized Lipase *Biochemical Engineering Journal* **39** 516–520

[6] Chetpattananondh P and Tongurai C 2008 Synthesis of high purity monoglycerides from crude glycerol and palm stearin *J. Sci. Technol.* **30** 515 – 521

[7] Wibowo T Y, Abdullah A Z, and Zakaria R 2011 Effect of Calcination Temperature and Tmaoh on Catalytic Activity of Basic Clay *Journal of Applied Science* **11** 3619-3624

[8] Yang Y C, Vali S R, and Ju Y H 2003 A process for synthesizing high purity monoglyceride *Journal of the Chinese Institute of Chemical Engineering* **34** 617-623

[9] Shende D and Sidhu G K 2014 Methods Used For Extraction Of Maize (Zea Mays, L.) Germ Oil-A Review *Ind. J. Sci. Res. and Tech.* **2** 48-54

[10] Hambali E and Mujdalipah S 2008 *Biodiesel Analysis* (Bogor: Surfactant and Bioenergy Research Center) p 11 – 18.

[11] Ma F and Hanna M A 1999 Biodiesel production: a review Bioresource Technology **70** 1 – 15

[12] Atadashi I M, Aroua M K, Aziz A R A, and Sulaiman N M N 2012 The effects of water on biodiesel production and refining technologies: A review Renewable and Sustainable Energy Reviews **16** 3456– 3470

[13] Kombe G G, Temu A K, Rajabu H M, Mrema G D, Kansedo J, and Teong Lee K T 2013 Pre-Treatment of High Free Fatty Acids Oils by Chemical Re-Esterification for Biodiesel Production—A Review *Advances in Chemical Engineering and Science* **3** 242-247

[14] Ding J, He B, and Li J 2011 Biodiesel Production from Acidified Oils via Supercritical Methanol *Energies* **4** 2212-2223

[15] Dasari S R and Goud V V 2013 Comparative Extraction of Castor seed Oil Using Polar and Non polar Solvents *International Journal of Current Engineering and Technology* **3** 121 – 123.

[16] Saxena D, Sharma S K S and Sambi S S 2011 Comparative Extraction of Cottonseed Oil by n-Hexane and Ethanol *ARPN Journal of Engineering and Applied Sciences* **6** 84 – 89.

[17] Kumoro A C 2012 Experimental and Modeling Studies of the Reaction Kinetics of Alkaline-Catalyzed used Frying Oil Glycerolysis using Isopropyl Alcohol as a Reaction Solvent *Research Journal of Applied Sciences, Engineering and Technology* **4** 869-876

[18] Sayyar S, Abidin Z Z, Yunus R, and Muhammad A 2009 Extraction of Oil from Jatropha Seeds-Optimization and Kinetics American *Journal of Applied Sciences* **6**1390-1395

IOP Conf. Series: Materials Science and Engineering **128** (2016) 012026 doi:10.1088/1757-899X/128/1/012026

Liquid volume monitoring based on ultrasonic sensor and Arduino microcontroller

M Husni*, D O Siahaan, H T Ciptaningtyas, H Studiawan and Y P Aliarham

Department of Informatics, Faculty of Information Technology,
Institut Teknologi Sepuluh Nopember, Indonesia

*Corresponding author: husni@its-sby.edu

Abstract. Incident of oil leakage and theft in oil tank often happens. To prevent it, the liquid volume insides the tank needs to be monitored continuously. Aim of the study is to calculate the liquid volume inside oil tank on any road condition and send the volume data and location data to the user. This research use some ultrasonic sensors (to monitor the fluid height), Bluetooth modules (to sent data from the sensors to the Arduino microcontroller), Arduino Microcontroller (to calculate the liquid volume), and also GPS / GPRS / GSM Shield module (to get location of vehicle and sent the data to the Server). The experimental results show that the accuracy rate of monitoring liquid volume inside tanker while the vehicle is in the flat road is 99.33% and the one while the vehicle is in the road with elevation angle is 84%. Thus, this system can be used to monitor the tanker position and the liquid volume in any road position continuously via web application to prevent illegal theft.

1. Introduction

Oil theft by taking small amount of the liquid inside the tank often happen, the latest case happened in Riau. The incident causes loss about IDR 1.3 trillion. The oil theft is done by creating small hole in the tank so the fluid can leak and distributed to the other party. This illegal act caused loss to the company (Pertamina) and also Indonesia [1].

To overcome this problem, Pertamina uses ATG (*Automatic Tanks Gauge*) but it only calculates the liquid volume in the gas station not in the oil tank[1]. To calculate the liquid volume inside tank, fluid velocity sensor which is embedded in the oil tap can be used. This method can be implemented if the leakage comes from the oil tap [1]. But, if it comes from the other hole, this method won't success.

The other method is using fuel level sensor inside the oil tanker and GPS sensor to monitor the oil tanker [2][3]. The monitoring result is already well presented by using maps, but, the fuel sensor will fail if the oil tanker is on the tilted road. To overcome the disadvantage, fuel level sensor mounted on the fuel tanker with arm on floater can be used [4]. But this approach can have less accuracy since it only has 1 sensor. It also only send the OK data in a plain teks so it's hard for the user to monitor the location and the fuel volume.

Aim of this study is to calculate the liquid volume inside oil tank on any road condition and send the volume data and location data to the user. This research contribution is the system can calculate the fuel volume even in a tilted road with more accurate than previous study. The advantage of this system is user doesn't have to request the data. The system will send the data regularly to the server, save it on the database, and displayed the record in the user's browser in a visual way.

This system uses ultrasonic sensors to monitor the fluid height, Bluetooth modules to sent data from the sensors to the Arduino microcontroller, Arduino microcontroller to calculate the liquid volume, GPS / GPRS / GSM Shield module to get location of vehicle and sent it to the Server. From the server, the data will be processed and the output is displayed in a web browser. It will display a map and show the location of the oil tank and the liquid volume.

2. Literature Study

The literature study explains the object used for this research.

2.1. Ultrasonic Sensor

Ultrasonic sensor is electronic device that change the electrical energy into mechanical energy in the form of ultrasonic sound waves. Ultrasonic sensors are most commonly used due to its simplicity and low cost. The sensor consists of a series of ultrasonic transmitter and receiver. The accuracy of the measured distance is dependent on the separation between the ultrasonic transmitter and receiver [5].

Example of ultrasonic usage is distance measurement of an object in the path of a person, equipment, or a vehicle, stationary or moving is used in a large number of applications such as robotic movement control, vehicle control, blind man's walking stick, medical applications, binary gas mixture fraction, etc [6][7][8].

This tool is used to measure the ultrasonic waves. Ultrasonic waves are mechanical waves which have the characteristic longitudinal and typically have a frequency above 20 kHz. Ultrasonic waves can propagate through the solid, liquid or gas. Ultrasonic waves are waves of energy propagation and mechanical momentum that propagate through the third element as the interaction with the molecule and the nature of the medium inertia path [9].

2.2. Arduino

The Arduino integrated development environment (IDE) is a cross-platform application written in Java. Arduino programs are written in C or C++ [10]. The Arduino Uno Boards are uploaded with specific programs that enable them to perform their required operation [11]. Arduino is flexible (offers a variety of inputs, interface, and output), inexpensive, and can communicate with software running on your computer [12].

2.3. GPS

GPS (Global Positioning System) is a radio-based navigation system that provides information about the coordinates of the position, velocity, and time to users worldwide. The use of GPS satellites service is not charged. Users only need a GPS receiver to determine the coordinates of the location. The accuracy of location coordinates depending on the type of GPS receiver. GPS consists of three parts: satellites orbiting the Earth (The GPS satellites orbit the earth twice a day), control and monitoring stations on Earth, and the GPS receiver [13]. Example of GPS implementations are: land vehicle navigation [14], human activity [15], and tracking system [16][17].

GPS output data format is defined by the NMEA (National Marine Electronics Association). Example of GPS data is '$ GPGGA, 061 648, 0.276513, S, 112.791692, E, 1, 09, 0.8, 70.0, M, 1.5, M,, * 5E'. The explanation of the code can be seen in table 1 [18].

Table 1. GPS Data Description

GPS output	Description	GPS output	Description
'GPGGA'	Sentence identifier	'1'	position (0 = invalid, 1 = valid)
'061 648'	Time	'09'	Number of satellites
'0.276513'	Latitude	'0.8'	Precision horizontally
'S'	Indicator (S = South, N = North)	'70 M'	Altitude
'112.791692'	Longitude	'1.5 M'	Height
'E'	Indicator (E = East, W = West)	'5E'	Checksum

2.4. GPS/GPRS/GSM Shield

GPS / GSM / GPRS Shield version 3.0 is a product of DF Robot. The shield supports Quad-band GSM / GPRS frequency at 850 MHz, 900 MHz, 1800 MHz, and 1900 MHz. The shield is also equipped with a GPS antenna for satellite navigation purposes which enable system to send data location over GSM networks [19][20].

This shield needs specific command called AT command. Examples of AT commands which used in this system can be seen in table 2 [21][22][23].

Table 2. AT Command on GPS/GPRS/GSM Shield V3.0

AT Command	Description	AT Command	Description
AT+CGPSPWR	turn on the GPS	AT+HTTPINIT	HTTP mode Initiation
AT+CGPSRST	Set the GPS Mode	AT+HTTPTERM	HTTP mode Termination
AT+CGPSINF	Obtain location information	AT+HTTPPARAM	Set the URL to be visited
AT+CGREG=?	Checking the network status		

2.5. Bluetooth Module HC 05 & HC 06

Bluetooth technology has been considered as a cheap, reliable, and power efficient replacement of cables for connecting electronic devices in a short range [24][25]. Bluetooth is a great protocol for wireless communication because it's capable of transmitting data at nearly 1MB/s, while consuming 1/100th of the power of Wi-Fi [26].

HC 05 and HC 06 is a Bluetooth module that is compatible with Arduino microcontroller. This module serves as a plug in on Arduino. It act as the media data recipient or sender through the air (wireless). Both modules use the AT command. Examples of AT commands which used in this system can be seen in Table 3 [27].

Table 3. AT Command on Bluetooth HC 05 and HC 06

AT Command	Description	HC 05	HC 06
AT + RESET	Reset module	√	-
AT + RMAAD	Disconnect the connected device	√	-
AT + INIT	Initialize module	√	-
AT + LINK	Connecting to other devices with the MAC address parameter	√	-
AT + ROLE	Set mode (slave / master)	√	-
AT + NAME	Rename the device	√	√

3. Methodology

This research creates a monitoring system of liquid volume inside the fuel tank by combining some components: Ultrasonic sensors, Arduino microcontroller, Bluetooth module and GSM/GPRS/GPS Shield v3.0. These components are designed to form a simple wireless network for obtaining liquid volume data in the tank and get the location of the tank. The system sends the data to the server, processed by the server, and displayed via web-based application.

To monitor an oil tank, the vehicle, capacity, the route, and the liquid volume should be registered before it leaves the base. As long as the vehicle has not reached the destination, the components which are embedded in the vehicle will continuously send liquid volume data and vehicle location data every 10 seconds. In the web application, the vehicle icon will continuously move according the coordinates of the monitored vehicle. The vehicle icons contained information on the percentage of the volume of the tank on the monitored vehicle. This system uses leaflet.js API for drawing the map and the vehicle in the browser.

3.1. *Workflow Process*

There are 3 workflow processes that occur in the system, the first is the process flow of the child node (figure 1), the second is the process flow of the centre node (figure 2), and the third is a process flow of the monitoring application (figure 3).

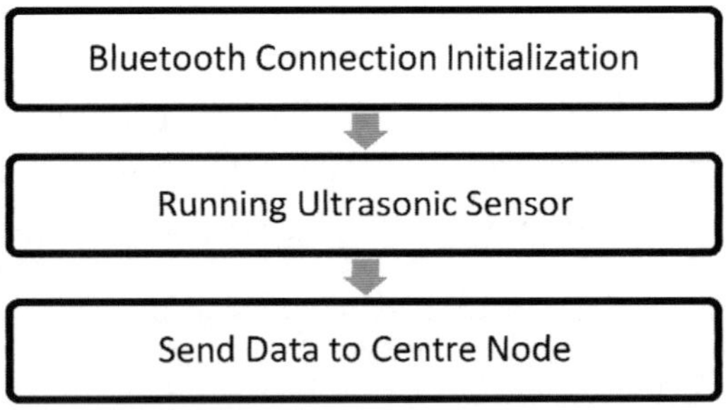

Figure 1. Child node process

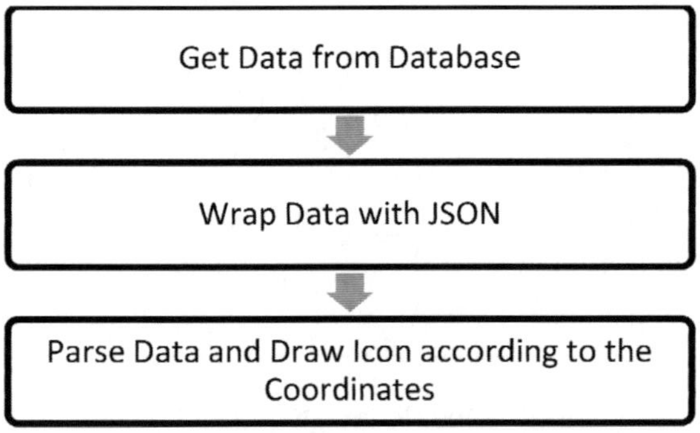

Figure 2. Monitoring Process

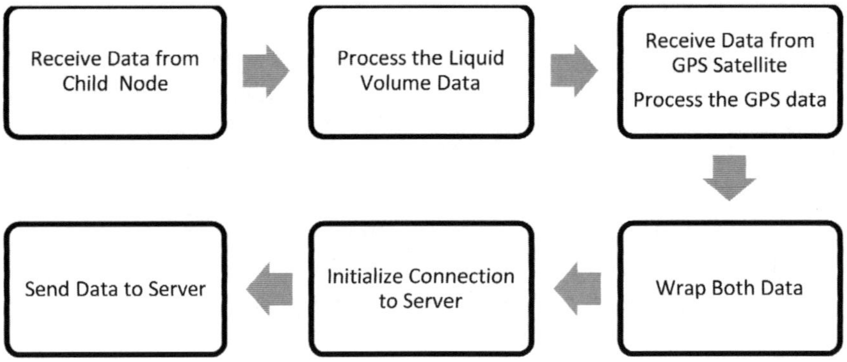

Figure 3. Centre node process

3.2. *System Architecture*

The ultrasonic sensor used to capture the depth of the liquid in the tank. To increase the data validity, this system uses two ultrasonic sensors. Create integration system between those sensors is necessary, so we set up a small network wireless using Bluetooth module as a medium of communication between nodes. An additional GPS/GSM/GPRS Shield module is needed to capture location data and send it to the server. The system architecture in figure 4 shows that there are two child nodes to capture liquid level data in the tank. While the centre node assigned to receive sensor data from child node, receives location data from the GPS, and transmits the data to a server. Users can monitor the liquid volume via web-based application, so it can be viewed anywhere and anytime. Overview of the monitoring process can be seen in figure 5.

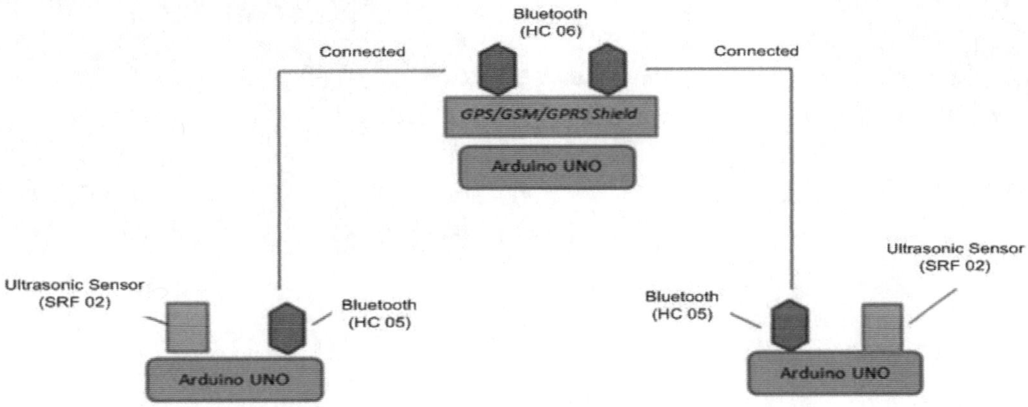

Figure 4. System Architecture

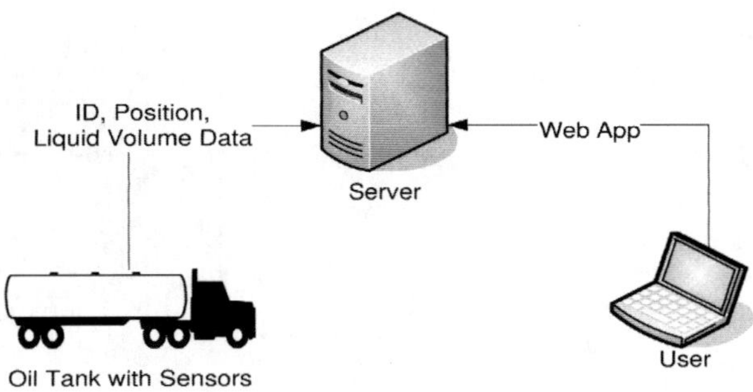

Figure 5. Monitoring Process

4. Implementation

Child node consists of Ultrasonic sensor, Bluetooth module (HC 05), and Arduino microcontroller. Centre node consists of Bluetooth module (HC 06), GPS/GSM/GPRS Shield module, and Arduino microcontroller. Implementation of components child node and centre node can be seen in Figure 6 and Figure 7 for the diagram (figure 6.a and figure 7.a) and also the implementation (figure 6.b and figure 7.b). Server specification has Intel Core i3 processor, 4 GB RAM, 64-bit Windows Operation System. The tank prototype is a box container with height x width x length = 48.5 x 32.7 x 65.4 cm and the liquid height = 15 cm.

IOP Conf. Series: Materials Science and Engineering **128** (2016) 012026 doi:10.1088/1757-899X/128/1/012026

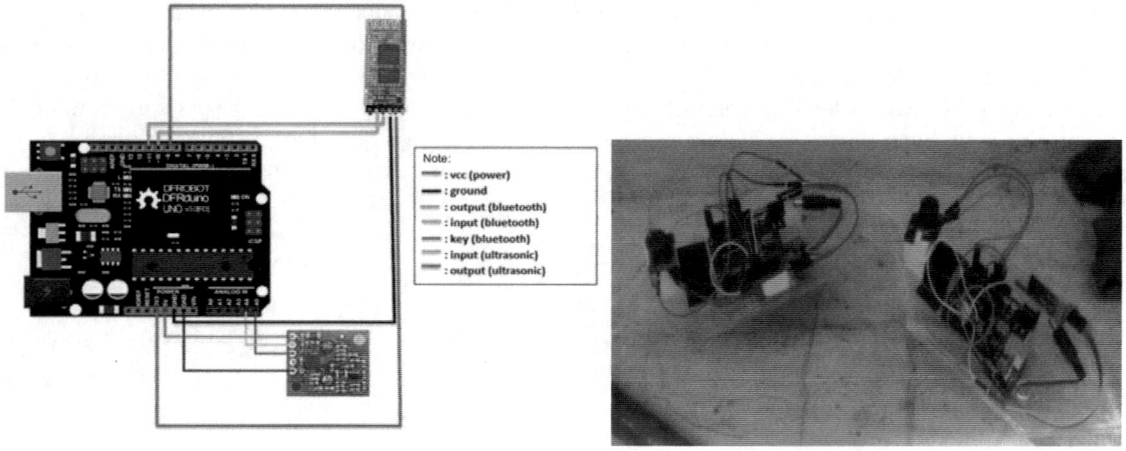

(a) (b)

Figure 6. Child Node: (a) Diagram; (b) Implementation

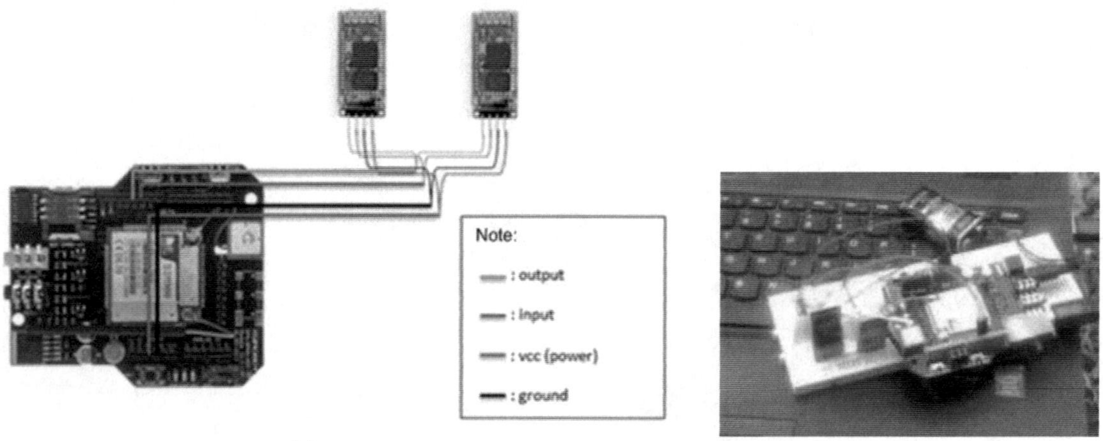

(a) (b)

Figure 7. Centre Node: (a) Diagram; (b) Implementation

5. Experimental Result

To determine the accuracy level of sensor components when detecting liquid volume in the tank, we create three test scenarios. Those tests are performed on a prototype tank contains water. The last scenario is to check the web based application.

5.1. First Scenario

The first testing scenario is to check the component accuracy under normal circumstances. The components and the tank prototype can be seen in figure 8. This scenario is performed for 70 seconds, and managed to send 5 data to the server. Graph accuracy rate sensor in the first scenario can be seen in figure 9. The blue line shows real volume of the liquid while the red one shows the detected volume. The accuracy rate of the system in normal circumstances is 99.33%. It is almost the same with previous study [2][3].

Figure 8. First Scenario

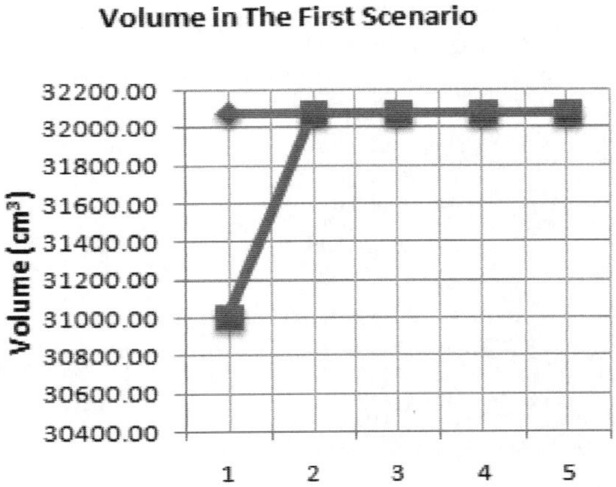

Figure 9. Liquid volume in the First Scenario

5.2. *Second Scenario*

The second testing scenario is to check the component accuracy when the tank is tilted 10^0. The components and the tank prototype can be seen in figure 10. This scenario is performed for 75 seconds, and managed to send 5 data to the server. Graph accuracy rate sensor in the second scenario can be seen in figure 11. The blue line shows real volume of the liquid while the red one shows the detected volume. The accuracy rate of the system in tilted circumstances is 84%.

Figure 10. Second Scenario

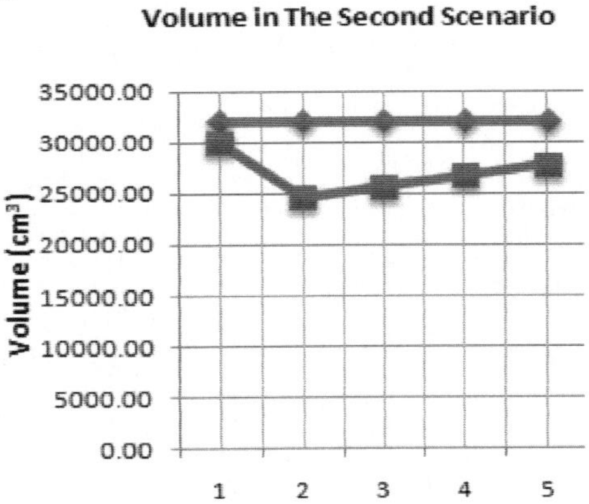

Figure 11. Liquid volume in the Second Scenario

5.3. *Third Scenario*

The third testing scenario is to check the component accuracy when the tank leakage happens in a flat road. The components and the tank prototype can be seen in figure 12. This scenario is managed to send 10 data to the server. Graph of the leakage in the third scenario can be seen in figure 13. The blue line shows the liquid height. When the liquid height decreases gradually over a certain period in different locations, it can be suspected as leakage or illegal theft.

Figure 12. Third Scenario

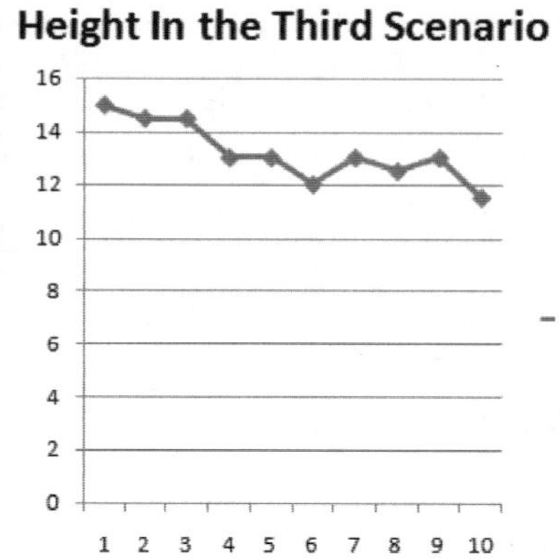

Figure 13. Liquid height in the Third Scenario

5.4. Fourth Scenario

The fourth testing scenario is to check the registration process and output of monitoring system in a web based application. The result can be seen in figure 14 and figure 15. The interface for monitoring fuel inside is better than just text based [4] and give more details than the other research [2][3].

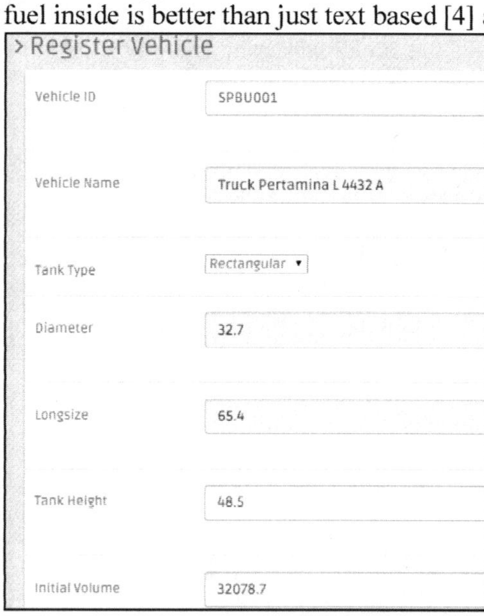

Figure 14. Registration Process

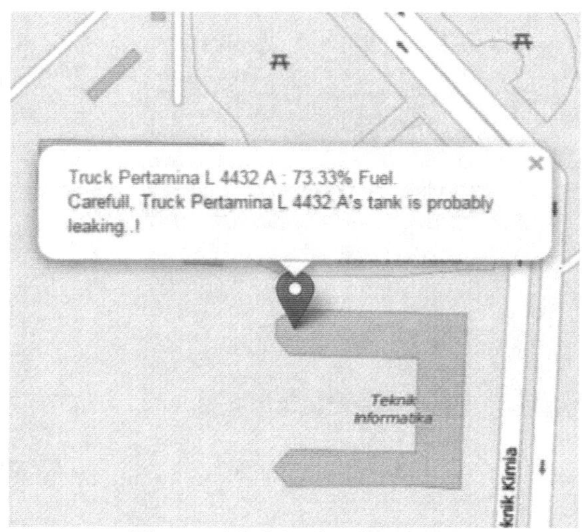

Figure 15. Monitoring System

6. Conclusion

From the experimental result, we can draw conclusions as follows:

1. This system that consists of ultrasonic sensors, Arduino microcontroller, GSM/GPRS/GPS Shield modules and Bluetooth module can monitor the liquid volume inside tank and detect the leakage via web based application continuously.
2. The road condition (flat or tilt) is affecting the accuracy rate of the sensor. In the normal road the accuracy rate is 99.33% while in tilted road the accuracy rate drop until 84%. This system can be implemented in any road condition.

Acknowledgments
This research was supported by PUPT Lab Research Grant 2015 at Institut Teknologi Sepuluh Nopember (ITS), Surabaya with Research ID 003246.45/IT2.11/PN.08/2015.

References

[1] Husni M, Djanali S, Ciptaningtyas H T, Wicaksono H 2014 *Aplikasi Pemantauan Pendistribusian Bahan Bakar Minyak Menggunakan GPS pada Perangkat Android dan Sensor Kecepatan Air Berbasis Mikrokontroler Arduino* Proc. Seminar Nasional Teknologi Terapan - SNTT 2014 (Yogyakarta, Indonesia, 15 November 2014) ISBN: 978-602-1159-06-4.

[2] Sachin S A and Kokate R D 2012 *Fuel Monitoring and Vehicle Tracking Using GPS, GSM, and MSP430F149* J. International Journal of Advances in Engineering & Technology – IJAET vol. 4, Issue 1, pp 571-578.

[3] Daniel O G, Dayo O and Anne O O 2011 *Monitoring And Controlling Fuel Level Of Remote Tanks Using Aplicom 12 GSM Module* J. Asian Research Publishing Network – ARPN vol. 6, No. 1, January 2011 pp 56-60.

[4] Obikoya G D 2014 *Design, construction, and implementation of a remote fuel-level monitoring system* J. EURASIP Journal on Wireless Communication and Engineering 2014:76.

[5] Shrivastava A K, Verma A and Singh S P 2009 *Effect Of Variation Of Separation Between The Ultrasonic Transmitter And Receiver On The Accuracy Of Distance Measurement* J. International Journal of Computer science & Information Technology (IJCSIT), Vol 1, No 2, November 2009 pp 19-28.

[6] Shrivastava A K, Verma A, and Singh S P 2010 *Distance Measurement of an Object or Obstacle by Ultrasound Sensors using P89C51RD2* J. International Journal of Computer Theory and Engineering, Vol. 2, No. 1 February, 2010 pp 64-68.

[7] Bates R, Battistin M, Berry S, et.al 2014 *Implementation of Ultrasonic Sensing for High Resolution Measurement of Binary Gas Mixture Fractions* J. Sensors, Vol. 14 pp 11260-11276.

[8] Tahir M B and Abdullah M 2012 *Distance Measuring (Hurdle detection System) for Safe Environment in Vehicles through Ultrasonic Rays* J. Global Journal of researches in engineering Automotive engineering Vol.12(1) Ver.1 February 2012 pp 15-21.

[9] Carullo A and Parvis M *An Ultrasonic Sensor for Distance Measurement in Automotive application* J. Sensor Vol. 1(2) August 2011 pp 145.

[10] Anusha D, Sarma P M, and SandhyaRani M N 2013 *Appliance Remote Control Using Arduino* J. International Journal of Latest Trends in Engineering and Technology (IJLTET) Vol. 2(4) July 2013 pp. 35-41.

[11] Purkayastha A, Prasad A D, Bora A, Et Al. 2014 *Hand Gestures Controlled Robotic Arm* J. Journal Of International Academic Research For Multidisciplinary Vol. 2(4) May 2014 pp. 234-240.

[12] Kumar P and Kumar P 2013 *Arduino Based Wireless Intrusion Detection Using IR Sensor and GSM* J. International Journal of Computer Science and Mobile Computing - IJCSMC Vol. 2(5) May 2013 pp 417-424.

[13] Ahamed S S R 2009 *Technological Strategy Of Using Global Positioning System: An Analysis* J. International Journal of Engineering Science and Technology Vol.1(1) 2009 pp 8-16.

[14] Abbot E and Powell D 1999 *Land-Vehicle Navigation Using GPS* Proc. IEEE Vol. 87 (1) January 1999 pp 145-162.

[15] Spek S V D, Schaick J V, Bois P D, et al. 2009 *Sensing Human Activity: GPS Tracking* J. Sensors No. 9 pp. 3033-3055.

[16] Al-Khedher M A 2011 *Hybrid GPS-GSM Localization of Automobile Tracking System* J. International Journal of Computer Science & Information Technology (IJCSIT) Vol 3 (6) December 2011 pp. 75-85.

[17] Verma P and Bhatia J S 2013 *Design And Development Of GPS-GSM Based Tracking System With Googlemap Based Monitoring* J. International Journal of Computer Science, Engineering and Applications (IJCSEA) Vol. 3(3) June 2013 pp. 33-40.

[18] DFRobot 2014 *DFRobot NMEA Reference Manual Rev 2.7 Dec07* http://www.dfrobot.com/image/data/TEL0051/3.0/NMEA%20Reference%20Manual-Rev2.1-Dec07.pdf Accessed December 2014.

[19] Selvapriya C, Sathya P S2, Abdulrahim M, et al. 2013 *LPG Leakage Monitoring and Multilevel Alerting System* J. International Journal Of Engineering Sciences & Research Technology - IJESRT Vol. 2(11) November 2013 pp. 3287-3290.

[20] Zaghloul M S 2014 *GSM-GPRS Arduino Shield (GS-001) with SIM 900 chip module in wireless data transmission system for data acquisition and control of power induction furnace* J. International Journal of Scientific & Engineering Research - IJSER Vol. 5(4) April 2014 pp. 776-780.

[21] Libellium *Cooking Hacks* https://www.cooking-hacks.com/projects/arduino-realtime-gps-gprs-vehicle-tracking Accessed March 2015.

[22] Shanghai SIM 2011 *SIM 908 AT Command Manual_V1.01.*

[23] DFRobot *GSM Module V3* http://www.dfrobot.com/wiki/index.php/GPS/GPRS/GSM_Module_V3.0_(SKU:TEL0051) Accessed December 2014.

[24] Minar N B I and Tarique M 2012 *Bluetooth Security Threats And Solutions: A Survey* J. International Journal of Distributed and Parallel Systems - IJDPS Vol. 3(1) January 2012 pp 127-148.

[25] Chanana A, Singh H, Miglani D, et al. 2012 *Research on Bluetooth Technology* J. International Journal of Innovative Technology and Exploring Engineering - IJITEE Vol. 1(6) November 2012 pp 36-41.

[26] Singh P, Sharma D, and Agrawal 2011 *A Modern Study of Bluetooth Wireless Technology* J. International Journal of Computer Science, Engineering and Information Technology - IJCSEIT Vol. 1(3) August 2011 pp 55-63.

[27] Instructables Team *AT Command Mode of HC05* http://www.instructables.com/id/AT-command-mode-of-HC-05-Bluetooth-module/ Accessed March 2015.

The material performance of HSS (high speed steel) tools and its relation with chemical composition and carbide distribution

B Darmawan*, M Kusman and R A Hamdani

Department of Mechanical Engineering Education, FPTK
Universitas Pendidikan Indonesia (UPI)

*Corresponding author: bamdarsap@gmail.com

Abstract. The study aims to compare the performance of two types of material HSS (High Speed Steel) are widely used. It also will be the chemical composition and distribution of carbide particles therein. Two types of HSS are available in the market: HSS from Germany (Bohler) and HSS from China. This research employed the pure experimental design. It consists of two stages. The first, aims to test/operate lathe machines to determine the lifetime and performance of tools based on specified wear criteria. The second, characterization of microstructure using SEM-EDS was conducted. Firstly, grinding of toolss was done so that the toolss could be used for cutting metal in the turning process. Grinding processes of the two types of toolss were done at the same geometry, that is side rake angle (12°-18°), angle of keenness (60°-68°), and side relief angle (10°-12°). Likewise, machining parameters were set in the same machining conditions. Based on the results of the tests, it is found that to reach 0.2 mm wear point, toolss made of HSS from Germany needed 24 minutes, while toolss made of HSS from China needed 8 minutes. Next, microstructure tests using SEM/EDS were done. The results of the SEM tests indicate that the carbide particles of HSS from Germany were more evenly distributed than the carbide particles of HSS from China. Carbide compounds identified in HSS from China were Cr23C6 and Fe4Mo2C. Oxide impurity of Al2O3 was also found in the material. On the other hand, in HSS from Germany, no impurity and other carbide compounds were identified, except Cr23C6 and Fe4Mo2C, also Fe4W2C, and VC or V4C3.

1. Introduction

In the process of forming and machining, cutting tools are used to change the shape of materials into particular desired shapes or certain products. The materials of the tools should ensure the continuity of the process of cutting/forming with optimum results. The materials used should have properties that are better or superior to the materials of workpieces to be formed into specific products [1]. The advantages of the tools can be achieved because the tools/toolss are made from materials which have certain characteristics, such as:

- Hardness: harder than the materials of workpieces, not only at room temperature but also at high temperature during machining process;
- Ductility: large enough to withstand shock loads occurring during the process of machining with termination/interruption or when cutting workpieces which have hard spots;
- Thermal shock load: required when significant temperature changes occur regularly;

- Low adhesive properties: to reduce the affinity of workpieces against the tools/toolss, reducing the wear rate;
- Low solubility of elements or materials of the tools/toolss: required to minimize the wear rate due to diffusion mechanism.

The material which is widely used because it has the characteristics described above is High Speed Steel (HSS) [2]. This material can easily be found in the market. It is affordable and can easily be sharpened when it undergoes the process of wear and tear. There are many manufacturers producing this type of cutting tool material. They offer various prices. The various prices reflect the quality of the material. There are different characteristics in terms of material performance if the material is used in machining processes. The performance of the material can be seen from the life of toolss/tools. It is interesting to explore the different characteristics of the material, which reflect different performances, by looking at the metallurgical features of the material through a micro approach/perspective by using SEM/EDS. The purposes of this study are:

- to get a description of the performance of various types of HSS material through machine testing with specified critical wear criteria;
- to get a description of chemical composition and the distribution of carbide as well as its size through the use of SEM/EDS.

2. Research Methods

The method used as shown in the following diagram:

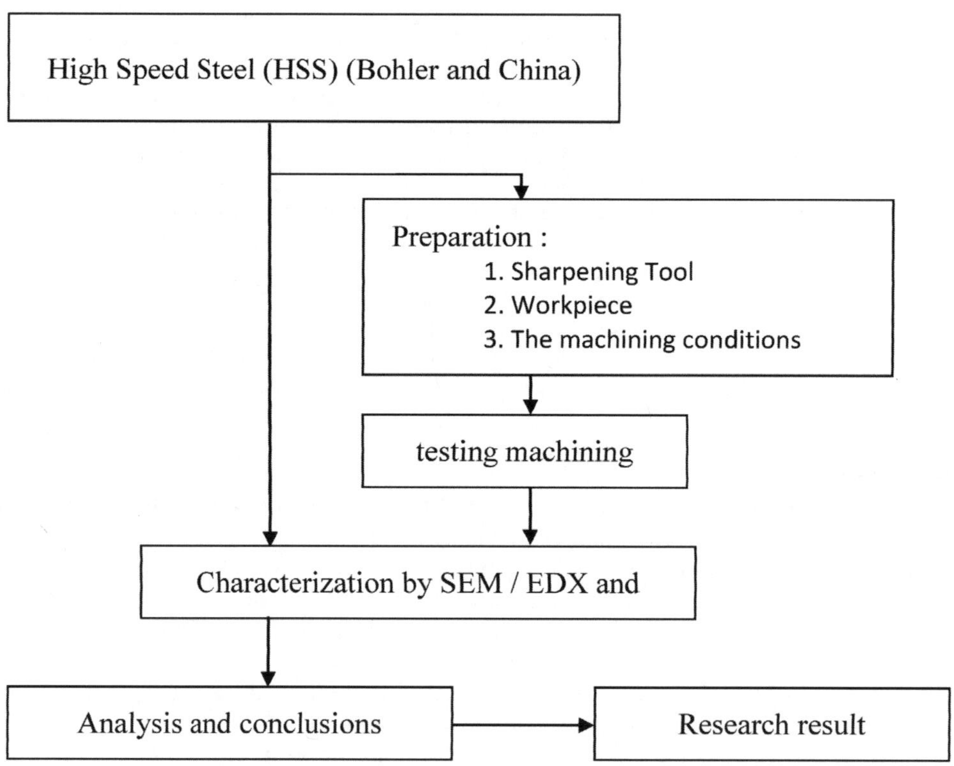

Figure 1. Method

IOP Conf. Series: Materials Science and Engineering **128** (2016) 012027 doi:10.1088/1757-899X/128/1/012027

3. Results and Discussions

3.1. Machine Testing

Machining processes to identify the life of cutting tools were done by way of stoppage; for a certain length of cut, the size of edge wear (VB) was measured to determine whether the tools reach the specified critical wear. The results of the machining processes can be seen in the following table:

Table 3. The wear of tools – HSS ½" Bohler Germany

No	Length (mm)	Duration (min)	Wear (mm)
1	400	4	0,05
2	400	4	0,05 - 0,1
3	400	4	0,1
4	400	4	0,1 - 0,15
5	400	4	0,15
6	400	4	0,15 - 0,2
Total	**2400**	**24**	**0,2**

Table 4. The wear of tools – HSS ½" China.

No	Length (mm)	Duration (min)	Wear (mm)
1	400	4	0,1
2	400	4	0,2
Total	**800**	**8**	**0,2**

From the comparison of tool life with critical wear VB = 0.2 mm, it is found that the life of the German HSS tools is longer (tc = 24 minutes), compared to that of the China HSS tools (tc = 8 minutes). The causes of this difference can be identified by looking at the metallurgical characteristics of materials of the toolss using SEM/EDS.

3.2. Examination using SEM/EDS

Examination using SEM/EDS was aimed to determine the microstructure and chemical composition of the materials, as well as to identify the distribution of carbide particles which commonly emerge from HSS tools materials [3]. Here are the results of the examination using SEM.

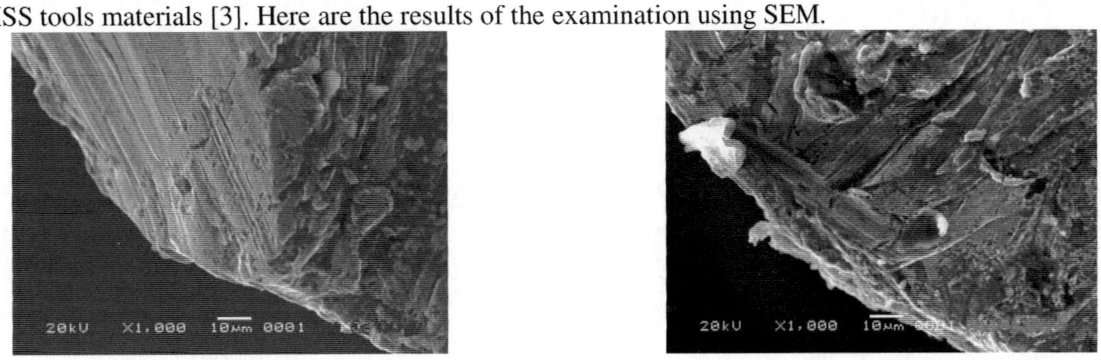

(a) China HSS **(b) German HSS**

Figure 2. Microstructures of the wear areas of HSS tools materials

Based on the two images, it can be explained qualitatively that the spread of carbide grains in German HSS is more even that that of China HSS. Furthermore, the grains of German HSS are finer. This differences affect the life of the toolss. Carbide is a very hard compound. If it is spread evenly, the hardness will also be evenly distributed, so that the tool life could be longer.

IOP Conf. Series: Materials Science and Engineering **128** (2016) 012027 doi:10.1088/1757-899X/128/1/012027

Meanwhile, in accordance with the theory of HallPecth, the small size of carbide grains will increase the strength of the material. This explains why German HSS toolss have longer service life. The following images indicate the microstructure of China HSS and German HSS.

Figure 3. The microstructure of China HSS

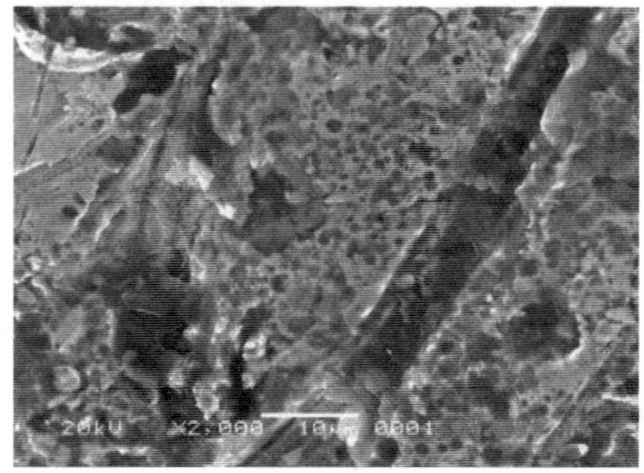

Figure 4. The microstructure of German HSS

3.3. Chemical composition analysis using EDS

The purpose of this analysis was to determine the chemical compositions of the two types of HSS. The technique used was Energy Dispersive Spectroscopy (EDS). The chemical composition analysis was applied to the same areas, which were around the wear of the toolss and distant form the wear. The morphological image and chemical compositions of the two toolss are presented in the following figure.

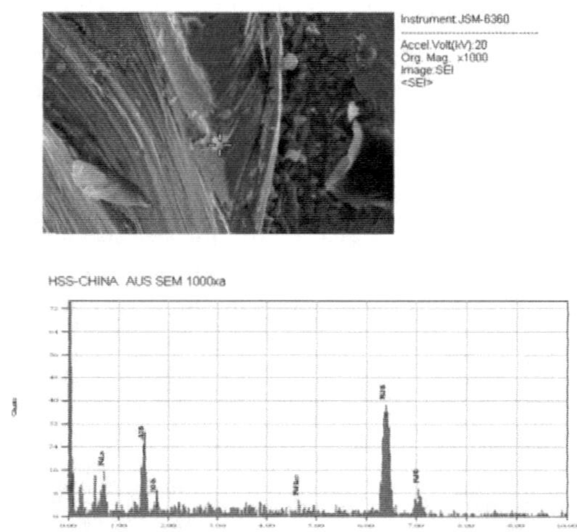

Element		(keV)	mass%	Error%	At%	Compound		mass%	Cation	K
O			27.82							
Al K		1.486	8.37	6.35	13.07	Al2O3	17.71	4.78	6.9448	
Si K	*	1.739	1.76	5.86	4.70	SiO2	3.76	0.86	1.5608	
Fe K		6.398	61.05	9.71	82.23	FeO	78.54	15.08	91.4943	
Total			100.00		100.00		100.00	20.74		

Figure 5. Chemical composition analysis of China HSS outside wear area

Based on the above figure and with reference to the literature on the formation pattern of compounds as a result of an element, it can be estimated that there is an element of Aluminium (Al) around the wear area of the China HSS tools. The Al allowed the oxidation of Al2O3, which is a detrimental impurity (able to decrease the material strength).

It is predicted that the Al element was from the walls of the melting furnaces (refractory) during the making of HSS. While Chromium (Cr) and Molybdenum (Mo) can form carbide compound which can increase the material strength. The pattern of carbide compound from Cr is M23C6, in accordance with the references, it can be predicted that carbide compound from Cr is Cr23C6. While the pattern of carbide compound from Mo is Fe4Mo2C.

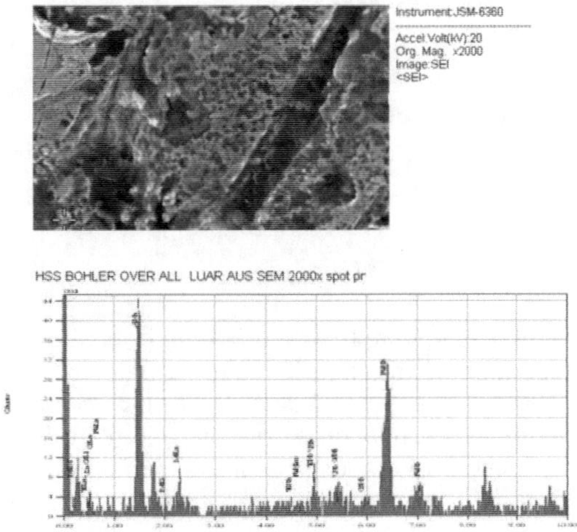

Element		(keV)	mass%	Error%	At%	Compound	mass%	Cation	K
Al K		1.488	26.92	4.44	44.20			17.3091	
Ti K	*								
V K		4.949	5.89	7.54	5.13			6.9856	
Cr K		5.411	2.41	8.16	2.05			3.0582	
Fe K		6.398	56.45	11.14	44.78			65.6649	
Mo L		2.293	8.33	10.78	3.85			6.9822	
Total			100.00		100.00				

Figure 6. Chemical composition analysis of German HSS outside wear area

Different from the results of the chemical composition analysis of China HSS, German HSS contains other elements such as Vanadium (V) and Cobalt (Co), besides Chromium (Cr) and Molibdenum (Mo). Such elements will form carbides in accordance with the patters of carbide compound: VC or V4C3. The addition of Cobalt (Co) to the HSS will increase the hardness of the material at high temperature and tenacity. The grains become finer; thus, the sharp end can be maintained during the heat treatment process at high temperatures.

4. Conclusions

Based on the results of the performance testing of the toolss, it can be concluded that:

- German HSS toolss have longer lifetime compared to China HSS toolss under the same machining parameters and critical wear criteria;
- The difference of lifetime, which reflect their endurance, is due to the different chemical compositions of the materials. Both German HSS toolss and China HSS toolss contain carbide compund; however, they have diffent compositions.

References

[1] Rochim T, 1993, Teori & Teknologi Proses Pemesinan, Proyek HEDS
[2] De Garmo E Paul, 1984, Material and processes in Manufacturing, MacMillan Publishing Company.
[3] Goldstein JI and Yakowiz H, P, 1977, Practical Scanning Electron Microscopy, Plenum Press, New York.

Enhancing the performance of the domestic refrigerator with hot gas injection to suction line

E T Berman*, S Hasan and Mutaufiq

Department Pendidikan Teknik Mesin, FPTK
Universitas Pendidikan Indonesia (UPI)

*Corresponding author: egatb@upi.edu

Abstract. The purpose of this study was to determine the increase in performance of a domestic refrigerator that uses hot gas injection (IHG) to the suction line. The experiment was conducted by flowing refrigerant from the discharge line to the suction line. To get performance data, measurements performed on the liquid brine as cooling load with various temperatures (range from 3^0C to -3^0C). The working fluid is used as a cooling medium is R-134a. The experimental results showed that the injection of hot gas to the suction line generates an increase in the coefficient of performance systems (COPs) of 7% and is able to lower the discharge temperature, causing the compressor to work lighter/easier, saving electric power needed by the refrigerator.

1. Introduction

The use of refrigerators in households increases every year. It is because people increasingly do their activities outside their houses; they prefer to purchase their needs at one time and store them in refrigerators. The role of refrigerators in today's modern era occupies a primary position in people's lives [1]. In general, refrigerators use a vapour compression cycle that requires energy input from the outside in the form of electrical energy to drive the compressor motor. The energy required by refrigerators is quite big [2], as the refrigerators work for 24 hours. Therefore, energy saving measures are necessary as it would significantly impact on energy conservation efforts in the world [3][4].

In order to solve the problem of energy saving of refrigerators, some researchers have implemented various ways to reduce energy consumption and improve the efficiency of the refrigeration system, including: (i) replacing the synthetic refrigerant with natural refrigerants [5][6][7][8]; and (ii) installing suction-liquid heat exchangers (SLHX) [9][10]. However, some problems are still found. The natural refrigerants that are made from hydrocarbons have flammable properties and the advantages of SLHX installation rely on the combination of operating conditions and the properties of the fluid [10].

In general, the technique of hot gas injection is used to defrosting process on the evaporator [11][12][13]. Difference with other report, the technique of hot gas injection in this study is used to increase the cooling capacity and lowering the discharge temperature of the refrigerant. When the refrigeration system off cycling, the pressure of both low side and high side will be the same. However, when the refrigeration system starts running both pressure values will differ significantly. Therefore, the technique of hot gas injection to the suction line is needed to reduce the pressure difference on both sides.

The purpose of this study was to determine the increase in performance of a refrigerator utilizing hot gas refrigerant from the compressor's discharge which was injected into the suction line. This method is adopted from the heat pump cycle [14][15], which useful to maintain the balance of the capacity of the evaporator and compressor when the pressure and temperature of refrigerant at the suction channel are low. Therefore, the application of this engineering method to domestic refrigerators is expected to reduce the load of the compressor when doing compression work; this will have an impact on the increased value of coefficient of performance systems (COPs) and the decrease of energy/power required as the driving force.

2. Materials and Experimental method

Figure 1 shows the layout of a domestic refrigerator used for the experiement. The tes unit was a Sanyo refrigerator and designed to work with R-134a refrigerant. The main components of the domestic refrigerator consist of evaporator, compressor, condenser and refrigerant control. Components of the hot gas injection line consist of a 2 m capillary tube and a check valve. Two pressure gauges (Robin Air, USA) and two temperature sensors (Lutron, Taiwan) were placed on the inlet and outlet of the compressor to measure the suction and discharge pressures. On the inside of the refrigerator, a container of brine (salt water) weighing 500g was kept, serving as cooling load. A thermometer was placed in the brine container to determine changes in temperature every 1^0C.

At the beginning of the research, the system was operated under a normal condition to obtain baseline data of the performance of the refrigerator. After that, hot gas injection line was installed to the suction line. Before the second data retrieval, the system was flushed using nitrogen to eliminate dirt, moisture and other substances in the system which might affect its performance. Then, the system was filled with R-134a weighing 100g in accordance with the recommendation of the manufacturer. Data capture started when the temperature of the brine was 3^0C, with the assumption that the system was steady. Next, changes were observed until the temperature of the brine reached -3^0C. During the study, the temperature of the environment was kept at $28^0C \pm 2^0C$. The research procedure was implemented 3 times to obtain the average value.

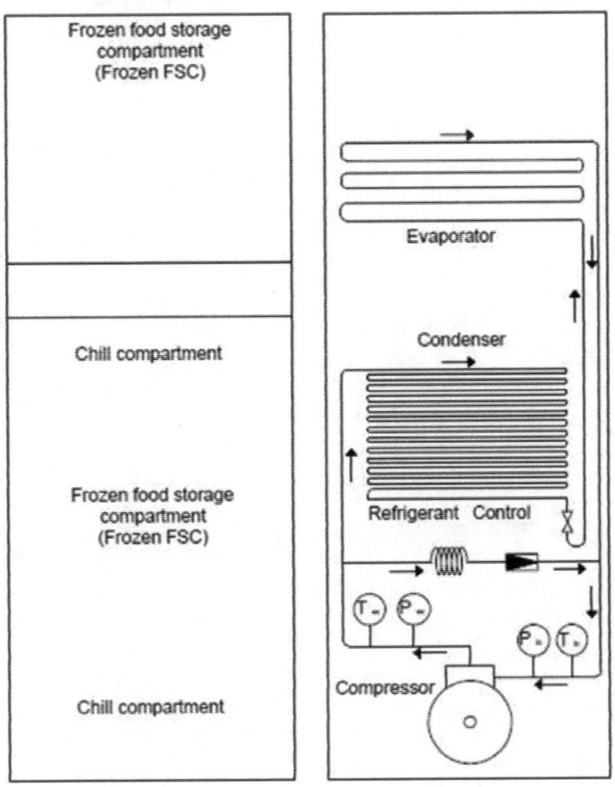

Figure 1. Layout of the domestic refrigerator

3. Results and Discussions

The results were obtained from the performance testing of the domestic refrigerator in two operating conditions: normal and hot gas injection. Tests were carried out at an ambient temperature of 28^0C in the refrigeration cycle operating mode under load conditions of 3^0C to -3^0C.

3.1. Work of compression

Figure 2 shows the results of the compression works of the domestic refrigerator under two testing conditions. In general, the compression work generated by the system using hot gas injection has a lower value than that of the normal system. In the normal system, the compression generated when the brine was 3^{0}C was 26.27 Btu/lb. Then, this increased gradually to 26.57 Btu/lb when the temperature of the brine was -1^{0}C. After that, it feel back to to 26.48 Btu/lb when the temperature of the brine was -3^{0}C. On the other hand, in the system which used hot gas injection, the compression work generated when the brine was 3^0 C was 25.25 Btu/lb. Further, this gradually decreased to 25.12 Btu/lb when the temperature of the brine was -3^{0}C. Based on the results, there had been a 5% decrease of compression work in the system using hot gas injection. The effects of the use of the hot gas injection on the domestic refrigerator result in a smaller ratio of discharge and suction pressures, causing the compression work which compresses vapour from condensing pressure to evaporating pressure becomes shorter. Such results/values will contribute to the energy consumption of the compressor, which becomes smaller.

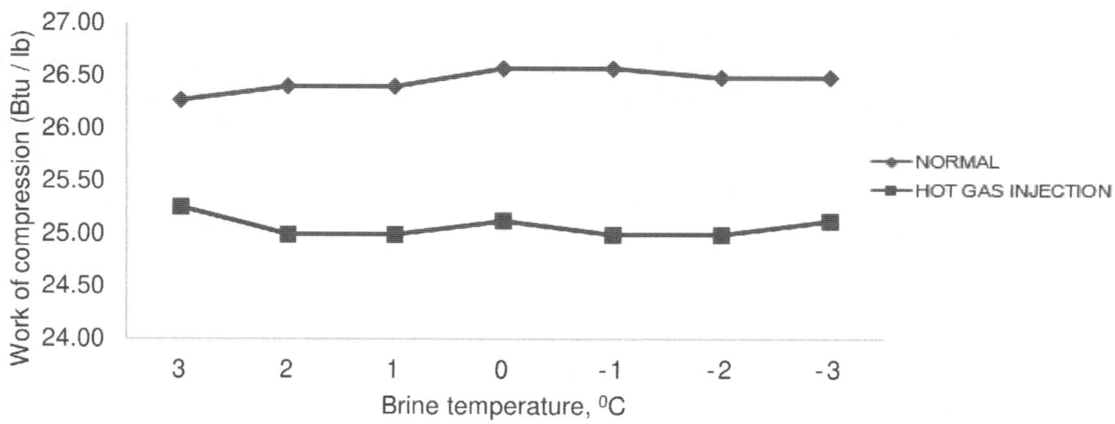

Figure 2. The work of compression based on temperature changes of the brine

3.2. Coefficient of performance

Figure 3 presents coefficient of performance data (Cop) of two conditions of the domestic refrigerator testing. In the normal system, the Cop decreased along with the decreasing temperature of the brine. At first, the obtained Cop was 2.15 when the temperature of the brine was 3^{0}C; the Cop of 2.12 was obtained when the brine temperature was -3^{0}C.

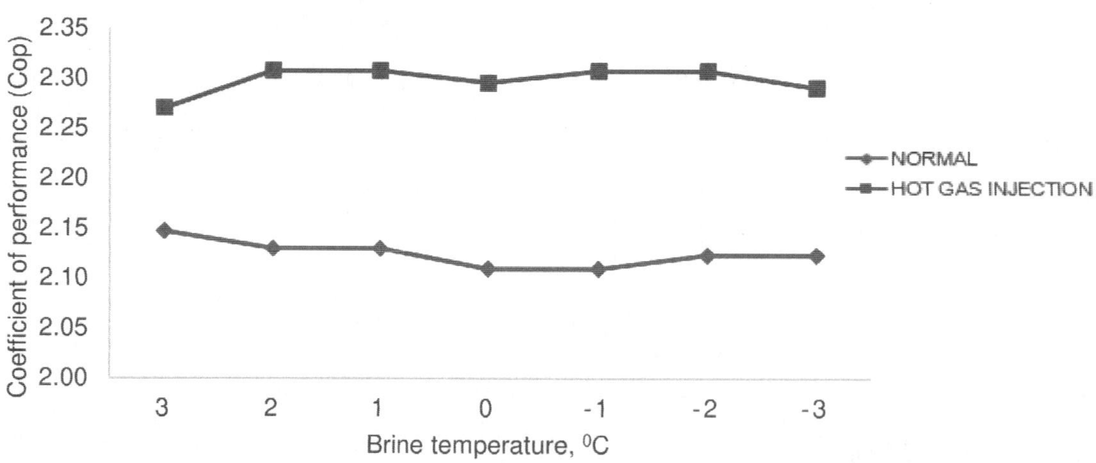

Figure 3. Acquisition Cop value based on changes in temperature brine

Further, when the system employed hot gas injection, Cop increased significantly. When the temperature of the brine was 3^0 C, the obtained Cop was 2.27. Then, the value fluctuated to 2.29 when the brine temperature was -3^0 C. These results indicate that the use of hot gas injection on the refrigeration system results in the increase value of Cop, which is 7%. The great value of Cop indicates that the system worked properly. The value of Cop is influenced by the amount of heat which can be absorbed by the refrigerant flowing in the evaporator (refrigeration effect), and the compression work performed by the compressor when it compressed the low temperature-pressured refrigerant vapour to high temperature-pressured refrigerant vapour. If the value of the refrigeration effect is big and the value of compression work are small, the Cop obtained will be large. The bigger the value of the refrigeration effect and the smaller the value of the compression work, the value of Cop obtained will be better.

4. Conclusions

Research on the use of hot gas injection to improve the performance of domestic refrigerators has been conducted. To demonstrate the performance improvement of refrigerators, we tested the refrigeration system in two different conditions. The results indicate that hot gas injection to the suction line generates an increased value of coefficient of performance (Cop) and lowers the temperature discharge. This has a significant implication for the lighter work of the compressor and low electrical power needed by the refrigerator.

Acknowledgement

The author would like to thank the Rector of UPI for funding this research through PPDKI program 2011-2015, on research scheme *penguatan kompetensi*.

References

[1] Sari O, Balli M. From conventional to magnetic refrigerator technology. International Journal of Refrigeration. 2014. p. 8–15.
[2] Geppert J, Stamminger R. Do consumers act in a sustainable way using their refrigerator? The influence of consumer real life behaviour on the energy consumption of cooling appliances. Int J Consum Stud [Internet]. Blackwell Publishing Ltd; 2010;34(2):219–27.
[3] Ueno T, Sano F, Saeki O, Tsuji K. Effectiveness of an energy-consumption information system on energy savings in residential houses based on monitored data. Appl Energy. 2006;83(2):166–83.
[4] Hasanuzzaman M, Saidur R, Masjuki HH. Effects of operating variables on heat transfer and energy consumption of a household refrigerator-freezer during closed door operation. Energy. 2009;34(2):196–8.
[5] Tashtoush B, Tahat M, Shudeifat MA. Experimental study of new refrigerant mixtures to replace R12 in domestic refrigerators. Appl Therm Eng. 2002;22(5):495–506.
[6] Sekhar SJ, Lal DM, Renganarayanan S. Improved energy efficiency for CFC domestic refrigerators retrofitted with ozone-friendly HFC134a/HC refrigerant mixture. Int J Therm Sci. 2004;43(3):307–14.
[7] Wongwises S, Chimres N. Experimental study of hydrocarbon mixtures to replace HFC-134a in a domestic refrigerator. Energy Convers Manag. 2005;46(1):85–100.
[8] Sattar MA., Saidur R, Masjuki HH. Performance investigation of domestic refrigerator using pure hydrocarbons and blends of hydrocarbons as refrigerants. Waset. 2007;23(2002):223–8.
[9] Mastrullo R, Mauro AW, Tino S, Vanoli GP. A chart for predicting the possible advantage of adopting a suction/liquid heat exchanger in refrigerating system. Appl Therm Eng. 2007;27(14-15):2443–8.
[10] Klein SA, Reindl DT, Brownell K. Refrigeration system performance using liquid-suction heat exchangers. Int J Refrig. Elsevier Science Ltd; 2000;23(8):588–96.
[11] Cho H, Kim Y, Jang I. Performance of a showcase refrigeration system with multi-evaporator during on-off cycling and hot-gas bypass defrost. Energy. 2005;30(10):1915–30.
[12] Dopazo JA, Fernandez-Seara J, Uhía FJ, Diz R. Modelling and experimental validation of the hot-gas defrost process of an air-cooled evaporator. Int J Refrig. 2010;33(4):829–39.
[13] Yaqub M, M. Zubair S, Khan J-R. Performance evaluation of hot-gas by-pass capacity control

schemes for refrigeration and air-conditioning systems. Energy. 2000. p. 543–61.
[14] Dutta AK, Yanagisawa T, Fukuta M. An investigation of the performance of a scroll compressor under liquid refrigerant injection. Int J Refrig. 2001;24:577–87.
[15] Hu B, Wang X, Cao F, He Z, Xing Z. Experimental analysis of an air-source transcritical CO2 heat pump water heater using the hot gas bypass defrosting method. Appl Therm Eng [Internet]. 2014;71(1):528–35.

Typology of building shading elements on Jalan Sudirman corridor in Pekanbaru

G Faisal* and P Aldy

Program Studi Arsitektur Universitas Riau

*Corresponding author: gunfaisal@gmail.com

Abstract. In 2013, temperature in Pekanbaru was between 22.60°C and 34.6°Cwith humidity 79.14 percent. This condition has increase the concern of energy utilization to building comfort. Buildings have the biggest energy consuming due to the use of air conditioner in Pekanbaru. One effort to reduced energy is shading devices application. Application of air conditioner need huge energy, replaced natural circulation with architecture elements to reduced building thermal. This research study about system and building shading devices types that influence building thermal in Pekanbaru so that knowing characteristics and elements form. This study aims to determine and identify of systems and building elements types in Pekanbaru, which the element forms to conquer in climate condition. Qualitative method with rationalistic-paradigm has used to identify typology of building shading devices on Jalan Sudirman corridor. The research orientation on typology theory, thermal theory, and building shading device to identification of building shading device types on Jalan Sudirman corridor. Based on the survey result, there are 2 type of building shading devices on Jalan Sudirman Pekanbaru which is based on forms and quantity of shading. The types are building shading devices based on shading quantity and building shading devices based on shading forms.

1. Introduction

Pekanbaru is the biggest city in Riau Province According to [1] temperature in that city was between 22.60° and 34.6° with humidity 79.14 percent. While Rainfall was 256.45 mm with the highest in December reached 614 mm. The high temperature has in impact to increase of energy utilization for building comfort.

Air Conditioner is the largest contributor in energy consumption. So are buildings in Pekanbaru almost entirely using air-conditioner. Energy efficient architecture is the architecture with the lowest possible energy requirement can be achieved by reducing the amount of resources that makes sense [2]. Thus, energy-efficient architecture is based on the idea to minimize energy use without limit or alter the building function, comfort, and productivity of its users. One effort to reduce energy use is the utilization of the thermal shading elements of the building. *Air Conditioner* takes a lot of energy can be replaced by natural air using architectural elements that can reduce building heat.

This study will discuss about how the system and the type of building shading elements that affect the thermal building in Pekanbaru, so that we determine the character and shape of the elements used. This study aims to determine and identify the system and the type of building elements in Pekanbaru, which those elements are ways of building climate conditions.

1.1 Identify of Problems

1. What types of building shading elements on Jalan Sudirman's corridor in Pekanbaru?
2. What kind of shading elements shape and function that affect building thermal comfort on Jalan Sudirman corridor in Pekanbaru?

1.2 Study Purposes

The purpose of this study is to identify, learn and study shading elements usage, in terms of both form and function. In addition, this study aims to observe of usage shading elements building on Jalan Sudirman corridor in Pekanbaru as reducing energy for building thermal comfort used.

2. Typology

Typology derived from the word "type" which defined as a concept to describe a group of objects that have similar characteristics of formal structure [3]. Typology is the study of the types of elements that are not reducible anymore. The word in the context of the type of architecture into Architype and then the typology which is an order of paradigms and tools owned by the architecture to put the position of architecture as a science [4]. Meanwhile, according to [5]typology is an activity or activities generating the same type to classify and categorize. According to [6] typology is an approach that separates attributes of coherence architectural, and to identify it as a characteristic, in an aim to comparing with attributes abstracts from other contexts, and to define similarities or differences. [7] in *Pr`ecis of the Lectures on Architecture* in architectural typology divides into three, namely: traditional typology, modern typology and functional typology.

While [8] argues, the typology is a concept describing the object groups based on imilarity of the basic properties that are trying to sort or classify the diversity and similarity of species. In this case, the typology is the elaboration result of architectural characteristics, which is composed of various local and foreign cultural elements specific to a classification structure, either functions, geometric, or style or styles classification.

Furthermore, [9] said that, typology studies in architecture world means studies in an effort to sorting, classification, to be revealed the diversity and similarities in architectural products with one another. Basically, the typology is a concept that describes the object groups on the basis of the basic properties similarity.

Therefore, it was concluded that the typology is a science or a method which involves the study of grouping or classifying an observation object that based on certain criteria to produce specific types of the observed objects.

3. Shading Device

Solar radiation adjustment which entered into building is a major step in passively cooling process to get thermal comfort condition. Sunlight modulation in setting process can be achieved with regard: Orientation and aperture ometry; Shading devices; Property of opaque and transparent surfaces. [10].

A strategy in passive design of building envelope which can be applied in order to reduce the externalheat gain in buildings is shadow elements usage. Shadow elements are many variations. Principally, external shading element divided into horizontal type *(overhang),* vertical *(side fin),* and *an eggcrate.* Despite having different characteristics, the core of shadow elements usage is to minimize the angle of solar radiation that hits the glass field.

Orientation and aperture are closely related aspects in controlling the thermal comfort of the building. Building with sunlight radiation high levels requires needs aperture and orientation. Instead of building with sunlight radiation low levels requires needs to avoid aperture and direct sun orientation. By arranging the shape and orientation towards direct solar radiation can be adjusted in accordance with the position of the equator location.

Shading devices usage is the next step after controlling the orientation and aperture. If the orientation and aperture cannot be tolerated because of design needs, then the shading devices are very important elements. Shading devices can be designed in accordance with the position and direction of solar radiation arrival so that aperture can be shielded from direct radiation.

According to [10], there are two classifications of shading devices namely fixed shading elements and adjustable/rectracable shading elements. Fixed shading devices categories are external form position, overhangs, vertical fins, horizontal and vertical combinations (egg-crate type), and balcony. Among other internal position are light-shelves form and louvre above the window. While adjustable shading elements categories are element external position, tents, awning, blinds, pergola, and the internal ones such as curtains, rollers and venetian blinds. There are external factors of the building that can serve as shading devices, namely vegetation around buildings. Vegetation near a window can provide shading effect and reduce sunlight's direct radiation.

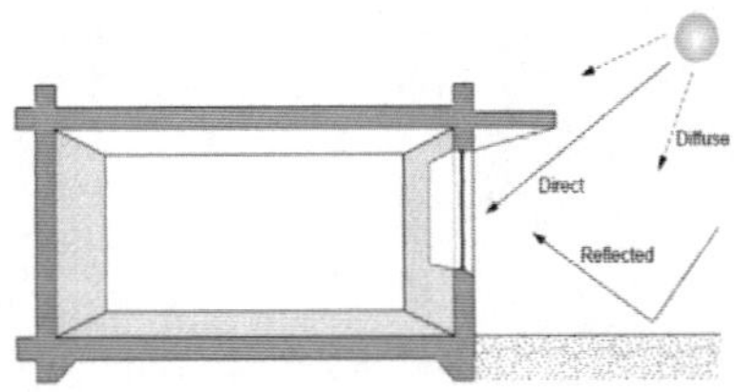

Figure 1. Direct and Indirect Solar Radiation

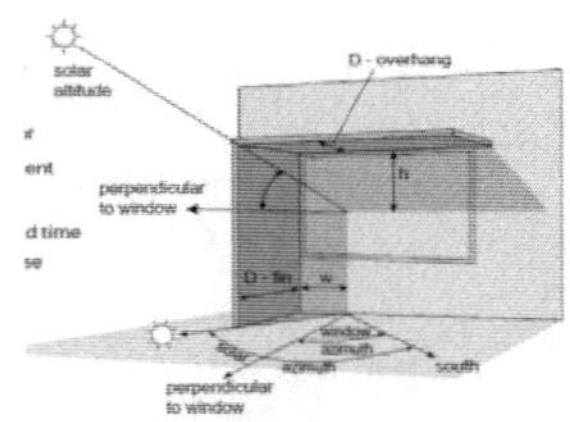

Figure 2. Shadowing area by overhang and side fin

Shading device is able to block direct solar radiation and reduce the influence of indirect illumination that is the sky reflection and the ground level. The position of the sun determines the sunlight's angle of incidence on the glass surface (*penetration).*When the sunlight's angle of incidence equals the angle of the shadow which formed shading device that direct sunlight can be driven effectively. Therefore, sunlight's position and the depth of shading device becomes a very important factor.

According to [11], shading devices divided into two types based on application, internal shading and external shading.

- Internal shading is interior element on inside windows. Internal shading devices generally can adjustable. Thus shading can be blinds, curtains and others.
- External shading is the shading element which is placed on building external. External shading devices can reduced thermal energy effectively to anticipate and reduced radiation before glass surfaces.

  External shading devices divided into 2 types, Fixed Devices and Adjustable Devices.
  - Adjustable Devices generally used indoor or outdoor or between inside and outside of windows (double glazing system) and generally used in internal shading systems.
  - Fixed Devices generally used outdoor and influence façade building. Building orientation was the point.

Some of shading devices were generally used is explained on Table 1

IOP Conf. Series: Materials Science and Engineering **128** (2016) 012029 doi:10.1088/1757-899X/128/1/012029

Table 1. External Shading Devices

	3D-View	***Best Orientation***
Overhang		south, west, east
Overhang Horizontal Louver s	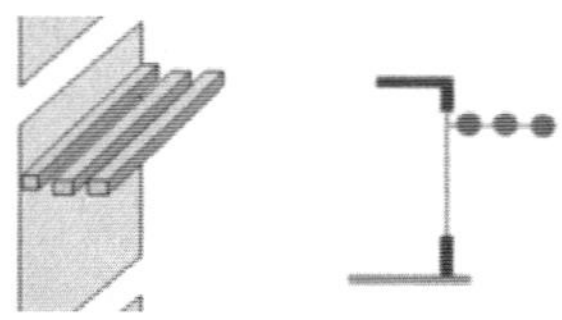	south, west, east
Overhang Multiple Blades	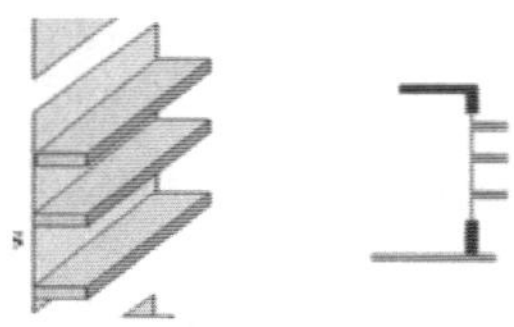	south, west, east
Overhang Vertical panel	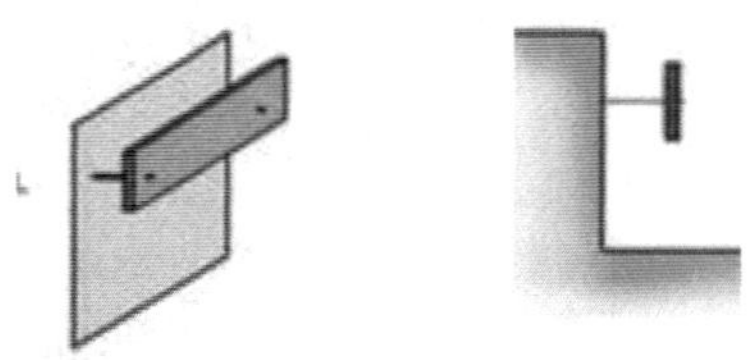	south, west, east
Vertical Fin	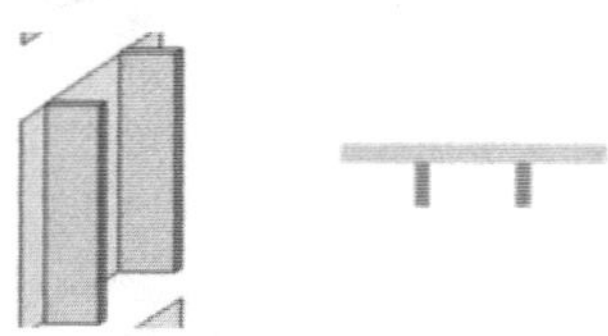	west, east, north
Slanted Vertical Fin	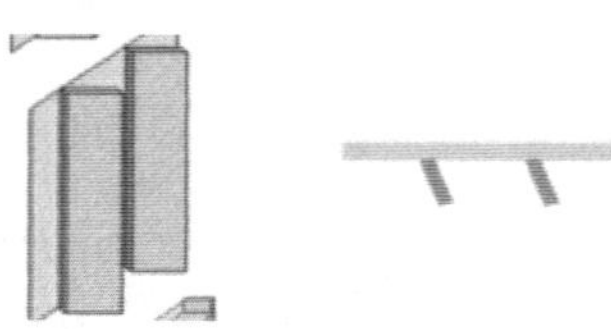	west and east
Eggcrate	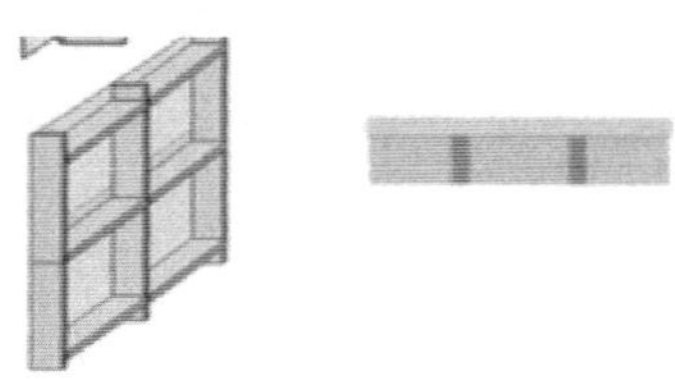	west and east

4. Method

This research used qualitative method with rationalistic-paradigm. According to [12], the purpose of this method was to identification of government building shading device types on Jalan Sudirman, Pekanbaru. The research orientation on typology theory, thermal theory, and building shading device to identification of building shading device types on Jalan Sudirman. The object of this research was building on Jalan Sudirman corridor. Observation, interview, documentation, and document study was used for data collection. And deductive-qualitative was used to data analysis.

IOP Conf. Series: Materials Science and Engineering **128** (2016) 012029 doi:10.1088/1757-899X/128/1/012029

5. Results and Analysis

Observation research divided into 3 segments area and each segment divided into 2 parts, west and east, namely segment 1A and 1B, segment 2A and 2B, and segment 3A and 3B. Total of buildings in west segments were 70 buildings which are consists of 23 buildings on segment 1A, 20 buildings on segment 2A, and 27 buildings on segment 3A. Whereas total buildings in east segments were 102 buildings which are consists of 34 buildings on segment 1B, 36 buildings on segment 2B and 32 buildings on segment 3B.

Figure 4. Observation Segments

Figure 5. 1A Observation Segment

Figure 6. 2A Observation Segment

Figure 7. 3A Observation Segment

Figure 8. 3B Observation Segment

Figure 9. 2B Observation Segment

Figure 10. 1B Observation Segment

Based on research results, there were 2 types of building shading devices on Jalan Pekanbaru, as follow:

1. Building shading devices based on quantity shading device.
    a. 1st type, shading devices with 1 form of shading element (71 samples or 41%)
    b. 2nd type, shading devices with 2 forms of shading elements (52 samples or 30%)
    c. 3rd type, shading devices with 3 forms of shading elements (33 samples or 19%)
    d. 4th type, shading devices with 4 forms of shading elements (8 samples or 5%)
2. Building shading devices based on shading device type.
    a. Overhang type (136 samples or 44%)
    b. Vertical Finn type (49 samples or 16%)
    c. Eggcrate type (51 samples or 16%)
    d. Vegetation type (67 samples or 21%)

In this research, a few of samples were reduction. Reductions reasons consist of the shading was in the building (Light Shekves type), 7 samples building without shading element, 1 sample building with 5 form of shading elements and 8 samples building used water as shading elements.

6. Conclusion and Recommendation

There were two types of building shading device on Jalan Sudirman in Pekanbaru:

1. Building shading devices according to quantity of shading device: 1st type with 1 form of shading device, 2nd type with 2 forms of shading device, 3rd type with 3 forms of shading device, and 4th type with 4 forms of shading device.
2. Building shading device according to shading type, consist of overhang, vertical-finn, eggcrate and vegetation.

Re-arrange building shading device elements which is one of solving aspect in climate condition issues and alternative design of shading device for energy saving efforts.

IOP Conf. Series: Materials Science and Engineering **128** (2016) 012029 doi:10.1088/1757-899X/128/1/012029

References

[1] Badan Pusat Statistik Kota Pekanbaru. 2014. *Statistik Daerah Kota Pekanbaru 2014.* Katalog BPS, Pekanbaru.
[2] Abel, Enno. 1994. Low-Energy Building, *Energy and Building Science Journal Vol. 21*, Elsevier Science S.A. p. 74-169.
[3] Moneo, Rafael. 1979. Oppositions Summer On Typology, *A Journal for Ideas and Criticism in Architecture*, vol. 13, The MIT Press, Massachusetts, h. 23-45
[4] Johnson, P A. 1994. *The Theory of Architecture*, Van Nostrand Reinhold Company, New York.
[5] Francescatto, Guido. 1994. *Type and the Possibility of an Architecture Scolarship*, Ordering Space, Types in Architectural and Design, Karen A. Franck, Lynda H. Schneekloth (ed). Van Nostrand Reinhold, NY.
[6] Pfeifer, G.; P. Brauneck. 2008. *Courtyard Houses–A Housing Typology.* Birkhauser Verlag AG, Germany.
[7] Durand, Jean Nicolas Louis. 2000. *Pr`ecis of the Lectures on Architecture*, The Getty ResearchInstitute, Los Angeles, h. 146-147
[8] Amiuza, C. 2006. Tipologi Rumah Tinggal Administratur P.G. Kebon Agung di Kabupaten Malang. *Jurnal RUAS*. IV (1): 1-22.
[9] Tjahjono, R. 1992. *Studi Karakteristik Arsitektural Pada Hunian Masyarakat Berbahasa Madura di Malang Selatan (Studi Kasus: Desa Ganjaran, Gondanglegi)*. Jurusan Arsitektur Universitas Brawijaya, Malang.
[10] Santamouris, M, dan Asimakopoulos, D. 1996. *Passive Cooling of Buildings*. James & James, Universitas Michigan.
[11] Olgyay, Victor. 1963. *Design With Climate*. Princeton University Press, New Jersey.
[12] Noeng, Muhadjir. 1996. *Metode Penelitian Kualitatif Edisi III*. Rake Sarasin, Yogyakarta.

Exploration of making date seed's flour and its nutritional contents analysis

M Wahini*
Program Studi Tata Boga, Jurusan Pendidikan Kesejahteraan Keluarga, Fakultas Teknik, Universitas Negeri Surabaya
Surabaya-Indonesia

*Corresponding author: medawahini@unesa.ac.id

Abstract. The date palm (Phoenix dactylifera L) is one of the oldest fruit plants that identical with people's lives in the Middle East including The Kingdom of Saudi Arabia since ancient times.The date palm is known and consumed by most of people in the form of fruit flesh, while its seed is discarded tough it is rich in nutrient. Therefore, need to be explored the potential of date seed through product innovation of foodstuffs with a high nutritional value. The aims of this study were to 1) know how to make flour from date seed, and 2) determine nutritional content of date seed's flour. This study was experiment and conducted in July, 2015 at the laboratory of food technology, Family Welfare Education department. Chemical analysis was used to determine nutrients content. The results showed that 1) the flour of date seed produced from the process of washing, soaking, flushing, boiling, draining, drying, grinding, and sieving; 2) the flour of date seed's flour have a macro nutrients value. This study explains that date seed is regarded as rubbish, in fact, it has a high value that can be an alternative substitution of wheat flour.

1. Introduction

Date (the date palm/dates) has been known widely by the society that related to their's life in the Middle East and The Kingdom of Saudi Arabia. In almost all areas of this region, it is consider as part of the most people diet because of its benefits are good for health. A people there used to eat date as main dish. However, in Indonesia the date fruit is mostly found in the market in the Ramadan untill Eid Hajj's month that its consumption tend to increase. Date is generally consumed, either eaten immediately or in various types of food and beverage, such as cake, pancake, pie, sus, pudding, sauce, ice cream, date's stew, syrup, jam, tea, and juice [1] and cookies [2]. On the other hand, the date palm is useful to take care of eyes, bones and teeth's health, lung cancer, cancer and colon prevent. In the terms of nutritional value, it has 23 calories, 9-13 of soluble fiber, 2.5-4.3% crude fiber, 1.4-1.7 grams/100 grams protein of date flesh (0.3-0.5% 100-800mg/100g wet weight, fatty acids and potassium amounted)[3]. People generally know that the date palm is consumed fruit flesh only. Such views are not always true. Because, it can be used as a whole, including seed that is considered as a rubbish. Whereas its seed contains of 71.9 to 73.4 % of carbohydrate, 5 to 6.3 % protein, 9.9 to 13.5 % fat, 6.4 to 11, 5 % fiber, and 1 to 1.8 of ash. Date seed can be useful materials for flour [4]. It is assumed with semi-processed form of the equivalent protein value of wheat flour, which means it can be used date seed into various preparations such as cakes, breads, pastries and others. Therefore, the date palm can be used for foods which have a history and religion value.

The date palm had a history value in the world. This plant thought to have come from the plains of Mesopotamia, Palestine, or about northern Africa (Morocco) around 4000 BC. It spread into areas of Egypt, Africa, Central Asia, and the surrounding areas since 3000 BC. In terms of religious aspect, the date palm is quite special to get a place. The palm leaves as a symbol of longevity; the fruit is a symbol of the presence of the god Amun ra' and the fertility goddess Hathor. In the ancient Egyptian nation, the date fruit is used as a dish in the morgue, and is found as one element in the process of purification corpse preservatives.

The date palm is from family of Arecaceae which have seeds with one institution (monocots). Date seed do not have a smell or odorless and has a slightly bitter taste bland. In general, it has a light and dark brown. Today, a various studies regarding date seed have been published in order to determine the functional properties of date seed is used for food and non-food items such as thermal properties [5], treatment and diet [6][7], the composition of macro and micronutrients [8], the composition of phenolic acids [9], as an ingredient of bread [3], and protein solubility [10]. Date seed are highly recommended for use in foods and dietary supplements. Because, it is a very good source of dietary fiber [9]. Total mineral content found in single seed is comparable to the mineral content of barley. The minerals contained in date seed include sodium, potassium, calcium, iron, copper, magnesium, manganese, zinc, phosphorus, lead, cadmium. This suggests that a single seeds is a good source of minerals, and also can be used to replace barley in food products [11]. The amount of dietary fiber found in date seed about 58 % of which 53 % soluble dietary fiber as hemicellulose, cellulose and ligni [7][9]. In comparison, the higher dietary fiber was detected in other studies conducted in three different varieties of date seed, the number of fibers between 65 to 69 percentage. This shows the high content of lignin and resistant starch [10].

On the other hand, protein is also found in date seed with a sizeable amount. Albumin, globulin, prolamin and glutelin are soluble protein which found in seed of the current date, the number of 5-6 percentage of the total protein level [10]. Total phenolic content which is found in date seed is 48.64 mg / 100 g. Phenolic acids contained in date seed that the gallic acid, acid protocatechuic acid p-hydroxybenzoic acid vanilla, caffeic acid, p-coumaric acid, ferulic acid, acid m-coumaric and o-coumaric acid [12]. Based on Boundries [1], indicated that the highest antioxidant content is in that seed, included penolic,

Thus it can be further developed into new product or improve existing product. Various studies indicate that date seed can be used as a good source of dietary fiber, phenolic components and natural antioxidants. Date seed actually have a more complex function as a food source that can improve the health of the body. It is necessary to further research to explore the way of making flour made from date seed with emphasis on the different drying process.

2. Materials and Methods

This research was conducted at the laboratory of food technology, Family and Welfare Education Departement, Faculty of Engineering, in July, 2015. Equipment used in this study is the basin, siever, pan, hammer, and grinder. The materials are used for flour production included the date palm seed of Tunisia's type which collected from producers who make cookies by using the date fruit in ied Mubarak. It chosen because this is one of the preferred types of date that many people of Indonesia, as well as the largest importer of dates for this type [1]; Natrium bisulfite. This material is used to maintain color and to prevent browning reaction during the heating process, and tap water for cleaning the date seed. These presented in the table 1 as follow.

Table 1. Materials of flour

Materials	Total
Date seed	2500 gram
Natrium bisulfite	500 ppm
Tap water	2 l

The methods were used are 8 steps, namely 1) washing, This step aims to clear up residual and dirt in seed by means washed the seed with tap water; 2) soaking. When seed has been cleaned, then soaking it into natrium bisulfite ($NaHSO_3$) for 24 hours This aims to maintain the color and prevent the profusion of browning reaction during warming process; 3) flushing, This step aims to clear up the seed after soaking process; 4) boiling, The next step is boiled seed in the water with 80-90°C of temperature during 10 minutes, it is mean to soften the texture of date seed; 5) draining, After that drained it to make drain faster before drying process; 6) drying. The drying process in this research are using by two treatments, namely drying under he sun for two days, and drying by oven for 24 hours at 50-60°C. This process aims to dry date seed to allow in the grinding process; 7) grinding. The grinding have two steps. The first step is date seed is hammered. This is done because the outer skin of the date seeds are very hard, and to get a rough flour that seed have to destroye with a hammer. The next step is grinding by using machine. This aims to get the ideal texure of date seed's flour before performed by siever; 8) the last method is sieving. The aim is to get smooth of flour texture is sieving by using stainless siever. The explanation of methods of making date seed's flour present on picture 1 as follow:

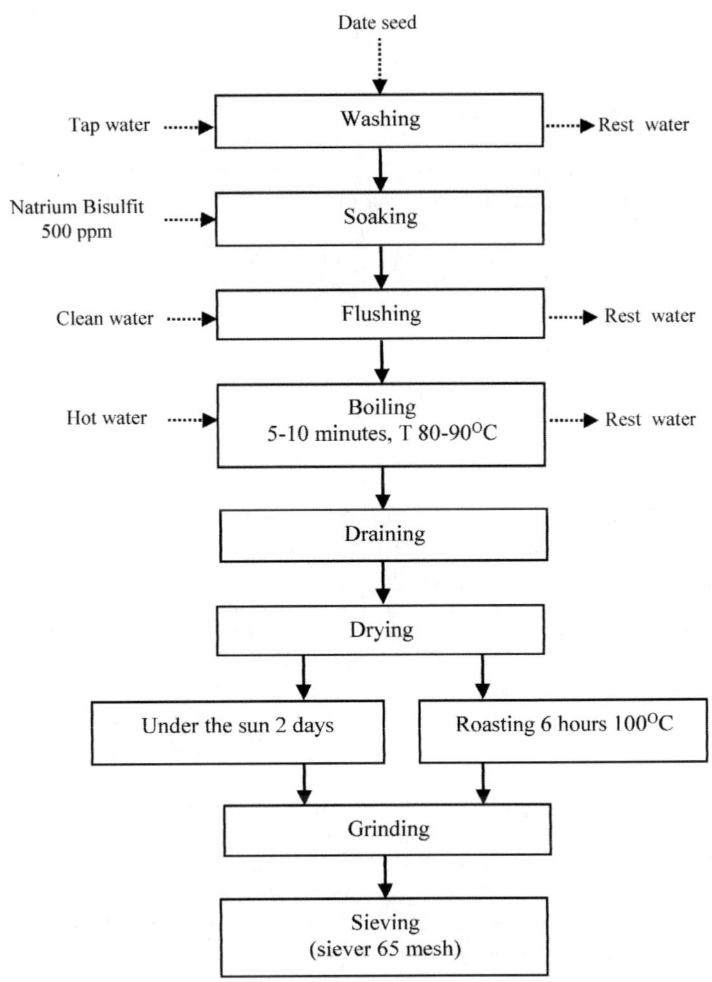

Picture 1. Scheme of date seed flour's made

3. Results and Discussion

3.1. The flour of date seed

Based on the results of the experiment can be explained that the flour of date seed which made from Tunisia's type is conducted through eleven stages, namely washing, soaking, flushing, boiling, draining, drying, grinding, and sieving. It supports the research that conducted by [4], the seed's flour of date seed obtained by washing, soaking, boiling, drying, grinding, and sieving.

A difference in the two studies are located at boiling treatment by soaking. In this research, date seed is boiling at 80-90 degrees Celsius for 5-10 minutes which aims to soften date seed; While the research [2] proved that boiling process is not carried out with a high temperature and a long time, but by using blanching technique. It uses blanching techniques and the addition of natrium bisulfite by giving effect to the bulk density and whiteness in flour [13]. Referring to the research of Kusumawati that the recommended temperature to produce color and density of grain flour is not more than 70 degrees Celsius. While in this study date seed boiling process is used too high temperatures (80-90 degrees Celsius). Thus, the boiling temperature of date seed in this study is not recommended to be used as a reference because it will affect the color and density of date seed grain flour. Although the temperature of boiling had no effect on bonding compounds contained in date seed's flour.

The most important stages in the process of making flour from the seeds of these date palm are soaking by using natrium bisulfite. The aims of it to defend the colors of the materials and prevent non-enzymatic browning reactions as well as enzymatic, as well as to inhibit microbial growth [14]. The other importance stages is the drying process in two ways, namely solar thermal and heat oven. The aims of this two-way warming is intend heating with solar heat have better results in terms of the quality of the material, it is supported by research [15] that heating with sun better than a hot oven, although the weakness it takes a longer time.

3. 2. Nutrients content of date seed's flour

Results of the chemical test analysis on date seed's flour with different drying process is presented in Table 2 below. In this study, date seed's flour dried by the sun heating has better nutritional value compared with the nutrients that are heated in the oven. The flour of date seed with sun dried process contains 33.84% of carbohydrates; 26.52% of Protein; 31.54% of Fat; Minerals (102.42 mg /100gr of Ca, 81.05 mg/100gr of P, 16.54 mg/110gr of Fe), also vitamins (3.08 mg/100gr of B, 23.55 mg/100gr of C). Whereas the nutritional value ofdate seed's flour through a drying oven was 5.03% of protein; 12.37% of fat; Minerals (64 mg/100gr of CA, 62 mg/100gr of P, 16.54 mg/100g of Fe), and vitamins (5% of A, no C, 2% K). The amount of 26.54% protein contained in the date seed's flour with sun heat as much as protein in date seed's flour with oven heating only one-fifth of the protein date seed's flour with sun heating that is as much as 5.03%. The carbohydrate content of the date seed's flour with sun heat as much as 33.84%. The results of protein content in this study are better than the results of previous studies which have been carried out by [16] that protein in the flour contribute to improve water absorption of flour as the manufacture of cakes dough by using flour with a high protein material can withstand the gas allowing the developer and the dough can swell up like a balloon. Results of the cake with this dough will have a cavity structure is smooth and uniform, and the texture is soft and elastic [17]. High protein flour (hard wheat) has a total protein content of 12-13%, while the protein content in date seed flour with better solar heating is 26.54%. This proves that date seed's flour can be an alternative food substitute for wheat flour. Whereas the nutritional value of date seed's flour through a drying oven was 5.03% of protein; 12.37% of fat; Minerals (64 mg/100gr of CA, 62 mg/100gr of P, 16.54 mg/100g of Fe), and vitamins (5% of A, no C, 2% K).

Table 2. Nutrients content of date seed's flour

Nutrition component	Sun heat (%)	Oven heat (%)
Carbohydrate	33,84%;	25,64%

Protein	26,54 %	15,03 %
Fat	31,54%	17,37%
Mineral Ca	102,42 mg/100gr	64 mg/100gr
Mineral P	81,05 mg/100gr	62 mg/100gr
Mineral Fe	16,54 mg/100gr	16,54 mg/100gr
Vitamin A	7%	5%
Vitamin B	3,08 mg/100gr	2,03 mg/100gr
Vitamin C	23,55 mg/100gr	17,55 mg/100gr
Vitamin K	2,9%	2%
Water content	-	-
Ash content	-	-
Fiber	-	-

4. Conclusions

4.1. The flour product of date seed which made from Tunisia's type is conducted through eigh stages, namely washing, soaking, flushing, boiling, draining, drying, grinding, and sieving.

4.2. Date seed's flour dried by the sun heat has better nutritional value compared with the nutrients that are heated in the oven. The amount of 26.54% protein contained in date seed's flour with as much as the protein in date seed's flour with oven heating only one-fifth of the protein palm seeds flour with sun heat that is as much as 5.03%. The carbohydrate content in date seed's flour with sun heat as much as 31.54%.

References

[1] Satuhu, Suyanti. 2010. Kurma Khasiat dan Olahannya Jakarta: Penebar Plus+.

[2] 2Setyiono, Luthfi. 2011. Pemanfaatan Biji Kurma (Phoenix Dactyliferal L.) Sebagai Tepung dan Analisis erubahan Mutunya Selama Penyimpanan. Skripsi Tidak Untuk Dipublikasikan. Bogor: IPB.

[3] Almana, H. A. and Mahmoud, R. M. 1994. Palm date seeds as an alternative source of dietary fiber in Saudi bread. Ecology of Food and Nutrition 32: 261-270.

[4] Rahmadi A. 2010. Kurma.Samarinda: Food Technologist, Neuro-biologist, and Pharmacologist, University of Mulawarman.

[5] Rahman, M. S., Kasapis, S., Al-Kharusi, N. S. Z., AlMarhubi, I. M. and Khan, A. J. 2007. Composition characterization and thermal transition of date pits powders. Journal of Food Engineering 80: 1-10.

[6] Hussein, A. S., Alhadrami, G. A. and Khalil, Y. H. 1998. The use of dates and date pits in broiler starter and finisher diets. Bioresource Technology 66: 219-223.

[7] Aldhaheri, A., Alhadrami, G., Aboalnaga, N., Wasfi, I. and Elridi, M. 2004. Chemical composition of date pits and reproductive hormonal status of rats fed date pits. Food Chemistry 86: 93-97

[8] Habib, H. M. and Ibrahim, W. H. 2009. Nutritional quality evaluation of eighteen date pits varieties. International Journal of Food Sciences and Nutrition 60:99-111

[9] Al-Farsi MA dan Lee CY.2008b. Optimization of phenolics and dietary fibre extraction from date seeds. Food Chemistry 108: 977-985.

[10] Hamada, J. S., Hashim, I. B. and Sharif, F. A. 2002. Preliminary analysis and potential uses of date pits in foods. Food Chemistry 76: 135-137.

[11] Ali-Mohamed, A. Y. and Khamis, A. S. H. 2004. Mineral ion content of the seeds of six cultivars of Bahraini date palm (Phoenix dactylfera). Journal of Agriculture and Food Chemistry 52: 6522-6525..

[12] Al-Farsi MA dan Lee CY. 2008. Nutritional and functional properties of dates: a review. Critical Reviews in Food Science and Nutrition48(10): 877 - 887.

[13] Kusumawati. 2012. Pengaruh Perlakuan Pendahuluan Dan Suhu Pengeringan Terhadap Sifat Fisik, Kimia, Dan Sensori Tepung Biji Nangka (*Artocarpus Heterophyllus*). Jurnal Teknosains Pangan Vol 1 No 1. Surakarta: Universitas Sebelas Maret.

[14] Rizal, Saifur. 2013. Pengaruh Konsentrasi Natrium Bisulfit Dan Suhu Pengeringan Terhadap Sifat Fisik-Kimia Tepung Biji Nangka (*Artocarpus heterophyllus*). Jurnal Bioproses Komoditas Tropis Vol. 1 No. 2. Malang: Universitas Brawijaya.

[15] Masduqi, Ahmad. 2014. Efek Metode Pengeringan Terhadap Kandungan Bahan Kimia Dalam Rumput Laut *Sargassumpolycystum*. Buletin Anatomi dan Fisiologi Volume XXII, Nomor 1. Semarang: Fakultas Sains dan Matematika Undip.

[16] Al-Qarawi AA, Ali BH, Al-Mougy SA and Mousa HM (2003) Gastrointestinal transit in mice treated withvarious extracts of date (Phoenix dactylifera L). Food & Chem. Toxicol. 41, 37-39

[17] Koswara, Sutrisno. 2009. *Seri* Teknologi Pangan Populer (Teori dan Praktek). eBookPangan.com.

IOP Conf. Series: Materials Science and Engineering **128** (2016) 012032 doi:10.1088/1757-899X/128/1/012032

First principles molecular dynamics simulation of graphene growth on Nickel (111) surface

R Arifin* [1,2], Y Shibuta[3], K Shimamura[1] and F Shimojo[1]

[1]Department of Physics, Kumamoto University, 2-39-1 Kurokami, Chuo-ku, Kumamoto 860-8555 Japan
[2]Fakultas Teknik, Universitas Muhammadiyah Ponorogo, Jl. Budi Utomo No. 10, Ponorogo 63471 Indonesia
[3]Department of Materials Engineering, The University of Tokyo, 7-3-1 Hongo, Bunkyo-ku, Tokyo 113-8656 Japan

*Corresponding author: rarifin@umpo.ac.id

Abstract. The mechanism of the graphene growth on the nickel (111) surface is studied using first principles molecular dynamics (MD) simulation. To investigate the effects of growth temperature and concentration of carbon atoms on the nickel (111) surface, the systems consisting 50%, 66.7%, and 100% of carbon concentration are annealed from 800 K, 1000 K, 1200 K, and 1500 K until room temperature. It is found from the MD simulation that the optimum temperature for the graphene growth is approximately 1000 K, although the graphene can also be grown at lower temperatures (~800 K) in a high concentration of carbon atoms which is in good agreement with the experiment.

1. Introduction

Graphene has attracted a lot of attentions due to its desirable electronic, optical, and mechanical properties [1][2][3]. Since several years ago, the number of publications and patents on graphene field has been increased significantly [4]. Graphene can be obtained by several ways. One of the most prospective method for large scale production of graphene is known as epitaxial growth on the particular substrate, such as silicon carbide [5][6] or metal substrate [7][8]. Nickel is one of metals that are widely used as the substrate for the graphene growth process via chemical vapor deposition (CVD) method [8]. It has been known that nickel exhibits high carbon solubility [9], which means that carbon atom can easily diffuse into the nickel substrate.

The previous experimental research of carbon behaviour on the nickel (111) surface has reported that specific formations of carbon are observed in particular temperature range. At the temperature more than 1200 K, carbons are diluted in the surface. Carbon atoms segregate to the surface to form monolayer graphene at the temperature between 1100 K and 1200 K. At the temperature below 1100 K, multilayer graphenes are formed by the precipitation of carbon atoms on the nickel surface due to the reduced carbon solubility of nickel bulk at low temperature [10].

So far, many theoretical studies have been performed to investigate the evolution of isolated carbon atoms on the nickel (111) surface. Meng and co-workers have employed the reactive force field (ReaxFF) potential as the interatomic potential within classical molecular dynamics (MD) simulation [11]. Other MD studies of the graphene growth on the nickel (111) surface have been carried out by means of density functional tight binding (DFTB) method. In the initial stage of their simulations, the

IOP Conf. Series: Materials Science and Engineering **128** (2016) 012032 doi:10.1088/1757-899X/128/1/012032

carbon atoms are distributed randomly in the interstitial state of the nickel subsurface. During the MD simulation at 1180 K, the carbon atoms nucleate underneath, and then segregate to the nickel surface. They obtained many pentagonal (5-membered) and hexagonal (6-membered) with few heptagonal (7-membered) carbon rings from the MD simulation. Additionally, some species of amorphous carbon cluster are also found in the nickel subsurface [12][13].

A number of theoretical studies of the graphene growth on the nickel (111) surface have been carried out recently. However, so far we are unaware of any MD studies of carbon nucleation on the nickel surface employing pure DFT method. In this study, we aim to reveal the mechanism of graphene growth on the nickel (111) substrate in various concentrations of carbon atoms and growth temperatures using first principles MD simulations. This simulation method has been employed in our previous studies of the hydrocarbon dehydrogenation on the nickel (111) surface [14][15][16].

2. Methods

This research has been conducted by using first principles MD simulation. The electronic states are calculated by the projector-augmented-wave (PAW) [17][18] method. The Perdew-Burke-Ernzerhov (PBE) [19] generalized gradient approximation (GGA) is employed for the nonlocal exchange-correlation effect. The empirical correction of the van der Waals interaction is described by the DFT-D approach [20]. The plane wave cutoff energies are 30 Ry and 250 Ry for the electronic pseudo-wave function and pseudo-charge-density, respectively. The energy functional is minimized iteratively using a preconditioned conjugate-gradient method [21]. The Brillouin zone is sampled using Γ point. Projector functions are generated for the $3d$, $4s$, and $4p$ states of nickel and the $2s$ and $2p$ states of carbon.

Three simulation systems are generated as the initial configurations. The systems include three layers of the nickel (111) plane, containing 30 nickel atoms in each layer, with (*i*) 50%, (*ii*) 66.7%, and (*iii*) 100% of carbon concentration, *i.e.* 15, 20, and 30 carbon atoms, respectively. These carbon atoms are distributed randomly on the nickel (111) surface. In order to hinder the carbon diffusion into the nickel subsurface, all carbon atoms on the nickel surface are connected with the bonds. The atoms in the bottommost layer of the nickel substrate are fixed to mimic the infinite thickness of substrate. The dimensions of supercell are $12.94 \times 12.95 \times 20.0$ Å^3 in the x-, y-, and z-directions, respectively. The Nosé-Hoover thermostat [22][23] is employed in the canonical-ensemble at the finite temperatures. The equations of motion are numerically solved via an explicit reversible integrator [24] with a time step of 0.968 fs. In each MD simulation process, the systems are equilibrated for 5 ps at particular growth temperatures, and then annealed until room temperature by the cooling rate 200 K/ps.

3. Results and Discussion

In this section, the results of the MD simulation of graphene growth on the nickel (111) surface are reported. These include the investigation on the effect of the annealing temperatures and the concentrations of isolated carbon atoms on the nickel surface as well.

3.1. Effects of the annealing temperature on the graphene growth on nickel (111) surface.

The growth temperature affects the quality of the product material. It has been known from the experimental report that the carbon atoms on the nickel surface exist in the different forms depending on the temperature, either they dissolve into the nickel subsurface or segregate to the nickel surface to form the nickel carbide or the graphene, respectively [10].

The optimum temperature for the graphene growth on the nickel (111) surface is investigated in our study. For this purpose, the MD simulations at four different annealing temperatures, *i.e.* 800 K, 1000 K, 1200 K, and 1500 K, are performed for the system with 66.7 % of carbon concentration in which the initial configuration is shown in figure 1(a).

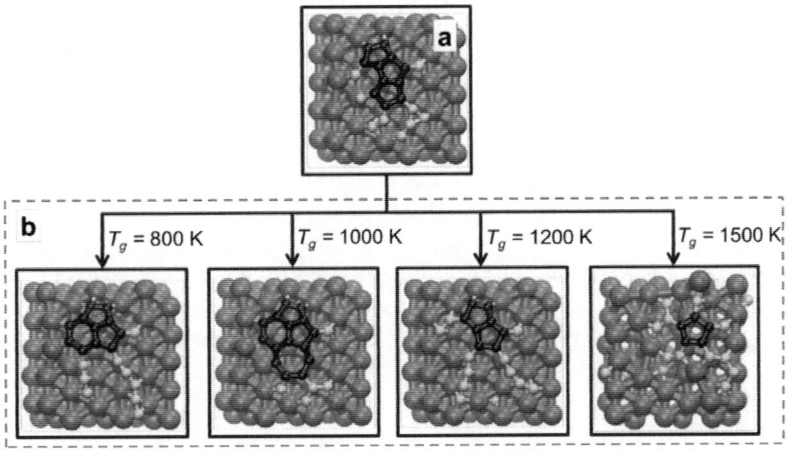

Figure 1. Atomic configuration of the graphene growth process in the system with 66.7 % of carbon concentration on the nickel (111) surface. (a) Initial, and (b) final configurations which are annealed from growth temperature (T_g) to the room temperature. Red, cyan, and black spheres represent nickel atoms, carbon atoms, and carbon chains, respectively.

Several new carbon rings are formed at the growth temperatures of 800 K and 1000 K. In this study, the carbon ring is defined as the pentagonal, hexagonal and heptagonal ring which is observed in the grain boundaries of graphene. Figure 1(b) shows that one new hexagonal carbon ring is formed at 800 K of growth temperature. A new hexagonal and a new heptagonal carbon rings are found at 1000 K of growth temperature in addition to two pentagonal carbon rings which are not broken during growth process. Since the hexagonal carbon ring is the main component of graphene, its formation indicates the existence of graphene on the system. At the growth temperatures of 1200 K and 1500 K, one and two pentagonal rings are deformed from three pentagonal rings found in the initial configuration. At the growth temperature 1500 K, some carbon atoms are dissociated from carbon chains and buried in the nickel subsurface arising from the disorder of nickel atoms at the surface. This result is in agreement with the experimental fact that carbon atoms are diluted in the surface at the temperature higher than 1200 K [10].

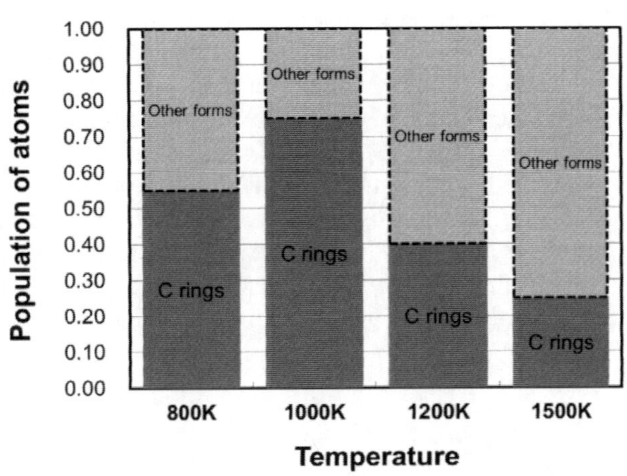

Figure 2. Population of carbon rings and other forms of carbon on the nickel (111) surface for the system with 66.7% of carbon concentration as the function of growth temperatures.

Figure 2 shows the population of carbon rings and other forms of carbon atoms at various growth temperatures. Carbon rings are formed by only 40 % and 25 % of the total carbon atoms on the nickel (111) surface at 1200 K and 1500 K, respectively. It is found that the carbon atoms as much as 55 % are in the form of carbon ring at growth temperature of 800 K. The highest population of carbon chains are obtained at the growth temperature of 1000 K thus considered as the optimum temperature for the graphene growth. This is in reasonable agreement with the recent experimental finding that the monolayer graphene is grown within the temperature range between 753 K and 923 K [25].

3.2. Effect of the carbon concentration on the graphene growth on nickel (111) surface

In this subsection, the effect of different carbon concentration on the graphene growth on the nickel (111) surface is discussed. For this purpose, three different systems containing 50 %, 66.7 %, and 100 % of carbon concentration are annealed from 1000 K to the room temperature.

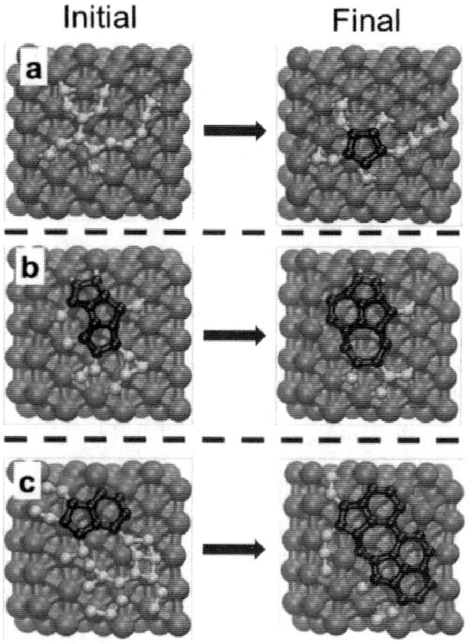

Figure 3. Initial and final configurations of the systems with (a) 50 %, (b) 66.7 %, and (c) 100 % of carbon concentration on the nickel (111) surface. All the systems are annealed from 1000 K to the room temperature. Red, cyan, and black spheres represent nickel atoms, carbon atoms, and carbon chains, respectively.

In the system with 50 % of carbon concentration, the carbon chain in the initial configuration makes a bond with another carbon on the surface to form a pentagonal carbon ring as shown in figure 3(a). When the concentration of the carbon atoms on the nickel (111) surface is increased to 66.7 %, one new hexagonal and one new heptagonal carbon rings are formed from carbon chains and a broken pentagonal carbon rings that exist in the initial configuration as it is seen in figure 3(b). For the system with 100 % of carbon concentration on the nickel (111) surface, the carbon atoms form one new pentagonal, two new hexagonal, and one new heptagonal carbon rings in addition to one pentagonal and one hexagonal carbon rings found in the initial configuration (figure 3(c)). These results are similar with the graphene growth process in the experiment that the number of hexagonal carbon rings is increased when more carbon atoms are supplied to the nickel (111) surface from the dehydrogenation reactions of carbon precursor molecules.

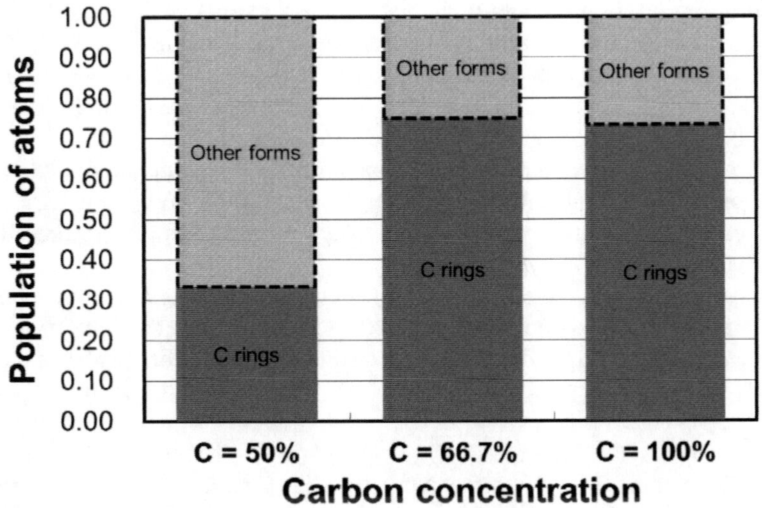

Figure 4. Populations of carbon rings and other forms of carbon as the function of carbon concentration on the nickel (111) surface after the annealing process from 1000 K until room temperature.

It can be seen from figure 4 that only 35 % of carbon population forms the carbon rings in the system with 50 % of carbon concentration on the nickel (111) surface. On the other hand, the populations of carbon atoms which form carbon rings increase to 75 % for the system with higher concentration of carbon atoms on the nickel (111) surface.

3.3. Graphene growth on nickel (111) surface in the high concentration of carbon atoms

In this subsection, we demonstrate the graphitization process of the system with 100 % of carbon concentration on the nickel (111) surface at growth temperatures of 800 K and 1000 K.

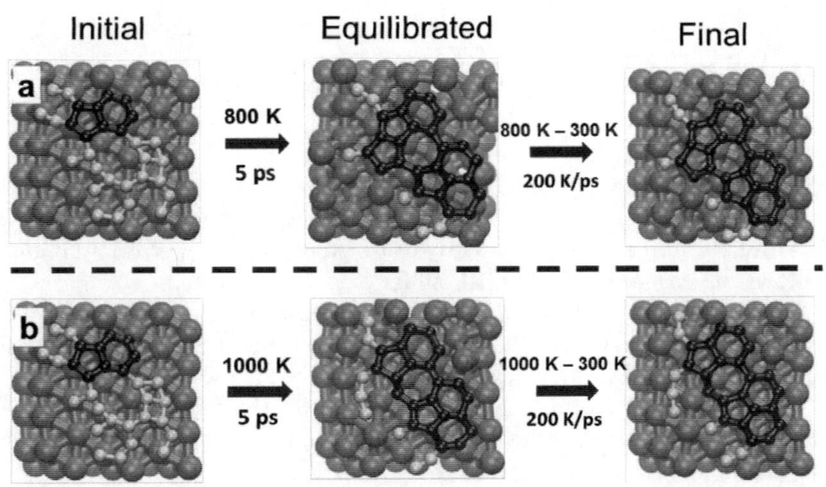

Figure 5. Initial, equilibrated, and final configurations of the graphene growth processes at (a) 800 K and (b) 1000 K using the system with 100 % of carbon concentration. Red, cyan, and black spheres represent nickel atoms, carbon atoms, and carbon chains, respectively.

In the initial configuration shown in figure 5, the carbon structures on the nickel (111) surface consist of one pentagonal and one hexagonal carbon rings and the other structures. The new formations of two pentagonal, two hexagonal, and one heptagonal carbon rings are observed after the system are equilibrated at 800 K for 5 ps as shown in figure 5(a). The number of carbon rings does not change when the temperature of the system is decreased until room temperature.

It can be seen from figure 5 (b) that the carbon atoms form one new pentagonal, two new hexagonal, and one new heptagonal carbon rings after the equilibration at 1000 K. When the temperature is lowered to 300 K, the new carbon ring is not formed. The same mechanism is also observed in the graphene growth process at 800 K of annealing temperature.

This simulation result supports the experimental finding that the monolayer graphene can be grown on the nickel (111) surface at low temperature (~800 K) using high concentration of carbon atoms [25][26]. It is also confirmed from this result that the graphitization processes occur during the equilibration.

4. Conclusion

We have investigated the graphene growth mechanism on the nickel (111) surface using first principles MD simulations. The MD simulations have been carried out for various concentrations of carbon atoms and for different annealing temperatures. The optimum temperature for the graphene growth process on the nickel (111) surface is estimated as 1000 K. It has also been confirmed that the higher concentration of carbon atoms on the nickel (111) substrate provides the better graphene formation, as well as the capability of the graphene growth at low temperature (~800 K). Furthermore, the graphitization process occurs during the equilibration of the system.

Acknowledgement

The authors acknowledge the Supercomputer Center, Institute for Solid State Physics, The University of Tokyo and the Research Institute for Information Technology, Kyushu University for the use of their facilities.

References

[1] Geim A K and Novoselov K S 2007 The rise of graphene *Nat. Mater.* **6** 183
[2] Schwierz F 2010 Graphene transistor *Nat. Nanotechnol.* **5** 487
[3] Geim A K 2009 Graphene: status and prospects *Science* **324** 1530
[4] Shapira P, Youtie J and Arora S 2013 Early patterns of commercial activity in graphene *J. Nanoparticle Res.* **14** 811
[5] Poon S W, Chen W, Wee A T S and Took E S 2010 Growth dynamics and kinetics of monolayer and multilayer graphene on a 6H-SiC(0001) substrate *Phys. Chem. Chem. Phys.* **12** 13522
[6] Jakse N, Arifin R and Lai S K 2011 Growth of graphene on 6H-SiC by molecular dynamics simulation *Condens. Matter. Phys.* **14** 43802
[7] Bae S, Kim H, Lee Y, Xu X, Park J-S, Zheng Y, Balakrisnan J, Lei T, Kim H R, Song Y Il, Kim Y-J, Kim K S, Özylmaz B, Ahn J-H, Hong B H and Iijima S 2010 Roll-to roll production of 30-inch graphene films for transparent electrodes *Nat. Nanotechnol.* **5** 574
[8] Kim K S, Zhao Y, Jang H, Lee S Y, Kim J M, Kim K S, Ahn J-H, Kim P, Choi J-Y and Hong B H 2009 Large-scale pattern growth of graphene films for stretchable transparent electrodes *Nature* **457** 706
[9] Portnoi V K, Leonov A V, Mudretsova S N and Fedotov S A 2010 Formation of nickel carbide in the course of deformation treatment of Ni-C mixtures *Phys. Met. Metallogr.* **109** 153
[10] Shelton J C, Patil H R, and Blakely J M 1974 Equilibrium segregation of carbon to a nickel (111) surface: a surface phase transition *Surf. Sci.* **43** 493
[11] Meng L, Sun Q, Wang J and Ding F 2012 Molecular dynamics simulation of chemical vapor deposition graphene growth on Ni (111) surface *J. Phys. Chem. C* **116** 6097

[12] Li H-B, Page A, Wang Y, Irle S and Morokuma K 2012 Sub-surface nucleation of graphene precursors near a Ni(111) step-edge *Chem. Commun.* **48** 7937

[13] Wang Y, Page A, Li H-B, Qian H, Jiao M, Wu Z, Morokuma K and Irle S 2014 Step-edge self-assembly during graphene nucleation on a nickel surface: QM/MD simulations *Nanoscale* **6** 140

[14] Shibuta Y, Arifin R, Shimamura K, Oguri T, Shimojo F and Yamaguchi S 2013 Ab initio molecular dynamics simulation of dissociation of methane on nickel(111) surface: unravelling initial stage of graphene growth via a CVD technique *Chem. Phys. Lett.* **565** 92

[15] Arifin R, Shibuta Y, Shimamura K, Shimojo F and Yamaguchi S 2015 *Ab initio* molecular dynamics simulation of ethylene reaction on nickel (111) surface *J. Phys. Chem. C* **119** 3210

[16] Arifin R, Shibuta Y, Shimamura K and Shimojo F 2015 First principles calculation of CH_4 decomposition on nickel (111) surface *Eur. Phys. J. B* **88** 303

[17] Blöchl P E 1994 Projector augmented-wave method *Phys. Rev. B* **50** 17953

[18] Kresse G 1999 From ultrasoft pseudopotentials to the projector augmented-wave method *Phys. Rev. B* **59** 1758

[19] Perdew J P, Burke K and Ernzerhof M 1996 Generalized gradient approximation made simple *Phys. Rev. Lett.* 77, **18** 3865

[20] Grimme S 2006 Semiempirical GGA-type density functional constructed with a long-range dispersion correction 2006 *J. Comput. Chem.* **27** 1787

[21] Shimojo F, Kalia R K, Nakano A and Vashishta P 2001 Linear-scaling density-functional-theory calculations of electronic structure based on real-space grids: design, analysis, and scalability test of parallel algorithms *Comput. Phys. Commun.* **140** 303

[22] Nosé S 2006 A molecular dynamics method for simulations in the canonical ensemble *Molec. Phys.* **52** 255

[23] Hoover W 1985 Canonical dynamics: equilibrium phase-space distributions *Phys. Rev. A* **31** 1695

[24] Tuckerman M, Berne B J and Martyna G J 1992 Reversible multiple time scale molecular dynamics *J. Chem. Phys.* **97** 1990

[25] Lahiri J, Miller T S, Ross A J, Adamska L, Oleynik I I and Batzill M 2011 Graphene growth and stability at nickel surfaces *New J. Phys.* **13** 025001

[26] Addou R, Dahal A, Sutter P and Batzill M 2012 Monolayer graphene growth on Ni(111) by low temperature chemical vapor deposition *Appl. Phys. Lett.* **100** 021601

Utilization of industrial dairy waste as microalgae cultivation medium : a potential study for sustainable energy resources

S Nurmayani*, Y Sugiarti and R H Putra

Fakultas Pendidikan Teknologi dan Kejuruan, Universitas Pendidikan Indonesia

*Corresponding author : sari.nurmayani@student.upi.edu

Abstract. Microalgae is one of biodiesel resources and call as third generation biofuel. Biodiesel is one alternative energy that being developed. So study about resource of biodiesel need a development, for the example is development the basic material such as microalgae. In this paper we explain the potential use of dairy waste from industry as a cultivation medium of *microalgae* for biodiesel production. Dairy waste from dairy industry contains 34.98% protein, 4.42% lactose, 9.77% fiber, 11.04% fat, 2.33% calcium, 1.05% phosfor, and 0.4 % magnesium, meaning that the dairy waste from dairy industry has a relatively high nutrient content and complete from a source of carbon, nitrogen and phosphorus as macro nutrients. The method in this paper is literature review to resulting a new conclusion about the potency of waste water from dairy industry as microalgae cultivation medium. Based on the study, the dairy waste from dairy industry has potency to be used as cultivation medium of *Botryococcus braunii* in the production of biodiesel, replacing the conventional cultivation medium.

1. Introduction

Along with economic growth and population growth, Indonesian energy needs has increased every year. So that the demand and consumption of fuel oil are increase. It become a problem because fuel oil is not renewable, and the sources has been decrease. According to data from the Ministry of Energy and Mineral Resources (ESDM), the energy needs of Indonesia in 2010 approximately 3.3 million barrels of oil equivalent per day and will increase two-fold in 2025, when oil production Indonesia for several years has decreased 3- 5% in the last 6 years.

Then base on the rule from Presidential Regulation No. 5 year 2006 on National Energy policy has describes renewable resources such as biofuels as an alternative fuel. One example of biofuels as an alternative energy source to replace fossil fuels are biodiesel. Biodiesel is an alternative fuel made from vegetable oils and animal fats. One of potential feedstock for biodiesel is microalgae. Microalgae has a high oil contain so it is the reason for developing microalgae as biodiesel material basic. Biodiesel from microalgae has been developed by countries in Europe and called as third generation biofuel.

Generally the method of biodiesel production process have three stage, they are preparation, microalgae harvesting and extraction. Preparation stage is a very important stage because in this stage microalgae will cultivated and need a appropriate cultivation medium. So in harvesting time will resulting a large number of microalgae biomass.

Based on the energy requirements which is continuously increasing, biodiesel from microalgae as third-generation biofuel is one of the alternative sustainable energy and needs a good manufacturing process from the raw material, cultivation medium until the extraction process. Hopefully the manufacturing process can be environmental friendly.

Indonesia is one of country with a number of ever-increasing milk production per year. Along with the increase in milk production, the waste generated from the production of milk increased. National milk production in 2014 is 1.24 million tons and milk production in 2015 is expected to reach 1.53 million tons (Antaranews: 2014). Budi and Karyono (2008) reported that the volume of wastewater generated from the dairy base is 3.9 ltr / kg of milk products. In a report of industry practice in the PT Indolakto, Cicurug, the amount of dairy waste that must be processed per day is 2000 m3 / day. (Aji, 2013)

Based on a study conducted by Insani (2006)dairy waste water has the physical characteristics with BOD value + 4000 mg / L, COD + 2000 mg / L and levels of suspended solids (TSS) dairy waste water is + 800 mg / L. Meanwhile, according to KEP-51/ MENLH/ 10/1995 ON wastewater quality standard, the maximum allowable pollution load of dairy waste water is BOD 50 mg / L, COD of 100 mg / L and TSS of 40 mg / l with discharge of waste production is 3,5 liters per kg of milk solids.

Dairy industry is one of developing industry in Indonesia and they have increased production. Along with the production, the waste from the industry has increased. The waste from the industry have two types they are sludge and waste water. The waste water from the industry which is have a high organic compound is much and polluted the environment. So that, the waste need a treatment before flowed to river body. One kind of the treatment is bioremediation with plant or decomposers microorganism. Microalgae is one of decomposers microorganism, so it is potentially to live in the waste.

Based on the explanation before, the problem in this research is how the potency of waste water from dairy industry to use as microalgae cultivation medium and how the potency of microalgae living process to remediate the waste water as a bioremediation treatment. Then the purpose of this paper is describe the potency of potency of waste water from dairy industry to use as microalgae cultivation medium and describe the potency of microalgae living process to remediate the waste water as a bioremediation treatment.

2. Materials and Method

Marlina (2007) has reported that waste water from dairy industry has high organic compound content. Here is the content of the waste water:

Table 1. Chemical Composition of sewage sludge dairy industry

crude protein	34.98%
lactose	4.42%
crude fiber	9.77%
crude fat	11.04%
Calcium	2.33%
phosfor	1.05%
Mg	0.4%

Source: Marlina 2007

From the table above it can be seen that the dairy waste water contains high nutrients that enough to be the sources of carbon, nitrogen and phosphorus as macro nutrients and supported with minerals and vitamins as a micro nutrients so it can be used as nutrients for living organisms such as plants and decomposing microorganisms.

Botryococcus braunii is a type of microalgae are included in*clorophyta* the group of algae that has chlorophyll or green substance as a main pigment. *B. braunii* is a single-celled green plants, many found in the waters of the lake, pond or brackish waters to the sea (Metzger & Largeau, 2005)

B. braunii is one type of microalgae that has a high number of akinet with hydrocarbon content as oil. The high oil contain is one of reason for developing microalgae biodiesel as third generation biufuels. The composition of fatty acids in the microalgae are very varied so that the characteristics of biodiesel produced also varies. Compared with other oil sources microalgae, *B. braunii* has a higher oil content as in the following table:

Table 2. Comparison of Oil Content In Microalgae

Species of microalgae	oil content (% dry weight)
Scenedesmus dimorphus	16-40
Chlorella vulgaris	56
Dunaliella bioculata	8
Spirulina maxima	4-9
Botryococcus braunii	* 75

Source: * Banerjee et al., 2002; Gouveia and Oliveira 2009 in Amini, Sri and Susilowati, Rini (2010)

The method used in this research is the study of literature with a variety of sources to support the ideas presented by the authors. The big picture of this research showing bellow:

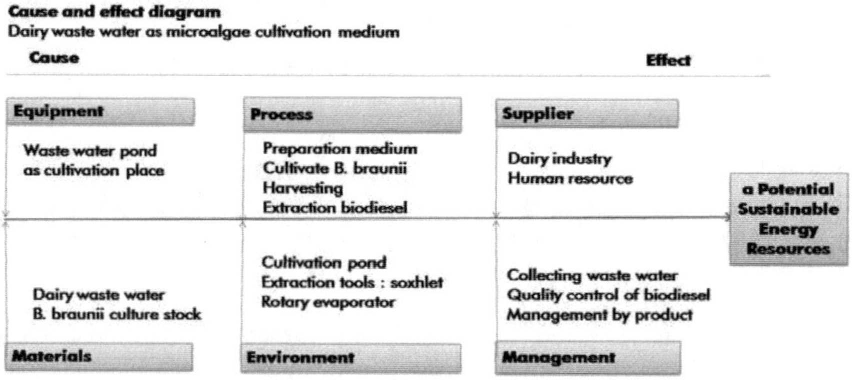

Figure 1. Cause and Effect Diagram

3. Result and Discussion

Based on the case study, the authors have an innovative idea in the form of:

3.1. *Biodiesel Production Process B. braunii*

Generally production process of biodiesel from B. braunii have here stage, they are preparation, harvesting and extraction biodiesel. The preparation stage is begin with cultivate B. braunii culture in agar medium with indoor cultivation. Then scaling up the culture with a gradual process from small scale with indoor cultivation on an agar medium, followed by a semi-indoor cultivation in the bioreactor with a capacity of 40-1000 L and outdoor as large scale. The next stage is the harvesting of biomass, one of the most efficient ways is using a chemical flocculant to precipitate cultivars *B. braunii* up to 80%. Then the final stage is the extraction of oil that can be done by several methods such as mechanical method, chemical solvents, super critical fluid, and osmotic shock. But the type of extraction method that is frequently and easily used mechanical method with pressing, its can extract oil up to 75% (Andrews, 2008) and the chemical solvents method with n-hexane as solvent, it can extract up to 95% (McMichens, 2009) so it can produce biodiesel oil from *B. braunii* as an isoprenoid triterpenes with formula C_nH_{2n-10} derived from fatty acids. n has a value of 30-37 as biodiesel range elements of hydrocracking gasoline hydrocarbon type. (Frenz et al., 1989 in Amini, Sri and Susilowati, Rini: 2010).

3.2. *Potency of dairy waste water as cultivation medium of B. braunii for producing biodiesel*
Based on studies in the previous section, dairy waste water contains 34.98% crude protein, lactose 4.42%, 9.77% crude fiber, crude fat 11.04%, 2.33% calcium, 1.05% phosfor and Mg of 0.4% based on dry matter (Marlina, 2007) so that dairy waste water has a relatively high nutrient content and complete from a source of carbon, nitrogen and phosphorus as macro nutrients. By paying attention to the data, the waste water from dairy industry has the potency to be used as B. braunii cultivation medium to produce biodiesel, replacing the conventional cultivation medium. In addition the abundance of waste water from dairy industry in Indonesia is so much and need a treatment to reducing pollutant effect.

3.3. *Cultivation of B. braunii as bioremediation efforts to dairy waste water*
According to the KEP -51/MENLH/10/1995 on Standards of Quality of wastewater has been decide that maximum allowable pollution for waste water from dairy industry describe as BOD value 50 mg/L, COD value 100 mg/L and TSS value is 40 mg/L and waste production rate is 3, 5 L/kg of milk solids. Meanwhile, according to a study from the Human (2006), the waste water from dairy industry has the physical characteristics with BOD value more than 4000 mg/L, COD value more than 2000 mg / L and TSS value more than 800 mg/L. It show that waste water from dairy industry has polluted environment and need a treatment such as bioremediation so available to flowed into river.

With the utilization of dairy waste water as a medium of cultivation *B. braunii*, the value of BOD, COD and TSS can be reduced, because the content of organic compounds as a determining factor TSS has been used by *B. braunii* as a living factor. Then *B. braunii* can synthesize oxygen and increase the levels of dissolved oxygen in the waste water. So that the levels of BOD and COD decreased and is expected less than the applicable environmental standards.

3.4. *Other findings: potency of producing bioethanol frombiomass B. Braunii as by-product*
B. braunii cell structure not only contains oil, but also contains other carbon sources such as cellulose biomass which is a by-product of biodiesel production during the extraction stage. up to now its used as an organic fertilizer. However, based on their chemical composition which is largely a polysaccharide, its very potential as raw material for bioethanol. Because the materials containing starch or cellulose can be hydrolyzed to be monosaccharide then fermented into ethanol. (Wahyudi: 2010).

Utilization of by-products into bioethanol could be an innovation in the field of renewable energy. In practice, the process of producing bioethanol from biomass *B. braunii* can be done shorter than the process of producing bioethanol from other materials that have complex stages. The process is carried out with the a fermentation system called SSF/Simultaneous saccharification and Fermentation which includes the step of pretreatment of the process to separate the hemicellulose and cellulose, then the SSF with microorganisms hydrolyzes complex carbohydrates such as *Bifidobacterium bifidum* and yeast such as *Saccharomyces cereviseae* to convert glucose into bioethanol and final distillation to purify the ethanol has been formed. (Nurmayani, *et al*, 2015)

4. Conclusion

Based on the discussion in the previous section, it can be concluded that the waste water from dairy industry has a relatively high nutrient content and complete from a source of carbon, nitrogen and phosphorus as macro nutrients. So that the dairy waste water has the potency to be used as *B. braunii* cultivation medium to producing biodiesel. Furthermore, the utilization of dairy waste water as *B. braunii* cultivation medium is one of bioremediation treatment to reduce environmental pollution from this waste. Because the content of organic compounds has been used by *B. braunii* as living factor. Then *B. braunii* can synthesize oxygen and increase the levels of dissolved oxygen in the wastewater dairy industry so that the levels of BOD and COD decreased and is expected to or less than the applicable environmental standards. In addition the process will resulting a by-product. It is B. barunii cellulose biomass which is largely a polysaccharide, its very potential as raw material for bioethanol. Because the materials containing starch or cellulose can be hydrolyzed to be monosaccharide then fermented into ethanol.

Therefore, research and technological development must be continued to updating procedures cultivation, harvesting, extraction and processing by-products as the center of attention. So resultobtained kept on improving quality and becoming a real solution to energy needs and became solution to environmental problems.

References

[1] Aji, Santana, G. H. (2013). *Laporan Praktik Lapangan Pengolahan Air Limbah Di PT Indolakto Kabupaten Sukabumi, Jawa Barat*. Bogor : Institut Pertanian Bogor.
[2] Amini, Sridan Susilowat, Rini (2010) Produksi Biodiesel Dari Mikroalga*Botryococcus braunii*. *Squalen Vol. 5 No. 1, Mei 2010*
[3] Insani, Yun. dkk (2006). *Panduan Inspeksi Penaatan Pengelolaan Lingkungan Industri Pengolahan Susu*. Jakarta : Asisten Deputi Urusan Pengendalian Pencemaran Agroindustri, Deputi MENLH Bidang Pengendalian Pencemaran, Kementerian Negara Lingkungan Hidup.
[4] Karyono, Wagini dan Budi, Agus S. (2008) Pengolahan Limbah Cair Industri Susu. Yogyakarta : Pusat Lingkungan Hidup Universitas Gajah Mada.
[5] Kementrian Energi dan Sumber Daya Mineral (2012). Laporan Akhir Tahun 2012
[6] Keputusan Menteri Negara Lingkungan Hidup Nomor : KEP-51/MENLH/10/1995 Baku Mutu Limbah Cair Bagi Kegiatan Industri
[7] Marlina, E.T. (2007). Kandungan Gizi Lumpur Susu PT Indomilk Laboratorium Nutrisi Ternak Ruminansia dan Kimia Makanan Ternak. Sumdang : Fakultas Peternakan Universitas Padjajaran
[8] McMichens, R.B. 2009. *Algae as a Source for Biodiesel.* Paper of University of Maryland, College Park library.

[9] Nurmayani, Sari, Putra, Rosdiana H. dan Tri Adi, S. Yohanes (2015) Rekayasa Produksi Bioetanol Limbah Agrokompleks Industri Teh Sebagai Energi Terbarukan, Ramah Lingkungan dan Berkelanjutan Untuk Mendukung KetahananEnergi Indonesia. Universitas Pendidikan Indonesia

[10] Wahyudi dan Supriyanto, Tri. (2010) Proses Produksi Etanol Oleh SaccharomycesCerivisiae Dengan Operasi Kontinyu Pada Kondisi Vakum. [Undergraduatethesis], Jurusan Teknik Kimia Fakultas Teknik Universitas Diponogoro.

[11] Metzger, P., Berkaloff, C., Casadevall, E., and Coute, A. 1985. Alkadeine-and botryococcene-producing races of wild strains of *Botryococcus braunii. Phytochemistry.* 24: 2305–2312.

[12] Peraturan Presiden Republik Indonesia Nomor 5 Tahun 2006 tentang Kebijakan Energi Nasional

[13] Subagyo. (2014). *Produksi susu tahun ini ditargetkan 1,24 juta ton.* [online]tersedia di http://www.antaranews.com/berita/436968/produksi-susutahunini-ditargetkan-124-juta-ton. diakses pada tanggal 18 september 2015.

The land-use of Bandung, its density, overcrowded area and public facility toward a compact city

B Paramita*

Jl. Dr. Setiabudhi no.229, Bandung 40154

*Corresponding author: betaparamita@upi.edu

Abstract. The concept of a compact city has been introduced since 1973. It is a utopian vision largely driven by a desire to see more efficient uses of resources. In 1980s, the reconfiguration of the physical urban form of metropolitan areas was increasingly debated by both theorists and practitioners. Recently, the concept of a compact city has been more focused on developed countries in which the population tends to decrease. However, in Asia, except Japan which contains many dense cities, it has become a concept which promotes relatively high residential density with mixed land uses, though rather only in population and density. This paper addresses the land-use of Bandung that having the density over 14,000people/km2, which has been so much potential toward a compact city. Somehow, unprepared ness of urban planning and regulation, the city seemed overwrought to serve its inhabitants. This condition is shown from the demographic condition, especially population density in Bandung based on its sub areas of the city (SWK). The stack of public facilities in a certain district has led the concentration of density and activity, which finally raising the slum and overcrowded settlement. Finally, this paper explores the implications of land use management and describes challenges faced and possible approaches, especially in land-use management strategies to be implemented in Bandung.

1. Introduction

The number of urban areas, especially in developing countries in the coming years has been predicted to be more widespread. One of the most serious problems in the 21st century is no other than the global population growth, and consequent urban expansion, particularly in developing countries. Based on Demographia (2015), 57% population live in Asia continent, which most of them are developing country. The concern of the urban form, infrastructure as well as the density is considered as an important factor in understanding how cities function. Density then becomes one of the demography problems, which related to urban spaces has risen, such as overcrowded settlements, mostly inhabited by squatters, spread throughout the downtown area. The unprepared ness of spatial planning and law enforcement to keep implementation on track seems to become a big problem for many urban spatial planners in some cities in Indonesia. Urban environmental problems tend to increase exponentially. The idea of the compact city which requires the population density and optimal land use, it seems only managed in developed countries, mainly because of the number of inhabitants slowly decreasing. In contrast, in developing countries, there are many dense cities that promote a high population with their inhabitants living in the mixed land uses area, but somehow the compact city becomes just a term to describe an amount of density, rather than a well-managed city.

There is the issue of how best to design and implement compact city, these are mostly formulated for existing cities, each with its specific context, a locality needs to adapt its compact city strategies to fit its particular circumstances. In other words, no single comprehensive compact city model is applicable to all cities. The implication of compact city in a developing country somehow is interesting to discuss here. Thus, the challenge of compact city concept in developing country, such as in Indonesia is interesting to explore, in what extent this concept can be applied to cities in developing countries. Bandung then, has been chosen as a area study, as the capital city of West Java Province which is the most densed province in Indonesia.

2. High density vs Compact city

2.1. High Density in Developing Country

Density, as employed in land use planning and related applications, appears to be a simple concept. However, amounts of density are often misinterpreted and distorted due to the numerous technical complexities involved in coming up with them. Therefore, it is significant that the scales of geographical references be explicitly limited in density calculations, otherwise the comparison of density measures will be unmanageable. Nevertheless, there is no standard measure of density; there are only measures that are more widely used than others[1].

Densities in developing country cities are much higher than in developed countries, especially in the core city. Many explanations have been given: [2]

a. Higher rates of population and urban growth
b. Lower incomes have meant much smaller dwelling sizes and tiny lots,
c. In terms of the 'modern' sector in developing countries, housing preferences have favored high-rise apartments rather than single family housing.
d. The prevalence of mixed-use structures and neighborhoods has resulted in residential densities becoming higher the closer they are to the city centre.
e. Lax planning regulations, and either the absence or the inadequacy of building codes, have led to uncontrolled increases in densities
f. There remains considerable scope for further increases in densities in low income settlements

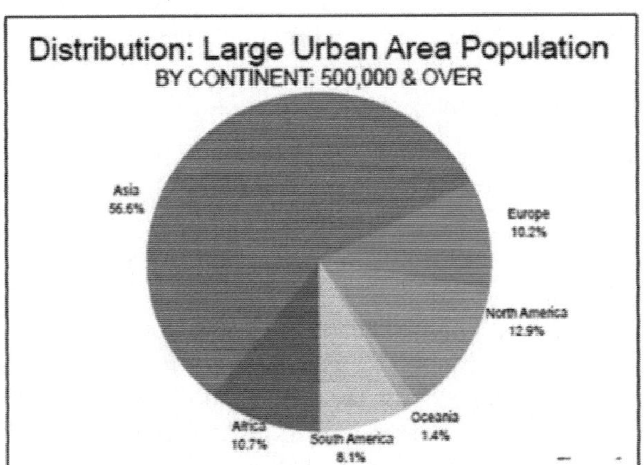

Figure 1 shows the continent with the largest urban area population in the world is Asia, which has a population of more than 500,000 people. The next table 1 describes the top 10 cities in Southeast Asia based on population density.

Table. 1. Largest Urban Areas in Southeast Asia, based on population density

1	Metro Manila [3]	11,855,975	638.55	18,567
2	Bandung + Cimahi [4]	3,017,681	208.03	14,506
3	Jabodetabek (metro) [5]	17,220,076	1345.82	12,795
4	Yogyakarta [6]	394,012	32.5	12,123
5	Surakarta [5]	520,061	44.03	11,812
6	Yangon	5,998,000	598.75	10,018
7	Surabaya [5]	3,314,700	374.78	8,844
8	Singapore [7]	5,535,000	719.1	7,697
9	Kuala Lumpur [8]	1,627,172	243	6,696
10	Bangkok [9]	8,280,925	1,568.74	5,279

As the 4th most populous country in the world, Indonesia still dealing with uneven population distribution, there are 12 cities with high populations, from a total of 510 cities spread throughout 34 provinces. These cities have population densities of more than 10,000/km2, and all of them are located on Java Island. The urban form characteristic exibits the overcrowded neighborhoods with relatively low FAR but incontrary high dwelling unit. Most informal developments have a similar density measurement profile. Moreover, consider its high population, it does not give each person a sizeable amount of personal living space.

Thus, the density here, except in Singapore hardly to say that cities with urban form characteristics above are the result of some prescient ground design. Rather, their compactness has emerged spontaneously, if not chaotically.

2.2. Compact city in Developing Country

Discusses how the application of compact city in the developing country, is inseparable from term of : ''What is sustainable urban form?'' Furthermore, the term sustainability is currently quite vague and has not been defined clearly. However, the following is likely to be the common discussion dimension in recent years; that has three aspects [10].

a. To keep a stable economic growth with restructuring the productive system for saving recourses and energy, (Economic sustainability)
b. To maintain safe and comfortable living environment through zero emission, (Environmental Sustainability)
c. To secure social equity in distribution of wealth and social services (Social Sustainability)

Anyhow, to achieve the terms of ''sustainability'' as mentioned above, it seems the road is still long and left a lot of homework to be completed. The very basic problems in the higly densed city in developing country are often described mostly in living environment are not in good condition. The traffic congestion, noise and air pollution are something that is constantly faced every day. The overcrowed settlement which then an embryo of slum and squatter settlement with poor sanitary condition due to lack of tap water and sewerage. These settlements are common to find in the cluster of commercial area.

Focus on compact city implementation then is the intensification of the use of space in the city. This proposal is sharpening the distinction between density and intensity of development, for they are not the same thing. Density is a quantitative measure of number within a prescribed area, whereas intensity reflects a more subjective measure of built-up-ness or urbanity. Density itself is of little importance unless it is related to built form. Compact is meaningless unless it is related to some facts and figures [11]. With higher residential densities and centralization, and they write that 'planners should aim for compactness and integration of land uses, for some degree of "self-containment". Furthermore, Newman and Kenworthy also demand more intensive land use, centralized activity and higher densities. [12]

3. Study Case: Bandung, a Capital of Densest Province in Indonesia

Bandung is the second largest metropolitan area in Indonesia, after the capital city of Indonesia, Jakarta. With 2,470,802 citizens recorded inhabited in the land of 167.67 km^2 bringing the density of Bandung into 14.736 / km^2. The population growth rates in Bandung from 2005 to 2010 are reported 1.15%.[1] Bandung is divided into six development areas[2], shown at figure.2.

In previous research it is found that there is no correlation between population and overcrowded areas, nor the population density to overcrowded areas [13], the correlation between three of them are shown by figure. 3a,3b, and 3c.

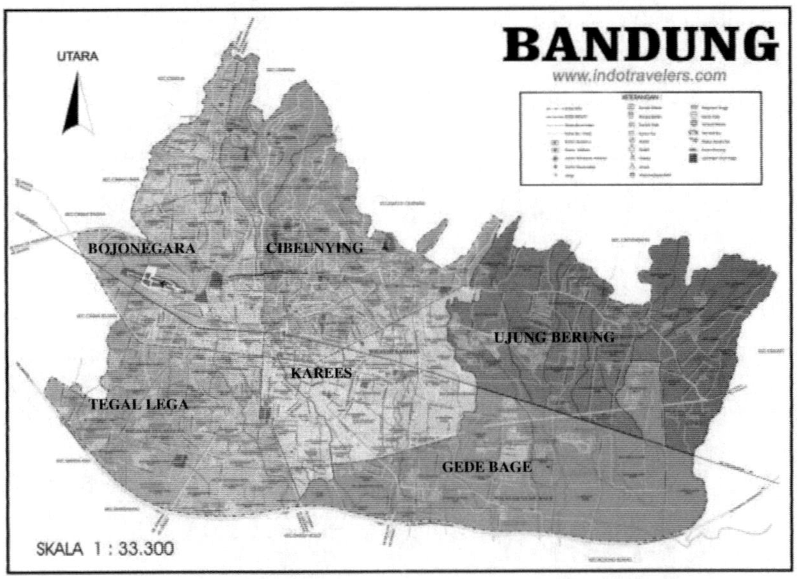

Figure 2. Development areas (SWK) in Bandung, 2007

[1] http://bandung.go.id/

[2] From 2011 this development area is change into 8 areas, but the supporting data are not compatible enough, so it was decided to take the data in 2007

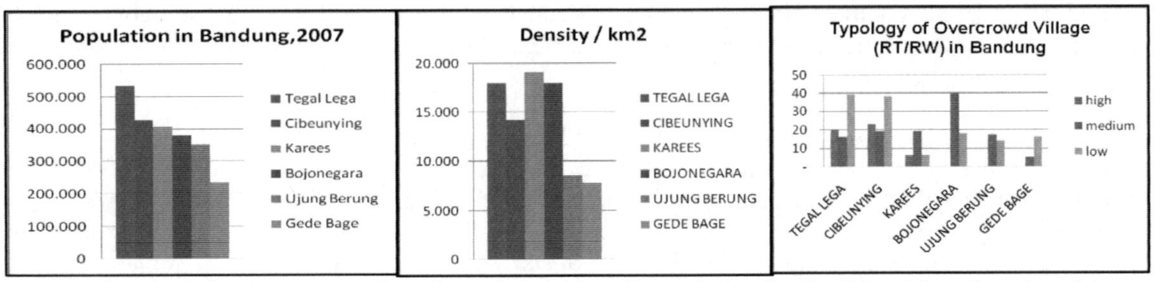

Figure 3. Correlation between number of population (a) with population density (b) and its urban slum (c)

As shown in the above figure, the most overcrowd area is found in Cibeuying district, meanwhile the highest population is in Tegal Lega district and the most densest area is in Karees district.

From figure.4, it emphasizes the statement that the overcrowded area as an urban slum and squatter does not occur as a result of the density.

3.1. Bandung Land-use

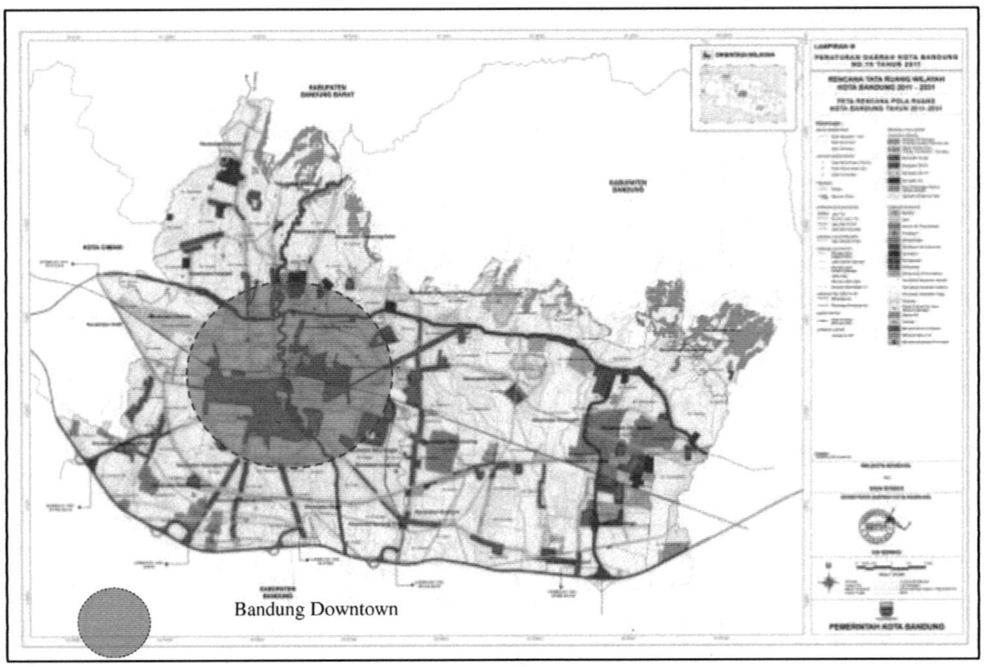

Figure 4. Bandung Land-use Master-plan 2011-2030
Retrieved from: http://bapeda.bandungkab.go.id/

The condition of Bandung density becomes a dilution that restrains from a compact city. Furthermore, this condition will be explained by giving the spread of Bandung Master Plan of land-use (figure.4).

Bandung downtown is located in the intersection of area between Cibeunying, Karees, Tegal Lega and Bojanegara. This area is called high economic productivity of land, which accounted for regional economic growth, and also giving the negative effects of the impact of urban growth.

The elements that affect to urban growth which influence on the percentage of land use is shown in figure.5.

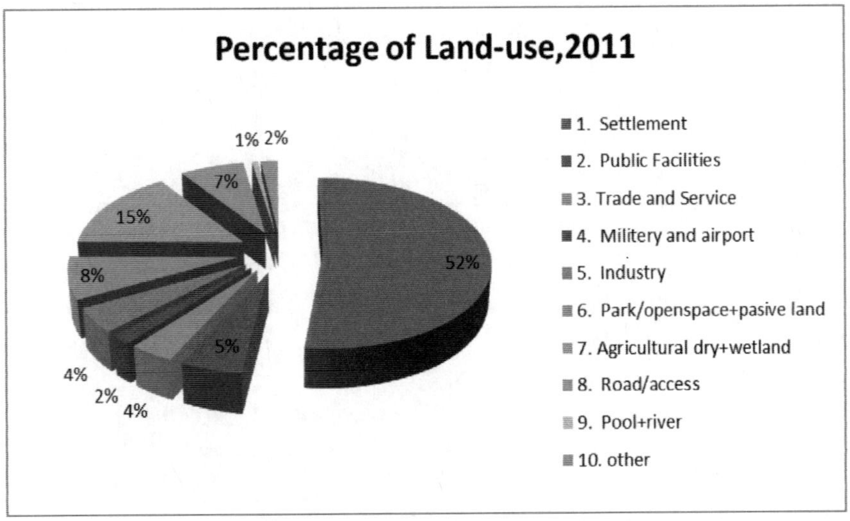

Figure 5. Percentage of Land-use
Processed from Bandung in figure, 2011

The key points of a compact city as mentioned before are described as:

- promoting the 'intensification of the use of space in the city' with higher residential densities and centralization;
- planning for compactness and integration of land uses, for some degree of "self-containment";
- demanding more intensive land use, centralized activity and higher densities.

Thus, physical elements, such as a group of buildings with their activities and less-movement of an inner city, have a strong influence to induce the success of a compact city.

Based on the land-use percentage in Bandung, the physical elements of the impact of urban growth can be categorized into:

- Service (education, government, health)
- Settlement
- Industry
- Commercial and recreation

The problem then occurs when these physical elements are stacking in a clustering area. As mentioned, it is in the downtown area, which is increasing the movement from one district to other districts.

The main problem of this movement is because of the imbalance of public facilities, especially the quality of education and health facilities.

The top five good schools are stacking in Cibeuying area, from elementary level to high school. This is compounded by the existence of state universities which are also located in Cibeunying.

3.2. Land-use Assessment

Land-use assessment is intended to determine the level of use of the land used by the physical elements (building function) that generates movement in the city.

The selected elements are:

- Schools (divided from elementary to higher education)
- Government
- Health Facilities
- Commercial
- Industry
- Settlement

Table 2 Scoring of Land-Use

no.	Physical Element	Indicator	Element land-use	Tegal Lega	Cibeunying	Karees	Boja-negara	Ujung Berung	Gede Bage
1	SCHOOL (top 3 from elementary school to University)	Favorite	SDN Sabang		v				
			SDN Merdeka		v				
			SDN Banjarsari		v				
		Passing Grade	SMP 2		v				
			SMP 5		v				
			SMP7		v				
			SMAN 3		v				
			SMAN 5		v				
			SMAN 8			v			
			ITB		v				
			UNPAD		v				
			UPI				v		
2	GOVERN MENT	Main office	Cityhall and mayor's office		v				
			governor office		v				
			BAPPEDA		v				
3	HEALTH	Main Hospital	RSHS				v		
			RS. BOROMEUS		v				
			RS. ADVENT				v		
4	COMMER CIAL	Biggest Mall	Cihampelas walk		v				
			BIP		v				
			BSM			v			
5	Industry	trading traffic	Cibaduyut			v			
			Binong Jati			v			
			Cihampelas		v				
6	Settlement	Higest price	Dago		v				
			Setiabudhi				v		
			Batununggal			v			

Diagram of Land-use scoring as shown in figure 6 below:

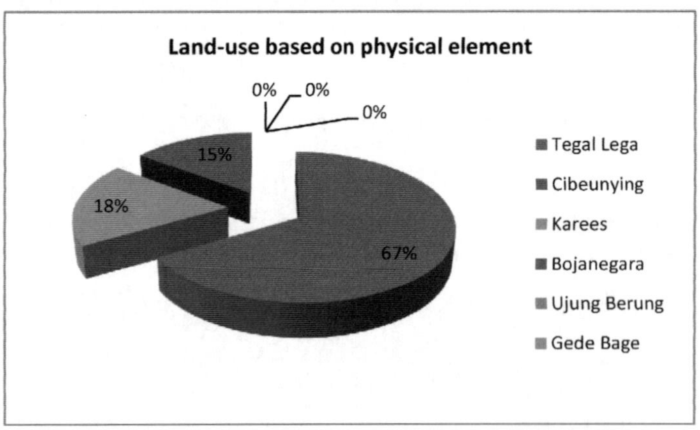

Figure 6. Stack of Land-use distribution in Bandung

The diagram beside explain that SWK Cibeunying is the densest land use for educational, governmental, health, commercial, industrial and residential. Previous study about SWK Cibeunying has found that the change of land-use in this SWK has started in 2000. The change of housing function into rental office has changed in 31%, from 60% of total land-use has changed in Cibeuying area. [14]

The trend of land-use changing, especially penetrating the land-use of settlement shows there is lack of supervision by the city government to implement the Master Plan 2013. Furthermore, in his article, Tanuwidjaya giving the analysis that ''Master Plan 2005 was failed to realise the carrying capacity of Bandung (natural and social). Due to rapid urbanization, the Master Plan 2013 had to increase the population capacity of Bandung from 750'000 persons ("Karsten Plan") to 3 million persons. Further, the integrated land use and transportation planning approach in Master Plan 2013 were failed to be implemented due to land and development speculation by private sectors. The current commercials are mostly designated mostly in the city centres, while affordable housings are located in the suburban area. The separation actually will decrease accessibility, city vibrant life and livelihood issues. And this would worsen the city traffic jams. '' [15].

4. Conclusions

The description of statistical data on land use in Bandung and the influence of its physical elements indicate that the compact city cannot be achieved only by relying on the sheer density. The fundamental difference of compact cities in developing countries lies in the management of land use that has not been well managed and proportional; the main problem is the quality and uneven infrastructures.

SWK Cibeunying then, becomes the most vulnerable SWK, and seriously urgent to make intensification land-use in order to avoid a decline in the environmental quality and quality of life of city residents.

References

[1] Edward Ng, e.a., Designing High-Density Cities for Social and Environmental Sustainability. 2010, Sterling, USA: Earthscan.

[2] Burgess, M.J.a.R., *Compact Cities: Sustainable Urban Forms for Developing Countries*. 1st ed. 2000, New York: Spon Press.

[3] Phillipines_Statistics_Authority, *2000 Census-Based Population Projections*. 2010.

[4] BPS_Jabar, *West Java in Figure (Indonesian : Jabar dalam angka)*. 2014.

[5] BPS, *National Census 2010 (Indonesian Sensus Penduduk 2010)*, in *5 years*. 2010, Badan Pusat Statistik: Jakarta.

[6] BPS_Jogjakarta, *Jogjakarta in Figure*. 2012.

[7] Departement_of_Statistic_of_Singapore, *Population of Singapore*. 2015.

[8] Dept_of_Statistics_of_Malaysia, *Population and Demography Malaysia*. 2015.

[9] NSO_Thai, *Population and Society*. 2011, National Statistics Office

[10] Hideki KAJI, H.K., Ken-ichi ISHIBASHI, Nobuhiro HARA, *Compact City and Developing Countries*, in *Open Meeting of the Global Environmental Change Research Community*. 2003: Montreal, Canada.

[11] Vale, E.S.a.B., *How Compact is Sustainable—How Sustainable is Compact?*, in *The Compact City:A Sustainable Urban Form?*, E.B.a.K.W. Mike Jenks, Editor. 1996, Spon Press: Oxford, UK.

[12] Mike Jenks, E.B.a.K.W., *The Compact City: A Sustainable Urban Form?*, ed. E.B.a.K.W. Mike Jenks. 1996: Spon Press.

[13] Beta Paramita, H.F. *Passive Design through Solar Envelope for Vertical Settlement*. in *IEEA*. 2012. Singapore: IEEA.

[14] Anugrahani, I.K.d.M., *perubahan pemantaatan lahan perumahan ke perkantoran: implikasinya terhadap pengendalian pemanfaatan ruang kota.* Jurnal PWK, 2000. **11**(1).

[15] Gunawan Tanuwidjaja *the city of bandung and review of bandung spatial planning strategies in 2005*. 2005. DOI: https://greenimpactindo.wordpress.com/2010/03/25/the-city-of-bandung-and-review-of-bandung-spatial-planning-strategies-in-2005/.

IOP Conf. Series: Materials Science and Engineering **128** (2016) 012035 doi:10.1088/1757-899X/128/1/012035

Effect of modified yam (*Dioscorea esculenta*) flour on some physicochemical and sensory properties of synbiotic yoghurt

M N Handayani*, D Cakrawati and S Handayani

Agro-industrial Technology Education Program, Faculty of Technology and Vocational Education, Universitas Pendidikan Indonesia

*Corresponding author: mustika@upi.edu

Abstract. The aim of the study were to know characteristics of yam modified flour; to know the effect of modified yam flour on some physicochemical and sensory properties of synbiotic yoghurt and to determine the concentration level of modified yam flour to produce symbiotic yoghurt preferred by panelists. The reasearch was conducted using one factor complete randomized design. Modified yam flour was added to yoghurt at concentration of 2%, 4%, 6%. The effect of physical modification were investigated. Proximate analysis showed modified yam flour consist of 7.66% moisture content, 1.42% ash content, 10.16%, dietary fiber, 7.49% inulin, and 71.78% total starch content. Result obtained that modified yam flour has yield of 10.54%. the modified yam flour showed solubility and water absopsion of 77,63% and 136,65 respectively. The addition of modified yam flour on yoghurt resulted significantly difference effect on texture, but did not have significantly difference on colour, flavour and aroma. Modified yam flour added yoghurt thickness because it was gelatinized when added to yoghurt at 40°C. Sensory analysis conducted with hedonic test showed synbiotic yoghurt added with 2% of modified yam flour most preferred by panellists. Synbiotic yoghurt with 2% of modified yam flour has pH number of 4,8 and total acid tirated of 1,7%.

1. Introduction

Currently yogurt promoted massively through various media as functional beverages, to promote healthy digestion. Yoghurt consumption has increased among consumers especially women since they believed that yoghurt could help them to maintain body slimness, improve skin beauty and prevent premature aging [1]. Increasing in yoghurt consumption is expected to continue along with the increasing of public awareness on functional food importance. Functional food not only fulfill the body basic needs of nutrients but also give benefit health effect. According to [2] Indonesia was one of the world's largest yoghurt consumers. Yoghurt is present in Indonesia in a variety of brand by offering a variety of health-related advantages. One of them is the latest variant of synbiotic yoghurt.

Synbiotic yoghurt contains probiotic bacteria and prebiotic component [3]. Probiotics and prebiotics may provide synergistic benefits to health to provide the balance of microflora in the digestive system. Prebiotics can stimulate the growth of good bacteria in the digestive tract as well as during the process of milk fermentation [4]. Thus, the presence of probiotics and prebiotics in the form yoghurt synbiotic can inhibit the growth of microbial pathogens and maintain the balance of intestinal micro flora. They improve absorption of minerals, prevent incidence of diarrhea, and optimize the assimilation of nutrients [5]. Prebiotic component can be obtained from several sources of agricultural products, one of them was inulin contained in yam (Dioscorea esculenta) tuber. Inulin from yam tubers can maintain the growth of probiotic bacteria. Inulin prebiotic activity was affected by incubation time, where the longer incubation can enhance the value of yam tuber inulin prebiotic activity. Therefore, inulin from yam tuber can be used as a prebiotic source and added to food products [6].

Addition of yam flour in yoghurt were expected to provide a prebiotic effect because of its inulin content [7]. Yam flour can not be added directly to yoghurt since yoghurt has low pH and using heat in its processing. Thus, yam flour required modification so it would produce desired yoghurt characteristics. In addition, modification yam flour can increase the levels of resistant starch in the flour [8]. Resistant starch is resistant maltodextrin belonging to the oligosaccharides that can be fermented in the intestine often referred to as dietary fiber. In other words starch modification can increase fiber content of food. Dietary fiber needed by the body as a prebiotic. Therefore, research is needed on the addition of yam flour modified in yoghurt products. The aim of this study was to determine characteristics of modified yam flour, determine effect of modified yam flour addition on some physicochemical and sensory properties of synbiotic yoghurt and determine the concentration level of yam flour modified to produce symbiotic yoghurt most preferred by panelists.

2. Materials and Methods

Yam (*Dioscorea esculenta*) tubers were obtained from farmers, whole milk, yoghurt cultures consist of *Lactobacillus bulgaricus, Streptococcus thermophilus* and *Lactobacillus acidophilus.* Yam flour was modified physically according to [9]. Proximate value were determined including moisture content, ash, dietary fiber and inulin.

The research was conducted by several stages: (1) Making of modified yam flour; (2) Analysis of characteristics of modified yam flour; (3) producing synbiotic yoghurt with modified yam flour addition; (4) Analysis of physicochemical and sensory properties of symbiotic yoghurt.

The research method was using complete randomized experimental design. The treatment factor was concentration of modified yam flour with three concentration levels namely 2%, 4%, 6%. Each treatment was repeated 2 times to minimize experimental error. The analysis was performed using Anaysis of variance (ANOVA). The effect of modified yam flour addtion to yoghurt chareceristic were investigated. Analysis were carried out on chemical characteristics included pH and total acid titrated. Sensory analysis conducted using hedonic test for texture, colour, aroma and flavour. The flowchart of diagram of modified yam flour processing can be seen in Figure 1.

Manufacture of synbiotic yoghurt with the addition of yam flour modified, consists of several stages, i.e. the main raw material preparation of fresh whole milk and then pasteurized at a temperature of 65 C for 15 minutes. Then, yam flour modified is added to the various treatments which is 2%, 4% and 6% then stirred until homogeneous. Subsequently cooled to a temperature of 40 C for inoculation with the starter yogurt as much as 3% consisting of *Lactobacillus bulgaricus, Streptococcus thermophillus and Lactobacillus acidophyllus* as probiotic bacteria. Anaerobic incubation yoghurt conducted at a temperature of 37C for 12 hours.Here is a diagram of processingsynbiotic yoghurt with the addition of yam flour modified (figure 2).

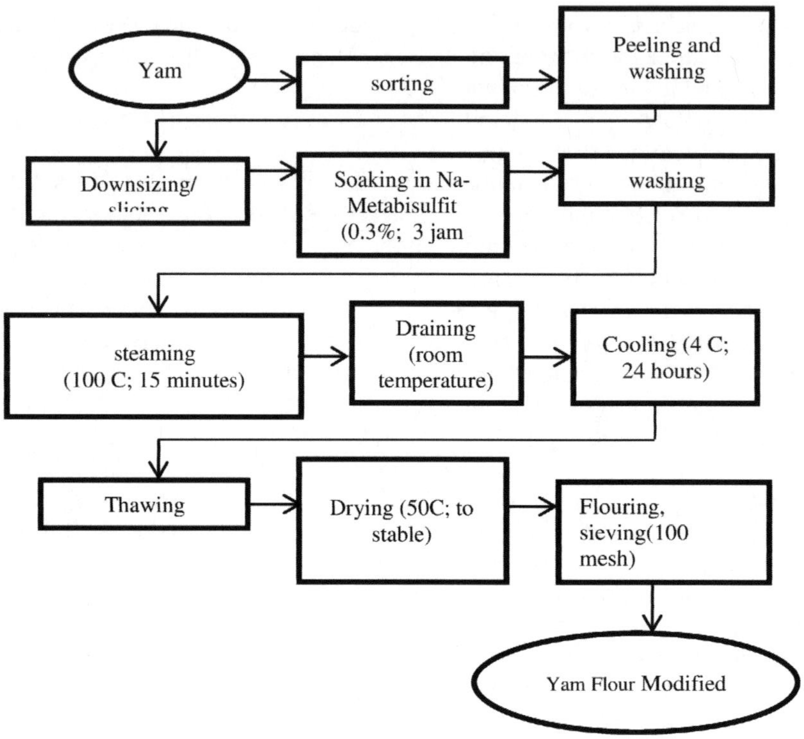

Figure 1. Diagram processing of yam flour modified

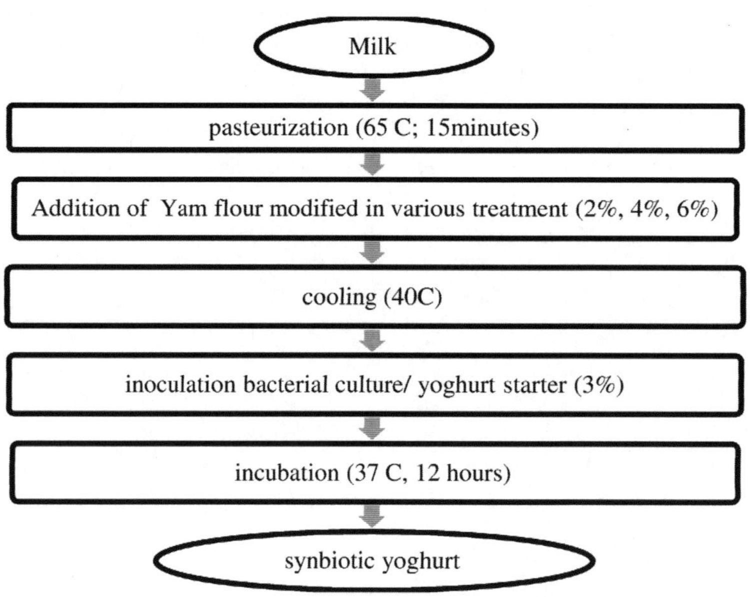

Figure 2. Diagram processing of synbiotic yoghurt with addition of yam flour modified

3. Results and Discussion

3.1. The Characteristics of Modified Yam Flour

Modification of starch and flour can be chemically or physically modified. In this research, floue modification using physical treatment which was steaming. Research conducted by [8] shows the process of pre-cooking (combined heating and cooling) can increase the levels of resistant starch in bananas. Yam has white tuber flesh color so that it produce white color flour. Shortly after peeled, yam tubers were immersed in 0.3% sodium meta-bisulfite solution. The maximum limit use of sulfites in dried foods by the FDA (food drug administration) is 2000-3000 ppm. Immersing is intended to prevent browning reactions that may occur during the process of peeling and drying. Yam flour modification produced a white, powdery substance with a degree of fineness of 100 mesh, has a normal aroma and taste. Sensory characteristics of yam flour is influenced by raw material used (yam tuber) and various treatments during the flour making.

The yield of yam starch modification obtained in this study is 10.54%. Yield is the percentage of flour weight compared to yam tubers intact, meaning that 100 grams of yam tubers can be obtained 10.54 grams of yam flour modified. Physicochemical characteristics of modified yam flour are presented in Table 1.

Table 1. Characteristics of yam flour modified

Characteristics	Yam (*Dioscorea esculenta*) flour modified
Yield	10.54 %
Solubility	77.63 %
Water absorption	136.66 %
Swelling power	618.87 %
Moisture	7.66 %
Ash content	1.42 %
Total starch content	71.78 %
Total dietary fiber content	10.16 %
Inulin Content	7.49 %

Water absorption test was conducted to determine how much flour to absorb water which would affect the treatment process when the flour applied to food processing. Physical modification yam flour has a high water absorption 136.66%. This value is influenced by heat treatment in the process of making physical modifications flour. Yam starch in the flour has undergone gelatinization, so it is no longer in amylose starch granules; it diffuses out with amylopectin. Amylose which is no longer in granules, it will be easier to form hydrogen bonds and more easily absorb water.

Swelling power value characterizes the strength of the flour to inflate derived from the ratio of the weight of sediment pasta starch (supernatant) with a dry weight of wheat flour to form a paste. Yam flour modified has swelling power high value i.e. 618.87 %. This happens because of heat treatment which causes the starch in the flour gelatinization previous experience or a so-called pre-cooked so that the flour has a greater force expands.

Solubility is the ability of the material to dissolve in water. This solubility is a trait that is quite important for the usefulness of starch in food processing. Yam flour modified has high solubility values 77.63%. This is possible because the starch in the flour so that the experience gelatinization constituent molecules such as amylose and amylopectin starch diffuses out of the starch granules, the solubility will increase.

Yam flour modified obtained in this study is a dry material that has moisture content 7.66% so including foodstuffs non-perishable and this value meets the standards of quality flour in general which, according to the Indonesian National Standard (SNI) the maximum moisture content of flour about 14%. Yam flour modified has ash content 1.42%. Ash content indicates mineral content in a food [10]. Mineral contained in a food can be in the form of salts of organic and inorganic salts. According [11] all commercial starch derived from cereals and tubers contain small amounts of inorganic salts can be derived from the material itself or from the water during treatment.

Starch contained in yam flour modified based on this research is 71.78%, This shows that the starch content owned still high enough that can be utilized in further food processing applications. As well as total dietary fiber content in the yam flour modified is quite high at 10.16 % derived from the formation of resistant starch as a result of heating-cooling treatment when modifications yam flour. This makes the value of yam flour physical modifications can be categorized as a source of dietary fiber because the food can be claimed as a source of dietary fiber if it contains dietary fiber of 3-6 g / 100 g[12]. In addition, high fiber food in the yam flour modified can be used as a prebiotic. This is similar to the one proposed by [13] that ingredient food can not be digested will provide beneficial effects for the body because it supports the activities of microflora of the digestive tract, so it can provide healthful effects on the body.

Yam flour modified in this study had higher inulin content at 7.49% where the value is in line with the statement Winarti, et al [14] regarding the content of inulin contained in tubers group Discorea spp. which ranged from 2.8 to 14.7%. Inulin is not digested by the human digestive enzymes, which can stimulate the growth of good bacteria in the large intestine (Lactobacillus sp and Bifidobacterium sp). Inulin is fermented by microflora present in the colon making is prebiotic inulin.

3.2. Characteristics of synbiotic yoghurt with yam flour modified addition

3.2.1. pH of synbiotic yoghurt

World nutrition standards of yoghurt containing a maximum pH of 4.5. The low pH of yoghurt is caused by activity of lactic acid bacteria are *L.bulgaricus* and *S.thermophillus*. *Streptococcus thermopilus*had first role in yoghurt fermentation to decrease pH to about 5.0 and then *Lactobacillus bulgaricus*decrease again until it reaches a pH of 4.0. In this study, the pH of yoghurt sinbiotik with the addition of yam flour modified ranged from 4,38 to 4,56.

Table 2. pH of synbiotic yoghurt with the addition of yam flour modified

Concentration level yam flour modified	pH synbiotic yoghurt
Without added yam flour modified	4.68
Added yam flour modified 2 %	4.48
Added yam flour modified 4 %	4.38
Added yam flour modified 6 %	4.56

Based on these data (table 2), it can be seen that the addition of yam flour modified 4%as a source of prebiotic effect on a decrease in pH 4.38. Whereas when the addition of flour to 6% an increase in the pH to 4.56. As revealed in previous studies conducted [15] by using tubers of arrowroot, more concentration of flour is added, lower the pH value. Yam flour modified addition 4% to 3% starter alleged to provide optimum results for microbes to ferment for a comparison of flour and yoghurt working culture that is not too different. Inulin and oligosaccharides as nutrients more bacteria present in the flour will be faster to lower pH than starch which is only slightly contains inulin and oligosaccharides [15].

3.2.2. Lactic acid levels of synbiotic yoghurt

Lactic acid is the result of bacterial activity *L.bulgaricus* and *S.thermopilus* that breaks lactose during yoghurt fermentation so that both bacteria called lactic acid bacteria. The content of lactic acid in yogurt is one indicator of the quality of yoghurt. Yogurt lactic acid levels that meet the quality standards according to national standards of Indonesia [16] ranged between 0.5% - 2% (w / w). Yogurt lactic acid levels according to international standards (codex) fermented milk product that is at least 0.6% (w / w) and a minimum of 0.9% (w / w) according to the USDA.

Table 3. Lactic acid levels of synbiotic yoghurt with the addition of yam flour modified

Concentration level yam flour modified	Lactic acid levels of synbiotic yoghurt
Without added yam flour modified	1.2 %
Added yam flour modified 2 %	1.7 %
Added yam flour modified 4 %	2.02%
Added yam flour modified 6 %	1.59%

In this study, sinbiotik yogurt contains lactic acid that meet the standards (table 3), whether ISO, the international standards (Codex) and the USDA. Lactic acid levels are highest are owned by yoghurt sinbiotik yam flour modified with the addition of as much as 4%. The lactic acid produced from the fermentation of milk into yoghurt will lower the pH value of the growth environment and cause a sour taste [17]. Aside from the influence of the number of microorganisms, the yoghurt sinbiotik also influenced by the composition of prebiotics in the flour.

3.2.3. Sensory properties of Synbiotic Yoghurt: color attributes

The color of yoghurt is influenced by raw materials and components that are added also during the fermentation process. At sinbiotik yogurt, milk as a raw material has a yellowish white color. Milk white color on the effect of the spread of colloidal droplets of fat, calcium caseinate (a colloidal dispersion opaque) while the yellowish color influenced the content of carotenoids and riboflavin [18]. Yam flour modified added in the manufacture of yoghurt sinbiotik has a white color so resulting white yoghurt. The results showed (table 4) that the differences in the addition of yam flour modified did not show significant color differences according to panelists at the organoleptic assessment.

Table 4. Color attributes of synbiotic yoghurt with the addition of yam flour modified

Concentration level yam flour modified	Color attribute scores of synbiotic yoghurt
Without added yam flour modified	2.30 [a]
Added yam flour modified 2 %	2.35[a]
Added yam flour modified 4 %	2.05 [a]
Added yam flour modified 6 %	1.75 [a]

Description: treatmentthat is characterized bythe samesmall letterare not significantly differentat 5%

3.2.4. Sensory properties of Synbiotic Yoghurt: Aroma attributes

Yoghurt has a distinctive aroma that is affected by bacterial activity L.bulgaricus and S.thermopilus during the fermentation process. Starter cultures bacteria that is used is the main responsible in the formation of flavor compounds in the aroma of yoghurt [19]. In addition, the aroma of the yoghurt is also influenced by the other components are added.

Table 5. Aroma attributes of synbiotic yoghurt with the addition of yam flour modified

Concentration level yam flour modified	Aroma attribute scores of synbiotic yoghurt
Without added yam flour modified	2.25 [a]
Added yam flour modified 2 %	2.90[a]
Added yam flour modified 4 %	2.35 [a]
Added yam flour modified 6 %	2.65 [a]

Description: treatmentthat is characterized bythe samesmall letterare not significantly differentat 5%

The results showed (table 5) the addition of yam flour modified at various concentration levels contributed to the increase panelist evaluation to yoghurt aroma. It is suspected that the addition of yam flour modified become prebiotic source for the bacteria that make the bacteria more active and produce more flavor compounds in yogurt.

3.2.5. Sensory properties of Synbiotic Yoghurt: Flavour attributes

Yoghurt has a distinctive flavor that is product of lactic acid bacteria fermentation. *L. bulgaricus*had bigger role in the formation of aroma, whereas *S. thermophilus*had bigger role in the formation of yoghurt flavor. Results of both bacteria fermentation which produces the characteristic flavor of yogurt is lactic acid, acetaldehyde, acetic acid and diacetyl. A sense of sensory attributes that determine the final decision of consumers to accept or reject a food product [20].

Flavor is a combination of the overall perception of the sense of smell, feelings and visions play an important role in determining consumer food as desired [21]. The results showed that yogurt without addition yam flour modified has the highest score in the organoleptic evaluation by panelists. However, with the addition of yogurt synbiotik yam flour modified is acceptable panelists with scores did not differ much, and did not differ significantly based on analysis of variance (table 6).

Table 6. Flavour attributes of synbiotic yoghurt with the addition of yam flour modified

Concentration level yam flour modified	Flavor attribute scores of synbiotic yoghurt
Without added yam flour modified	3.20 [a]
Added yam flour modified 2 %	2.70[a]
Added yam flour modified 4 %	2.70 [a]
Added yam flour modified 6 %	2.75 [a]

Description: treatmentthat is characterized bythe samesmall letterare not significantly differentat 5%

3.2.6. Sensory properties of Synbiotic Yoghurt: Texture attributes

Yogurt textures is different with milk as raw material. Yoghurt has a somewhat viscous texture with a viscosity greater than milk. According to the Indonesian National Standard (SNI yoghurt) the consistency of yogurt is viscous/semi-solid. Creamy texture of yogurt that is influenced by both the fermentation of lactic acid bacteria. During fermentation, the bacteria L.bulgaricus and S.thermopilus produce acids that lower the pH, so that coagulates milk proteins and make the consistency of yogurt thickens.

In this study, yoghurt synbiotik with yam flour modified addition have significant effect on the texture of yogurt organoleptic test (table 7). The addition of yam flour modified as much as 6% have the highest score, which means that most panelists liked the yogurt sinbiotik with the addition of yam flour modified.

Table 7. Texture attributes of synbiotic yoghurt with the addition of yam flour modified

Concentration level yam flour modified	Texture attribute scores of synbiotic yoghurt
Without added yam flour modified	2.80[b]
Added yam flour modified 2 %	1.70[a]
Added yam flour modified 4 %	3.35[c]
Added yam flour modified 6 %	3.75[d]

Description: treatmentthat is characterized bythe samesmall letterare not significantly differentat 5%

The texture and taste of yoghurt is influenced by the concentration of prebiotics. The higher concentration of addition flour to yogurt will be more viscous [15]. Yam also contains starch that is high enough that when heated in water to give a viscous texture through gelatinization process. Gelatinization of starch is starch granule swelling of the starch granules due to the increase in volume that occur in water at temperatures between 55°C to 65°C. If the starch granules are heated in water, the thermal energy will cause hydrogen bonding is interrupted, and water into the starch granule. Water entering subsequently form hydrogen bonds with amylose and amylopectin. Water seep into granules cause swelling of the starch granules. Granule size will increase to a certain extent before the starch granules burst [22]. Rupture granules cause parts of amylose and amylopectin diffuse out. The process of entry of water into the starch granules swell and eventually cause rupture called gelatinization. It is why the higher the concentration the addition of arrowroot, yoghurt more viscous texture and taste will be more acidic [15].

4. Conclusion

The results showedthatyam (*Dioscorea esculenta*) flour modified characteristics are follows 7.66% moisture content, ash content of1.42%, totaldietary fibercontent of10.16%, 7.49%inulincontent, totalstarchcontent of71.78%, solubility is 77.63%, water absorption136.65%, 618.86% swellingpower, andyield of10.54%. The addition of yam flour modified in produce synbiotic yoghurt providea significantly different effectonhedonicquality test(texture), butdoes notgivea significantly different effecton thecolor, flavourandaroma. Based on theanalysis ofhedonicquality, synbiotic yoghurt with the addition ofyamflour modified 2% is the most preferredcharacteristics ofpanelists fromthe aspectof color, aroma, flavor, texture. It has apHvalueof 4.48, lactic acid levels is1.7%.

References

[1] Euromonitor. 2014. *Yoghurt and sour milk products in Indonesia*. Available on: http://www.euromonitor.com/yoghurt-and-sour-milk-products-in-indonesia

[2] Yin. 2014. *Countries that consume the most yogurt.* Available on: : http://www.yogurtinnutrition.com/11-countries-that-consume-the-most-yogurt

[3] Shaghaghi, Pourahmad, Adeli . 2013. Synbiotic Yoghurt Production by using prebiotic compounds and probiotic lactobacilli.*International Research Journal of Applied and Basic Sciences* Vol, 5 (7): 839-846 Science Explorer Publications

[4] Chen Ming-Ju, Kun-Nan Chen, Chin-Wen Lin. 2003. Optimization of the growth rate of probiotics infermented milk using genetic algorithms and sequential quadratic programming techniques.*Asian-Aust J Anim Sci* 16 (6): 894-902

[5] Buterikis, Matusevikcius P, Januskevicius A, Jankowski J, Mikulski D, Blok J, Cadieux P, Wind A, Sommer P, Schaefer L, Crowley K, Britton RA, Reid G. 2008. Evaluation of reuterin production in urogenital probiotic Lactobacillus reuteri RC-14.*Applied and Environmental Microbiology* 74 (15): 4645-4649

[6] Winarti, S. 2014. *Karakterisasi dan evaluasi sifat prebiotik inulin umbi gembili (dioscorea esculenta*). Disertasi UGM

[7] Yuniar, D.P. 2010. Karakterisitk beberapa umbi uwi (Dioscorea spp) dan Kajian Potensi Kadar Inulin. Fakultas Teknologi Industri. Universitas Pembangunan Nasional Veteran.Surabaya.

[8] Rosyida, Marsono,Y., dan Haryadi 2001. *Tepung Pra-Masak Tinggi Pati Resistant: Pengaruhnya Terhadap Sifat Fisik dan Kimia Digesta Tikus Wistar*. Himpunan Makalah Seminar Teknologi Pangan. PATPI. Semarang

[9] Wulan, S., et al. 2006. Modifikasi Pati Sederhana dengan Metode Fisik, Kimia, dan Kombinasi Fisik-Kimia yang Dibuat dari Jagung, Kentang, dan Ubi Kayu. *Jurnal Teknologi Pertanian.* Vol 7 (1): 1-9

[10] Suarni, Upe A, Harlim T. (2005). Karakteristik Fisik dan Kandungan Nutrisi Bahan Setengah Jadi dari Jagung. *Jurnal Pascapanen*. 4:101-106

[11] Wijayanti, Y. R. (2007). *Substitusi Tepung Gandum (Triticum aestivum) Dengan Tepung Garut (Maranta arundinaceae L) Pada Pembuatan Roti Tawar.* Yogyakarta: Universitas Gajah Mada

[12] Hasanah, R. U. (2007). *Pemanfaatan Rumput Laut (Gracilaria sp.) dalam Peningkatkan*

Kandungan Serat Pangan Sponge Cake. Institut Pertanian Bogor, Bogor

[13] Gibson GR, HM Robert, JAE Van Loo dan MB Roberfroid. (2004). *Dietary Modulation of The Human Colonic Microbiota: Updating The Concept of Prebiotics*. Nutr Res Rev, 2004; 17: 257-9

[14] Winarti, S. E. Harmayani, R. Nurismanto. 2011. Karakteristik Dan Profil Inulin Beberapa Jenis Uwi (*Dioscorea spp.*). *Jurnal Agritech* Vol. 31 no. 4 (on-line) http//www.jurnal-agritech.tp.ugm.ac.id(diunduh tanggal 18 Januari 2013

[15] Rosa Navila. 2010. Pengaruh Penambahan Umbi Garut (Maranta Arundinaceae L) Dalam Bentuk Tepung Dan Pati Sebagai Prebiotik Pada Yoghurt Sebagai Produk Sinbiotik Terhadap Daya Hambat Bakteri *Escherichia coli*. Artikel Penelitian, universitas Dipenogero. Semarang

[16] SNI: Standar Nasional Indonesia (SNI). 2009. SNI 2981:2009. Yogurt. Badan Standarisasi Nasional (BSN), Jakarta

[17] Buckle, K. A. et al. 2007. Ilmu Pangan. UI-Press, Jakarta

[18] Walstra, Wouters. 2006. Dairy Science and Technology. CRC Press. Boca Raton NewYork

[19] Antarini, A.A.N. 2011. Sinbiotik Antara Prebiotik dan Probiotik. Jurnal Ilmu Gizi, Volume 2 Nomor 2, Agustus 2011 : 148-155

[20] Yulisti, Maharani. 2000. Pengaruh Konsentrasi Garam dan Lama Penjemuran Terhadap Mutu Produk Fermentasi Usus Teripang Pasir Holothuria scabra. Institute Pertanian Bogor. Bogor

[21] Soekarto ST. 1995. Penilaian Organoleptik Untuk Industri Pangan dan Hasil Pertanian. Bharata Karya Aksara. Jakarta. 78-81, 90-97

[22] BeMiller, J. N. dan West Lafayette. (1997). *Starch Modification: Challenges and Prospects*. USA. Review 127-131

IOP Conf. Series: Materials Science and Engineering **128** (2016) 012036 doi:10.1088/1757-899X/128/1/012036

An evaluation of authentication methods for smartphone based on users' preferences

P K Sari*, G S Ratnasari and A Prasetio

Faculty of Economic and Business, Telkom University, Bandung, Indonesia

*Corresponding author: puspitakencana@telkomuniversity.ac.id

Abstract. This study discusses about smartphone screen lock preferences using some types of authentication methods. The purpose is to determine the user behaviours based on the perceived security and convenience, as well as the preferences for different types of authentication methods. Variables used are the considerations for locking the screens and the types of authentication methods. The population consists of the smartphone users with the total samples of 400 respondents within a nonprobability sampling method. Data analysis method used is the descriptive analysis. The results showed that the convenience factor is still the major consideration for locking the smartphone screens. Majority of the users chose the pattern unlock as the most convenient method to use. Meanwhile, fingerprint unlock becomes the most secure method in the users' perceptions and as the method chosen to be used in the future.

1. Introduction

In 2014, smartphone users in Indonesia increased for more than 70% compared to the previous period [1]. Smartphone is used not only as a communication device (to make phone calls and send messages) but also for storing personal data, some of which are sensitive to be accessed by others. Therefore, the smartphone manufacturers provide some security features to protect smartphones from unauthorized accesses. The operating systems on smartphones such as Android, Blackberry OS, iOS, and Windows Phone provide the unlock screen mechanism (as the security system) with various types of authentication methods that are different on each platform [2].

Concerns about the security and privacy of personal data have been increased, but it is not offset by the behavior of the smartphone users. In a previous study, it was found that more than half of the respondents said they did not use a password or PIN to lock their smartphones or tablets, even 44 % of the users who did not lock their mobile devices stated that using a password was "too difficult" [3]. It is also supported by another study which discovered that the users did not lock their devices because of the matter of convenience, in addition the users also felt that protecting the data stored on their smartphones are not necessary [4]. Another previous study found that the most prevalent authentication mechanism used in smartphones was the PIN even though the users did not perceive PIN as a highly secure method [5]. Despite this fact, they will still use this method in the future since it is a relatively convenient authentication method [5]. Those studies demonstrate that convenience beats security. Convenience is still more being a priority than the security because users tend to use a product or service if they feel convenient in using it [3].

In the other hand, there are differences in the behaviors and characteristics of users for different smartphones. Companies wanting their products and services to be widely used would need more information on the use of smartphones and the users' behaviors in order to be able to create strategies in a variety of mobile devices [6]. The use of the authentication methods on smartphones is important because among the factors which are considered when selecting a handset is the security features which occupy the second place after battery life [5]. By knowing the characteristics of the users, smartphone companies will be able to customize the types of authentication methods offered to the users based on their needs.

Based on those backgrounds, it is important to describe users' response regarding the selection and preferences of the authentication method. Therefore, the main research questions are:

1. What are the users' perceptions regarding the security and convenience in selecting the authentication methods?
2. What are the users' preferences among several authentication methods to be used currently and in the future?

2. Literature Review

2.1. Information Security

The first and probably the most important aspect of information security is the security policy. If the information security were a person, the security policy would be the central nervous system [7]. To protect the information resources, some controls that support the organization security objectives are needed. Security controls can be grouped into four categories, namely physical control, access control, communication control, and application control [8]. If the security policy was the central nervous system of information security, the access control would be the skin [7]. This research only discusses about the access control that limits the unauthorized individuals from using the information resources by applying several authentication methods.

2.2. Authentication Method

Authentication is the process for determining whether a person or a certain device is allowed to access a system, application, and specific data contained in a device or not [2]. Authentication schemes can be broadly classified into four main categories [7][8]:

1. Knowledge-based authentication; based on the knowledge of "what you know" (single factor). Example: PIN and Password
2. Token-based authentication; based on the ownership of "what you have" (two-factor). Example: Credit Card
3. Biometric authentication; based on the physical state of a certain "what you are" (three factors). Example: fingerprint, face unlocks
4. User doing-based; based on the physical acts or "what you do". Example: keystroke pattern, voice unlock

This study does not consider token-based methods because the hardware constraints of mobile phones do not favor them at the moment [5]. It does not consider mobile phones used as tokens as well.

2.3. User Perceptions for Using Authentication Method

There are two possibilities of behaviour in using the security features on a smartphone, doing locking screen and not-doing locking screen. This locking behaviour includes a variety of reasons. Based on the previous studies about security and privacy, some popular reasons for using or not using a screen lock consists of [4]:

- Prevent strangers from using smartphone
- Control when other people (friends or family) use the smartphone
- It is easy to do, never thought about why I use one
- Other people I know use a security lock
- In an emergency, others can use smartphone
- It's too much of a hassle

- No one would care about what's on my smartphone

The security mechanism is considered as a tool that provides protection against threats [9]. Like other products or services, when using this mechanism, saving time and effort also become the key factors that determine the convenience of usage [10]. Preference category (by comparing the authentication methods) in the study by Ben-Asher et al. [5] consists of three variables:

a. Perceived Security : users' perceptions about the protection to their personal information in the smartphone that is provided by an authentication method
b. Perceived Convenience : users' perceptions about the easiness, simplicity and robustness of an authentication method
c. Possible Future Use : likelihood of using an authentication method in the future

3. Methodology

This study employs a descriptive approach with the smartphone users as the population. Data was collected using questionnaires which were distributed to the users via online and offline. Non-probability samples used in this research is the convenience sampling. Samples were taken in accordance with the population and at the most easily accessible by researchers. The total samples used were 400 smartphone users that had responded to the questionnaire. After the data was collected, it was edited and organized in the forms of tables and charts.

The questionnaire consists of several sections. In the first section, the respondents must fill in with the demographic data. Then, those who lock their smartphone screens have to answer the question: "what method do you use to unlock your smartphone screen?". They also have to include the reasons why they use such method by choosing multiple choice questions provided with these options:

a. because the method provides sense of security
b. because the method is convenient to use
c. both
d. I do not know

The final section asks all respondents (both with the screen locked and unlocked) to choose one type of the authentication methods that is considered to be the most secure, most convenient and a method to be used in the future.

4. Results and Discussions

4.1. Respondents Profile

This study uses a non-probability sampling, so the proportion in each category is not specified. The result is calculated based on responses from 400 respondents. Majority of the respondents were female (58,5%), aged between 18 to 24 years old (72%) and high school graduates (46%). Most of the respondents were students (51%) and using Android as the operating system (75%).

The survey found that 329 respondents (82%) chose to lock their smartphone screens, and 71 respondents (18%) chose not to lock them. Most of the respondents who did not lock their phone screens were the Android OS users (70%). However, if it is clustered into its operating system, the Android users who did not lock the screen were only about 18%. Meanwhile, the BlackBerry OS users had the highest ratio (33%) and the iOS users had the lowest (9%). It can be interpreted that among the respondents who used the same platform, the BlackBerry users is the group which uses the least of authentication methods to protect their smartphone's screens. Another result shows the respondents of 18-24 years of age are the largest group (46 of 71 users) that did not protect their phones since they are also the majority group in this research. On the other hand, the users of >45-year-old group who did not lock their phone's screens have the highest ratio (85%). The result details are shown in Table 1.

Table 1. Demographics of Respondents

Factor	Variable	Total of Respondents
Gender	Male	166
	Female	234
Age	18-24	288
	25-34	85
	35-44	20
	above 44	7
Education	High School	184
	Diploma	50
	Bachelor	147
	Master	17
	Doctor	2
Occupation	Students	204
	Private Employees	112
	Teacher/Lecturer	10
	Civil Servants	24
	Entrepreneur	21
	Unemployment	15
	Others	14
Operating System	Android	300
	iOS	68
	BB OS	24
	Windows	8

4.2. Motivation For Locking (or not) Smartphone Screen

In this research, the respondents were also asked about the reasons why they chose to use or not use the unlocking screen methods in their smartphones. Some options provided in the questionnaire are based on the popular reasons for users in locking their smartphones referring to the previous study by Egelman [4]. The result is shown in Figure 1 and Figure 2.

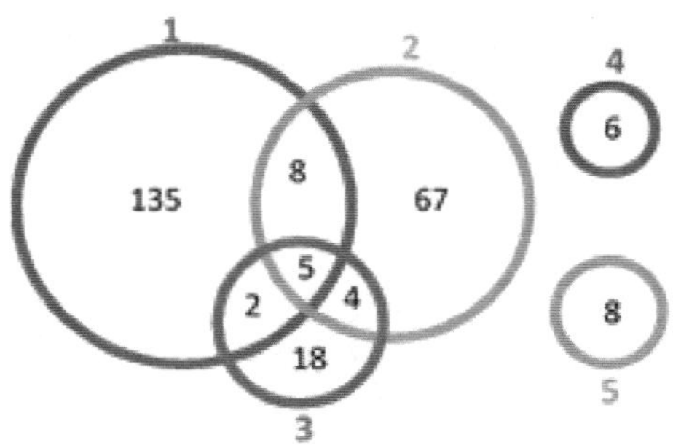

Figure 1. Motivation for Locking Screen

Legend:

1. Red circle: Prevent strangers from using my smartphone (if it is stolen, lost or accessed by unknown people)
2. Green circle: Control when other people use my smartphone (from people surrounding like family and friends)
3. Blue circle: It's easy to do, I've never thought about why I use one
4. Purple circle: Other people I know use a security lock
5. Orange circle: None of the above

As shown in Figure 1, there are intersections of answer sets because the respondents could choose more than one option. The result shows that most of the users apply an authentication method for unlocking their smartphone's screens to prevent strangers from using their phones and also to control other people who use them. Both of these answers have something in common which is to restrict the access of others to the mobile phones. This result is similar with the previous study that the main reason why users did the screen locking was to prevent unauthorized access from strangers [4]. Furthermore, there are a lot of users who use an authentication method because the feature is easy to use, instead of protecting their data. The least chosen answer is because people around them are using such methods.

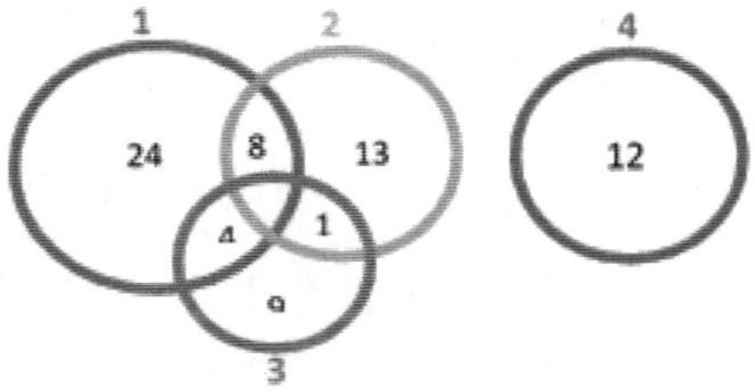

Figure 2. Motivation for Not Locking the Screen

Legend:
1. Red circle: In an emergency situation, others can use my smartphone
2. Green circle: It's too much of a hassle
3. Blue circle: I don't care about what's on my smartphone
4. Purple circle: None of the above

People who did not use any authentication method for unlocking their device also have some reasons as shown in Figure 2. The result indicates that some respondents still prioritize convenience while using their smartphones without having to repeatedly perform complicated authentication process. In this research, the reason "as a precaution in emergencies" becomes the most chosen reason followed by the inconvenience problem. This result is slightly different from the previous study by Egelman [4] that found the problem of emergency was in the last rank of reasons chosen by the users who did not lock their phone's screens. Some users still consider that using the method of authentication is too complex. There are also a few people who have other reasons outside the given options. Furthermore, only about 10% from 400 respondents who do not really care about protecting the data in their smartphones, either those who use an authentication method or not.

These results illustrate that smartphone users are more aware about information security issues. Nevertheless, the convenience problem still becomes one of the considerations that drives the users to use any access control to their smartphones, especially in locking the screen.

4.3. *Consideration of Security Factor and Convenience Factor*

In this section, the total respondents to be further analyzed are only 329 of the 400 respondents. 71 other respondents did not lock their smartphone screens so that the respondents could not provide feedbacks on the authentication methods used and the reasons for using such methods. Result about the reasons why the users apply the authentication methods is shown in Figure 3.

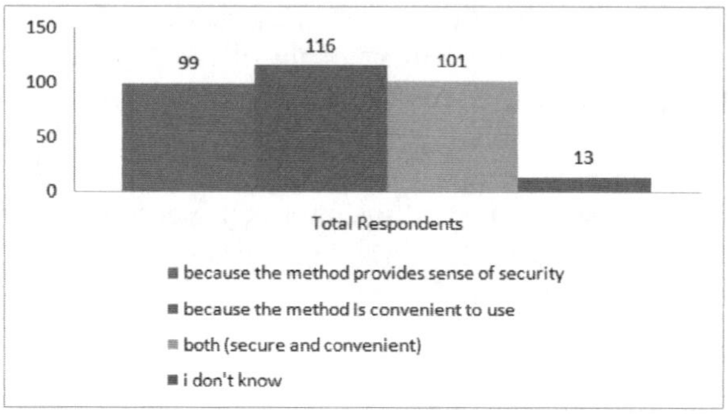

Figure 3. Reason for Using Authentication Methods

In general, most respondents (116 people) use an authentication method to lock their smartphone screens for the reason that it provides convenience (only). However, the result also shows that giving security is another important reason that received high response. This indicates that the respondents understand that choosing a method which can provide data protection is also a major goal for locking the smartphone screens.

The high response to both aspects of the security and convenience is contributed by 101 respondents who chose this reason. It indicates that 30.7% of the respondents felt that the use of authentication methods provided convenience in use and sufficient protection for their data at the same time. Meanwhile, only 13 of the 329 respondents did not know the reason for choosing the authentication method being used. It indicates that only a few users do not really understand the importance of safety and convenience when choosing a method of authentication. It is also an indication that the authentication method being used did not provide convenience and protection of data (security) as how the users expected.

Based on the results, the reasons for using a method do not have the same proportion to each type of authentication method. In other words, although convenience was the most selected reason by the users, not every method chosen was for the matter of convenience. For more details it can be seen in Table 2. The reason for using PIN and password is because these methods provide a sense of security, while the reason for using pattern and face unlocks is because they are more convenient to use. And, the reasons why the users apply fingerprint and voice unlocks are because these methods are both secure and convenient.

Table 2. The Reason use of Authentication Methods

Authentication Methods	The Reason Use of Such Methods				Total
	Because the method provides sense of security	Because the method is convenient to use	Both (secure and convenient)	I don't know	
PIN	33%	31%	30%	6%	100%
Password	22%	23%	8%	8%	100%
Pattern	22%	43%	33%	2%	100%
Face Unlock	50%	50%	0%	0%	100%
Voice Unlock	0%	0%	100%	0%	100%
Fingerprint	25%	25%	45%	5%	100%

4.4. User Preferences for Authentication Methods

This section will explain the results of user preferences for different types of authentication methods. All respondents were asked to select one type of authentication methods that is considered to be the most secure, convenient, currently used and will be used in the future. The results shown in Figure 4 and Figure 5.

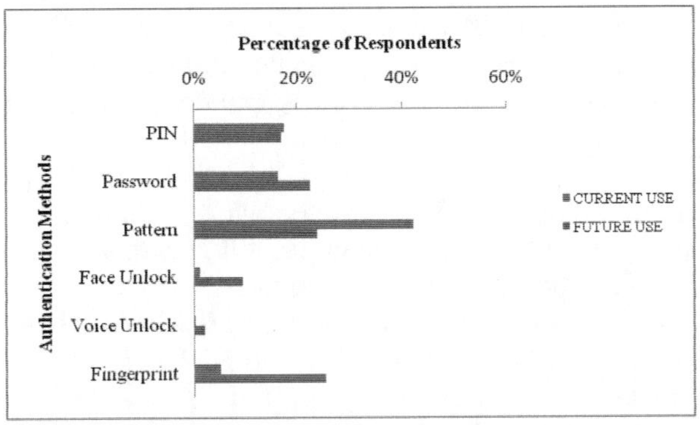

Figure 4. Current Use and Future Use of Authentication Methods

Figure 4 shows that the most widely used authentication method currently is pattern, while PIN and password are the second and the third preferences. This finding is different from the previous study that mentioned PIN and password were the most common methods used [5]. Although they gave different choices, both of them have become the preferred methods by considering convenience as the reason. The least preferred method in this research is voice unlock, whether for the current or future use. Slightly different from the voice unlock, face unlock is more favorable for the future use although it is less used today. Furthermore, fingerprint is not very popular in the current use but it is more considered to be used in the future and so are pattern and password.

Based on the assessment of the respondents shown in Figure 5, fingerprint becomes the most secure authentication method and also a preferred method to be used in the future. It can be said that the respondents (users of smartphones) have already begun to realize that the authentication methods used to lock the smartphone's screens should be prioritized on the security issues to avoid security threats such as accessing users' information without permission.

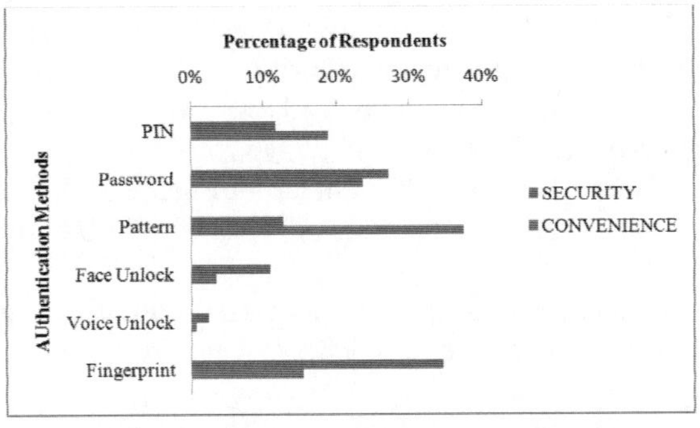

Figure 5. User Preferences Based on Perceived of Security and Convenience

If the choice of authentication method widely used today is the pattern unlock, then it follows the user's needs for convenient use. It is supported by the reasons that are included in the selection of authentication methods that many users use such method because of its perceived convenience. The result of this research also reveals that the pattern unlock has become the most convenient authentication method.

Unlike the previous research [5], fingerprint was the most favorable method based on the perspectives of security and convenience. There was also a similar method like pattern unlock and it became the least selected authentication method that would be used in the future. Yet, in this study it is found to be the second most preferred method for use in the future after the fingerprint. The least chosen authentication method is voice unlock. Based on observations, it probably happens because only a few platforms provide this method. The possibility that makes it different might be the mismatch between the types of the method with the characteristics of the respondents and the different time of research.

5. Conclusion

Indirectly, the security and convenience affect the users when deciding to use authentication methods based on various reasons. The convenience factor is still the major consideration in selecting the authentication method to be used by users.

Fingerprint is selected as the most secure type of authentication method in providing protection for data on smartphones and also as the method which would be mostly used by users in the future. Meanwhile, the pattern unlock is chosen by the majority of users as the most convenient method. This method is also the most widely used today. It is in line with the reason of convenience as the biggest factor in selecting a method. The future expectation of the users is that they can use the most secure method to protect their data.

Manufacturers of smartphones are expected to use this results as a consideration in providing the authentication methods for many users, especially for the future use. It is expected that more and more users would use authentication methods for locking their smartphones in order to increase the level of security for critical data and information in their smartphones. Based on the results, smartphone manufacturers should also reconsider to provide face unlock and voice unlock as the choices of the methods because they are less popular to be used now and in the future.

References

[1] Wijaya K K 2015 Android dan browser Opera dominasi pengguna mobile Indonesia selama 2014 [Online] Available: http://id.techinasia.com/android-opera-dominasi-smartphone-indonesia-2014/

[2] Sametinger J and Roland S 2012 Secure and Usable Authentication on Mobile Devices *MoMM, 10th International Conference on Advances in Mobile Computing & Multimedia*, 257-262

[3] Confident Technologies Inc. 2011 Survey Shows Smartphone Users Choose Convenience Over Security [Online] Available: http://confidenttechnologies.com/news_events/survey-shows-smartphone-users-choose-convenience-security/

[4] Egelman S *et al* 2014 Are you ready to lock? Understanding user motivations for smartphone locking behaviors *Proc. CCS , November 2014*, 750-761

[5] Ben-Asher N *et al* 2011 On The Need For Different Security Methods on Mobile Phones. *Proc. 13th Int'l Conf. Human Computer Interaction with Mobile Devices and Services (MobileHCI)*, 465-473

[6] Setiamanah A 2013 Strategi Marketing Berbasis KarakteristikPengguna Smartphone. [Online] Available: http://www.frontier.co.id/strategi-marketing-berbasis-karakteristik-pengguna-smartphone.html

[7] Peltier T R 2013 Information Security Fundamental (2nd ed.) Boca Raton: CRC Press Taylor& Francis Group

[8] Rainer R K and Turban E 2009 Introduction to Information Systems: Enabling and Transforming Business (2nd ed.)", New Jersey: John Wiley & Sons.

[9] Ang C S and Zaphiris P 2009 Human Computer Interaction: Concepts, Methodologies, Tools, and Applications. USA: IGI Global

[10] Hsu H and Chang Y 2013 Extended TAM Model: Impacts of Conveniencce on Acceptance and

Use of Moodle *US-China Education Review* **3:4**, 211-218

Comparison of the calculation QRS angle for bundle branch block detection

L Goeirmanto*, R Mengko and T L Rajab

Bandung 40132, Sekolah Teknik Elektro dan Informatika, Institut Teknologi Bandung

*Corresponding author: id.l.goeirmanto@ieee.org

Abstract. QRS angle represent condition of blood circulation in the heart. Normally QRS angle is between -30 until 90 degree. Left Axis Defiation (LAD) and Right Axis Defiation (RAD) are abnormality conditions that lead to Bundle Branch Block. QRS angle is calculated using common method from physicians and compared to mathematical method using difference amplitudos and difference areas. We analyzed the standard 12 lead electrocardiogram data from MITBIH physiobank database. All methods using lead I and lead avF produce similar QRS angle and right QRS axis quadrant. QRS angle from mathematical method using difference areas is close to common method from physician. Mathematical method using difference areas can be used as a trigger for detecting heart condition.

1. Introduction

The electrocardiogram (ECG) is a tool to record electrical activity of the heart. ECG result can be printed or showed in the monitor is commonly used to to non-invasively evaluate the cardiac activity of a patient [1]. ECG data is representative of cardiac physiology, used to analysing cardiac condition. The condition of cardiac health is usually seen in the shape of ECG waveform and heart rate. Human activities, lifestyles and food consumption have direct impact to condition of the heart [2].

ECG data is recorded in different forms according to the location of electrodes (leads) [3]. Standard ECG device records activity of the heart from six frontal plane leads and six horizontal plane leads. Frontal plane leads contain lead I, II, III and augmented lead avF, avL, avR. Horizontal plane leads contain lead V1, V2, V3, V4, V5 and V6. QRS axis is a spatial visualization of two frontal leads. It shows condition of the heart whether normal or abnormality conditions which lead to Bundle Branch Block. Previous research algorithm to detect condition of heart is slow and better method is need [4].

2. Theoretical Review

There are five parameters in normal ECG data which are P, Q, R, S and T as can be seen in figure 1. P wave is a recording of depolarization through the atrial myocardium from start to finish, the QRS complex is reflecting depolatization of ventricular myocardial and T wave represents ventricular repolarization [5]. A normal QRS complex is consisting of a small Q wave, a large R wave and a small S wave. Physicians usually analyse four intervals in ECG waveform. PR interval typically is analysed in lead II of the 12 leads ECG consisted of P wave and the straight line connecting it to QRS complex. QRS interval is visualized using frontal plane leads. ST interval is used to examine Myocardial Infraction. QT interval is usefull for heart rate correction.

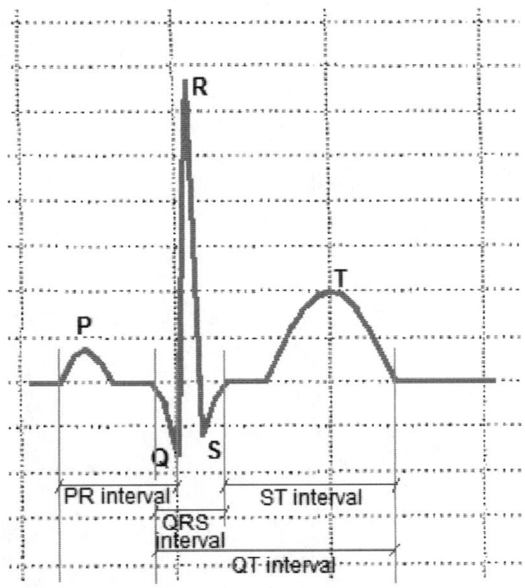

Figure 1. Normal ECG waveform, parameters and intervals

Physicians usually analyse a pair of leads from frontal plane leads which are perpendicular. The pair leads are lead I – lead avF, lead II – lead avL and lead III – lead avR.Resultant of the pair leads is visualized in the QRS axis. The visualization shows conditions of the heart especially in ventricular area. This research used 12 lead electrocardiogram data from MITBIH physiobank database. We compared three methods of calculation QRS angle in the QRS axis which are normal physician method, calculation directly from ECG data using different peaks and different areas.

Noise removal and baseline wander are done as preporcessing. Algorithm for QRS detection is complied to get values of QRS complex. Mathematical formulation to calculate area of surfaces in the QRS complex area is used.

3. Experimental Analysis

The 12 lead ECG data from MITBIH is converted to .mat extention in order to match MATLAB software requirement. All prepocessing and mathematical formulation is done using MATLAB. Three methods are compared to know difference of QRS angles.

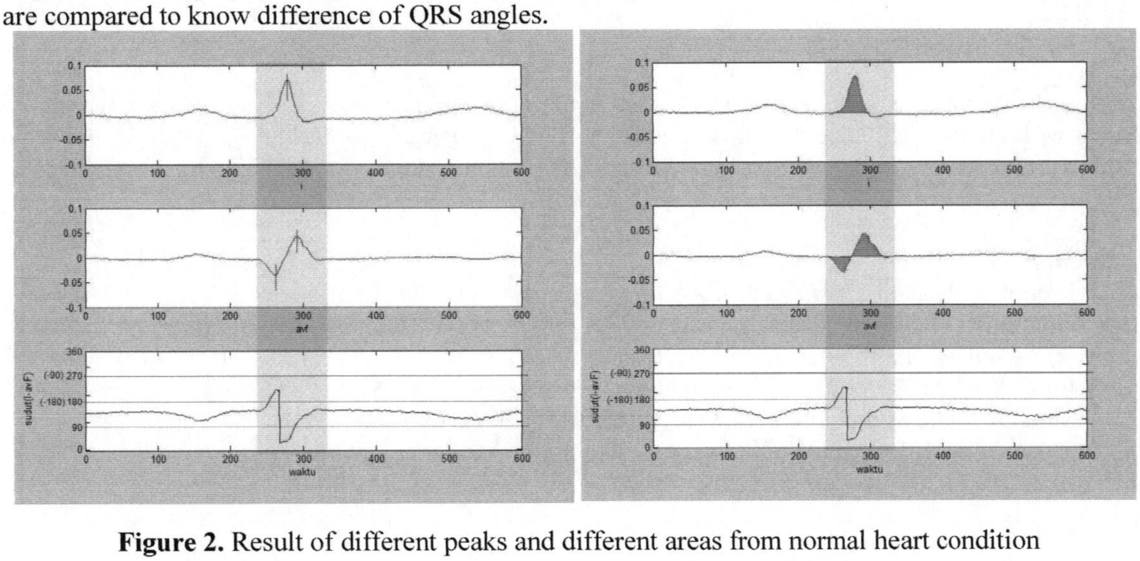

Figure 2. Result of different peaks and different areas from normal heart condition

Figure 2 shows two visualizations from normal condition of ECG data using lead I – lead avF. QRS angle from physician interpretation is 8°. Result of calculation from different peaks is 5.7884° and from different areas is 6.3604°. All results of QRS angle are in the normal quadrant.

There are two visualizations from abnormal heart condition. QRS angle from physician interpretation is -51°. Result of calculation from different peaks is -42.8548° and from different areas is -52.1493°. All results of QRS angle are in the left axis deviation (LAD) quadrant which lead to Left Bundle Branch Block.

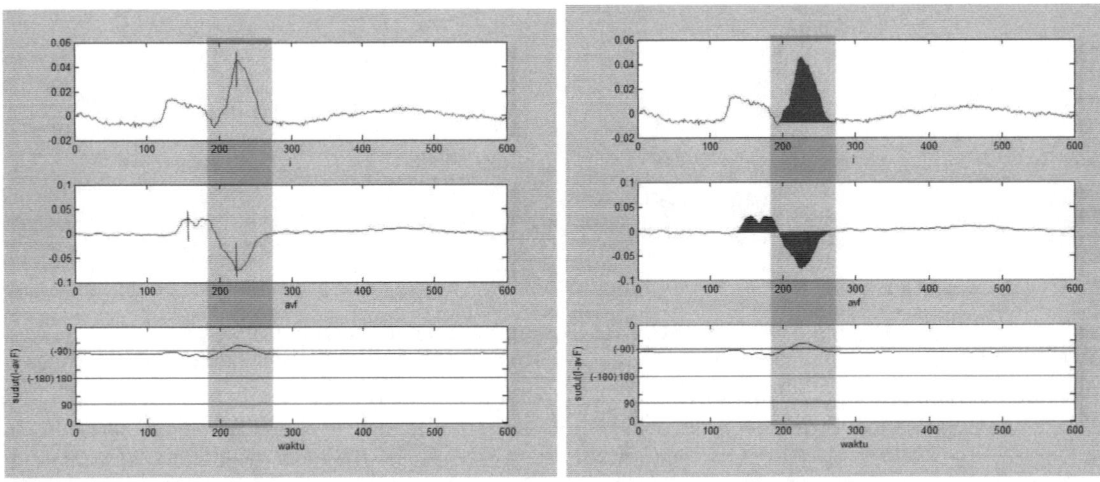

Figure 3. Result of different peaks and different areas from abnormal heart condition

Result of the experiment with normal heart conditions are calculation using different peaks and different areas lead to normal axis. Difference between calculation from data and interpretation from physician is 1.0215° to 3.2250°. Experiment with LAD heart condition also show same quadrant of QRS axis between calculation from data and physician's interpretation. The difference is 1.1135° to 18.8105°. These result lead to Left Bundle Branch Block condition. Other abnormal heart condition which is Right Bundle Branch Block also can be detected using three methods because the result lead to right axis deviation (RAD) quadrant. Comparing to interpretation of physician, the difference is 1.2147° to 17.4415°.

4. Conclusion

Calculation QRS angle using different peaks and areas give similar result to physician's interpretation. These methods were calculated using computer software which can produce QRS angle in the same quadrant. Like other computer result in reference, it is cost advantages by use of existing computer [6].

References

[1] Hampton JR 2003 *The ECG Made Easy* New York: Churchill Livingstone
[2] Goeirmanto L 2013 *Proc. International Conference on Economic and Information System Management* 50-54
[3] Luna ABD 2012 *Clinical Electrocardiography* West Sussex: Wiley-Blackwell
[4] Sutjiady F and Mengko R 2013 *Proc. IEEE ICICI BME* 157-160
[5] Thaler MS 2012 *The Only EKG Book* Philadelphia: Wolters Kluwer
[6] Bausseljot R and Kreiseler D 1998 *Computers in Cardiology* 25 349-352

Tunning PID controller using particle swarm optimization algorithm on automatic voltage regulator system

M F Aranza*, J Kustija, B Trisno and D L Hakim

Department of Electrical Engineering, Universitas Pendidikan Indonesia
Jl. Dr. Setiabudhi No. 207 Bandung, Indonesia

*Corresponding author: m.falah11@yahoo.com

Abstract. PID Controller (Proportional Integral Derivative) was invented since 1910, but till today still is used in industries, even though there are many kind of modern controllers like fuzz controller and neural network controller are being developed. Performance of PID controller is depend on on Proportional Gain (Kp), Integral Gain (Ki) and Derivative Gain (Kd). These gains can be got by using method Ziegler-Nichols (ZN), gain-phase margin, Root Locus, Minimum Variance dan Gain Scheduling however these methods are not optimal to control systems that nonlinear and have high-orde, in addition, some methods relative hard. To solve those obstacles, particle swarm optimization (PSO) algorithm is proposed to get optimal Kp, Ki and Kd. PSO is proposed because PSO has convergent result and not require many iterations. On this research, PID controller is applied on AVR (Automatic Voltage Regulator). Based on result of analyzing transient, stability Root Locus and frequency response, performance of PID controller is better than Ziegler-Nichols.

1. Introduction

PID controller is one of the controllers of the most widely used in industries. Although there are controllers modern, recently being developed for example, fuzzy controller, but controller PID is still used till today [1]. The performance of PID controller is depend on Gain Proportional (Kp), Gain Integral (Ki) and Gain Derivative (Kd), the selection of those values are not appropriate can make the system become unstable [2][3]. There are several methods to find the value of the gains such as Ziegler-Nichols (ZN), gain-phase margin, Root Locus, Minimum Variance and Gain Scheduling. However, these methods were considered less than optimal for a system that non-linear and has high orde [4][5][3], besides several of these methods in calculation relative hard. To solve these obstacles the algorithm Particle Swarm Optimization (PSO) is proposed which is assigned to find the value of Kp, Ki and Kd are optimized, where controller PID is applied on automatic voltage regulator (AVR). PSO is proposed because it has the results of convergance and do not require a number of iterations, so in relative calculation relative quick.

2. Model AVR System, Controller PID, Performace Estimation of PID Controller

2.1. Model AVR System

AVR role in the electric power system is to withstand the magnitude of the terminal voltage synchronous generator at a certain rate. An AVR system consists of four main components, namely amplifiers, exciter, generators and sensors. The transfer function of the four components as shown tabel 1 [6].

2.2. PID Controller

PID controller is an automatic controller that compares the actual value of the output of a system at the desired price and generates a control signal to minimize the error value [7]. As the name implies PID controller consists of three basic types, namely controller proportional, integral and derivative that can be used separately or together depending on what we need. Each controller has karakteristrik respectively.

Table 1.

Model	Fungsi Alih	Keterangan
Amplifier	$\frac{Vr(s)}{Vs(s)} = \frac{Ka}{1 + Ta_s}$	$10 \leq Ka \leq 400$ $0.002 \leq Ta \leq 0.1$
Exciter	$\frac{Vf\ (s)}{Vr\ (s)} = \frac{Ke}{1 + Te_s}$	$10 \leq Ka \leq 400$ $0.5 \leq Te \leq 1$
Generator	$\frac{Vt\ (s)}{Vf\ (s)} = \frac{Kg}{1 + Tg_s}$	$0.7 \leq Kg \leq 1.0$ $1.0 \leq Tg \leq 2.0$
Sensor	$\frac{Vs\ (s)}{Vt\ (s)} = \frac{Ks}{1 + Ts_s}$	$0.01 \leq Ts \leq 0.06$

Characteristics of a proportional controller is determined by Kp (Constant Proportional) [8]. Kp value is too small to generate a response rise time is slow, increase the value of Kp will increase the response faster, but when the value of Kp is too large will create an oscillating output. Characteristics of the integral controller can improve response while eliminating the steady-state error, but the selection of Ki (Integral Constants) which may cause high transient response, which can cause system instability. Selection of very high Ki can also cause the output to oscillate. Characteristics of derivative controllers can not work alone because it is improve the transient response with an error predicting what will happen. Selection of the value of Kd (constant Derivative) is appropriate can improve system stability and reduce overshoot [3].

The output of the PID controller is given as follows: [9]

$$(t) = Kp\, e(t) + Kd \frac{de(t)}{dt} + Ki \int_0^1 e\,(t)\, dt \tag{1}$$

Where *u* is the control signal, and *e* is the error value. Then the transfer function:

$$\frac{U\,(s)}{E\,(s)} = Kp + \frac{Ki}{s} + Kd_s \tag{2}$$

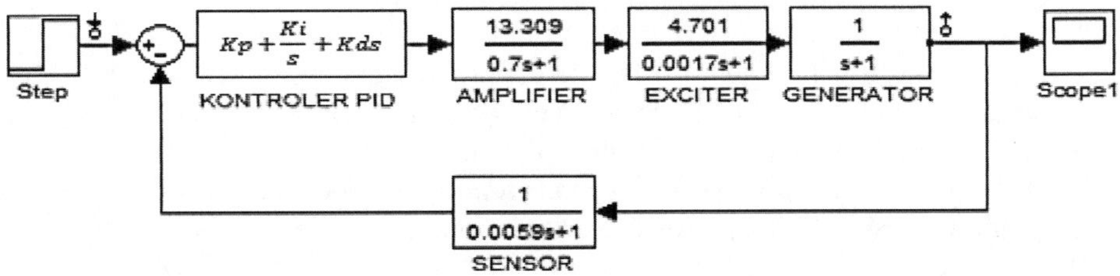

Figure 1. Block Diagram of AVR System with PID Controller

From figure 1 we got transfer function AVR system with PID controller, as shown below :

$$\frac{Vref}{Vout} \underset{0}{=} \frac{3.69Kds^3+(3.69\ Kp+\ 62.56\ Kd)s^2+(3.69Ki+62.56\ Kp)s+62.56Ki}{(3).000011s^5+0.00605s^4+0.7046s^3+(1.7017+62.56Kd)s^2+(1+62.56Kp)s+62.56Ki} \quad (3)$$

2.3. Performace Estimation of PID Controller

In some literature many techniques have been proposed to evaluate the performance of PID controllers. Evaluation of the performance of the following criteria: Integrated Absolute Error (IAE), Integral Time Absolute Error (ITAE) and Integral of Timing-Weghted-Squared-Error (ITSE). Third performasi index has advantages and disadvantages of each. For example, weakness IAE is able to generate a response with overshoot ralatif small, but long setling time , although the ITSE can overcome the shortcomings of the criteria IAE, persamaanya calculation process is complex and takes a relatively longer time. ITAE equation, IAE and itse shown in the equation below [10] :

$$ITAE = \int_0^t t\,|v_r - v_t|\ dt \quad (4)$$

$$IAE = \int_0^t |v_r - v_t|\ dt \quad (5)$$

$$ITSE = \int_0^t t\,(v_r - v_t)^2\ dt \quad (6)$$

Where Vt is the terminal voltage, a reference voltage Vr and t is the time span from smiulasi [3].

3. Paricle Swarm Optimization (PSO)

PSO algorithm was first introduced by Eberhart and Kennedy in 1995 [11]. The origin of the PSO terinipirasi of the behavior of a flock of birds or a school of fish while searching for prey [12].

Demonstrating how the Particle Swarm Optimization, to take the example of a number of patikel (in PSO, individuals are often referred to patikel), *N* moving together in a search space S. Each particles of *i* is kanidat settlement and expressed by the vector *xi.* Each particle has a position and velocity and will move based on experience and information from the social environment and the current position and the particle. Experience particle *i* expressed as *pi* best position ever achieved by these particles. Information from the environment is represented by particles that have the best position *g*, in the collection of the particles, whereas, the current position of particle *i* is expressed by *xi (t-1).* Change the speed of the particle and particle position (*vi, xi*) was determined based on two equations below as follows [13].

$$v_{i=}\ v_i\ (t-1) +\ \varphi c_1(p_i - x_i(t-1)) +\ \varphi c_2(g - x_i(t-1)) \quad (9)$$

Where *xi*

$$x_i = x_i\,(t-1) + v_i\,(t) \tag{10}$$

Random vector φ has a value range [0,1]. Meanwhile, *c1* and *c2* are two positive constants called cognitive learning and social learning. Each particel speed limited by [Vmin, Vmax] [14].

Selection of the proper w inertia weight provides a balance between global exploration and local exploration, so do not require many iterations in searching optimal solution. *W* always decline linearly approximately ranging from 0.9 to 04 for the calculation. general inertia weight w is set berdasarakn persasaam below [4].

$$w = w_{max} - \frac{w_{max} - w_{win}}{iter_{max}}\ x\ iter \tag{11}$$

Frist step in PSO is initialization is to determine the number of iterations, the number of population (n) inertia weight (w) and cognetif learning and social learning (c1 and c2), The next step aroused the population in the form of a random matrix with a range of values [0,1] that the dimension (dimensi_masalah.xn). Generation population by typing sinkaks matlab rand (dimensi_masalah, n). . After that, the initialization speed and position. In this step makes the value of the velocity and position of a particle to be equal to zero, Than Calculate the error (Vref-Vout), The amount used in this case is a power unit (pu) are worth one. Description 1 pu equals yout reference and a generator terminal voltage value, then error = | 1-yout |. The next step is to calculate the value of fitness or function to be optimized. In this paper ITAE (Integrated Time of Weighted Absolute Error) proposed, and continue with update velocity and update position. This process contoinue till iteration maximum. After finshing calculate till iterasi maximum, finally chek whether the result of calculating already convergence ? if it already convergance excecute if no try again and or change the intialtization (back to frist stap). To understand easyly, Flowchart PSO is givent at figure 2.

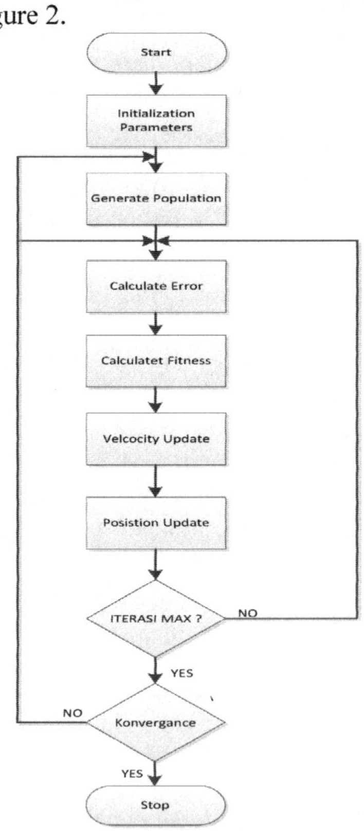

Figure 2. Flowchart

IOP Conf. Series: Materials Science and Engineering **128** (2016) 012038 doi:10.1088/1757-899X/128/1/012038

4. Result and Discussion

In the flowchart PSO is explained that first step is initialization. Proper initialization can produce maximum results. Initialize the form of information and the price used in penilitian this time presented in Table 2.

Table 2. Initialization Parameter PSO

Keterangan	Nilai
Jumlah populasi (n)	50
Jumlah iterasi (Iterasi_maksimum)	50
Dimensi permasalahan (v)	3
Berat *inertia* maksimum (w_max)	0.9
Berat *inertia* minimum (w_min)	0.4
Cognitive learning dan *social learning* $c_1 = c_2$	0.8

After calculating as many as fifty iterations (50) result of value Kp, Ki and Kd. Kp = 1,068; Ki = 0:04 Kd = 0.4011. These value is chosen not only already reach the calculations as much as fifty times (50 iterations), but also the tendency of convergance. Figure 3 shows after the 25th iteration of the convergent tendencies and maintain maximum results until the 50th iteration.

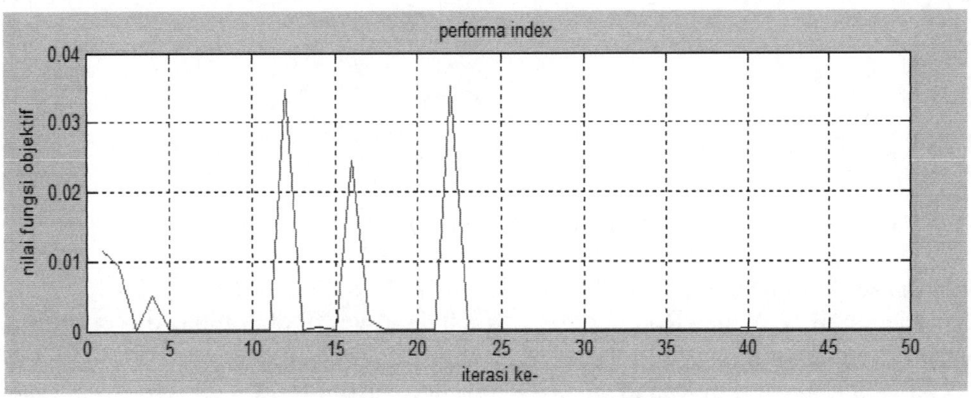

Figure 3. Result of Trendency Convergance

Table 3. Value Kp Ki & Kd Based on Two Methods

PSO-PID	ZN-PID
Kp= 1.068 Ki = 0.04 Kd = 0.411	Kp = 0.1457 Ki = 0.05901 Kd = 0.01644

4.1. The Result Of Transient Analysis

Transient analysis is done by providing a signal on the system. The best system is a system that has the input and output signals are identical. The results of transient analysis using the algorithm PSO has a value of overshoot is smaller 15.9% of the system with the ZN-PID and less 65% of the system without PID controller, for systems with PSO-PID value of peak-time 20.3% faster than the system without the PID and 73.8% faster compared to ZN-PID system. Values rise time for PSO-PID system with 37.01% and 7.61% faster than a system with ZN-PID and without PID system. Value settling time for the system to PSO PID faster 398.9% (almost 4x faster) than the system without the PID and 239.1% (2x faster) than a system with ZN-PID and final value error steady state for the system with the PSO-PID is 0 %.

Table 4. The Result Of Transient Analysis

Kondisi	*Overshoot*	*Peak Time*	*Rise Time*	*Settling Time*	*error steady state*
Tanpa PID	74.8 % (1.72 pu)	0.33 s	0.118 s	4.37 s	1.60%
ZN-PID	15.9 % (1.16 pu)	0.865 s	0.412 s	2.46 s	0%
PSO-PID	0.18 % (1.001 pu)	0.127 s	0.0419 s	0.0691 s	0%

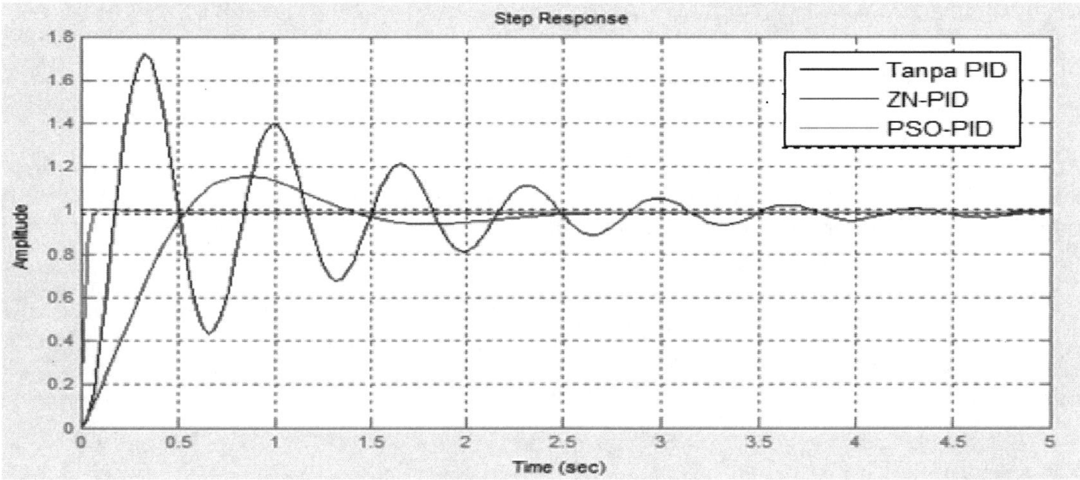

Figure 4. Step Response

4.2. The Result of Root Locus

Results of the stability of the Root Locus analysis showed that for the three conditions are conditions without the PID system, the system with the ZN-PID and the PSO-PID system are stable because they have a pole and a zero to the left of the field s. However, a system with PSO-PID has the location of the lie away from axis *s* and do not have the critical point that indicates the system is very stable, plus a system with PSO-PID had a price damping ratio ζ is greater 45.8% compared with the ZN-PID and 85.1% greater than the system without PID.

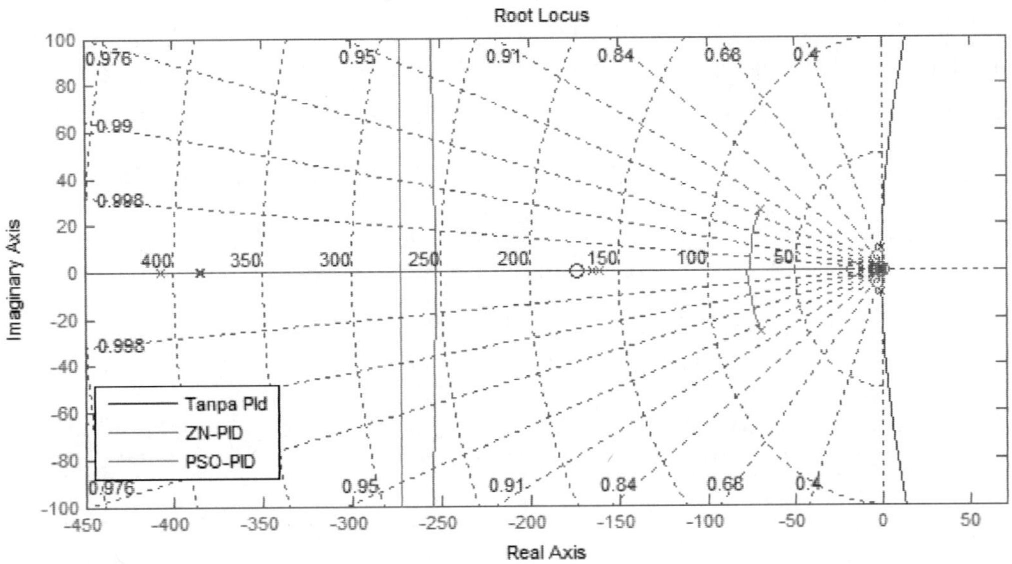

Figure 5. Curve Root Locus

Table 5. The Result of Root Locus Analysis

Tanpa PID		ZN-PID		PSO-PID	
Pole	*Damping Ratio*	Pole	*Damping Ratio*	Pole	*Damping Ratio*
-384	1	-386	1	-407	1
-164	1	-161	1	-2.68	1
-0.833+9.54i	0.087	-1.76+3.21i	0.48	-70+25.81i	0.938
-0.833-9.54i	0.087	-1.76-3.21i	0.48	-70-25.81i	0.938

4.3. Frequency Response

Systems that have the widthest bandwidth and has the greatest resonance frequency values indicated have the best performance because these values show a rapid system response. The stability of the system can be seen from the gain margin and phase margin. The system is said to be stable if the price gain margin is lower than 0 dB and phase margin is above -180^0. Bode diagram of the PSO-PID system, ZN-PID and systems without the PID controller is shown in Figure 7 and Table 6 shows the results of a frequency response analysis that PSO-PID has the widthest bandwidth and the greatest resonance frequency thay indicate, PSO-PID has the best performance.

Table 6. The Result of Response Frequency

Kondisi	Fasa Margin	*Gain* Margin	*Bandwidth*	Frekuensi Resonansi
Tanpa PID	12.1 (deg)	16 dB	14.784 (rad/s)	9.5 rad/s
ZN-PID	142.95 (deg)	inf	4.2597 (rad/s)	2.34 rad/s
PSO-PID	83.8 (deg)	inf	344.289 (rad/s)	61.5 rad/s

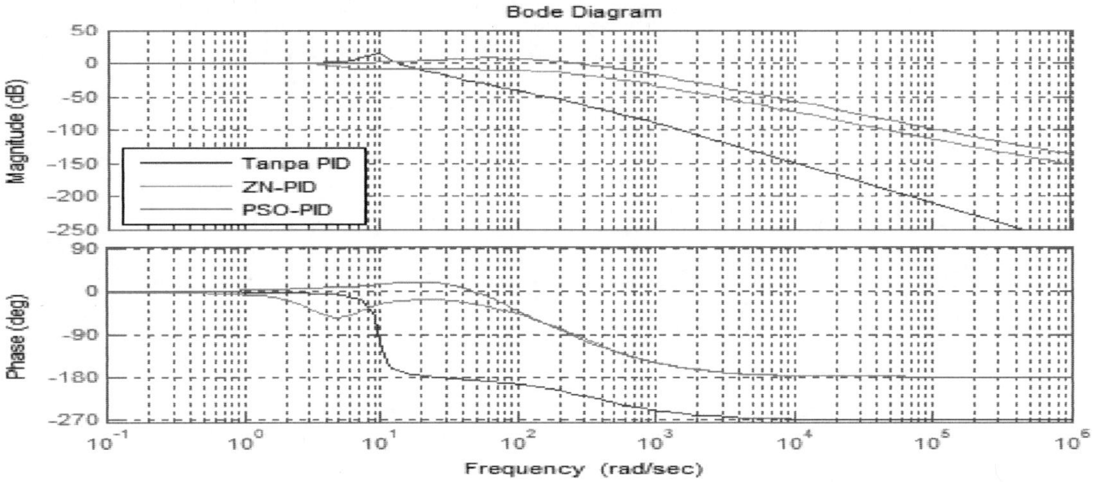

Figure 6. Bode Diagram

5. Conclusion

The main purpose of the application of PSO algorithm in PID controller is to determine the price of Kp, Ki and Kd are optimal, so it can improve the performance of PID controllers. Results of the application of PSO algorithm will compare the results with the method ZN (Ziegler-Nichols) and without PID system. Results of the transient analysis shows that the PSO-PID overshoot and settling time have the lowest compared to ZN-PID and PID Without a system. Based on Root Locus damping ratio owned PSO-PID has the largest value even close to one and is based on the frequency response using Bode plots PSO-PID has a bandwidth of the widest and the price of the resonant frequency of the greatest, so that both the transient analysis, Root locus and frequency response of the system by PSO-PID has the best performance among ZN-PID and without PID system.

References

[1] K. H. A. K. H. Ang, G. Chong, and Y. L. Y. Li, "PID control system analysis, design, and technology," *IEEE Trans. Control Syst. Technol.*, vol. 13, no. 4, pp. 559–576, 2005.

[2] Y.-B. W. Y.-B. Wang, X. P. X. Peng, and B.-Z. W. B.-Z. Wei, "A new particle swarm optimization based auto-tuning of PID controller," *2008 Int. Conf. Mach. Learn. Cybern.*, vol. 4, no. July, pp. 12–15, 2008.

[3] S. Panda, B. K. Sahu, and P. K. Mohanty, "Design and performance analysis of PID controller for an automatic voltage regulator system using simplified particle swarm optimization," IET *J. Franklin Inst.*, vol. 349, no. 8, pp. 2609–2625, 2012.

[4] Z.-L. L. Gaing, "A Particle Swarm Optimization Approach for Optimum Design of PID Controller in AVR System," *IEEE Trans. Energy Convers.*, vol. 19, no. 2, pp. 384–391, 2004.

[5] D. D. B. Selvabala, "Real-coded genetic algorithm and fuzzy logic approach for real-time tuning of proportional – integral – derivative controller in automatic voltage regulator system," vol. 3, no. February, pp. 641–649, 2009.

[6] Saadat, H. (1999). *Power Stability Analysis. Third Edition.* Mcgraw-Hill:New York
[7] Ogata, Katsuhiko. (1985). *Teknik Kontrol Automatik Jilid 1.* Alih Bahasa oleh Edi Laksono. Jakarta: Erlangga.
[8] C. Wong, S. Li, and H. Wang, "Optimal PID Controller Design for AVR System," vol. 12, no. 3, pp. 259–270, 2009.
[9] B. K. Sahu, P. K. Mohanty, and N. Mishra, "system using Pattern Search algorithm," *IEEE Int. Conf. Power Electron. Device Energy Syst. 2012*, 2012.
[10] N. Madinehi, K. Shaloudegi, M. Abedi, and H. A. Abyaneh, "Optimum design of PID controller in AVR system using intelligent methods," *2011 IEEE PES Trondheim PowerTech Power Technol. a Sustain. Soc. POWERTECH 2011*, pp. 1–6, 2011.
[11] Kimiyaghalam, A., & Ashouri, A. (2008). *Advanced Particle Swarm Optimization-Based PID Controller Parameter Tuning. Proceedings Ofthe 12th* IEEE *International Multitopic Conference*, 429–435.
[12] J. C. J. Cao and B. C. B. Cao, "Design of Fractional Order Controllers Based on Particle Swarm Optimization," *2006 1ST IEEE Conf. Ind. Electron. Appl.*, pp. 775–781, 2006.
[13] Purnomo, H. (2014). *Cara Mudah Belajar Metode Optimasi Methahuirstik Menggunakan Matlab.* Gave Media : Yogyakarta.
[14] M. R. AlRashidi and M. E. El-Hawary, "A Survey of Particle Swarm Optimization Applications in Electric Power Systems," *IEEE Trans. Evol. Comput.*, vol. 13, no. 4, pp. 1–6, 2009.

IOP Conf. Series: Materials Science and Engineering **128** (2016) 012039 doi:10.1088/1757-899X/128/1/012039

Effects of MgO buffer annealing on optical and electrical quality of P-MBE grown ZnO films on c-sapphire

A Setiawan*[1] and T Yao[2]

[1]Department of Mechanical Engineering Education, Universitas Pendidikan Indonesia, Jl. Dr. Setiabudhi 229 Bandung 40154, Indonesia
[2]Center for Interdisciplinary Research Tohoku University, Aramaki aza Aoba 6-3, Aoba-ku, Sendai 980-8578, Japan

*Corresponding author: agus_setiawan@upi.edu

Abstract. Zinc oxide (ZnO) has been attracting much attention because of its potential applications in photonic and optoelectronic devices. In this present study, we investigated the effect of MgO buffer annealing on the optical and electrical quality of P-MBE grown ZnO films on c-sapphire with MgO buffer layer. The optical quality was observed by low-temperature PL (photoluminescence) measurement in the near band edge emission region measured at 10K and at 77K. The emission line located at 3.368eV dominates the spectrum in both samples (ZnO with and without MgO buffer annealing) at 10K and 77K. This emission can be divided into two peaks, 3.367eV and 3.363eV and assigned as I2 (ionized donor bound excitons emission) and I4 (Hydrogen donor related emission), respectively. The relative intensity of these donor bound exactions to free exaction emission of the sample without MgO buffer annealing is greater than that of the sample with MgO buffer annealing. Comparison of the PL spectra of ZnO with and without annealing revealed that the intensity of free exciton emission from the sample with MgO buffer annealing is twice of that from the sample without annealing. We also found that the intensity of deep-level broad emission is reduced by about 1/3 by MgO-buffer annealing. Hence, the decrease of deep level emission intensity and the increase of free exciton emission intensity by annealing of MgO buffer corresponds to the reduction of defects of the ZnO film. The PL properties also suggest that there are fewer nonradiative recombination centers in ZnO layers with MgO buffer annealing than those in ZnO layers grown without MgO buffer annealing. The electrical quality was measured by room temperature Hall measurements. We found that the samples have a background n-type carrier concentration. The ZnO samples with MgO buffer annealing has a carrier concentration of 1.17x1017 cm-3 and Hall mobility of 120 cm2/V.s, while the ZnO sample without MgO buffer annealing has a carrier concentration of 2.63 x 1016 cm-3 and Hall mobility of 105 cm2/V.s. The improvement of electron mobility of ZnO films by MgO buffer annealing is due to a decrease in dislocation density. We conclude that annealing of MgO buffer layer increases the optical and electrical quality of the ZnO films. These results agree with the structural quality as observed by HRXRD.

1. Introduction

ZnO is a direct band gap semiconductor (Eg=3.37 eV at RT) with a wurtzite structure. A unique property of ZnO is its large exciton binding energy, 60 meV, which is about three times larger than that of ZnSe or GaN. Reports on the lasing mechanisms of ZnO have shown that ZnO is a promising photonic material for exciton devices in the wavelengths ranging from blue to ultraviolet [1][2][3].

High- quality ZnO is a prerequisite for ZnO-based optical device applications. Up to now, challenges of moving toward device applications are the difficulty in achieving high crystal quality of ZnO, as well as the difficulty in controlling p-type conductivity. On the other hand, controlling the crystalline defects in naturally n-type ZnO is the critical issue to obtain p-type ZnO.

Low price high-quality c-sapphire has been extensively used as a substrate for ZnO epitaxy or deposition in many growth techniques. Among growth techniques, P-MBE has shown high controllability to grow high-quality ZnO layers [4][5]. Because of the large lattice misfit between ZnO and c-sapphire (18%) and the formation of 30o rotated domains, ZnO layers grown on c-sapphire showed rough surface morphology and poor crystalline quality [5][6].

The use of a buffer layer is the most widespread technique used in heteroepitaxy. To overcome the problems caused by the large mismatch between ZnO and a c-sapphire substrate, the growth of double buffer layers consisting of low- temperature MgO and ZnO buffer followed by high-temperature annealing has been utilized with success by P-MBE [7][8]. Consequently, the growth of high temperature (HT) ZnO is dominated by layer-by-layer growth, the formation of 30o-rotated domains is completely suppressed, and the dislocation density is reduced. However, the dislocation density still remains high, in particularly edge dislocation (higher than 5x109 cm-2) [9]. Therefore, further reducing of the ZnO dislocation density is crucial for device applications. Buffer growth parameters such as thickness, growth temperature, and its annealing, and growth rate should be precisely controlled. So far, most studies on the quality of ZnO films have been addressed on the structural quality of the ZnO film. But only a few studies focussed on the optical and electrical quality of the ZnO films.

2. Experiment

In this experiment, we prepared ZnO sample grown on c-sapphire with annealed and un-annealed low-temperature MgO/ZnO buffer layer before ZnO growth. The c-sapphire substrate was degreased in acetone and methanol in an ultrasonic cleaner followed by rinsing in deionized water. The substrate was then chemically etched in an H2SO4 (96%): H3PO4 (85%) = 3:1 solution. Prior to growth, the substrate was thermally cleaned at 750 oC in the preparation chamber for 1 hour. The substrate was then treated in an oxygen plasma at 650 oC for 30 minutes in the growth chamber to produce an oxygen-terminated c-sapphire surface. The sample structure is ZnO/ZnO-buffer/MgO-buffer/c-sapphire. The buffer layers were grown under optimum growth temperature; at 490 oC for MgO buffer and 500 oC for ZnO buffer. ZnO buffer was annealed at 750 oC for 5 minutes. High-temperature ZnO layers were then grown at 700 oC. In the case of ZnO layers grown with MgO buffer annealing, the MgO buffer was annealed at 800 oC for 25 minutes. ZnO layers were then characterized by high-resolution x-ray diffraction (HRXRD), Photoluminescence (PL), and Hall measurements.

3. Result and Discussion

ZnO film has been successfully grown on c-sapphire with MgO/ZnO double buffer layer. AFM measurement revealed that that the ZnO films show a smooth surface morphology. Structural quality addressed by HRXRD measurement also shown that the ZnO films have a good crystalline quality. The surface roughness of the ZnO layers is 0.6 nm and 0.3 nm while step height (terraces size) average 0.2 nm (94 nm) and 0.3 nm (245 nm) for the ZnO films without and with MgO buffer annealing, respectively. Here we note that annealing of LT-MgO buffer at high temperature enhanced the surface migration of adatoms, leading to the formation of larger terraces on the surface and smoother surface morphology of ZnO layer. [10]

The crystalline quality of ZnO layers was addressed by HRXRD. FWHM values of (0002) □ ((10-11)□□ rocking curves (in arcsec) are 18 (1076) and 22 (843), respectively for ZnO layers grown on an LT-MgO buffer and on an annealed LT-MgO buffer. Nearly the same low FWHM values of (0002) □ scans for these two samples implies low screw dislocation density in both [10]. A much broader (10-11) □ reflection than (0002) reflection implies the present of a high density of edge-type dislocations. Note that all types of dislocations (edge, screw, and mixed) broaden the (10-11) reflection, whereas the (0002) reflection is only sensitive to screw and mixed type dislocations[11][12]. Furthermore, the

FWHM values of the (10-11) □ scan of the ZnO layer grown on annealed MgO buffer is smaller than that grown on an LT-MgO buffer, indicating much lower edge dislocation density.

3.1. Optical Quality

For further insight into the quality of the ZnO layers, we investigated the optical properties by PL measurement. Figure 1 shows low-temperature PL spectra of the samples with and without MgO-buffer annealing in the near band edge emission region measured at 10K (a) and at 77K (b). The spectra were normalized by the free exciton-A emission intensity for clarity. The emission line located at 3.368eV dominates the spectrum in both samples at 10K. This emission can be divided into two peaks, 3.367eV and 3.363eV and assigned as I2 (ionized donor bound excitons emission) and I4 (Hydrogen donor related emission), respectively [13]. Note that these lines are clearly resolved into two lines at 77K in the sample with annealing as shown in Figure 1 (b). The intensity of I2 is greater than that of I4 in both samples. In the donor bound exactions region [9], another relatively weak emission which is assigned as I8 or I9 (Gallium or Indium donor related emission) [13] was also observed at around 3.358eV. It can be seen that the relative intensity of these donor bound exactions to free exaction 96 emission of the sample without MgO buffer annealing is greater than that of the sample with MgO buffer annealing. It might be related to different impurity incorporation or formation of native defects in the samples.

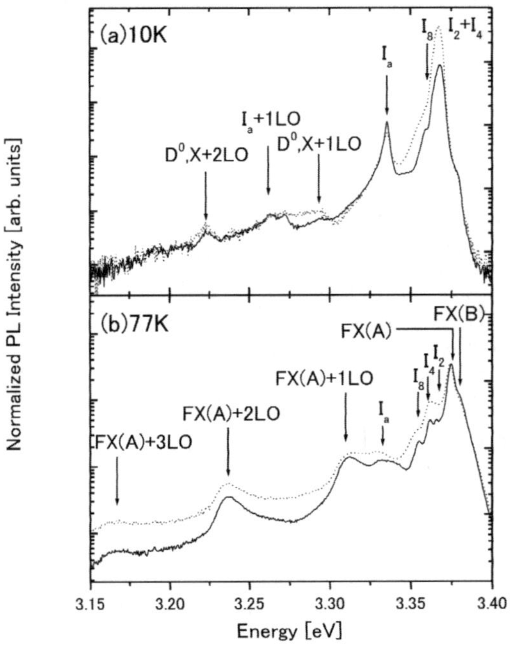

Figure 1. Photoluminescence spectra at low temperature. (a)T=10K (b)T=77K. Solid line: ZnO with MgO-buffer annealing; dotted line: is ZnO without MgO-buffer annealing. The spectra are normalized with respect to FX(A) emission.

At the lower energy region, a sharp emission located at 3.335eV and longitudinal optical (LO) phonon assisted bound exaction emission were observed in both samples. For the temperature dependence of PL spectra (data is not shown here), the intensity of the emission line at 3.335eV reduces like donor bound exactions with increasing temperature. Accordingly, this line can be attributed to exaction nature. Kato et al. [14] reported the origin of this emission is neutral acceptor bound exaction (Ian) emission from PL spectra of epitomical layers on a-sapphire grown by MBE. On the other hand, Laves et al. [15] assigned the exaction emission line at 3.335eV to exaction bound to structural defects

(DBX) emission from CL in experiments on as-grown and annealed bulk samples. In the present case, the intensity of the 3.335eV line in ZnO film with MgO buffer annealing is greater than that without annealing, that is, the intensity increases even though the dislocation density is lower. Considering this point, it is appropriate to assign this line to neutral acceptor bound excitons, in this case, though the origin of the acceptor is still unknown. At 77K, free exciton-A emission located at 3.374eV becomes dominant instead of donor bound excitons. The free exciton-B emission appears at 3.380eV with 6 meV energy spacing to the free exciton-A emission. LO-phonon replica of the free exciton emission clearly appeared at 3.312eV, 3.239eV, and 3.166eV. As each spectrum of excitonic emission lines can be resolved, the emission lines from the sample with MgO buffer annealing have narrower spectral width compared to those of ZnO without MgO buffer annealing. This indicates better crystal quality of ZnO film with MgO-buffer annealing [16].

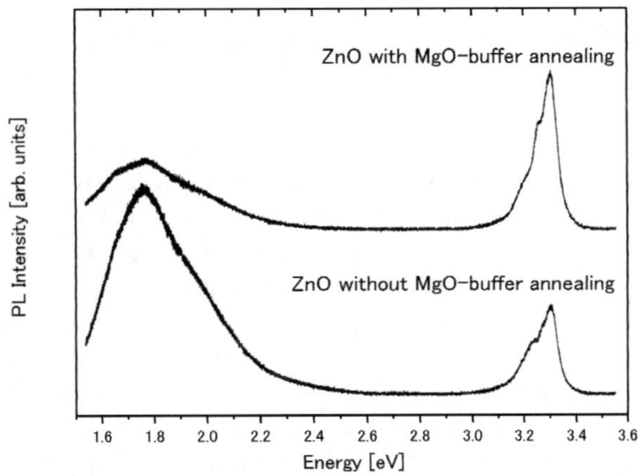

Figure 2. Room temperature PL spectra of ZnO layer grown with and without MgO buffer annealing

Figure 2 shows room temperature PL spectra of ZnO films grown with and without MgO-buffer annealing. The free exciton emission at 3.31eV and a broad deep-level emission at around 1.75eV were observed in both the samples. Comparison of the PL spectra of ZnO with and without annealing revealed that the intensity of free exciton emission from the sample with MgO buffer annealing is twice of that from the sample without annealing. We also found that the intensity of deep-level broad emission is reduced by about 1/3 by MgO-buffer annealing. Generally, the deep level luminescence of ZnO originates from oxygen vacancies and some impurity related complexes in ZnO films [12]. Hence, the decrease of deep level emission intensity and the increase of free exciton emission intensity by annealing of MgO buffer corresponds to the reduction of defects of the ZnO film. This agrees with the structural characterization which shown a reduction of dislocation density of the ZnO films from 5.3x109 cm-2 to 1.9x109 cm-2 by MgO-buffer annealing [10]. The PL properties suggest that there are fewer nonradiative recombination centers in ZnO layers with MgO buffer annealing than those in ZnO layers grown without MgO buffer annealing.

In order to gain further insight into the correlation between the structural and optical properties, Figure 3 provides a plot of structural properties represented by XRD against optical properties represented by PL intensity for ZnO samples grown on c-sapphire with and without MgO buffer. Broadening of 11 10 Ω scan can be correlated to the presence of an edge, screw, and mixed dislocations while PL intensity can be correlated to the 99 presence of non-radiative channels generated by the defects in the crystal. Figure 3 clearly shows that decreasing FWHM of 10-11 Ω scan is followed by increasing PL intensity. This feature directly implies a correlation between the structural and optical quality of the ZnO films.

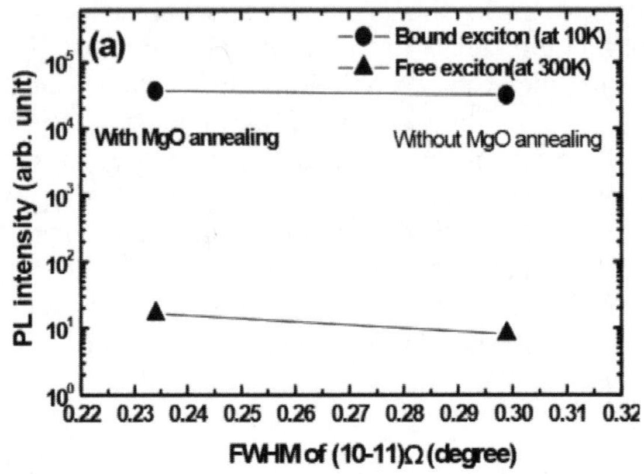

Figure 3. decreasing FWHM of 10-11 Ω scan is followed by increasing PL intensity

3.2. Electrical Quality

Room temperature Hall measurements show that the samples have a background n-type carrier concentration. The ZnO samples with MgO buffer annealing has a carrier concentration of 1.17×10^{17} cm^{-3} and Hall mobility of 120 cm^2/V.s while the ZnO sample without MgO buffer annealing has a carrier concentration of 2.63×10^{16} cm^{-3} and Hall mobility of 10^5 cm^2/V.s. The improvement of electron mobility of ZnO films by MgO buffer annealing is due to a decrease in dislocation density. Here, the edge dislocations introduce acceptor centers along the dislocation line, which capture electrons from the conduction band of an n-type semiconductor [17]. The dislocation line becomes negatively charged, and a space charge is formed around it which scatters electrons, thus reducing their mobility [18]. It should be noted here that the mobility of ZnO decreases as carrier concentration (n) decreases, in contrast to the behavior in most another semiconductor, in which mobility continues to rise as n falls. The same phenomenon was also observed in GaN grown on Al2O3 with the concentration below about 10^{17}-10^{18} cm^{-3} [19][20]. This phenomena can be explained by considering charged dislocation scattering, in which below a certain n, dislocation scattering dominates, while above a value of n, ionized-impurity (or defect) scattering is more important [21].

4. Conclusion

We have investigated the effect of MgO buffer annealing on the optical and electrical quality of P-MBE grown ZnO films on c-sapphire with MgO buffer layer. Optical quality observed by low-temperature PL measurement in the near band edge emission region at 10K and 77K revealed that the emission line located at 3.368eV dominates the spectrum in ZnO samples with and without MgO buffer annealing at 10K and 77K. This emission can be divided into two peaks, 3.367eV and 3.363eV and assigned as I_2 (ionized donor bound excitons emission) and I_4 (Hydrogen donor related emission), respectively. Comparison of the PL spectra of ZnO samples revealed that the intensity of free exciton emission from the sample with MgO buffer annealing is twice higher than that from the sample without annealing. We also found that the intensity of deep-level broad emission is reduced by about 1/3 by MgO-buffer annealing. which corresponds to the reduction of defects of the ZnO film. The PL properties suggest that there are fewer nonradiative recombination centers in ZnO with MgO buffer annealing than those in ZnO grown without MgO buffer annealing. The electrical quality measured by room temperature Hall measurements found that the samples have a background n-type carrier concentration. The ZnO sample with MgO buffer annealing has a carrier concentration of 1.17×10^{17} cm^{-3} and Hall mobility of 120 cm^2/V.s while the ZnO sample without MgO buffer annealing has a carrier concentration of 2.63×10^{16} cm^{-3} and Hall mobility of 10^5 cm^2/V.s. The improvement of electron mobility of ZnO films by MgO buffer annealing is due to a decrease in dislocation density. We conclude that annealing of MgO buffer layer increases the optical and electrical quality of the ZnO films. The results agree with the structural quality as observed by HRXRD.

Acknowledgment
The author would like to thank Dr. D. C. Oh and Mr. H. Goto, previous members of Yao-Lab. of Tohoku University Japan for their help for films characterizations and fruitful discussions.

References
[1] D.M. Bagnall, Y.F. Chen, Z. Zhu, T. Yao, M.Y. Shen, T. Goto, Appl. Phys. Lett. 73, 1038 (1998).
[2] H.J. Ko, Y.F Chen, T. Yao, K. Miyajima, A. Yamamoto, T. Goto, Appl. Phys. Lett. 77, 537 (2000).
[3] P. Ding, X.Pan, J. Huang, B. Lu, H. Zhang. Materials Letters. Vo.71. P.18-20. (2012).
[4] M. A. L. Johnson, S. Fujita, W. H. Rowland, Jr., W.C. Highes, J. W. Cook, Jr., J. F. Schetzina, J. Electron. Mater. 25, 855 (1996).
[5] Y. F. Chen, D. M. Bagnall, H. J. Ko, K. T. Park, K. Hiraga, Z. Zhu, T. Yao, J. Appl. Phys. 84, 3912 (1998).
[6] P. Fons, K. Iwata, A. Yamada, K. Matsubara, S. Niki, K. Nakahara, T. Tanabe, H. Takasu, Appl. Phys. Lett. 77, 1801 (2000).
[7] Y. F. Chen, H. J. Ko, S. K. Hong, T. Yao, Appl. Phys. Lett.76, 559 (2000).
[8] Y. F. Chen, H. J. Ko, S. K. Hong, T. Yao, Y. Segawa, J. Cryst. Growth 214/215, 87 (2000).
[9] Y. F. Chen, S. K. Hong, H. J. Ko, V. Kirshner H. Wenisch, T. Yao, K. Inaba, Y. Segawa. Appl. Phys. Lett. 78, 3352 (2001).
[10] A. Setiawan, H. J. Ko, S.K. Hong , Y. Chen, and T. Yao. Thin Solid Films 445 (2003) 213–218.
[11] J. P. Hirth J. Lothe, Theory of Dislocations, 2nd Ed. John Wiley & Sons, New York, (1982).
[12] B. Heying, X.H. Wu, S. Keller, Y. Li, D. Kapolnek, B. P. Keller, S.P. DenBaars, and J.S. Speck, Appl. Phys. Lett. 68, 643 (1996).
[13] B. K. Meyer, H. Alves, D. M. Hofmann, W. Kriegseis, D. Forster , F. Bertram, J. Christen, A. Hoffmann , M. Straßburg, M. Dworzak, U. Haboeck , and A. V. Rodina. Phys. Stat. Sol. (b) 241, No. 2, 231–260 (2004)
[14] H. Kato, M. Sano, M. Miyamoto, and T. Yao, J. Crystal Growth 237-239, 538 (2002).
[15] H. Alves, D. Pfisterer, A. Zeuner, T. Riemann, J. Christen, D.M. Hofmann, and B.K. Meyer, Optical Materials, 23, 33 (2003).
[16] V. Kumara, , V. Kumar, S. Som, A. Yaousif, N. Singh, O.M. Ntwaeaborwa, A. Kapoor, H.C. Swart. J. Of Coloid and Interface Science. Vol.428, p. 8–15.(2014).
[17] W. T. Read: Philos. Mag. 45 (1954) 775.
[18] K. Miyamoto, M. Sano, H. Kato, and T. Yao, Jpn. J. Appl. Phys. Vol. 41, L 1203–L 1205 (2002.
[19] N. G. Weimann, L. F. Eastman, D. Doppalapudi, H. Ng, and T. D. Moustakas, J. Appl. Phys. 83, 3656 (1998).
[20] H. M. Ng, D. Doppalapudi, R. Singh, and T. D. Mous-takas,Mater. Res. Soc. Symp. Proc. 482, 507 (1998); H. M. Ng, D. Doppalapudi, T. D. Moustakas, N. Weimann, and L. F. Eastman, Appl. Phys. Lett. 73, (1998).
[21] D. C. Look, J. R. Sizelove, Phys. Rev. Lett. 82, 1237 (1999).

Synthesis of silica particles from rice straw waste using a simple extraction method

A B D Nandiyanto*[1], T Rahman[1], M A Fadhlulloh[1], A G Abdullah[2], I Hamidah[3], and B Mulyanti[2]

[1]Departemen Kimia, Universitas Pendidikan Indonesia, Jl. Dr. Setiabudi no 229, Bandung 40154, Jawa Barat, Indonesia
[2]Departemen Pendidikan Teknik Elektro, Universitas Pendidikan Indonesia, Jl. Dr. Setiabudi no 229, Bandung 40154, Jawa Barat, Indonesia
[3]Departemen Pendidikan Teknik Mesin, Universitas Pendidikan Indonesia, Jl. Dr. Setiabudi no 229, Bandung 40154, Jawa Barat, Indonesia

*Corresponding author: nandiyanto@upi.edu

Abstract. The purpose of this study was to synthesize silica particles from rice straw waste using a simple extraction method. The experiment was conducted by heating and extracting rice straw waste into basic solution. To get silica particles, the extracted solution was then put into acid solution and heated to remove the remained solvent. The experimental results showed that the aggregated silica particles with sizes of about 200 nm were successfully produced. The XRD and FTIR analysis showed that the final product was silica and free of graphite. However, we found that some KCl component in the final product in which this was possibly from the use of KOH as the extraction agent. Therefore, further studies are still required to synthesize high purity silica particles from rice straw waste.

1. Introduction

Silica is one of the most important components and can be found in various applications such as biotechnology-related materials [1][2], environmental treatment components[3][4], raw materials for cement industry [5], composite materials [6][7], and medical-related materials [8][9][10]. The main reason for the use of silica is because of its excellent performances: chemically, physically, and thermally stable, compatible with various materials [6][7], largely available, and relatively inexpensive [1].

One of the potential silica raw materials is rice plant-related waste. This waste can be found in tropical countries, especially Indonesia. However, the use of rice straw waste as a silica raw material is still suboptimal. In fact, using rice straw as the silica raw material will be beneficial not only for economical aspect but also for environmental point of view. [11]

Many researches relating to extraction of silica from rice plant have been reported, including rice husk [12][13][14] and rice straw [10][15]. In general, to get silica, they used extraction agent (e.g. basic solution). Reports on the influence of mol ratios of extraction chemical and silica, extraction time, and process temperature on the successful production of silica have been reported [13][15][16]. Although their reports showed good for industrial applications, several disadvantages are found, including

IOP Conf. Series: Materials Science and Engineering **128** (2016) 012040 doi:10.1088/1757-899X/128/1/012040

complicated synthesis procedures, sophisticated apparatuses, high-temperature process, and time-consuming extraction process.

Here, the purpose of this study was to synthesize silica particles from rice straw waste using a simple extraction method. In short, our method was conducted by adding rice straw waste into basic solution. Then, the extracted solution was then put into acid solution to precipitate the silica. Different from other reports, our method is successful for extraction silica under atmospheric pressure. The synthesis method for silica particles is easily accomplished under relatively low-temperature process, while other methods must be done at a temperature of more than 80°C. Our extraction processing time is also relatively short (up to 2 hours), while other reports need more than 3 hours. To confirm our successful synthesis of silica materials, several characterizations have been conducted, including an X-Ray Diffraction (XRD), Scanning Electron Microscope (SEM), Thermo Gravity Differential Thermal Analysis (TG-DTA), Atomic Absorption Spectroscopy (AAS), and Fourier Transform Infra Red (FTIR).

2. Hypothetical analysis for the extraction of silica from rice straw waste

Rice straw contains cellulose (32-47 wt%), hemicellulose (19-32 wt%), lignin (5-24 wt%), and other components (13-20 wt%) [15][17]. Other components in the rice straw contains silica with composition about 60-80 wt% [15][18]. The chemical composition inside the rice straw depends on type of rice, planting time, climate, and geographical-related condition.[12] Since silica contains in the rice straw ash with high concentration, the rice straw can be classified as one of the prospective silica raw materials.

To extract silica from rice straw, a simple extraction method can be used. The reaction happening during the extraction process can be written as [14][19]

(1) $SiO_2 + 2KOH \rightarrow K_2SiO_3 + H_2O$

(2) $K_2SiO_3 + 2HCl \rightarrow SiO_2 + 2KCl + H_2O$

Then, for some cases, the reaction (2) can be written as [21]

(3) $K_2SiO_3 + H_2O + 2HCl \rightarrow Si(OH)_4 + 2KCl$

(4) $Si(OH)_4 + Si(OH)_4 \rightarrow n\,[(OH)_3Si - O - Si(OH)_3] + H_2O$

3. Experimental Method

3.1. Raw materials

We used rice straw waste (rice field in Subang, Indonesia), kalium hydroxide (KOH; technical grade; PT. Bratachem, Indonesia), and hydrochloric acid (HCl; technical grade; PT. Bratachem, Indonesia). Prior to using, rice straw waste was washed with de-ionized water, dried naturally for 3 days, and cut with sizes of about 1 cm. For KOH and HCl, these chemicals were used without further purification process.

3.2. Synthesis of silica particles

First, rice straw waste was heated with an electrical furnace at a specific temperature. Then, the heated rice straw waste (as ashes) was grinded, put into basic solution (KOH), and mixed with 600 rpm at temperature of 60°C for 2 hours. After the extraction process was finished, the extracted solution was filtered. The filtered ash is named as the residual ash, whereas the solution is the filtered solution. The filtered solution was then added to the acid solution (HCl). The process was conducted at pH = 2 to form clear gel. The gel was then aged for 1 night. Finally, the solution containing gel was filtered, washed with de-ionized water, and heated at 150°C for 30 minutes.

3.3. Characterization

We used several characterization analysis methods: X-Ray Diffraction (XRD; PANalytical X'Pert PRO; Philips Corp., Netherland), Thermo Gravity Differential Thermal Analysis (TG-DTA, DTG 60A TA 60 WS, Shimadzu Corp., Japan; operated at 5°C/min with 200 mL/min of carrier gas (oxygen gas)), Atomic Absorption Spectroscopy (AAS, Varian Spectra 240 FS, Varian Inc., Califonia), Fourier Transform Infrared (FTIR, FTIR-8400 dan Prestige 21, Shimadzu Corp., Japan), dan Scanning Electron Microscopy (SEM, JSM-6360LA; JEOL Ltd., Japan).

4. Results and Discussion

4.1. Physicochemical properties of rice straw waste

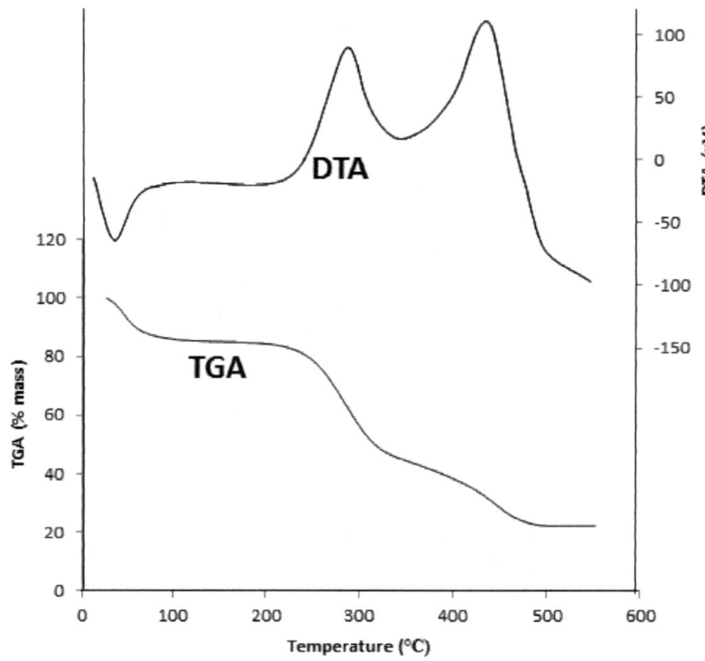

Figure 1. TG-DTA analysis result of rice straw waste

Figure 1 shows TG-DTA analysis results of rice straw waste. From this figure, we obtained four temperature zones, as discussed in the following:

(1) 20-100°C. The mass reduction (about 5 wt%) occurred because of evaporation of physically adsorbed water [6][7]. The evaporation process was confirmed by the detection of endotermic peak at 50°C.
(2) 250-320°C. The mass loss (48 wt%) happened due to the degradation of cellulose-related component to form carbon [11][16]. This is confirmed by the existence of exothermic peak at 299°C.
(3) 330-470°C. The mass reduction (25 wt%) was identified, in which this was due to degradation of lignin [11][16]. This was verified by the detection of exothermic peak at 442°C in the DTA analysis.
(4) More than 500°C. No mass loss was identified. The final mass was about 22 wt%. No change in mass was found because all organic compounds have been converted completely.

Based on the above TG-DTA results, the heat treatment is important to be added before the silica extraction process. The heat treatment can eliminate organic compound of up to 80 wt%, which is important to make the extraction of silica easily. [16]

To confirm the composition of silica in the heated sample, the AAS and the gravimetric analysis were conducted (**Table 1**). The result showed that the heated rice straw contained 85 wt% of silica. This result is in a good agreement with other reports [15][18]. This result also indicates that the present rice straw is one of the good candidates for silica raw materials.

Table 1. AAS and gravimetric analysis results of heated rice straw. Composition of silica was re-analysis using Gravimetric analysis method.

No.	Composition	Quantity (%)
1	SiO_2	84.60
2	K_2O	6.39
3	CaO	0.88
4	MgO	0.44
5	Fe_2O_3	0.16
6	Other	7.53

4.2. Synthesis of silica particles from rice straw waste

Figure 2. The photograph images (a-f) and the SEM analysis result of samples (g). Figure (a) is the initial rice straw waste. Figures (b)-(d) are the grinded samples after heat treatment process for 10 minutes with various temperatures: (a) 200, (b) 300, (c) 400, and (d) 500°C. Figure (d) is the sample after putting in the basic and acid treatment. Figure (f) is the product after final heating process. Figure (g) is the SEM analysis image of product shown in Figure (f). To make easy for observing visual observation, Figure (f) was put in the dark background, whereas Figures (a)-(e) was in the white background.

Figure 2 is the photograph and the SEM images of sample heated with various temperatures. **Figure 2a** shows rice straw sample after washed and cut into sizes of about 1 cm. After rice straw was heated and grinded, black powders were obtained (**Figures 2b-d**). The color of powders was getting brighter with increasing process temperatures. These figures confirmed that the transformation of rice straw into carbon from temperature of 200°C [20]. The phenomena of the brighter dark powder were due to the increases in silica composition in the ash [12]. This result is in a good agreement with TG-DTA analysis results in the above that increasing the temperature causes the decreases mass of organic compound. From these results, we conclude that the best heat treatment process is 500°C.

After sample was put into the basic solution, black solution (coffee-like solution) was obtained (not displayed). Then, filtering the solution, we got clear, transparent, and black solution. Adding this filtered solution into acid solution, the mild gel solution was obtained, in which the purified gel is shown in **Figure 2e**. To get silica nanopowder, the purified gel was heated (**Figure 2f**). As shown in Figure 2e, the crack in the sample was found. Silica starts to form smaller particles solitary [11]. This result was confirmed by the observation of the SEM image in **Figure 2g**, in which the product contained aggregated particles with sizes of about 200 nm. The white powder was obtained, indicating the potential material consisted in the product was silica [19].

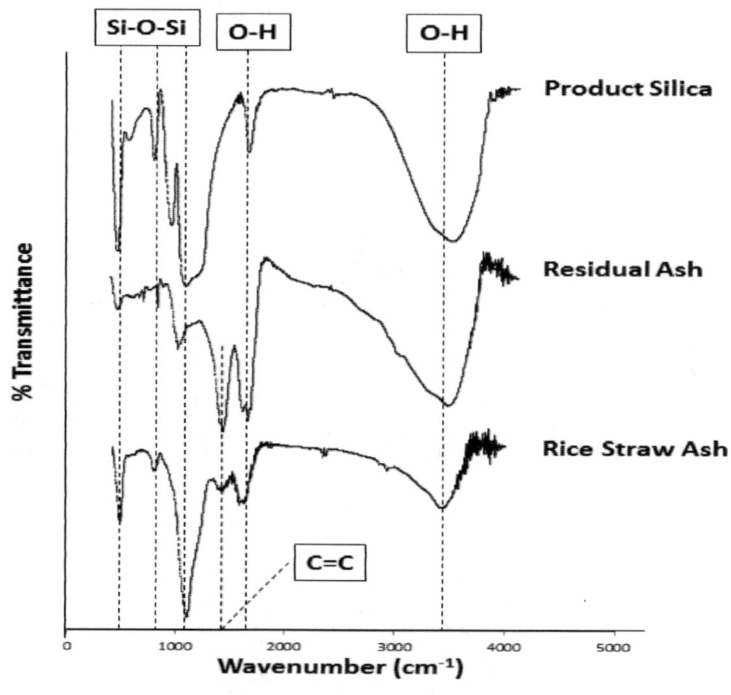

Figure 3. FTIR analysis result of rice straw ash, residual ash, and product samples

Figure 3 shows the FTIR analysis result of rice straw ash, residual ash, and product samples. All samples have some peaks at 460, 700, and 1000 cm^{-1}, in which these peaks are peaks for Si-O-Si bending [11], Si-O-Si symmetric stretching [12], and Si-O-Si asymmetric stretching [19], respectively. These results confirmed that all samples containing silica component. However, for rice straw ash and residual ash, we found a peak that was not found in the product sample. Specifically, we detected peak at 1400 cm^{-1} that indicated C=C aromatic from graphite. This result verified that the product was silica and free of graphite component. In the product sample, we also found a peak at 952 cm^{-1}, indicating that some water molecule (OH) was trapped inside the silica matrix [11]. Detailed analysis of peaks detected by FTIR is shown in **Table 2**.

Table 2. Peaks detected in the FTIR analysis results of rice straw ash, residual ash, and product samples

No.	Functional Group	Wavenumber (cm^{-1})		
		Rice Straw Ash	Residual Ash	Product
1	Si-O-Si bending	468	466	464
2	Si-O-Si symmetric stretching	796	756	798
3	Air dalam matrik silika	-	-	952
4	Si-O-Si assymmetric stretching	1099	1026	1083
5	C=C dari grafit	1421	1404	-
6	OH bending	1627	1651	1641
7	OH stretching	3423	3382	3427

Figure 4 shows the XRD analysis results of rice straw ash, residual ash, and product samples. As references, we also added several JCPDS result. The XRD of rice straw ash sample detects peaks of silica, $CaMgSi_2O_5$ (diopside), dan graphite. The appearance of diopside was different with AAS result in **Table 1**. AAS detected SiO_2, CaO and MgO, instead of diopside. AAS analysis uses flame that possibly changes the structure of diopside into their stable oxides, i.e. SiO_2, CaO, and MgO. The stability of diopside was also confirmed by the disappearance of diopside peaks in both residual ash and product samples.

Based on the XRD results of the product sample, the peak at 27° was not detected, confirming the product is free of graphite. The intensity of peaks of silica (JCPDS 27-1402) verified that the main product containing silica. However, in product, we found KCl peak, in which this was possibly from the KOH as the extraction agent. Therefore, further purification process is required for increasing the purity of silica.

Based on the above results, we conclude the hypothetical analysis for the synthesis of silica particles in **Figure 5**. Rice straw is prospective for silica raw source since the silica is contained in the rice straw (verified by TG-DTA, AAS, and gravimetric analyses). To get excellent process for isolating silica component from rice straw, additional heat treatment is the best choice. The heat treatment is effective to eliminate most of the organic components in the rice straw via the formation of ash. Indeed, by isolating silica from the rice straw ash, carbon porous materials can be produced; however, this will be performed in our future work. Then, to get silica component, combination of co-precipitation method (via acid etching process) and heat treatment can be used. Finally, the formation of silica particles in the final product can be observed directly by visual observation (white powder). Indeed, confirmation of silica particles can be achieved by conducting further physicochemical analysis such as SEM, XRD, and FTIR analysis.

5. Conclusions

We successfully synthesized silica particles from rice straw waste using simple extraction method. The experiment was conducted by heating and extracting rice straw waste into basic solution. To get silica particles, the extracted solution was then put into acid solution and heated to remove the remained solvent. The experimental results showed that the aggregated silica particles with sizes of about 200 nm were successfully produced. Although this method succeeded to form silica particles, the KCl in the product is still problems as impurities. The existence of KCl was possibly from the use of KOH as the extraction agent. Therefore, further studies on how to produce high purity silica from rice straw waste and how purify the product are still needed.

Acknowledgements

A. B. D. N. acknowledged Direktorat Jendral Pendidikan Tinggi (Dirjen DIKTI) for Program Unggulan Perguruan Tinggi Negeri (PUPTN). T. R. and M. A. F. acknowledged Direktorat Jendral Pendidikan Tinggi (Dirjen DIKTI) for Program Kreatifitas Mahasiswa - Penelitian Eksakta (PKM-PE).

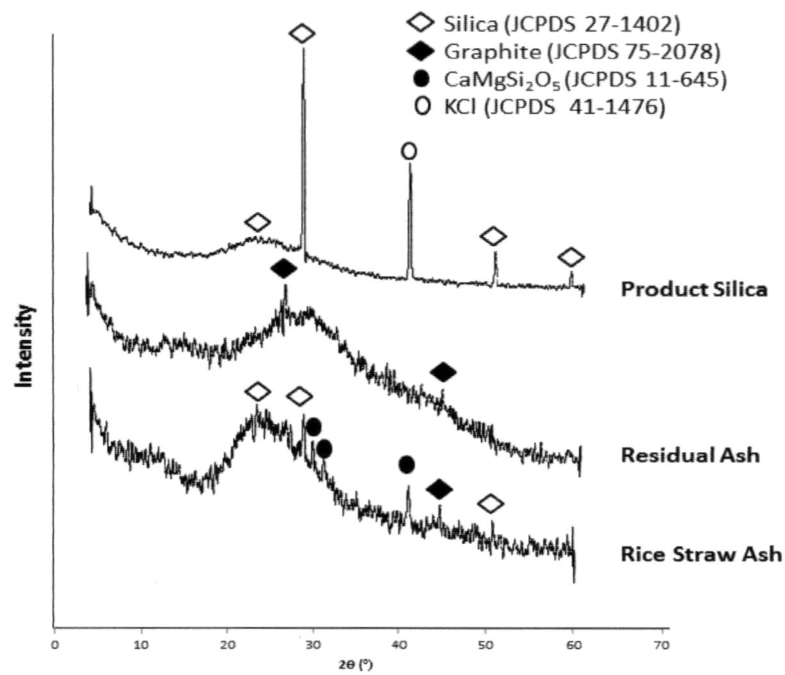

Figure 4. XRD analysis results of rice straw ash, residual ash, and product samples

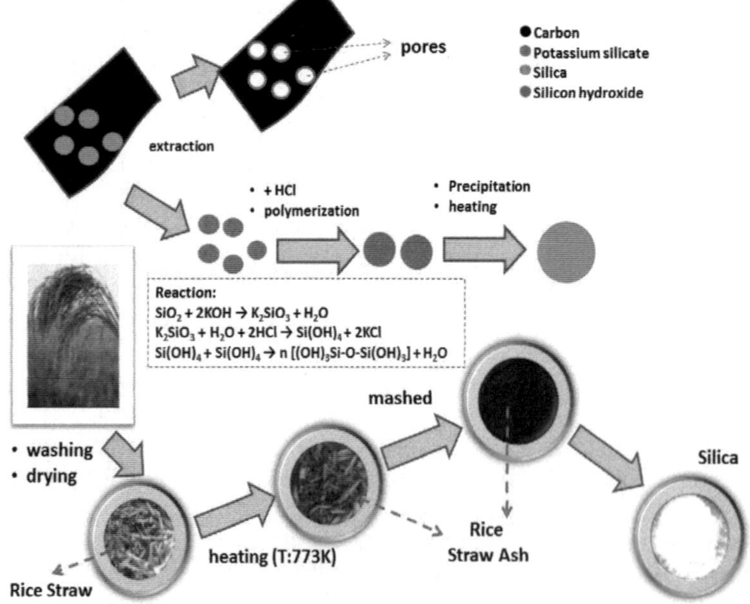

Figure 5. Hypothetical analysis for the synthesis of silica particles from rice straw waste

References

[1] Galliker, P.; Hommes, G.; Schlosser, D.; Corvini, P. F. X.; Shahgaldian, P., 2010 *J. Colloid Interface Sci.* **349** 1
[2] Cheang, T. Y.; Tang, B.; Xu, A.; Chang, G.; Hu, Z.; He, W.; Xing, Z.; Xu, J.; Wang, M.; Wang, S. 2012 *Int. J. Nanomedicine* **7** 1061.
[3] Liu, Y.; Tourbin, M.; Lachaize, S.; Guiraud, P., 2013 *Chemosphere* **92** 681.
[4] Le, V. H.; Thuc, C. N. H.; Thuc, H. H., 2013 *Nanoscale Res. Lett.* **8** 1.
[5] Singh, L. P.; Agarwal, S. K.; Bhattacharyya, S. K.; Sharma, U.; Ahalawat, S., 2011 *Nanometer Nanotechnol.*, **1**, 44.
[6] Nandiyanto, A.B.D.; Suhendi, A.; Ogi, T.; Umemoto, R.; Okuyama, K. 2014 *Chem. Eng. J.* **256** 421
[7] Nandiyanto, A.; Ogi, T.; Iskandar, F.; Okuyama, K. 2011 *Chem. Eng. J.* **167** 409
[8] Kwon, S.; Singh, R. K.; Perez, R. A.; Neel, E. A. A.; Kim, H. W.; Chrzanowski, W. 2013 *J. Tissue Eng.* **4** 1.
[9] Estevez, M. C.; O'Donoghue, M. B.; Chen, X.; Tan, W., 2009 *Nano Res.* **2** 448.
[10] Lu, J.; Liong, M.; Zink, J. I.; Tamanoi, F., 2007 *Small* **8** 1341.
[11] Lu, P.; Hsieh, Y.L. 2012 *Powder Technol.* **225** 149
[12] Rafiee, E.; Shahebrahimi, S.; Feyzi, M.; Shaterzadeh, M. 2012 *Int. Nano Letters* **2** 1
[13] Sousa, A.M.; Visconte, L.; Mansur, C.; Furtado C. 2009 *Chem. Chem. Technol.* **3** 321
[14] Kalapathy, U.; Proctor, A.; Shultz J. 2000 *Bioresource. Technol.* **73** 257
[15] Zaky, R.R.; Hessien, M.M.; El-Midany, A.A.; Khedr, M.H.; Abdel, E.A; El-Barawy; K.A. 2008 *Powder Technol.* **185** 31
[16] Junko, U.; Katsuyoshi, K. 2008 *Transactions* **37** 14
[17] Khandanlou, R.; Ahmad, M.; Shameli, K.; Kalantari, K. 2013 *Molecules* **18** 6597
[18] Sarnklong, C.; Cone, J.W.; Pellikaan, W.; Hendriks, W.H. 2010 *Asian Austral. J. Animal Sci.* **23** 680
[19] Soltani, N.; Bahrami, A.; Canul, M.I.P; Gonzalez, L.A. 2015 *Chem. Eng. J.* **264** 899
[20] Ugheoke I.B.; Mamat, O. 2012 *Maejo Int. J. Sci. Technol.* **6** 430
[21] Olalekan, A.P.; Dada, A.O.; Adesina, O.A. 2014 *J. Encapsul. Adsorp. Sci.* **4** 122

Effect of tungsten concentration on growth of *acetobacter xylinum* as a promising agent for eco-friendly recycling system

A B D Nandiyanto*, H S Halimatul, N H Rosyid and D B Effendi

Departemen Kimia, Universitas Pendidikan Indonesia, Jl. Dr. Setiabudi no 229, Bandung 40154, Jawa Barat, Indonesia

*Corresponding author: nandiyanto@upi.edu

Abstract. Effect of tungsten (W) concentration on *Acetobacter xylinum* growth was studied. In the experimental procedure, concentration of W in the bacterial growth medium containing pineapple peels waste was varied from 0.5 to 50 ppm. To confirm the influence of W, the bacterial incubation process was carried out for 72 hours. Spectrophotometer analysis showed that the growth rate of *Acetobacter xylinum* decreased with increasing concentration of W. The result from fourier transform infra red analysis showed a slightly change on the absorption peak intensities and informing the interaction of W ion and bacteria cell. The result confirmed that *Acetobacter xylinum* was able to uptake W concentration up to 15 ppm, indicating that *Acetobacter xylinum* might act as a promising agent for eco-friendly recycling system.

1. Introduction

Tungsten or Wolfram (W) is one of the rare earth elements and valuable metal groups. The properties of W are high thermo stability, strength, and abrasion resistance. Among metals, W has the highest melting point and the lowest coefficient of thermal expansion. For this reason, W is widely used for electric-lamp filaments, composite materials and polymers, semiconductor components, machine-part cutting tools, thermal shield plates, mill roles, and catalysts for de-nitration of exhaust gases and removal of organic materials.[1]

Due to its excellent performances, the exploitation of W was inevitable. Indeed, because of the demand of W material, waste problems gained from W cannot be avoided. For example, recently Hsu et. al. reported that semiconductor manufacturer waste in Taiwan contains 300 μg/L of W. This result is very high, compared to the world river average concentration of W (less than 0.10 μg/L).[2] Since W is suspected to be toxic (e.g. acute lymphocytic leukemia) and causing environmental damage, [3] finding process for solving this W waste problem is inevitable.

To solve the W waste, physical and chemical removal methods (*e.g.,* ion-exchange[1], solvent extraction[4], and alkaline solution [5]) were studied. However, these methods are economically and technologically difficult. Further, the process associates with the requirement in high cost and high energy, and relates to non eco-friendly waste product.

Biosorption is one of the alternative methods to remove metal from waste water. This method utilizes biomaterials as a main biosorbents. The effectiveness of biosorption has been well-known for the sequestration of heavy metals, rare earth elements, radio nuclides, and metalloids.[6][7]

Malekzadeh et al. reported a biosorption of W by *Bacillus sp.*, as well as the effects of parameters such as pH, initial W and cell concentration, live activity, and addition of other cations on the adsorption capacity.[8] However, the bacteria used in their study is only *Bacillus sp.* And, there is no information regarding transformation of structure and morphology of bacteria during the adsorption process. The mechanism during the interaction process of W with *Bacillus sp.* is also unclear.

Based on our experience in the use of W-related material,[9][10][11][12] the purpose of this study was to investigate the effect of W on the growth of *Acetobacter xylinum. Acetobacter xylinum* was used as a bacterial biosorbent model because it is largely available, inexpensive, and harmless. This bacterium also has a rapid cultivation rate. In this study, various W concentrations were tested. The biosorption ability of *Acetobacter xylinum* was also investigated. The results found that W influenced the bacterial growth of *Acetobacter xylinum* was stable when concentration of W of less than 15 ppm. This information indicate that bacterium can uptake W concentration of up to 15 ppm. We believe that *Acetobacter xylinum* might act as a promising agent for eco-friendly recycling system.

2. Experimental method

2.1. Raw materials
Acetobacter xylinum, commercial pure sugar, ammonium sulfate ($(NH_4)_2SO_4$) (Bratachem, Indonesia), ammonium tungstate(VI) pentahydrate (ATP; $(NH_4)_2WO_4.5H_2O$; Wako Chem. Co. Ltd., Japan), and de-ionized water. All purchased raw materials were used without further purifications. Prior to using, pineapple peel waste was washed with de-ionized water, blended, and boiled to get a pineapple peel waste juice under an aseptic condition.

2.2. Bacterial growth conditions
To prepare a medium, the pineapple peel waste juice was put into the aqueous solution (containing 7.50 %w/v of sugar and 0.50 %w/v of ammonium sulfate), mixed, and boiled. Then, the medium was cooled to room temperature and left for about 24 h to make sure all chemicals dissolved completely. Subsequently, *Acetobacter xylinum* (10%v/v) was inoculated into 250 mL Erlenmeyer flask containing 100 mL of the medium. Then, the flask was sealed with an absorbent cotton cap and incubated for 72 h under room temperature.

2.3. Preparation of W solution
A metal solution containing W ions was prepared by dissolving a known weight of ATP in de-ionized water. The concentration of W was varied from 0.5 to 50 ppm.

2.4. Sorption of W by Acetobacter xylinum
In a typical sorption experiment, W solution was added to the medium containing *Acetobacter xylinum.* The experiment was conducted at room temperature for 72 h. To follow the time course of the microbial W adsorption, aliquots of the mixture were periodically withdrawn and the concentration of bacteria in the solution was determined.

2.5. Characterization
The growth of *Acetobacter xylinum* with addition of various W concentrations was determined by an UV-visible spectrophotometer (UV-Vis; Shimadzu Corp., Japan). The morphology and structure of the samples were characterized by a transmission electron microscope (TEM; JEM-1400, JEOL Ltd., Japan; operating at 20 kV). Elemental and chemical compositions of the samples were evaluated by a fourier transform infrared spectroscopy analysis (FTIR, Shimadzu-8400, Shimadzu Corp., Japan).

3. Results and Discussion

Figure 1 shows the effects of various W concentrations on the bacterial growth rate. The concentration of bacteria in the medium without addition of W solution increased from 20 to 65 h. With addition of W concentration of up to 5 ppm, the bacterial growth pattern remained the same. In the sample with W concentrations of 10 and 15 ppm, the bacterial growth curve almost stable. However, the concentration decreased after 65 h. Regarding samples with W concentrations of more than 25 ppm, the bacterial growth rate decreased when the addition of W concentration was altered from 25 to 50 ppm. This results confirmed that the W concentration higher than 15 ppm must be considered the maximum adsorption of *Acetobacter xylinum*.

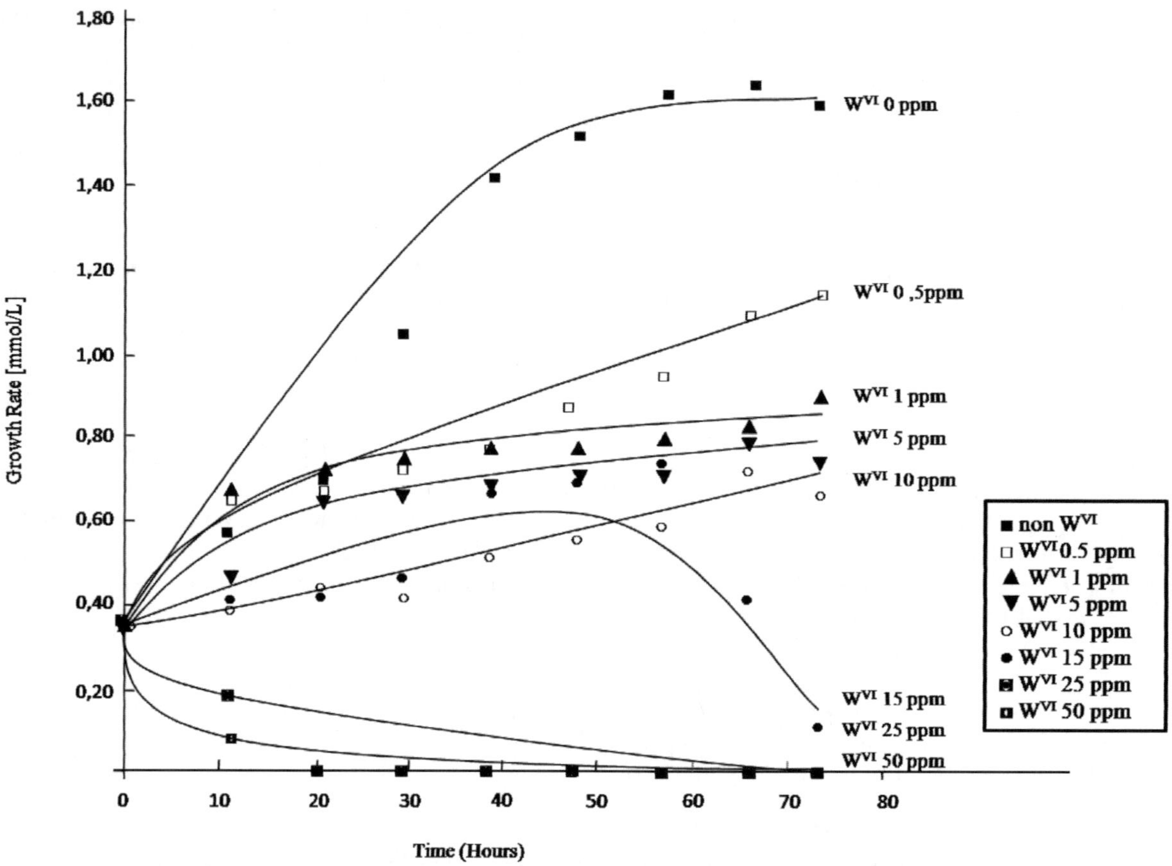

Figure 1. The curve of *Acetobacter xylinum* growth rate in the medium with various W concentrations.

Figures 2a and 2b shows the TEM images of samples before and after additional 50 ppm of W solution in the medium, respectively. TEM images indicated that the samples contained cell with bacillus structure. The sample with additional W solution had W-related particle (about 100 nm) on the outside of the body of the bacteria (**Figure 2b**), while these particles did not appear in the sample without W solution (**Figure 2a**). Based on these results, the main reason of the formation of particles can be described in the following: First, with the existence of W ion in the solution, bacteria use this ion for reduction and oxidation reactions.[12] The result of the reduction and oxidation reaction is the W-related particle. Some of the particles are adsorbed on the body, whereas the other is released to the medium. In addition, the possibility of the complete adsorption of W on the bacteria body depends on the charge of bacteria themselves.[13] Therefore, zeta potential analysis must be considered and will be done in our future study.

Figure 2c presents the FTIR analysis results of samples before and after additional 50 ppm of W solution. For the sample bacteria in the medium with addition of W solution, peaks at 3423, 2923, 1654, 1542, 1379, 1280, and 1068 cm^{-1}were detected. The detection of peaks for this sample is the same with that for sample bacteria in the medium without addition of W solution. There was only a slightly change on the absorption peaks intensity, which belonged to several functional groups of W-related component, such as carboxyl and phosphate groups (COOH and PO_4^{3-}).

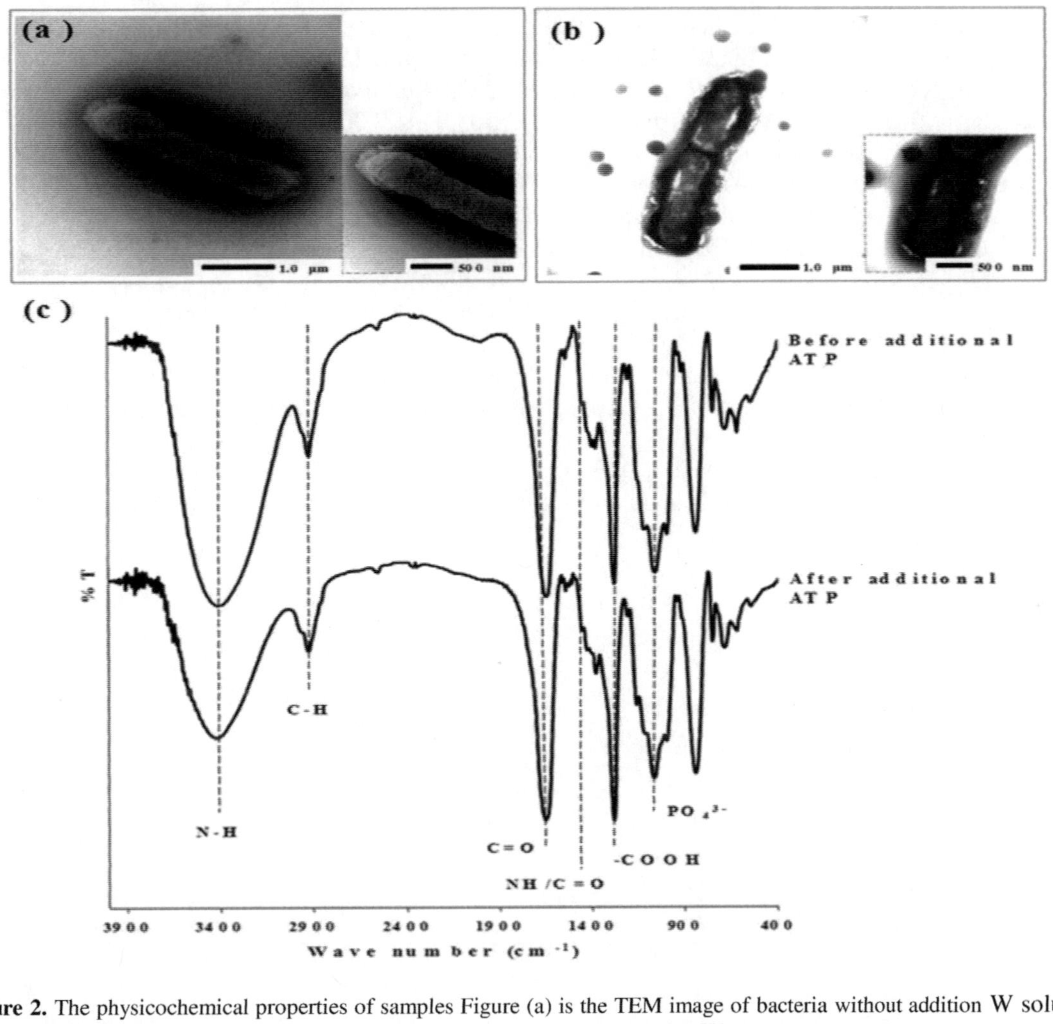

Figure 2. The physicochemical properties of samples Figure (a) is the TEM image of bacteria without addition W solution in the medium, whereas Figure (b) is the TEM image of bacteria with addition W concentration of 50 ppm in the medium. Figure (c) is the FTIR analysis results of samples.

4. Conclusions

We successfully investigated the effects of various W concentrations on the growth of *Acetobacter xylinum*. The *Acetobacter xylinum* was able to uptake W concentration of up to 15 ppm. The result from this study indicates that *Acetobacter xylinum* can be promising as an agent for eco-friendly recycling system.

Acknowledgements
A. B. D. N. acknowledged Direktorat Jendral Pendidikan Tinggi (Dirjen DIKTI) for Program Unggulan Perguruan Tinggi Negeri (PUPTN).

References
[1] Nataleand, F.D.; Lancia, A. 2007 *Ind. Eng. Chem. Res.* **46**, 6777.
[2] Hsu, S.–C.; Hsieh, H.–L.; Chen, C.–P.; Tseng, C.–M.; Huang, S.–C.; Huang, C.–H.; Huang, Y.–T.; Radashevsky, V.; Lin, S.–H. 2011 *J. Hazard. Mater.* **189**, 193.
[3] Strigul, N.; Koutsospyros, A.; Christodoulatos, C. 2010 *Ecotoxicol. Environ. Safety* **73**, 164.
[4] Gerhardt, N.I.; Palant, A.A.; Petrova, V.A.; Tagirov, R.K. 2001 *Hydrometallurgy* **60** 1
[5] Luo, L.; Miyazaki, T.; Shibayama, A.; Yen, W.; Fujita, T. 2003 *Mineral Eng.* **16** 665.
[6] Kaushik, P.; Malik, A. 2009 *Environ. Int.* **35** 127.
[7] Islam, M.S.; Tanaka, M. 2004 *Pollution* **48** 624.
[8] Malekzadeh, F.; Saeid, G.; Hossein, G.; Mohammad, R.S. 2007 *World J. Microbiol. Biotechnol.* **23** 905.
[9] Arutanti, O.; Nandiyanto, A.B.D.; Ogi, T.; Iskandar, F.; Kim, T.O.; Okuyama, K. 2014 *J. Alloys Compd.* **591** 121
[10] Arutanti, O.; Ogi, T.; Nandiyanto, A.B.D.; Iskandar, F.; Okuyama, K.; 2014 AIChE J. **60** 41.
[11] Nandiyanto, A.B.D.; Arutanti, O.; Ogi, T.; Iskandar, F.; Kim, T.O.; Okuyama, K. 2013 *Chem. Eng. Sci.* **101**, 523.
[12] Ogi T.; Sakamoto, Y.; Nandiyanto, A.B.D.; Okuyama, K. 2013 *Ind. Eng. Chem. Res.* **52** 14441.
[13] Nandiyanto, A.B.D.; Ogi T.; Okuyama, K. 2014 *ACS Appl. Mater. Interfaces* **6** 4418.

Synthesis of carbon nanoparticles from commercially available liquified petroleum gas

A B D Nandiyanto*, M A Fadhlulloh, T Rahman and A Mudzakir

Departemen Kimia, Universitas Pendidikan Indonesia, Jl. Dr. Setiabudi no 229, Bandung 40154, Jawa Barat, Indonesia

*Corresponding author: nandiyanto@upi.edu

Abstract. The aim of this study was to synthesize carbon nanoparticles (CNPs) from commercially available liquefied petroleum gas (LPG). In the research procedure, LPG was reacted with air to construct CNPs. To confirm the successful synthesis of CNPs, we conducted several sample analyses: Gas Chromatography-Mass Spectrometry (GC-MS), Transmission Electron Microscope (TEM), X-ray Diffraction (XRD), and Infrared Spectra (FTIR). We also varied LPG and oxygen mole ratios at 0.8; 2.4; 4.8; and 7.2. The GC-MS results indicated the composition of LPG was propane (58.90%), isobutane (18.35%), butane (22.26%), and butane, 2-methyl (0.48%). The TEM results showed that the particles were spheres with sizes of between 25 and 35 nm. The sizes of particles were controllable, depending on the mole ratio. The XRD results showed mole ratios of LPG and oxygen of 0.80 and 2.40 were natural graphite, whereas the mole ratios of 4.80 and 7.20 were hexagonal graphite. FT-IR results showed CNPs have absorption peaks at wave number (i) 752 (C-H bend sp^2); (ii) 835 (C=C); (iii) 1274 (C-O-C vibration); (iv) 1400 and 1600 (C-C stretch aromatic); (v) 2800 (C-H sp^2); (vi) 2900 (CH sp^3); (vii) 3100 (C-H aromatic); and (viii) 3400 cm^{-1} (O-H). From the FTIR analysis results, the sample contained allotrope graphite due to detection of peaks at 1400 and 1600 cm^{-1} (C-C stretch aromatic) and 3100 cm^{-1} (C-H aromatic).

1. Introduction

Carbon nanoparticles (CNPs) with the electron configuration [He] $2s^2$ $2p^2$ are one of the attractive materials because of their unique properties: excellent thermal conductivity, biocompatibility, and corrosion resistance [1][2[3]. CNPs have been found in various applications: super capacitor [4], high-performance electrode material in the battery [5], and photoluminescence materials [6].

Several methods for production CNPs have been reported: pyrolysis [7][8][9][10][11], microwave plasma chemical vapor deposition [12][13], molten salt electrolysis [14], graphitization method and microemulsion polymerization [15], laser vaporization of carbon pellets [16], and specific treatment in supercritical water [6][17]. Although their methods are good technique for the production of CNPs, some weaknesses have been found: (i) the energy used is relatively large, (ii) the resulting product is still in the laboratory scale and no confirmation for further scaling up production, (ii) the process must be conducted under complicated procedures.

Here, the purpose of this study was to report a rapid synthesis method of CNPs. Different from other methods, our CNPs were produced from a commercially available liquefied petroleum gas (LPG), in which this will be prospective for scaling up production. To confirm our successful synthesis of carbon nanoparticles, several characterizations have been conducted, including an X- Ray Diffraction (XRD), Transmission Electron Microscope (TEM), Gas Chromatography-Mass Spectrometry (GC-MS), and Fourier Transform Infra Red (FTIR).

2. Experimental method

To produce CNPs, commercially available LPG (PT. Pertamina, Indonesia) was used and burned using a home made burner that was connected with a flow of air (600 mL/min). The particles were collected using a filter paper (Whatman glass microfiber filters CAT No. 1822-055, Whatmann, Britain). The mole ratio of LPG and oxygen was varied: 0; 0.80; 2.40; 4.00; 4.80; 7.20; 9.60; and 12.

To confirm our study on the synthesis of CNPs, several characterization analyses were conducted: a Fourier transform infra red (FTIR, FTIR-8400, Shimadzu Corp., Japan), an X-ray diffraction analysis (XRD; PANalytical X'PertPRO; Philips Corp., The Netherlands), a gas chromatography mass spectrometry (GC-MS, GCMS-QP2010 Ultra, Shimadzu Corp., Japan), and an transform electron microscope (TEM, JEOL JEM-1400, JEOL Ltd., Japan).

3. Results and Discussion

Figure 1 shows the GC-MS analysis result of commercially available LPG. The result showed that the LPG consisted of propane (58.90%), isobutane (18.35%), butane (22.26%), and 2-methyl butane (0.48%). This result showed the LPG is prospective for carbon source.

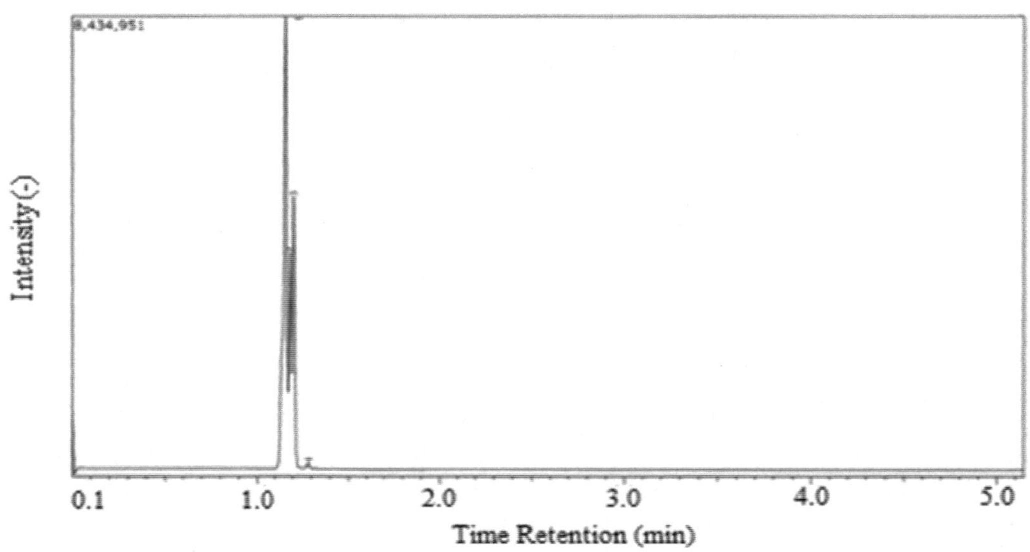

Figure 1. The GC-MS analysis result of commercially available LPG

Figure 2 shows the TEM images of CNPs prepared with various mole ratios of LPG and oxygen. The results showed that spherical CNPs were successfully produced in all variations. The outer diameter of CNPs was controllable depending on the ratio. Ferret sizes of CNPs produced with the ratio of 0.80; 2.40; 4.80; and 7.20 were 24; 26; 31; and 34 nm, respectively.

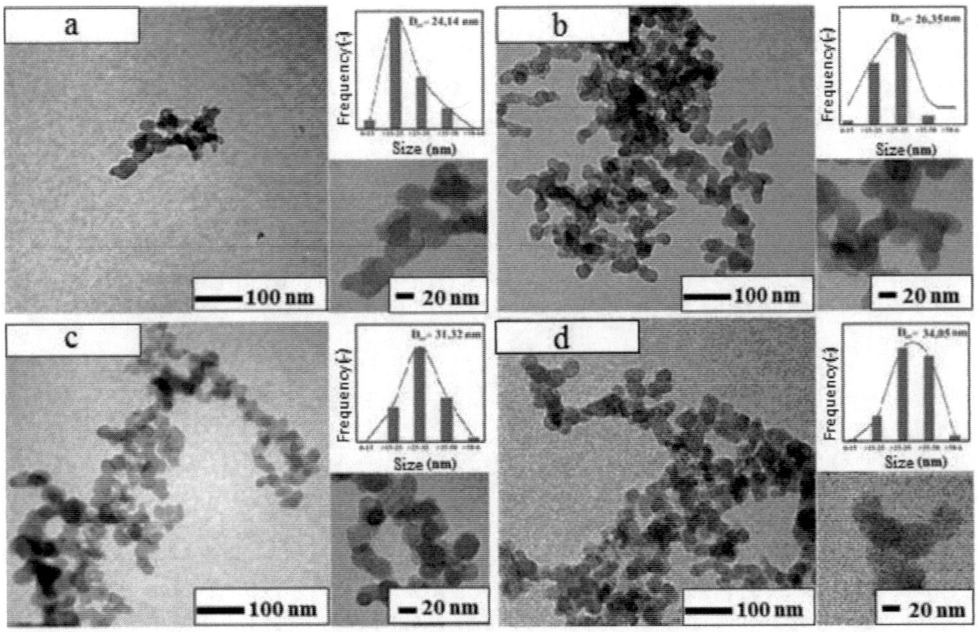

Figure 2. The TEM analysis images of samples prepared with various mole ratios of LPG and oxygen: (a)0.8; (b)2.4; (c)4.8; (d)7.2

To confirm that the prepared particles are CNPs, the FTIR and the XRD analyses were conducted, shown in **Figures 3a** and **b**, respectively. **Figure 3a** shows the detection of several absorption FTIR peaks, confirming the existence of chemical bonds in the sample: (i) 752 (C-H bend sp^2); (ii) 835 (C=C); (iii) 1274 (C-O-C vibration); (iv) 1400 and 1600 (C-C stretch aromatic); (v) 2800 (C-H sp^2); (vi) 2900 (CH sp^3); (vii) 3100 (C-H aromatic); and (viii) 3400 cm^{-1} (O-H). From the FTIR analysis results, the sample contained allotrope graphite due to detection of peaks at 1400 and 1600 cm^{-1} (C-C stretch aromatic) and 3100 cm^{-1} (C-H aromatic) [18]. The XRD result in **Figure 3b** identified that the mole ratios of LPG and oxygen of 0.80 and 2.40 were natural graphite, whereas the mole ratios of 4.80 and 7.20 were hexagonal graphite.

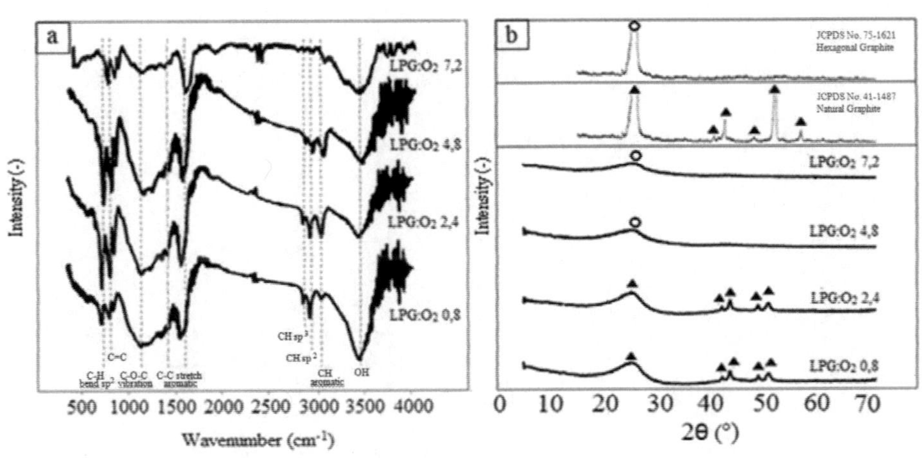

Figure 3. Physicochemical properties of the prepared particles prepared with various mole ratios of LPG and oxygen using the FTIR (a) and XRD (b) analysis methods.

4. Conclusions

We have successfully synthesized CNPs from the commercially available LPG. Using this method, the particles were spheres and the sizes of of CNPs were controllable from 20 to 40 nm depending on the mole ratio of LPG and oxygen. The results showed that our particles contained carbon graphite.

Acknowledgements

A. B. D. N. acknowledged Direktorat Jendral Pendidikan Tinggi (Dirjen DIKTI) for Program Unggulan Perguruan Tinggi Negeri (PUPTN). T. R. and M. A. F. acknowledged Direktorat Jendral Pendidikan Tinggi (Dirjen DIKTI) for Program Kreatifitas Mahasiswa - Penelitian Eksakta (PKM-PE).

References

[1] Shenderova, O.; Zhirnov, V.; Brenner, D. 2002 *Crit. Rev. Solid State* **27** 227.
[2] Wang, X.Z.; Hu, Z.; Chen, X.; Chen, Y. 2001 *Scripta Mater.* **44** 1567.
[3] Lu, J.; Yang, J.; Wang, J.; Lim, A.; Wang, S.; Loh, K.P. 2009 *ACS Nano* **3** 2367.
[4] Laforgue, A.; Simon, P.; Fauvarque, J.F.; Mastragostino, M.; Soavi, F.; Sarrau, J.F.; Lailler, P.; Conte, M.; Rossi, E.; Saguatti, S. 2003 *J. Electrochem. Soc.* **150** A645.
[5] Dominko, R.; Gaberscek, M.; Drofenik, J.; Bele, M.; Jamnik, J. 2003 *J. Electrochim. Acta* **48** 3709.
[6] Li, H.T.; He, X.D.; Liu, Y.; Huang, H.; Lian, S.Y.; Lee, S.T.; Kang, Z.H. 2011 *Carbon* **49** 605.
[7] Larciprete, R. Lizzit, S.; Botti, S.; Cepek, C.; Goldoni, A. 2002 *Phys. Rev. B* **66** 12140.
[8] Gherghel, L.; Kubel, C.; Lieser, G.; Rader, H.J.; Mullen, K.J. 2002 *J. Am. Chem. Soc.* **124** 13130.
[9] Ding, L.H.; Olesik, S.V. 2004 *Nano Lett.* **4** 2271.
[10] Néabo, J.R.; Gagné, S.R.; Carrière, C.V.; Morin, J.F. 2013 *Langmuir* **29** 3446.
[11] Ding, L.H.; Olesik, S.V. 2005 *Chem. Mater.* **17** 2353.
[12] Yu, J.; Ahn, J.; Q. Zhang, Yoon, S.F.; Rusli; Li, Y.J.; Gan, B.; Chew, K.; Tan, K.H.J.; 2002 *J. Appl. Phys.* **91** 433.
[13] Wang, Y.; Serrano, S.; Santiago-Aviles, J. 2002 *J. Mater. Sci. Lett.* **21** 1055.
[14] Hsu, W.K.; Terrones, M.; Hare, J.P.; Terrones, H.; Kroto, H.W.; Walton, D.R.M. 1996 *Chem. Phys. Lett.* **262** 161.
[15] Jang, J.; Oh, J.H.; Stucky, G.D. 2002 *Angew. Chem. Int. Ed.* **41** 4016.
[16] Asano, H.; Muraki, S.; Endo, H.; Bandow, S.; Iijima, S.J. 2010 *Phys. Condens. Matt.* **22** 1.
[17] Yang, X.G.; Li, C.; Wang, W.; Yang, B.J.; Zhang, S.Y.; Qian, Y.T. 2004 *Chem. Commun.* **3** 342.
[18] Hossain, M.A.; Islam, S. 2013 *ACS Nano* **1** 52.

IOP Conf. Series: Materials Science and Engineering **128** (2016) 012043 doi:10.1088/1757-899X/128/1/012043

Effect of processing techniques on color and active components amount of sweet potato (Ipomoea Batatas l) flakes

Y Rahmawati*, A Mahmudatussa'adah and S Yogha

Universitas Pendidikan Indonesia, Jl. Dr. Setiabudhi No. 229 Bandung Indonesia

*Corresponding author: yuliarahmawati56@gmail.com

Abstract. Sweet potato processing is limited, such as flour, snacks, cystic, or chips. Flakes as pre-cooked meals are made through the stages of making pasta and drying. The purpose of this study was to optimize the production of sweet potato flakes at the stage of making pasta and drying. Making the pasta is done through techniques steamed or baked. Pasta drying using tools a drum dryer or cabinet dryer. As an indicator of optimization is the total of monomeric anthocyanins, β-carotene and color the resulting flakes. The results showed that the amount of anthocyanin monomeric flakes by using steam, and drum dryer (3.83 ± 0.03 mg CYE / g db), flakes by the technique of steam, and cabinet dryer (3.03 ± 0.02 mg CYE / g db), flakes with techniques bake, drum dryer (2.49 ± 0.05 CYE mg / g db), flakes with bake technique, cabinet dryer (1.98 ± 0.03 mg CYE / g db). The Color of purple sweet potato flakes produced through steamed techniques bright purple, while the color purple sweet potato flakes produced through techniques roast give a brownish purple color. The amount of β-carotene yellow flakes sweet potato with stages of cooking steamed, drum dryer (152±0.5 mg / Kg db), grilled drum dryer (136±0.4 mg / Kg db), flakes of yellow sweet potato with stages of roasted and cabinet dryer (140±0.8 mg / Kg db), and grilled stage with cabinet dryer (122±0.3 mg / Kg db). In conclusion sweet potato flakes production techniques through the stages of steam process, and used drum dryers have a number of anthocyanins or β-carotene bigger and brighter colors than the baked flakes techniques and used cabinet dryer.

1. Introduction

The average productivity of sweet potato (Ipomoea batatas L) Indonesia 123.29 kw / ha, with a total production of 2.196.033 tons. The average productivity of sweet potato in West Java 153.73 ku/ha, with a total production of 429 378 tonnes (BPS 2011). Flesh color of sweet potato grown in West Java, beige, yellow, orange and purple. Purple sweet potato contains carbohydrates, minerals, vitamins, anthocyanins, dietary fiber and has a glycemic index (Glycemic Index) is low. In terms of its chemical composition, yams potentially be used as a source of carbohydrates, vitamins, minerals, and antioxidants such as phenolic acids, tocopherols, anthocyanins, and ß-carotene [1]. Colors and varieties of yams determine the level and profile of the content of phenolics such as anthocyanins and carotenoids. Phenolics present in sweet potatoes is chlorogenic acid which is an ester of cinnamic acid and acid quinat. Chlorogenic acid includes three isomers of mono caffeoylquinic acid (CQA): 3CQA (neochlorogenic acid); 4-CQA (cryptochlorogenic acid); and 5-CQA (chlorogenic acid); and three dicaffeoylquinic acids: 3,5-diCQA, 3,4-diCQA, and 4.5-diCQA (isochlorogenic acid) [2]. Anthocyanins are red dye, purple and blue fruits, vegetables, grains cereal / rice and flowers. Anthocyanin dissolves in water, the phenolic group of the group of flavonoids. Anthocyanins are dye on the flowers, leaves, tubers, fruits and vegetables that give ping color, red, blue, purple influenced by pH. Anthocyanins have

IOP Conf. Series: Materials Science and Engineering **128** (2016) 012043 doi:10.1088/1757-899X/128/1/012043

antioxidant activity [3][4][5][6][7]. Anthocyanin content in sweet potatoes were extracted at a temperature of 80 °C by using an acidified methanol (3.3% w / v) 186,1 highest of cyanidin, 3-glucoside / 100 g wb (wet base). Carotenoids is a dye found in plants that plays an important role as a pro-vitamin A. Sweet potatoes are yellow or orange contains ß- carotene higher than others [8][9][10]. Sweet potatoes are very prospective to serve as food for health (functional food) [11].

Some research suggests that the phytochemical content in sweet potatoes have antioxidant activity or has the ability to capture free radicals so as to provide a good influence on human health [12][1], Red sweet potato cultivars are grown in the Andean has antioxidant activity and phenolic content which is higher than blueberries, fruits with high antioxidant content [3]. The content of total phenolics of four varieties of sweet potatoes ranged from 45 to 103 mg gallic acid equivalent (GAE) / 100 g fresh weight [4]. The content of phenolics, anthocyanins, carotenoids yams correlated with antioxidant activity.

According to [13] cooking techniques can reduce the amount of phenolics content of for the o en, and for engineering fried oiled. Losing highest phenolic content sequentially boiled> fried> saute - steamed> microwave - oven roasted with. [14][15] found that the amount of anthocyanin reduced by nearly half when steamed and only slightly reduced when urned. Losing the content of trans ß-carotene with boiled technique more than urned. The content of trans ß-carotene is more abundant than the cis-ß-carotene were approximately 0.6 to 1.3% [16]. Sweet potato storage at cold temperatures cause increased phenolic content [17]. Caffeoyl-sucrose derivatives, 6-O-caffeoyl- (ß-D fructofuranosyl- (2-1)) - a-D-glucopyranoside (FCG), was found to increase during storage 15oC [2]. Carotenoid content decreased during storage due to autooksidasi ecause the pace of decline associated with ß-carotene content level of water and oxygen. Treatment processes can decrease the amount of anthocyanins present in sweet potatoes [18][15][19][6].

Sweet potatoes have a semi perishable nature. Within one month of storage, sweet potato sprout leaves. Yams are widely used as raw material for the manufacture of chips, dodol, flour and keremes. Utilization of sweet potato as a food inggridient still limited. Unlike the potato that has been processed into various food ingredient such as mash potato or potato flakes. Flakes potato is a potato-based food ingredient that is widely used both by households and catering industry, as materials for French fried potatoes, or potato chips [20]. The nature of sweet potato are relatively easily damaged, it needs to make other preparations which have a longer shelf life as sweet potato flakes. Sweet potato flakes made from raw sweet potato, cooked, prepared pasta and dried using a dryer.

Thus we see the need for a food ingredient made from yams that have a longer shelf life. In line with the diversification program of food, sweet potatoes can be processed into flakes of sweet potato. Sweet potato flakes are dried dosage form of flakes are made through the process of making pasta and drying. Manufacture of pasta can be done through a technique boiled, broiled, or grilled. According to [15] the amount of anthocyanin reduced by nearly half when steamed and only slightly reduced when baked. Thus in this study will be assessed regarding the optimization of the production of sweet potato flakes as diversification of sweet potato processing. The purpose of this study was to optimize the process of making sweet potato flakes, through the technique of making pasta with steamed or grilled and pasta drying technique using a drum dryer or dryer cabinet, as diversification of sweet potato processing.

2. Methods and Materials

2.1. Materials

The tools used UV / VIS Shimadzu UV-2450, Chromameter Minolta CR 310, Steamer, Oven, drum dryer and glassware. Materials used are purple sweet potato planting time - + 5 months of farmers in Bandung Regency Banjaran. Methanol, HCl, KCl, Sodium acetate PA from Merck.

2.2. Sample Preparation

To make sweet potato flakes made by way: as much as 5 kg sample of shelled, washed in running water, cut with a thickness of 3 cm. Pieces sample steamed for 15 minutes or baked at a temperature of 150 °

C oven for 60 minutes. Pieces of samples that have been steamed or baked entered grinder, homogenized, and then dried with a dryer drum 141 oC 3.75 bar rotation speed of 1 rpm. The dried samples (Flakes) formed a smaller size. Flakes sweet potatoes are stored in plastic bags and stored in the dark at room temperature up to further analysis [20].

2.3. Anthocyanin extraction [21][22][23]

As many 10g sweet potato powder was extracted with 320 mL of acid- methanol solution of 12% HCl (HCl, 1.5 mol / L in methanol) is inserted into the thermostatic water bath of 50°C for 60 minutes. Then centrifuged at a speed of 4000 rpm for 15 minutes. The supernatant was collected and filtered using a vacuum filter. The organic solvent is evaporated using a rotary evaporator temperature of 40°C to obtain the anthocyanin extracts concentrated. Concentrated extract stored in dark bottles stored at -20°C until used for analysis.

2.4. Analysis of total monomeric anthocyanin [23]

The content of total monomeric anthocyanin is measured by pH difference method [24] as modified by [23]. This method is based on the transformation of the structure of anthocyanin when the pH changes. 1 mL sample is introduced into a 5 mL volumetric flask and then added a solution of potassium chloride buffer (0.025 M) pH 1 up to 5 mL volume. A total of 1 mL of sample is introduced into a 5 mL volumetric flask else then added a solution of sodium acetate buffer (0.4 M) pH 4.5 up to 5 mL volume. Then the flask was placed in a dark place for 1 hour. Absorbance of each solution after berkesetimbangan measured using a spectrophotometer at a wavelength of maximum absorbance and 700 nm with distilled water blank. Absorbance is calculated as:

$$Abs:(A\lambda vis\text{-}max - A700)pH1.0-(A\lambda vis\text{-}max - A700)pH4.5, \qquad (1)$$

with a molar extinction coefficient of 26900 to sianidin-3-glucoside. Total monomeric anthocyanin is calculated and expressed as mg equivalent sianidin-3- glycoside / 100 grams of flakes. The equation used to calculate the number of monomeric anthocyanin is:

Antosianin monomerik (CyE, mg/L) : ....(2)

A: absorbance, BM: anthocyanin molecular weight (449.2 g / mol), FP dilution factor, ε: molar absorbance sianidin -3-glucoside (26900 L mol -1 cm -1), L length cuvette cell (1 cm). all measurements were performed 3 repetitions.

3. Result and Discussion

3.1. Production of Sweet Potato Flakes

Production of sweet potato flakes made through the stages of stripping the skin, making of sweet potato paste, and drying the pasta. Stripping the skin can be done by using a paring knife or abrasion. Sweet potato pasta-making is done by cooking fresh sweet potatoes with techniques steamed, boiled or roasted. According to [13] cooking techniques can reduce the amount of phenolics content of for the o en, and for engineering fried oiled. Losing highest phenolic content sequentially boiled> fried> saute - steamed> microwave - baked in an oven [14]. [15] found that the amount of anthocyanin reduced by nearly half when steamed and only slightly reduced when baked.

Figure 1 shows the color purple sweet potato flakes with the technique of making the pasta through the stages of steamed or roasted. The resulting paste is then dried using a drum dryer and the dryer cabinet. Figure 1A is an image of purple sweet potato flakes produced through the process of making pasta through stages pasta steamed and then dried with a drum dryer. Figure 1B is a purple sweet potato

IOP Conf. Series: Materials Science and Engineering **128** (2016) 012043 doi:10.1088/1757-899X/128/1/012043

flakes image generated through the process of making pasta through stages pasta steamed and then dried with a dryer cabinet. Figure 1C is a purple sweet potato flakes image generated through the process of making pasta with grilled stage and then dried pasta with a drum dryer. Figure 1D is a purple sweet potato flakes image generated through the process of making pasta with grilled stage and then dried pasta dryer cabinet. The water content of purple sweet potato flakes ranged from 6% - 7%. The yield of sweet potato flakes produced is ± 25%. The water content of fresh sweet potatoes about 67%, and the amount of sweet potato skin around 8-10%.

Figure 1 Flakes purple sweet potato (A: steam, drum dryer), (B: steamed, cabinet dryer, (C: grilled, drum dryer), and (D: grilled, cabinet dryer)

Fresh sweet potato cooking process is intended to get the texture of soft sweet potato to facilitate the making of pasta. Sweet potato paste can be made by using a grinder, mixer or crushed mixed manually using a hand. Sweet potato pasta drying process can be performed using an oven, dryer cabinet or with a drum dryer. Drying using a drum dryer produces purple sweet potato flakes subtle shine bright. Drying pasta using cabinet dryers produce purple sweet potato flakes darker. sweet potato flakes with the stages of cooking techniques and techniques steamed roast sequentially shown in Figure 1.

3.2. Total Anthocyanins of Purple Sweet Potato Flakes
Anthocyanin measurement method sweet potato flakes made by the method of pH differences. Anthocyanins are dye pink, red, blue, purple discolored influenced by differences in pH. At acidic pH conditions of the majority of anthocyanins are in the form of cations flavilium that gives the red color. At alkaline pH state of the majority of anthocyanin is in the form quinonoidal which gives a blue color. Besides anthocyanin can equilibrate form pseudobase or calcon giving colorless effect.

Anthocyanin extraction carried out in an acidic state. Anthocyanins are water soluble dye. From several studies optimum conditions for extraction of anthocyanin is to use methanol acidified with HCl 12% (1.5 M HCl in methanol), with a sample size - + 100 mesh with a ratio of solids: solvent 1: 32. Suspension samples with solvent stirred by using stirrer speed of 4, and the temperature of the room. The suspension is then stored in a rocking water bath at 50 ° C for 60 minutes. Then the sediment is separated between the supernatant by using sentrifius with a speed of 4000 rpm for 15 minutes at room temperature. Supernatant was then filtered through Whatman filter paper No. 1 and anthocyanin extract was concentrated by rotary evaporator 4 speed, 40oC. The yield of the concentrated extract of anthocyanin ± 23%. Each 1 mL of concentrated anthocyanin extract is inserted each into a 5 mL volumetric flask which is different, then each flask is added buffer CH3COONa KCl 0.4 M and 0.025 M. The solution was mixed, allowed to stand 1 hour then measured using a spectrophotometer at a wavelength of maximum (530 nm).

Measurement of the total number of monomeric anthocyanin in this study using a pH difference. The total amount of anthocyanin purple sweet potato flakes and β-Carotene of orange purple sweet potato showed in Table 1. Table 1 showed that stages of making pasta with steam and used drum dryer, total anthocyanin and β-carotene higher than stages of making pasta with bake and used cabinet dryer. According to [18] the amount of anthocyanins from purple sweet potato several genotypes ranged from 0.02 to 2.10 mg / g wet basis. According to [25] the highest of sweet potato anthocyanin 158 mg / 100 g wet base and 418 mg / 100 g wet basis [6]. The number of purple sweet potato anthocyanin 138 mg / 100 g dry basis. This research results contrary to the opinion of [15] which states roasting process (40-50 minutes the temperature of 200 oC) to lose a little of the total anthocyanin compared with steamed techniques (10 min 210 oC) which can lower half of the amount of anthocyanin. This difference is due to differences in sample preparation techniques and time, the temperature used in the different cooking techniques. Preparation of the samples in this study, yams undergone stripping the skin first, and conducted by [15] do not explain a process of stripping the skin.

Table 1 Quantity of anthocyanin monomeric and β-carotene of flakes sweet potato

Stages of making pasta	Tools of drying Pasta	Anthocyanin (mg CYE/g db)	β-carotene (mg/Kg db)
Steam	Drum dryer	3.83 ± 0.03	152 ± 0.5
Steam	Cabinet dryer	3.03 ± 0.02	136 ± 0.4
Bake	Drum dryer	2.49 ± 0.05	140 ± 0.8
Bake	Cabinet dryer	1.98 ± 0.03	122 ± 0.3

CYE : Cyanidin-3-glucoside Equivalent
db : dry basis

3.3. Total β-Caroten of Yellow Sweet Potato Flakes

The water content of sweet potato flakes with grilled engineering 7% and the water content of sweet potato flakes technique is 6% randemen steamed sweet potato flakes produced was 25%. The water content of fresh sweet potatoes about 67%, and the amount of sweet potato skin around 8-10%. From the results of this study, the amount of β-carotene and color flakes sweet potato yellow stages of ripening (A) steam dryer drum, (B) grilled drum dryer, flakes of sweet potato yellow stages of roasted dryer cabinet (C), and the grilled dryer cabinet (D) showed in Table 1 and Figure 2.

A B

C D

Figure 2. Flakes yellow sweet potato ripening stages (A) steam drum dryer, (B) grilled drum dryer, yellow sweet potato flakes roasted stages dryer cabinet (C), and bake stage cabinet dryer (D).

According to [13] cooking techniques can reduce the amount of phenolics content of for the oven, and for engineering fried oiled. Losing highest phenolic content sequentially boiled> fried> saute - steamed> microwave - oven roasted with. [14]. Losing the content of trans β-carotene with oiled technique more than urned. The content of trans β- carotene is more abundant than the cis-β-carotene which is about 0.6-1.3% [16]. Sweet potato storage at cold temperatures cause increased phenolic content [17]. Caffeoyl-sucrose derivatives, 6-O- caffeoyl- (β-D fructofuranosyl- (2-1)) - α-D-glucopyranoside (FCG), was found to increase during storage 15°C [2].

Carotenoid content decreased during storage due to autooksidasi because the pace of decline associated with β-carotene content level of water and oxygen. Sweet potatoes contain trypsin inhibitors from 0.2 to 43.6 IU / 100 g. Trypsin inhibitor can shut down the enzyme active site of trypsin that it can not bind with the substrate. Sweet potatoes contain stakiosa that can cause flatulence effects. Stakiosa fermented by the bacteria present in the gut to form H¬2 gas and CO2. Sweet potatoes contain trypsin inhibitors from 0.2 to 43.6 IU / 100 g. Trypsin inhibitor can shut down the enzyme active site of trypsin that it can not bind with the substrate.

Varieties of sweet potato with carotene content higher associated with a water texture, pumpkin flavor and orange color while the varieties of sweet potatoes with carotenoid content low have a characteristic sweet tasting, yellow color, creamy color, starchiness, hard texture , coarse texture and fibrous texture [26]. This is in line with the opinion reported that the varieties of sweet potatoes with orange flesh has a flavor that is approaching the yellow vegetables such as pumpkin.

4. Conclusion

The conclusion from this study is the technique of making the pasta is steamed in the best technique in the manufacture of sweet potato flakes which can maintain the amount of β-carotene and anthocyanin highest. Number antosinin purple sweet potato flakes highest is the technique of making pasta by means steamed and dried pasta using drum dryers. The amount of β-carotene is the yellow sweet potato flakes highest is the technique of making pasta by means steamed and dried pasta using drum dryers.

Acknowledgements

The authors would like to thank the Directorate General of Higher Education, Ministry of Research and Higher Education, Republic of Indonesia, for funding this research through Hibah Bersaing research grant 2014.

References

[1] Teow CC, Truong V, Feeters R, Thompson RL, Pecota KP, Yencho GC. 2007.
[2] Ishiguro K, Yahara S, Yoshimoto M. 2007. Changes in polyphenols content and radical-scavenging activity of sweetpotato (Ipomoea batatas L.) during storage at optimal and low temperatures. Journal of Agricultural and Food Chemistry 55: 10773–10778.
[3] Cevallos-Casals BA, dan Cisneros-Zevallos L. 2004. Stability of anthocyaninbased aqueous extracts of Andean purple corn and red-fleshed sweet potato compared to synthetic and natural colorants. Food Chemistry, 86, 69–77.
[4] Wu X, Beecher GR, Holden JM, Haytowitz DB, Gebhardt SE, Prior RL, 2004.
[5] Jiao Y, Jiang Y, Zhai W dan Yang Z. 2012. Studies on antioxidant capacity of anthocyanin extract from purple sweet potato (Ipomoea batatas L.) African Journal of Biotechnology : 11(27), 7046-7
[6] Burgos G, Amoros W, Munoa, Sosa P, Cayhualla E, Sanchez C, Dı´az C, Bonierbale M. 2013. Total phenolic, total anthocyanin and phenolic acid concentrations and antioxidant activity of purple-fleshed potatoes as affected by boiling. Journal of Food Composition and Analysis 30 : 6–12
[7] Mahmudatussa'adah, A., Fardiaz, D., Andarwulan, N., Kusnandar, F. 2 1 . Color characteristics and antioxidant activity of purple sweet potato anthocyanin. J.Tenol. dan Industri Pangan 25(2): 176-184.
[8] Juanda D, dan B. Cahyono. 2000. Sweet potatoes, cultivation and analysis of farming. Yogyakarta: Kanisius.
[9] Steed LE, Truong V-D. 2008. Anthocyanin content, antioxidant activity, and selected physical properties of flowable purple-fleshed sweetpotato purees. Journal of Food Science 73: S215–S221.
[10] Van den Berg H, Faulks R, Granado HF, Hirschberg J, Olmedilla B, Sandmann G, Southon S, Stahl W. 2000. The potential for the improvement of carotenoid levels in foods and the likely systemic effects. Journal of the Science of Food and Agriculture 80: 880–912.
[11] Ginting, E., J.S. Utomo, Yulifianti R., and M. Joseph. 2011. Potential purple sweet potato as a functional food. Science and Technology of Food Crops 6 (1): 116-138
[12] Rabah IO, Hou DX, Komine SI, & Fuji M. 2004. Potential chemopreventive properties of extract from baked sweet potato (Ipomoea batatas Lam. Cv. Koganesengan). Journal of Agricultural and Food Chemistry 23: 7152–7157.
[13] Jung J-K, Lee S, Kozukue N, Levin CE, Friedman M. 2011. Distribution of phenolic compounds and antioxidative activities in parts of sweet potato (Ipomoea batata L.) plants and in home processed roots. Journal of Food Composition and Analysis 24: 29–37.
[14] Bechoff A, Dhuique-Mayer C, Dornier M, Tomlins K, Boulanger R, Dufour D, West y A. 2 1 Relationship etween the kinetics of β-carotene degradation and formation of norisoprenoids in the storage of dried sweet potato chips. Food Chemistry 121: 348–357.
[15] Kim HW, Kim JB, Cho SM, Chung MN, Leen YM, Chu SM, Che JH, Kim SN, Kim SY, Cho YS, Kim JH, Park HJ, Lee DJ. 2012. Anthocyanin changes in the Korean purple-fleshed sweet potato, Shinzami, as affected by steaming and baking. Food Chemistry 130: 966–972.
[16] Thilsted. 2007. Effect of home preparation practices on the content of provitamin A carotenoids in coloured sweet potato varieties (Ipomoea batatas Lam.) from Kenya. Innovative Food Science and Emerging Technologies 8: 399–406.
[17] Padda MS, Picha DH. 2008. Effect of low temperature storage on phenolic composition and

antioxidant activity of sweetpotatoes. Postharvest Biology and Technology 47: 176–180.

[18] Truong VD, Deighton N, Thompson RT, McFeeters RF, Dean LO, Pecota KV, Yencho GC. 2010. Characterization of anthocyanins and anthocyanidins in purple-fleshed sweetpotatoes by HPLC-DAD/ESI-MS/MS. Journal of Agriculture and Food Chemistry 58: 404–410.

[19] Lachman J, Hamouz K, Orsak M, Pivec V, Hejtmankova K, Pazderu K, Dvorak P, Jaroslav. 2012. Impact of selected factors – Cultivar, storage, cooking and baking on the content of anthocyanins in coloured-flesh potatoes. Food Chemistry 133: 1107–1116.

[20] Lamberti M, Geiselmann A, Conde-Petit B, Escher F. 2004. Starch transformation and structure development in production and reconstitution of potato flakes. Lebensm.-Wiss. u.-Technol. 37: 417–427

[21] Fan G, Han Y, Gu Z, Chen D. 2008. Optimization conditions for anthocyanins extraction from purple sweet potato using response surface methodology (RSM). LWT Food Sci Technol 41:155–160.

[22] Huang CL, Liao WC, Chan CF, Lai YC. 2010. Optimization for extraction anthocyanin from purple sweet potato roots using response surface methodology. J. Taiwan Agric. Res. 59(3) : 143-150.

[23] Jie L, Xiao-ding L, Yun Z, Zheng-dong Z, Zhi-ya O, Meng L, Shao-hua Z, Shuo L, Meng W, Lu O. 2013. Identification and thermal stability of purple-fleshed sweet potato anthocyanins in aqueous solutions with various pH values and fruit juices. Food Chemistry 136: 1429–1434

[24] Lee J, Durst W, & Wrolstad RE. 2005. Determination of total monomeric anthocyanin pigment content of fruit juices, beverages, natural colorants, and wines by the pH differential method: Collaborative study. Journal of AOAC International, 88, 1269–1278.

[25] Truong VD. Hua Z, Thompson RL, Yencho GC, Pecota KV. 2012. Pressurized liquid extraction and quantification of anthocyanins in purple-fleshed sweet potato genotypes. Journal of Food Composition and Analysis 26 : 96–103.

[26] Tomlins K, Owori C, Bechoff A, Menya G, Westby A. 2012. Relationship among the carotenoid content, dry matter content and sensory attributes of sweet potato. Food Chemistry 131: 14–21

Implementation monitoring temperature, humidity and mositure soil based on wireless sensor network for e-agriculture technology

A Sumarudin*, A L Ghozali, A Hasyim and A Effendi

Department of Informatics State Polytechnic of Indramayu, Jalan Lohbener lama no 8 Lohbener Indramayu, west Java, Indonesia

*shumaru@polindra.ac.id

Abstract. Indonesian agriculture has great potensial for development. Agriculture a lot yet based on data collection for soil or plant, data soil can use for analys soil fertility. We propose e-agriculture system for monitoring soil. This system can monitoring soil status. Monitoring system based on wireless sensor mote that sensing soil status. Sensor monitoring utilize soil moisture, humidity and temperature. System monitoring design with mote based on microcontroler and xbee connection. Data sensing send to gateway with star topology with one gateway. Gateway utilize with mini personal computer and connect to xbee cordinator mode. On gateway, gateway include apache server for store data based on My-SQL. System web base with YII framework. System done implementation and can show soil status real time. Result the system can connection other mote 40 meters and mote lifetime 7 hours and minimum voltage 7 volt. The system can help famer for monitoring soil and farmer can making decision for treatment soil based on data. It can improve the quality in agricultural production and would decrease the management and farming costs.

Keywords: e-agriculture, monitoring soil, sensing, temperature, moisture, humidity.

1. Introduction

We propose e-agriculture system for monitoring soil. This system can monitoring soil status. The experimental setup and the preliminary results of the part of the application used for data measurement and data collection as well as initial concepts for improving the monitoring system. The proposed architecture is designed in order to improve the quality in agricultural production and would decrease the management and farming costs. The environmental parameters such as temperature, humidity and moisture should be continually monitored and controlled in order to monitoring plant conditions.

Monitoring system based on wireless sensor mote that sensing soil status. This sensor for monitoring utilize soil moisture, humidity and temperature. System monitoring design with mote based on microcontroler and xbee connection. Data sensing send to gateway with star topology with 1 gateway. Gateway utilize with mini personal computer and connect to xbee cordinator mode. On gateway, gateway include apache server for store data based on My-SQL. System web base with YII framework. System done implementation and can show soil status real time.

2. The Design System E-agriculture

2.1. System Design

The design of the system is an overview of the literature that has been studied and then presented to the timetable for implementation in the manufacturing process. In designing this system will be addressed needs that must be met to be able to operate the system in accordance with the plan that is as sender wireless media sensor data using XBee module.

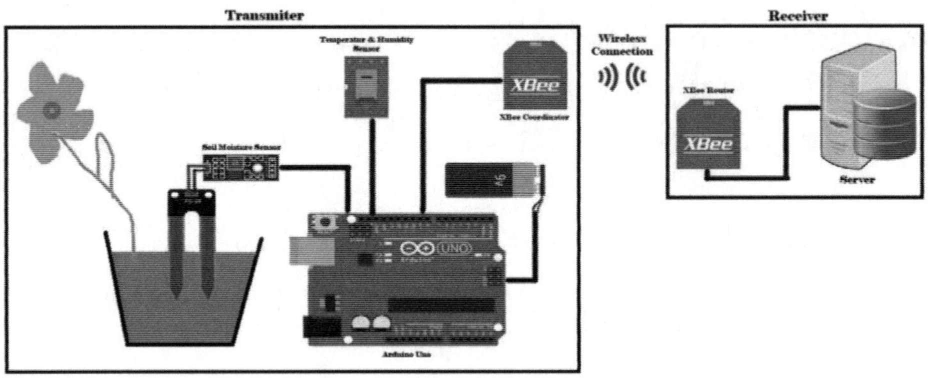

Figure 1. Monitoring System Design

3. The Method Design Implementation

For implementation the system, we development with microcontroller for data collection and controller, xbee connection utilize for communication mote end device to gateway system. The step implementation can show on figure 2.

In the working system is divided into two design, namely at the time of data transmission and storage of data and view data collection for sensor on web base interface, the figure 3 shows the phases that exist in both the current process is running.

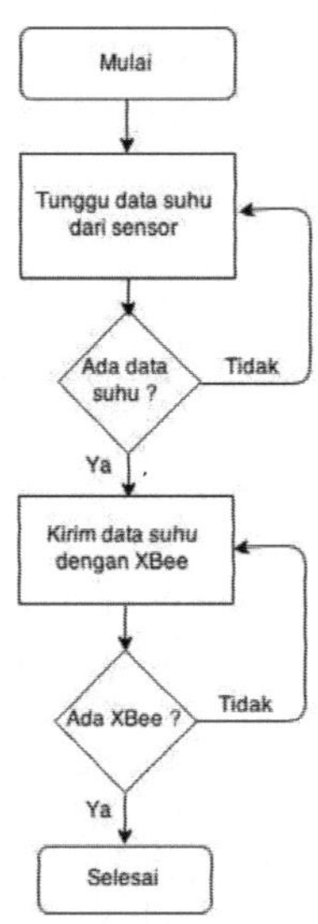

Figure 2. Flowchart of Data Delivery

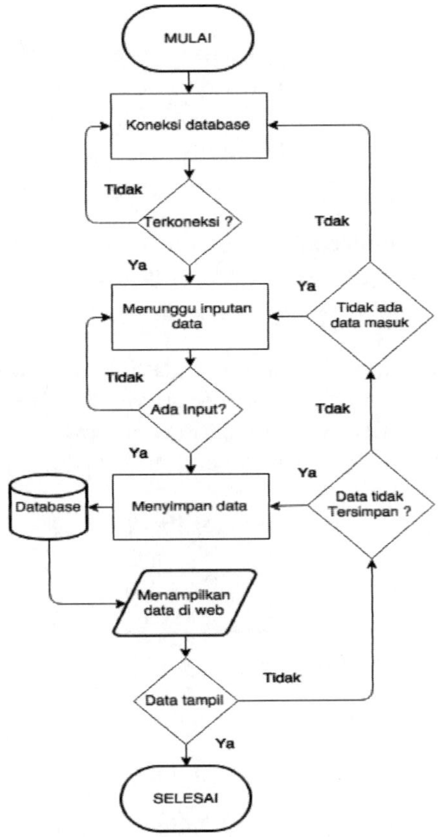

Figure 3. Flowchart Displaying Data

Figure 4. Implementation mote

4. Result and Discussion

This research produced a way to monitor the temperature and ambient temperature, also on a soil moisture levels. This device can facilitate in monitoring environmental conditions, because it does not have to come directly to check the environmental conditions. This device can also store the data of temperature or ambient temperature also soil moisture during the period desired, and from data stored tersbut be displayed into a website in the form of a line chart and table format with a full day or 24 hours.

IOP Conf. Series: Materials Science and Engineering **128** (2016) 012044 doi:10.1088/1757-899X/128/1/012044

4.1. Working Principle Transmitter Device

This monitoring system requires data obtained by the sensors are embedded. The sensors will tersus temperature and humidity transmit data obtained from the surrounding environment and sends it to a server PC wirelessly using a wireless module for 1 hour. This system is capable of running by getting resources from a 7,4V battery with capacity of 1300mAh.

4.2. Working Principle Receiver Devices

The data has been sent by the transmitter system will be accepted by the receiver system and then be converted into the form of a data string for the next tersbut will dismpan into a database.

4.3. Test Result

Tests carried one by one from the start of testing the communication between the XBee, temperature sensors and temperature (SHT), the humidity sensor (Soil Moisture), and testing to all monitoring tools, testing conducted on the ability of XBee communication and power required to run the transmitter, and the results tests outlined in the table below as follows :

Table 1. Table testing distance

No	distance	transmitter system	Receiver System	there barrier	Data Sent
		position	Position		
1	40m	Outdoor	Indoor	No	Yes
2	40m	Indoor	Indoor	No	Yes
3	40m	Outdoor	Indoor	Yes	No
4	10m	Indoor	Indoor	Yes	Yes
5	10m	Outdoor	Indoor	Yes	Yes

From the table above it can be concluded that the maximum distance to communicate with each other XBee is 40 meters without any obstacle, when the author of more than 40 meters to test the XBee can not communicate. Besides the barrier also interfere with communication between the XBee, XBee only able to communicate at a distance of no more than 10 meters when the two XBee blocked a doorway and surrounded by walls.

Table 2. Table testing battery power 7,4v

No	transmitter system	
	Battery Time	Data Sent
1	7 Volt	Yes
2	6 Volt	Yes
3	5 Volt	Yes
4	4 Volt	Yes
5	3 Volt	No
6	2 Volt	No
7	1 Volt	No

From table 4.3 can be authors conclude that the minimum power needed for XBee able to send the data is equal to 4V. The author tested using li - po battery 1300mAh 7,4V assisted a ubec capable of generating an output of 5V. The result is a system of the sender is able to work for ± 7 hours with the intensity of shipping 1 time for 1 hour, the system stops sending when the remaining battery is less than 4V or approximately 3,24V.

Table 3. Testing battery with 12 volt

No	transmitter system	
	Battery Time	Data Sent
1	12 Volt	Yes
2	11 Volt	Yes
3	10 Volt	Yes
4	9 Volt	Yes
5	8 Volt	Yes
6	7 Volt	Yes
7	6 Volt	No
8	5 Volt	No
9	4 Volt	No
10	3 Volt	Tidak
11	2 Volt	No
12	1 Volt	No

Table 3 is the result of a second experiment by using li - po battery 12V without using ubec, which at this second trial the sender system is only able to work for ± 4 hours with shipping intensity is the same as the first experiment.

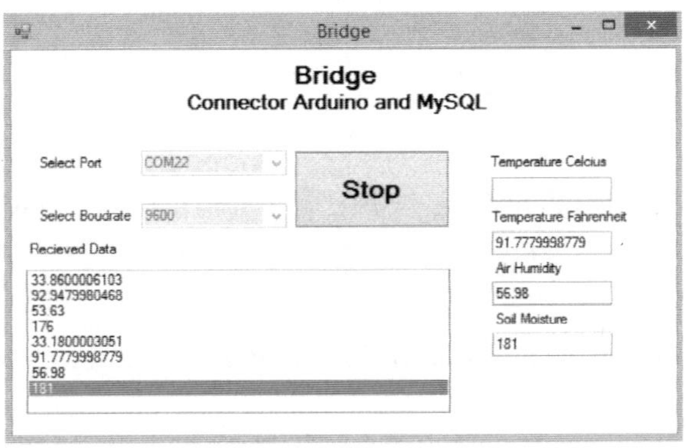

Figure 5. Application image link

Figure 2.4 is a picture which time the receiver receives the data, the application will download the user detection receiver port select along boudratenya. After the start button is pressed the data received will be converted into a string and stored into the database.

IOP Conf. Series: Materials Science and Engineering **128** (2016) 012044 doi:10.1088/1757-899X/128/1/012044

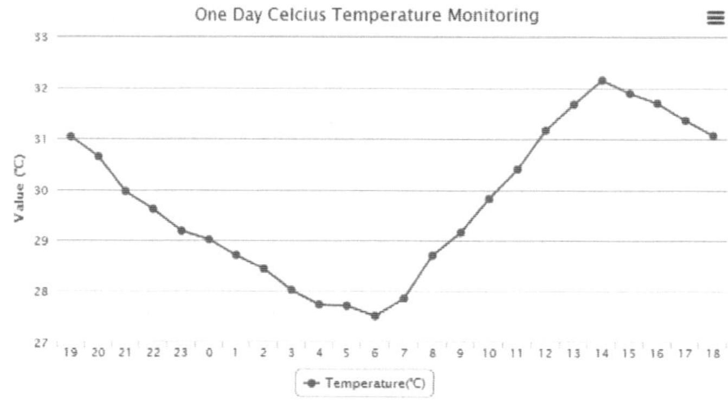

Figure 6. Display chart at website monitoring

In the Figure 7 displays the results of the data stored into the database in the form of a chart with a 24 -hour format, each data will be retrieved and displayed every hour.

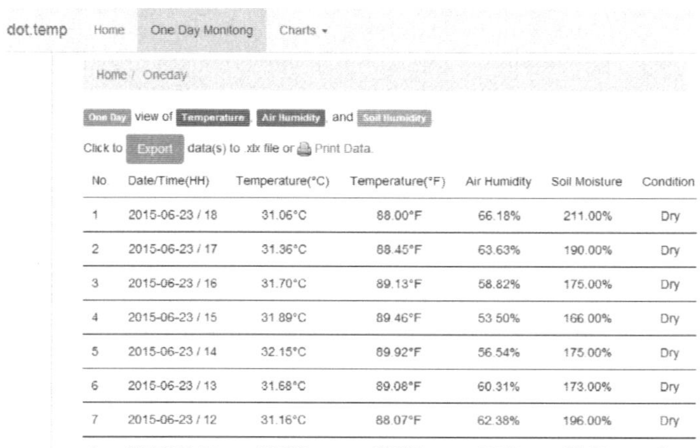
dot.temp Home One Day Monitong Charts

Home / Oneday

One Day view of Temperature, Air Humidity, and Soil Humidity

Click to Export data(s) to .xlx file or Print Data.

No	Date/Time(HH)	Temperature(°C)	Temperature(°F)	Air Humidity	Soil Moisture	Condition
1	2015-06-23 / 18	31.06°C	88.00°F	66.18%	211.00%	Dry
2	2015-06-23 / 17	31.36°C	88.45°F	63.63%	190.00%	Dry
3	2015-06-23 / 16	31.70°C	89.13°F	58.82%	175.00%	Dry
4	2015-06-23 / 15	31.89°C	89.46°F	53.50%	166.00%	Dry
5	2015-06-23 / 14	32.15°C	89.92°F	56.54%	175.00%	Dry
6	2015-06-23 / 13	31.68°C	89.08°F	60.31%	173.00%	Dry
7	2015-06-23 / 12	31.16°C	88.07°F	62.38%	196.00%	Dry

Figure 7. Display data in tabular form

In the figure 8 displays data in tabular form. Such as charts, tables display data in 24- hour format in the capture then displayed every hour.

5. Conclusion

Based on the above results obtained some conclusions from the manufacture of temperature monitoring system:

1. From the testing that has been done, it can be concluded that the design and implementation of monitoring temperature using the ambient temperature and XBee arduino media has been successful in terms of the accuracy of the sensor with the actual conditions, both when displaying the temperature and the temperature as well as at the time of data transmission.
2. The device temperature and ambient temperature monitoring is enabled and used to the environment temperature and the temperature must be monitored in real-time such as smart-home, smart-garden or server room.
3. Data that has been saved is displayed into a website in the form of line charts and tables, the data is displayed every hour for a total of 24 hours or a full day.

4. The process of monitoring the temperature and the ambient temperature is more efficient, in the setting of the time we can change the delay of delivery every 1 or 2 hours and so on, without spending time and can also monitor outcomes temperature changes and temperature in real-time so that we can know the temperature change that happened.

References

[1] Adminstrator. Connect Visual Basic .NET to MySql Database. http://www.visual-basic-tutorials.com. Diakses pada 25 Mei 2015
[2] Anonim. 2014. PIC32MX: XBee Wireless Round-trip Latency. http://hades.mech.northwestern.edu. Diakses pada tanggal 13 April 2015
[3] Apais. 2014. PART 1 - Send Arduino data to the Web (PHP/MySQL/D3.js). http://www.instructables.com. Diakses pada tanggal 14 April 2015
[4] Arduino. 2013. Arduino Wireless Shield with XBee Series 2 radios. http://www.arduino.cc. Diakses pada tanggal 10 April 2015
[5] Digi XBee. 2014. Digi XBee Examples & Guide. http://examples.digi.com. Diakses pada tanggal 15 April 2015
[6] Faludi, Rob. 2014. Basic XBee ZB ZigBee (Series 2) Chat. http://examples.digi.com. Diakses pada tanggal 16 April 2015
[7] Guo, Yanny. 2014. SHT10 Digital Temperature and Humidity Sensor for Arduino http://learn.linksprite.com. Diakses pada tanggal 11 April 2015
[8] JIMBO. 2013. Exploring XBees and XCTU. https://learn.sparkfun.com. Diakses pada tanggal 13 April 2015
[9] JIMBO. 2013. XBee Shield Hookup Guide. https://learn.sparkfun.com. Diakses pada tanggal 18 April 2015
[10] Jweymarn. 2013. MyHome - home automation with Arduino and XBee. http://www.instructables.com. Diakses pada tanggal 14 April 2015
[11] Kadir, Abdul. 2013. Panduan Praktis Mempelajari Aplikasi Mikrokontroler dan Pemrogramannya Menggunakan Arduino. Surabaya: Andi Publisher.
[12] Kautsar, Syamsial. 2013. Komunikasi Wireless Serial Meggunakan Xbee 2,4 GHz. http://kautsar-sam.blogspot.com. Diakses pada tanggal 15 April 2015
[13] Mortensen, Peter. 2013. Reading data over serial Arduino and XBee. http://stackoverflow.com. Diakses pada tanggal 2015
[14] Nikhil. 2013. Xbee – Part 7 – Led on/off using API mode. http://www.freemindscafe.com. Diakses pada tanggal 15 April 2015
[15] Rapp, Andrew. 2013. xbee-arduino. https://code.google.com. Diakses pada tanggal 15 April 2015
[16] Rives Horridge, Susan. 2014. XBee Tech Tip: Sending Serial Data From One XBee Wi-Fi to Another. http://www.digi.com. Diakses pada tanggal 17 April 2015
[17] Schuman, Milo. 2014. highchart. http://www.yiiframework.com. Diakses pada tanggal 15 Mei 2015
[18] Tshepang. 2014. How to send data from Arduino Xbee and receive it from C#. http://stackoverflow.com. Diakses pada tanggal 17 April 2015
[19] Wilson, Hannah Perner. 2013. XBEE SERIAL COMMUNICATION. http://www.kobakant.at. Diakses pada tanggal 16 April 2015

IOP Conf. Series: Materials Science and Engineering **128** (2016) 012045 doi:10.1088/1757-899X/128/1/012045

Modification of Cilembu sweet potato starch with ethanoic acid

A Mahmudatussa'adah*, Y Rahmawati and Sudewi

Departemen Pendidikan Kesejahteraan Keluarga, Fakultas Pendidikan Teknologi dan Kejuruan, Universitas Pendidikan Indonesia

*Corresponding author: aims@upi.edu

Abstract. Cilembu sweet potato harvest was abundant, its use was still limited. Starch was required by various industries. Starch is generally beige, and requires a long time for the drying process. The purpose of this research was to produce a modified starch with ethanoic acid. The method used in this study was the experimental method. The results showed acid modified starch yield was 18%, with the color characteristics of L*: 96.38 ± 0.82; a*: -0.70±0.02; b*: 2.70±0.03; C: 2.79±0.02. Native starch yield was 16%, with the color characteristics of L*: 93.55 ± 0.91; a*: -0.86±0.06; b*: 2.93±0.04; C: 3.05±0.03. The conclusion of this study was modified starch of Cilembu sweet potato using ethanoic acid have higher yield and more white bright than native starch.

1. Introduction

Starch is a food that is much needed in a variety of industries. Starch can be used as thickeners, binders, fillers, materials for liquid sugar, baby food ingredients, noodles and even material to make ethanol or sizing at the fabric factory. In Indonesia usually starch made from cassava, sago, mung beans, and corn. In China the main ingredient is sweet potato starch maker [1]. Production of sweet potato (Ipomoea batatas L) of West Java Indonesia 485 065 tons, with production centers at Garut and Kuningan District [2]. Sweet potato starch contains 50-80% of each sweet potato tubers in dry basis [3].

The process of making the starch is by way of sweet potatoes are cleaned, peeled, grated, added water, squeezed, water and starch suspension is allowed to stand until precipitation of starch, then decanted and the precipitate was dried. Starch yield and color will determine the economic value of the starch. The physicochemical properties of starch largely determine its designation in the industry. Thus in producing or isolating starch often be modified to affect yield, color, and physicochemical properties of starch.

Starch industry in China to make a starch using acid solution, centrifuged, or combine these two techniques. Manufacture of starch using acid solution produces fat, protein, K, P, solubility and L * higher than using centrifugation and setback, viscosity, lower crystalline [1]. Some previous researchers have used acid solution in the manufacture of starch mungbean [4][5][6], cassava [7], rice [8], corn [9], and sweet potato [1]. Researchers previously used acid solution is fermented a mixture of lactid acid, amino acids, peptides, protein, ash, monosaccharide, oligosaccharide, water and Lactococcus lactis bacteria [10]. There are reportedly about sweet potato modified starch manufacture using ethanoic acid. Ethanoic acid is cheap, readily available and food grade.

IOP Conf. Series: Materials Science and Engineering **128** (2016) 012045 doi:10.1088/1757-899X/128/1/012045

The purpose of this study was to observe the effect of ethanoic acid in the process of making sweet potato starch to the yield and the resulting color. The yield and the color of starch is important to note the commercialization efforts of sweet potato starch.

2. Materials and Methods

Research through the extraction stage, the calculations yield starch and starch color measurement.

2.1. Materials

Sweet potato (Ipomoea batatas L) Cilembu from the Cilembu village Pamulihan District of Sumedang district West Java Province Indonesia. The planting period of 5 months (November-March 2015), cream skin and yellow flesh. Samples used ± 100 Kg, good quality with relatively the same weight 150 - 200g. All sweet potatoes washed with running water, drained. Chemicals used ethanoic acid 70% was obtained from the store in the city of Bandung Indonesia.

The tools used scales, knives, a basin, a grater (rasper), abrasion peeler, cabinet dryer, 80 mesh sieve, and oven. Tools for the identification and analysis of the physical, chemical is analytical balance, colorimeter (Minolta CR 310).

Sample preparation

Extraction Starch

Sweet Potatoes washed with running water, abrasion peeler peeled, grated with a grater machines. Results grated fresh water added with a ratio of slurry sweet potato : water is 1: 5, stirred and then filtered using a filter cloth. Suspension (water, starch) allowed to stand 24 hours. Sediment, and the liquid is separated by means of filtered, then the precipitate was dried. Starch sediment ground, then sieved using a 80 mesh sieve. Siftings stored in the dry for further analysis. To make the starch modified with ethanoic acid, is added 5 mL of 70% ethanoic acid into each 1 L suspension of starch.

2.2. Analysis Method

2.2.1. Yield (Percent)

The yield of starch is the percentage of starch produced from fresh sweet potato, is calculated by weighing the weight of the starch resulting from the extraction process compared with the weight of fresh sweet potatoes multiplied by 100 percent Water Content [11].

Measurements of water content of flour and water content of fresh sweet potatoes done using the drying method. In this method, the sample is weighed as much as 5 grams, and then inserted into the cup which has been dried beforehand for 2 hours and weighed. The sample is dried in an oven at 110 ^{0}C for 24 hours or until dry weights obtained were fixed. Samples removed from the oven and then put in a desiccator until cool, then performed weighing using analytical balance. The water content of the samples was determined using the following formula;

$$\text{The water content (percent bb)}=\frac{(x-y)}{(x-\alpha)}\,x100 \qquad (1)$$

Description :

x = weight of dish and sample before drying (g)

y = weight of dish and sample after drying (g)

a = weight of empty cup

Starch levels (AOAC 1995)

Analysis of starch based methods Luff Schrool. Luff Schrool solution was prepared by 25 g $CuSO_4.5H_2O$ dissolved in 50 mL and citric acid in 50 mL of distilled water. A total of 388 g $Na_2CO_3.10H_2O$ dissolved in 400 mL of distilled water. Citric acid solution is added little by little to a solution of soda, then the mixture was added a solution terusi and diluted to 100 ml in volumetric flask. A total of 2 gram sample is introduced into a 500-mL Erlenmeyer, then added 200 mL of HCl 3 percent and the boiling stones. Erlenmeyer mounted on upright coolers and hydrolyzed for 3 hours. The solution is then cooled and neutralized with NaOH to the phenolphthalein indicator. The solution was put in a 500 mL volumetric flask, in calibration with distilled water, then filtered. A solution of 10 mL pipette into a 250-mL Erlenmeyer and Schrool Luff added a solution of 25 mL and 15 mL of distilled water. Shells made without the sample solution is analyzed. KI was then added a solution of 30 percent and 25 mL of H_2SO_4 25 percent. Once the reaction is complete, immediately titrated with $Na_2S_2O_3$ solution until the solution was less colored.

$$\text{starch content } (\%) = \frac{0{,}9\, x\, G\, x\, P}{g}\, x\ 100 \quad (2)$$

0.90 = a factor comparison of the molecular weight of the sugar units in the starch molecule
G = glucose equivalent mL $Na_2S_2O_3$ used for titration (mg) after sugars calculated
P = dilution
g = weight of sample (mg)

2.2.2. Color measurement

The color of the sample was measured with a colorimeter with CIELAB measurement system ([11]Truong et al. 2012). Parameters measured include L* (Lightness), a* = red (+) to green (-), b* = yellow (+) to blue (-). Chromacity (C) shows the intensity of the color which is calculated by the formula $\sqrt{a^{*2} + b^{*2}}$ 2). Hue angle (H*) calculated as tan-1 (b* /a*). Hue is expressed as degrees angle ranging from 0° - 360 °, where 0 ° (red) in the quadrant + a *, rotated counter clockwise 90° (yellow) to + b*, 180° (green) for -a*, 360° (blue) to -b*. Colorimeter calibrated with L* = 92.75, a* = - 0.76, b* = - 0.07. The sample is introduced into a petri dish and carefully pressed surface to remove air bubbles. The color parameter value calculated from the average of three replicates measurements.

3. Result and Discussion

The results showed the yield of acid modified starch 18%, with the characteristic color of L*: 96.38 ± 0.82; a*: -0.70 ± 0.02; b*: 2.70 ± 0.03 C: 2.79 ± 0.02. Native starch yield of 16%, with the characteristic color of L*: 93.55 ± 0.91; a*: -0.86 ± 0.06; b*: 2.93 ± 0.04; C*: 3.05 ± 0.03. The yield of starch with the addition of ethanoic acid in the water for the extraction of starch, higher than without the addition of ethanoic acid. The presence of acid can accelerate the deposition of starch [1]. According to [10][13] in the fermentation of starch suspension under acidic conditions grew dominant lactic acid bacteria is *Lactococcus lactis* and by [4] grew dominant *Streptococcus lactis*. The growth of lactic acid bacteria can change the properties of starch [13]. *Lactococcus lactis* bacteria have the ability to combine the starch granules into groups, so that the starch granule is separated from the protein and fiber. Starch granules flocking accelerate the deposition of starch [14].

Color is an important factor in determining the quality of starch. Starch is expected to have a high rate in the brightness (L*) and lower in the Chroma (C) [1]. Color affects consumer acceptance of starch [15]. Starch produced by depositing the starch under acidic conditions produces more starch white color L*: 96.38 ± 0.82, C: 2.79 ± 0.02; compared with no addition ethanoic acid, L*: 93.55 ± 0.91, C: 3.05 ± 0.03. It can be seen from the figures L* for the acid modified starch is higher than the number L* a native starch. Instead number C for the acid modified starch is lower than the number of native starch C. These results similar with the results of the study [1] that made sweet potato starch using acid solution and centrifugation. According to [4][6] mung bean fermented starch under acidic conditions can produce starch that is whiter and better tansmission. The fermentation process is expected to change the amorphous starch, thus affecting the physicochemical properties of starch produced. In addition it is estimated the lactic acid can inhibit the activity of peroxidase enzymes, so that the oxidation process of starch is inhibited [6][16][17]. Peroxidase enzyme responsible for the increased rates of chroma or browning.

4. Conclusion

The conclusion of this study was modified Cilembu sweet potato starch using ethanoic acid generating yield and white degrees higher than native starch. The yield of modified starch ethanoic acid 18%, with the characteristic color of L*: 96.38 ± 0.82; a*: -0.70 ± 0.02; b*: 2.70 ± 0.03; C: 2.79 ± 0.02. Native starch yield of 16%, with the characteristic color of L*: 93.55 ± 0.91; a *: -0.86 ± 0.06; b*: 2.93 ± 0.04; C: 3.05 ± 0.03.

Acknowledgements

The authors would like to thank the Directorate General of Higher Education, Ministry of Research Technology and Higher Education, Republic of Indonesia, for funding this research through Hibah Bersaing research grant 2015.

References

[1] Deng, F., Mu, T., Zhang, M. and Oluwaseyi Kemi Abegunde, O.K. 2013. Composition, structure, and physicochemical properties of sweet potato starches isolated by sour liquid processing and centrifugation. Starch/Sta¨ rke 65: 162–171.

[2] BPS. 2014. Sweet Potato Production. Central Bureau of Statistics and Director General of Food Crops Production. http:www.bps.go.id/sector/agri/pangan/table2.shtml.

[3] Zhu, F., Yang, X. S., Cai, Y. Z., Bertoft, E., Corke, H. 2011. Physicochemical properties of sweet potato starch. Starch/Sta¨ rke, 63, 249–259.

[4] Liu, W. J., Shen, F Q., 2007a. Studies on the physicochemical properties of mung bean starch from sour liquid processing centrifugation. Journal of Food Engineering, 79: 358–363.

[5] Li, Z. G., Liu, W. J., Shen, Q., Zheng, W., Tan, B. 2008. Properties qualities of vermicelli made from sour liquid processing centrifugation starch. Journal of Food Engineering, 86: 162–166.

[6] Liu, W. J., Shen, F. Q. 2007b. Structure analysis of mung bean starch from sour liquid processing centrifugation. Journal of Food Engineering, 79: 1310–1314.

[7] Numfor, F. A., Walter, W. M., Jr., Raleigh, S. J. S. 1995. Physicochemical changes in cassava starch and flour associated with fermentation: Effect on textural properties. Starch/Sta¨ rke, 47: 86–91.

[8] Lu, Z. H., Li, L. T., Cao, W., Li, Z. G., Tatsumi, E., 2003. Influence of spontaneous fermentation on physicochemical characteristics of rice noodles. International Journal of Food Science and Technology. 38: 505–510.

[9] Yuan, M. L., Lu, Z. H., Cheng, Y. Q., Li, L. T. 2008. Effect of spontaneous fermentation on the physical properties of corn starch rheological characteristics of corn starch noodle. Journal of Food Engineering. 85: 12–17.

[10] Liu,W. J. 2006. Studies on the mechanism of the mung bean starch agglutination by sour liquid and comparison of physicochemical properties of starch and vercimelli. Master Thesis. The Agriculture University of China, Beijing, China.

[11] AOAC. 1995. Official Methods of Analysis of the Association of Analytical Chemists, Washington D.C.

[12] Truong, V.D., Hua, Z., Thompson, R.L., Yencho, G.C., dan Pecota, K.V. 2012. Pressurized liquid extraction and quantification of anthocyanins in purple-fleshed sweet potato genotypes. Journal of Food Composition and Analysis 26 : 96–103.

[13] Gao,Q. Y. 1998. Systematic studies on properties of bean starches their application for starch silk, Ph. D. Thesis. South China University of Technology, China.

[14] Zhang, L. L., Xue, H., Li, X. H. 2010. Study on flocculating activity of the microorganisms in sweet potato acid steeping liquor. Sci. Tech. Food Industry 31, 172–175.

[15] Galvez, F. C. F., Resurreccion, A. V. A. 1992. Reliability of the focus group technique in determining the quality characteristics of mung bean (Vigna radiate (L.) Wilczek) noodles. Journal of Sensory Studies, 7: 315–326.

[16] Qin, L. K., Jiang, P., Zhang, Q., Shen, C. G., Qiu, S. 1997. Study on the effect of preventing

milled and filtered potato liquid from browning by sour liquor which was fermented naturally from pea. Journal of Food Science. 18, 32–37.

[17] Rani, V. S., John, J. K.,Moorthy, S. N.,Raja, K. C. M. 1998. Effect of pre-treatment of fresh Amorphophallus paeoniifolius tubers on the physicochemical properties of starch. Starch/Sta¨ rke. 50: 72–77.

Growth mechanism of Co:TiO_2 thin film deposited by metal organic chemical vapor deposition technique

A Saripudin*[1] and P Arifin[2]

[1]Study Program of Electrical Engineering, Dept.of Electrical Engineering Education, Faculty of Technology and Vocational, Universitas Pendidikan Indonesia
Jl. Dr. Setiabudi 207 Bandung 40154, West Java, Indonesia

[2]Physics of Electronics Material Lab., Study Program of Physics, Faculty of Mathematics and Natural Science, Institut Teknologi Bandung
Jl. Ganeca 10 Bandung 40132, West Java, Indonesia

*Corresponding author: asaripudin1604@gmail.com

Abstract. In this research, we investigated the growth mechanism of cobalt-doped titanium dioxide (Co:TiO_2) films. Thi Co:TiO_2 thin films were grown on the n-type silicon substrate. The films were grown by metal organic chemical vapor deposition method. The growth temperature was varied of 325°C – 450°C. The films were characterized by SEM. Using Arheniu's equation, it is known that the activation energy value of film growth is positive in the range of temperature of 325°C – 400°C and negative in the range of temperature of 400°C – 450°C. These results show that the decomposition rate in the range of temperature of 325°C – 400°C is due to diffusion phase of precursor gas. On the other hand, the decomposition rate decreased in the range of temperature of 400°C – 450°C because the precursor gas decreased, and the surface chemical reaction was high.

1. Introduction

TiO_2 is an interesting material. It has many properties that can be applied in various fields. TiO_2 has a high refractive index and dielectric constant [1]. TiO_2 is also chemically stable. Pigment, absorbent, catalyst supports, filters, coating, and dielectric material commonly used TiO_2 [2]. In the recent times, TiO_2 has a high potential for environmental purification, decomposition of carbonic acid gas, and hydrogen gas generation [2]. Besides as a semiconductor, TiO_2 doped by cobalt (Co:TiO_2) demonstrate ferromagnetic properties at room temperature. These ferromagnetic properties allow us to use Co:TiO_2 as sprintronics devices [3].

Many techniques can used to grow TiO_2 thin film. These techniques are atomic layer deposition, MBE (molecular beam epitaxy), PLD (pulsed layer deposition), DC and RF Sputtering, and MOCVD [4][5][1][6][7]. Each technique has special advantages, but MOCVD possible to grow epitaxial film, deposit film selectively, and control the growth parameter simplicity so MOCVD is a powerful technique [7].

The effect of growth temperature on the structural and morphological quality of Co:TiO_2 thin films deposited on Si(100) substrate by MOCVD technique has been done by Saripudin et al [8]. The

IOP Conf. Series: Materials Science and Engineering **128** (2016) 012046 doi:10.1088/1757-899X/128/1/012046

morphological quality and the growth rate of Co:TiO_2 thin film as a function of temperature was pictured. However, the growth mechanism of Co:TiO_2 thin films was not explained. The activation energy of film growth also was not evaluated. In this research, we focused on the growth mechanism of Co:TiO_2 thin film. The activation energy of film growth was then evaluated by Arheniu's equation.

2. Experimental Method

In this research, thin films of cobalt-doped TiO_2 were grown on Si(100) substrate by MOCVD method. A TTIP (Titanium Tetra Iso-Propoxide) was used as TiO_2 precursor, other while $Co(TMHD)_3$ was used as Co precursor. A THF (Tetra Hidrofuran) was used to make a Co solution.

TTIP of 20 mL was mixed by $Co(TMHD)_3$ of 900 mg and THF of 20 mL into the bubbler tube. The mixture of TTIP, $Co(TMHD)_3$, and THF in the bubbler tube was then heated at a constant temperature of 100°C. The vaporized TTIP and $Co(TMHD)_3$ of 0,3 kg/cm in partial pressure were driven by argon gas into the reactor chamber with 70 sccm in speed. A silicon substrate has been embedded on a heater disc in the reactor. The disc was heated until a constant temperature, that is, 325°C, 350°C, 375°C, 400°C, or 450°C. Each film was grown along 120 minutes with the total pressure of the reactor chamber was 2×10^{-3} torr. The details of the growth parameters of Co:TiO_2 thin films were listed at **Table 1**.

Table 1. The detail of the growth parameters of Co:TiO_2 thin film.

No.	Parameters	Value	Unit
1	Total pressure of the reactor chamber	2×10^{-3}	torr
2	Growth time	120	min
3	Argon gas (carrier) speed	70	sccm
4	Growth temperatures	325 - 450	°C
5	Bubbler tube temperature	100	°C
6	Pressure of the vaporized TTIP and $Co(TMHD)_3$	0,30	kg/cm^2

Each film was characterized by SEM (JEOL type JSM-6510LA) to know the film morphology and the film thickness. The thickness of each film was measured and, by knowing the growth time, we calculated the rate of film growth. To evaluate the energy activation of film growth, we used Arheniu's Equation, that is

$$r = Ae^{-\frac{E_A}{RT}} \tag{1}$$

where r is the rate of growth, A is the frequency factor for reaction, R is universal gas constant, T is temperature of growth, and E_A is activation energy.

3. Results and Discussion

SEM photograph of Co:TiO_2 films for various temperature growths resulting in this research is shown in **Figure 1**. The left-side of the photograph is SEM of film's surface, other while the right-side photograph is SEM of film's cross-section. These SEM photographs show, generally, that all film's surface look roughly with a relatively homogeneous grain.

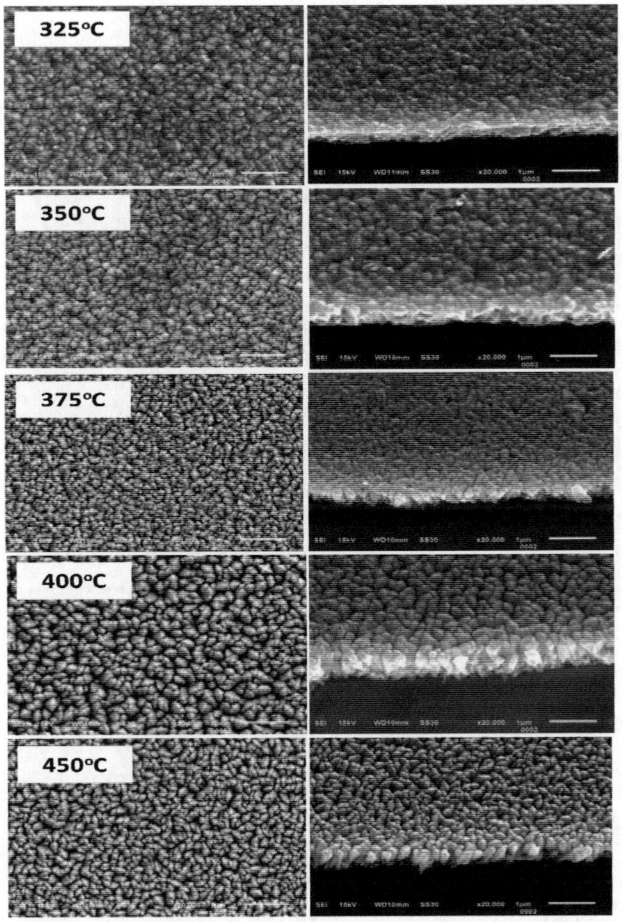

Figure 1. SEM photograph of films grown in various temperature.

SEM photograph of film's cross-section grown at 375°C – 450°C show that the film composed by like rod crystal grains. At the interface of film-substrate, crystal grains look random. These results were obtained because of competitively crystal growth so that the crystal grains formed depend on the diffusion process of surface's atomic precursor. The competitively crystal growing can be caused by various crystal grain orientations on the substrate's surface [9]. On the other hand, photographs of film's cross-section grown in the range of temperature of 325°C - 350°C show that the grain is in the random position and separated each other. It was caused by polluter covered the crystal surface that is growing.

Although the films were grown at the same growth time, the thickness of each film are different for different temperature. It means that the rate of film gowth is different. The film grown at temperature of 325°C is the lowest rate of growth, that is about 2.42 nm/min. On the other hand, the film grown at temperature of 400°C is the higest rate of growth, that is about 5.33 nm/min. This fact indicated that the rate of film growth depend on the temperature of film growth.

Graph of $\ln r$ versus $1/T$ for Co:TiO_2 films is shown in **Figure 2**. By linear fitting, we get the activation energy of film growth. It is revealed that the activation energy of film growth is positive, that is 32.2 kJ/mol, for the film grown in the range of temperature of 325°C – 400°C. This result shows that the rate of film growth increased as the temperature increased. This phenomenon caused by gas phase of precursor that higher than surface's chemical reaction [4]. In the other word, deposition of film is controlled by diffusion [7].

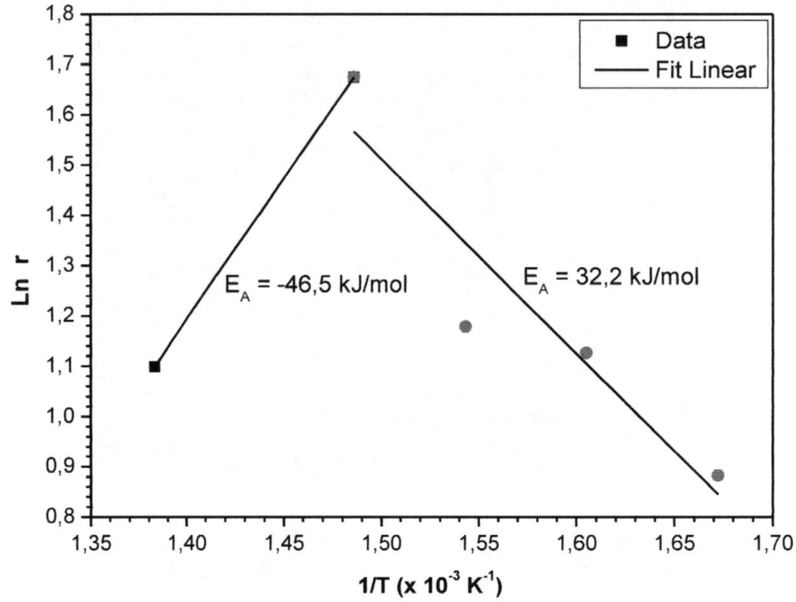

Figure 2. Graph of natural logaritm of film growth rate (ln r) versus inversely temperature (1/T).

A different result is shown by the film grown in the range of temperature of 400°C - 450°C. Its activation energy is negative, that is -46.5 kJ/mol. This result shows that the rate of film growth decreased as the temperature of film growth increased. The decreasing tendency of the rate of film growth at the higher temperature is commonly occur. In the high temperature, the decomposition rate of film and other surface chemical reaction becomes higher. This phenomenon decreased the gas diffusion of precursor. In that situation, film deposition is controlled by kinetically deposition [7].

4. Conclussion

We have grown a series of Cobalt-doped TiO_2 thin films on the n-type silicon substrate by MOCVD method in the range of temperature of 325°C – 450°C. SEM photograph of the film shows that all film's surface look roughly with a relatively homogeneous grain. The activation energy of film growth is positive in the range of temperature of 325°C – 400°C and negative in the range of temperature of 400°C – 450°C. These results show that the decomposition rate in the range of temperature of 325°C – 400°C is due to diffusion phase of precursor gas. On the other hand, the decomposition rate decreased in the range of temperature of 400°C – 450°C because the precursor gas decreased, and the surface chemical reaction was high.

References

[1] 8A. Saripudin, H. Saragih, Khairurijal, T. Winata, and P. Arifin, Effect of Growth Temperature on Cobalt-doped TiO_2 Thin Films Deposited on Si(100) Substrate by MOCVD Technique, Advanced Materials Research (2014) 896, 192

[2] Barna, P.B., and Adamik, M. (1998), Fundamental Structure Forming Phenomena of Polycrystalline Films and The Structure Zone Models, Thin Solid Films **317**, 27.

[3] 4C.H. Heo, S.B. Lee, J.H. Boo, Deposition of TiO2 thin films using RF magnetron sputtering method and study of their surface characteristics, Thin Solid Films 475 (2005) 183-188.

[4] 6C.K. Jung, B.C. Kang, H.Y. Chae, Y.S. Kim, M.K. Seo, S.K. Kim, S.B. Lee, J.H. Boo, Y.J.

Moon, J.Y. Lee, Growth of TiO2 Thin Films on Si(100) and Si(111) Substrates Using Single Molecular Precursor by High Vacuum MOCVD and Comparison of Growth Behavior and Structural Properties, Journal of Crystal Growth 235 (2002) 450-456.

[5] 2K. Thamapat, P. Limsuwan, and B. Ngotawornchai, Phase Characterization of TiO_2 Powder by XRD abd TEM, Kasetsart J. (Nat. Sci.) 42 (2008) 357-361.

[6] 1M.I.B. Bernardi, E.J.H. Lee, P.N. Lisboa-Filho, E.R. Leite, , E. Longo, and J.A. Varela, TiO2 Thin Film Growth Using the MOCVD Method, Material Research 4 (2001) 223-236.

[7] 5M.M. Hasan, A.S.M.A. Haseeb, H.H. Masjuki, and R. Saidur, Influence of Substrate Temperatures on Structural, Morphological, and Optical Properties of RF-Sputtered Anatase TiO2 Films, The Arabian Journal for Science and Engineering 35 (2010) 147-156.

[8] 7S.H. Nam, J.S. Hyun, J.H. Boo, Synthesis of TiO2 thin films using single molecular precursors by MOCVD method for dye-sensitized solar cells application and study on film growth mechanism, Materials Research Bulletin 47 (2012) 2717-2721.

[9] 3Y. Matsumoto, M. Murakami, T. Shono, T. Fukumura, M. Kawasaki, P. Ahmet, C. Chikyow, S. Koshihara, and H. Koinuma, Room-Temperature Ferromagnetism in Transparent Transition Metal-Doped Titanium Dioxide, Science 291 (2001) 854-856.

IOP Conf. Series: Materials Science and Engineering **128** (2016) 012047 doi:10.1088/1757-899X/128/1/012047

Type utilization of baked-smashed sweet potato and vegetables on patisserie product

Ana*, S Subekti, Sudewi, E N Perdani, F Hanum, T Suciani and V Tania

Home Economics Department, Universitas Pendidikan Indonesia

*Corresponding author: ana@upi.edu

Abstract. The research was an experimental study in Green Skill Patisserie Course using Project-Based Learning model. It aims to complete the project development of pie named guramnis rainbow pie. Several experiments were carried out to produce a pie dough crust mixed with baked-smashed sweet potato and added with vegetables extract as the food coloring. The experiment method in order to make a better appearance or an attractive shape and to have more nutrition. In addition, the pie was filled with a mixture of sweet and sour gurame as Indonesian traditional food. By applying an organoleptic test to 10 respondents, the result shows that pie dough recipe using flour substituted by baked-smashed sweet potato with 2:1 of a ratio. Coloring pie dough adding extract vegetables (carrots, beets and celery) as color . We found that pie dough has more interesting pie color (90%) and the texture of the pie with a quite level of crispness (60%). Moreover, the pie taste is fairly (70%) and tasty (70%). Nutritional analysis results show that per size, serving guramnis rainbow pie contains energy as much as 81.72 calories, carbohydrates 12.5 grams, fat 2.32 grams and 2.77 grams of protein. The main findings are the pie appearance and taste was different compared to the previous pies because of the pie was served with gurame asam manis as the filling and had flour and cilembu sweet potato as the basic ingredients. The color of guramnis rainbow pie was resulted not only from food coloring but also from vegetables extract namely carrot (orange), bit (red), and salary (green). Thus, it had many benefits for health and adds the nutrition. The researchers recommend a further study in order to make pie dough with baked sweet potato and vegetables extract having an optimal level of crispness.

1. Introduction

Student who joined the course of patisserie project based learning was required to accomplish a project under the theme of patisserie product development [1]. The study develops pie production using sweet potato and vegetables as additives in making pie dough. The pie filling used Indonesian food namely gurame asam manis (fried gurame fish with sour-sweet sauce) or guramnis.

Sweet potato (Ipomoea Batatas L) is an alternative food as a carbohydrate source besides rice, corn, and cassava. Sweet potato has a big potential to be used as an alternative ingredient of many snacks [2]. It grows relatively fast in all seasons and does not need fertile soil. The researchers made a modification of pie dough by using Cilembu sweet potato comes from Sumedang West Java. Nowadays, cilembu sweet potato can be found easily there. It has a different characteristic compared to other sweet potatoes. It has a natural sweet taste. Moreover, it consists of carbohydrate, beta carotene (pro vitamin A), minerals (Ca, P, Fe, and K), fibers, and a good color and smell. The smashed cilembu sweet potato in this study was used to enrich the pie dough nutrition. To make the pie dough more attractive, it was added with some natural colors from vegetables namely bit, carrot, and salary.

The modification of pie dough by using a mixture of sweet potato and vegetables was preferred to enrich the nutrition. Besides, a pie dough commonly had no variation in its color. The filling was also modified by using Indonesian food namely gurame asam manis (fried gurame fish with sour-sweet sauce) which is very popular and never been used before. A good pie has a tender, flaky crust that can top a variety of fillings. A pie can be sweet and served as a dessert. With various fruits as filling, a pie is a favorite way to add more fruit to your menus. It can be savory and filled with meat or vegetables [3]. Therefore, the researchers were interested to do an experiment to a patisserie product namely pie by making the dough from a mixture of cilembu sweet potato and vegetables and by filling it with an Indonesian food. The experiment included testing the product until the appropriate receipt formula was got, doing an organoleptic test, and measuring the nutrition facts.

2. Material and Method

2.1. Material
In the experiment, the ingredients of pie dough were cilembu sweet potato, moderate-protein flour, milk, egg, sugar powder, salt, green extract from salary, yellow extract from carrot, red extract from bit, and natural food coloring. For the filling, the researchers used gurame fish, pineapple, onion leaf, Bombay onion, garlic, onion, red paprika, yellow paprika, chili sauce, tomato sauce, salt, pepper powder, sugar, corn powder and cooking oil.

2.1.1. Method
To get good pie dough, we did the experiment three times by combining taro and cilembu sweet potato. The researchers used a pie basic recipe formula as the following:

Table 1. Pie dough basic formula

No	Ingredients	Amount
1	Wheat flour	200 g
2	Shortening	100 g
3	Salt	0.5 g
4	Water	10 ml

Then, the researchers developed the pie dough and the filling.

3. Result and Discussion
In the first experiment, the researchers developed a basic formula recipe of pie dough by adding 100 gr steamed taro. Besides, the amount of flour was lessened 50 g (25%), shortening was added 20 g (20%), egg yolk was added 25 g and water was replaced by milk. Taro was added into the pie dough to make it more nutritious and having a good taste. Moreover, the pie dough was given colors from bit for red, salary for green, and sweet corn for yellow. Table 2 shows the formula of the first experiment:

Table 2. The formula of pie dough receipt in experiment 1

No	Ingredient	Amount	Product Characteristic
1	Flour	125 g	The pie dough was less flaky crust, the color was too soft, the taste was plain
2	Taro	100 g	
3	Margarine	120 g	
4	Milk	10 ml	
5	Egg Yolk	25 g	
6	Salt	0.5 g	
7	Bit extract	5 ml	

8	Sweet corn extract	5 ml
9	Salary extract	5 ml

In making the pie dough, the researchers found many obstacles such as it became sticky, slick, and dropped when baking. For the filling in the first experiment, the researchers used rendang Padang but it was too oily and made the dough wet. As the result, when eating the rainbow rendang pie, the taste was not balanced for the filling was too dry. Meanwhile, the dough was tough but easily broken for it did not use sugar.

In the second experiment, the researchers tried to develop the pie dough by adding 100 gram baked cilembu sweet potato. The flour was lessened 50 g (25%), shortening was added 20 g (20%), egg yolk was added 25 g, and water was replaced by milk. The researchers also added 20 gram sugar powder to make the dough tastier and sturdier rather than the previous experiment. The sugar caramel when baking strengthened and united all dough ingredients; thus, it was not easily broken [4]. The filling used in this experiment was gurame asam manis (fried gurame fish with sour-sweet sauce) which was made from gurame fillet which was fried crispy with flour and egg and was poured with a sour-sweet sauce on top. The sour sweet sauce was preferred to balance the pretty sweet taste of the pie dough. The table 3 below shows the pie dough recipe formula in the second experiment.

Table 3. The pie dough recipe formula in the second experiment

No	Ingredient	Amount	Product Characteristic
1	Flour	125 g	The dough was pretty flaky crust, the color was too soft, and the taste was plain
2	*Cilembu* sweet potato	100 g	
3	Margarine	120 g	
4	Milk	10 ml	
5	Sugar powder	20 g	
6	Egg Yolk	25 g	
7	Salt	0.5 g	
8	Bit extract	20 ml	
9	Sweet corn extract	20 ml	
10	Salary extract	20 ml	

In the previous experiment, the pie dough color was less attractive. Therefore, the researchers added carrot extract to make it yellow, bit extract to make it red, and salary extract to make it green. Besides, they also added food coloring but the formula recipe of the dough and the filling was still the same. The following table 4 shows the recipe.

Table 4. The recipe formula of pie dough in the third experiment

No	Ingredient	Amount	Product Characteristic
1	Flour	125 g	The pie dough was flaky crust, the color was interesting, and the taste was good.
2	*Cilembu* sweet potato	100 g	
3	Margarine	120 g	
4	Milk	10 ml	
5.	Sugar powder	20 g	
6	Egg Yolk	25 g	
7	Salt	0.5 g	
8	Bit extract	20 ml	
9	Carrot extract	20 ml	
10	Salary extract	20 ml	

11	Food Color (green, red, yellow)	3 ml

In this experiment, the pie dough which previously used color from sweet corn extract was added with carrot extract for the color of sweet corn was less striking. The color of vegetable extract was soft, which made the dough pale. Therefore, the researchers added a small amount of natural food coloring to get a good color. They tried to perfect the experiment so the result can be maximal. For the technique of dough color fusion, the researchers piled up some dough with different colors and rolled them up. Thus, the dough colors were spread evenly, which made easy the process of molding the pie dough into pan. The researchers named it rainbow pie for the dough color was like rainbow. The figure 1 shows the process of dough making in the last experiment.

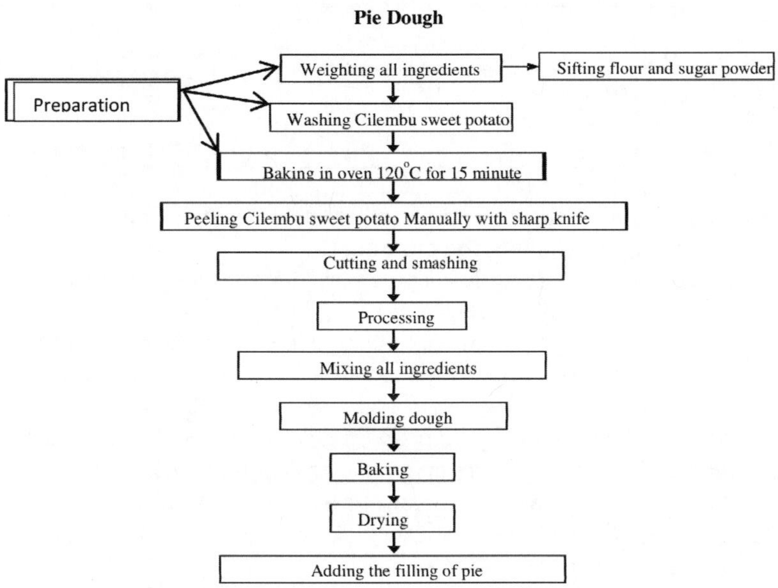

Figure 1. The process of pie dough making

In the last experiment, the researchers found the recipe formula of pie dough, analyzed the nutrition facts, and did an organoleptic test to 10 panelists. The recipe formula of pie filling is shown in table 5 below.

Table 5. The recipe of pie with gurame asam manis (fried gurame fish with sour-sweet sauce) filling

No	Ingredients	Amount
Basic Ingredient		
1	*Gurame* fish	500 g
Sauce		
1	Water	400 ml
2	Pineapple	250 g
3	Leek, slice aslant	2
4	Onion, slice round	½ clove
5	Garlic, slice	3 cloves
6	Red onion, slice	2 cloves
7	Red paprika	1
8	Yellow paprika	1

9	Chili sauce	4 spoons
10	Tomato	5 spoons
11	Salt	1 tea spoon
12	Pepper powder	1/2 tea spoon
13	Sugar	1/2 spoon
14	Cornstarch, dissolve	3 spoons
15	Cooking oil	100 g
Dressing		
1	Garlic	2 cloves
2	Lime	1
3	Pepper powder	1/2 teaspoon
4	Salt	1/2 teaspoon
Coating		
1	Flour	50 gram
2	Cornstarch	2 spoons
3	Pepper	1/2 teaspoon

Based on the recipe formula in table 5, the researchers made the pie filling from gurame asam manis. The following is the procedure of cooking the pie filling.

1. Discard fish scales, fillet, and cut into some parts. Wash them and also the head. Smear them with the dressing, keep for 30 minutes.
2. Roll the fish parts on the coating until spreading evenly. Pat them to avoid clumps of flour. Fry the fish in a hot cooking oil with a medium fire until the color is golden. Lift, drain, and set aside.
3. Heat two spoons of cooking oil, fry onion, garlic, red onion, and paprika until the smell is good. Pour water then add salt, sugar, and pepper. Add pineapple chunks and leek.
4. Pour tomato sauce and chili sauce into pan, stir well. Cook all ingredients until boiling then add cornstarch solution. Stir well until the sauce becomes thickened. Turn off the heat. Serve the fish with sour and sweet sauce as the dressing.

The experiment combined pie dough with gurame asam manis. For the next, the researchers named it guramnis rainbow pie. The result of organoleptic test of the product, which was tested to 10 panelists, involved some aspects namely pie dough color, pie dough flaky crust, taste, and aroma. Organoleptic method is comparing the food products at different variations with standard of the food products [5]. The following is the result of organoleptic test.

Table 6. The result of organoleptic test of the product

No	Aspect	Percentage (%)
1	Pie dough color	90
2	Pie dough flaky crust	60
3	Taste	70
4	Aroma	70

Serving size 100 g

According to the result of organoleptic test in table 6, it was known that the pie dough color, the taste, and the aroma were categorized as good. However, the dough flaky crust still needed improvement. The researchers found the product primacy as follows:

1. The pie appearance was different compared to the previous pies because of the rainbow colors which made the pie more beautiful.

2. The color of guramnis rainbow pie was resulted not only from food coloring but also from vegetables extract namely carrot (orange), bit (red), and salary (green). Thus, it had many benefits for health and adds the nutrition.
3. The taste of guramnis rainbow pie was different compared to the previous pies. Basically, a pie filling had a sweet taste such as vanila la, strawberry vla and so on. Nevertheless, guramnis pie was served with gurame asam manis as the filling and had flour and cilembu sweet potato as the basic ingredients. They made a new taste for this pie was never exist before, which may increase the selling price.

The nutrition of per serving size of guramnis rainbow pie was shown in table 7 below.

Table 7. The nutrition facts of guramnis rainbow pie

Energy	Carbohydrates	Fat	Protein
81.72 calories	12.5 grams	2.32	2.77 grams
Serving size 60 g			

Table 7 shows that the first fruit pie carp weighing 60 grams contains 81.72 calories of energy, carbohydrate 12.5 grams, 2.32 grams of fat, protein content by weight of 2.7 grams. Pie *guramnis* is a snack that can be eaten as a snack or a small meal that can be served as a complement to a cup of tea in the morning or afternoon. Presentation *guramnis* pie as a snack is very suitable, due to two pieces of pie *guramnis* is equivalent to two bowls of green bean porridge consisting of 25 grams of green beans + sugar + milk 13 grams and 50 grams = 175 calories can be eaten for 2 times in the morning and afternoon [6].

4. Conclusion

Guramnis rainbow pie is an innovative patisserie product which is resulted from a project-based learning. The innovation can be seen from the use of complementary ingredients namely *cilembu* sweet potato and vegetables extract for the pie dough. The study recommends a further experiment on pie dough flaky crust.

Acknowledgments

The authors would like to thank the Directorate of Higher-Education, Ministry of Education and Culture of the Republic Indonesia, which has provided funding of this research.

References

[1] Ana, Subekti, S., and Hamidah, S. (2015). The patisserie project based learning model to enhance vocational students' generic green skills. *Proceeding Atlantis Press* ISBN on line 978-94-62520-53-0.
[2] Zuraida, N. (2003). Sweet potato as an alternative food supplement during rice storage. Journal Lubang Pertanian. 22(4), pp. 150-155.
[3] National Food Service Management Institute. (2009). Culinary techniques for healthy school meals (2nd ed.). University, MS: Author.
[4] Bharathiar University . (2008) Food Production And Patisserie Rontline Institute Of Hotel Management Studies 14, Vallalar Salai, Pondicherry
[5] Gita Bisla,Shailza Choudhary, and Vijeta Chaudhary .(2014). Evaluation of the Nutritive and Organoleptic Values of Food Products Developed by Incorporated Catharanthus roseus (Sadabahar) Fresh Leaves Explore Their Hypoglycemic Potential. The Scientific World Journal Volume 2014 (2014), Article ID 304120, 5 pages
[6] Sunita Almatsier. (2005) *Penuntun Diet*. Instalasi Gizi Perjan RS Dr. Ciptomangun

Kusumo dan Asosiasi Dietesien Indonesia. Jakarta: PT Gramedia Pustaka Utama.

IOP Conf. Series: Materials Science and Engineering **128** (2016) 012048 doi:10.1088/1757-899X/128/1/012048

Entrepreneurial model based technology creative industries sector software through the use of free open source software for Universitas Pendidikan Indonesia students

B Hasan*, Hasbullah, W Purnama and A Hery

Department of Electrical Engineering Education, Universitas Pendidikan Indonesia
Jl. Dr. Setiabudhi No. 207 Bandung, Indonesia

*Corresponding author: bachtiarhasan55@yahoo.com

Abstract. Creative industry development areas of software by using Free Open Source Software (FOSS) is expected to be one of the solutions to foster new entrepreneurs of the students who can open job opportunities and contribute to economic development in Indonesia. This study aims to create entrepreneurial coaching model based on the creative industries by utilizing FOSS software field as well as provide understanding and fostering entrepreneurial creative industries based field software for students of Universitas Pendidikan Indonesia. This activity phase begins with identifying entrepreneurs or business software technology that will be developed, training and mentoring, apprenticeship process at industrial partners, creation of business plans and monitoring and evaluation. This activity involves 30 UPI student which has the motivation to self-employment and have competence in the field of information technology. The results and outcomes expected from these activities is the birth of a number of new entrepreneurs from the students engaged in the software industry both software in the world of commerce (e-commerce) and education/learning (e-learning / LMS) and games.

1. Introduction

Business and industrial development have risen with the globalization of the economy. The result range of products required to meet the needs and desires of consumers. Economic globalization and information development encourage the students who just graduate have competent entrepreneurial spirit. However, not all college graduates have an entrepreneurial spirit as desired by the employment opportunities. The fact shows that only a small proportion of college graduates who have an entrepreneurial spirit. On the other hand, the economic crisis caused the number of jobs was not growth, and even reduced due to bankruptcy. Under these conditions, the college graduates are required to not only be able to act as job seekers but also must be able to act as a job creator. Both require an entrepreneurial spirit. Therefore, in order that the college must be able to meet these demands, various innovations needed, include learning innovation to building technopreneurship generation today's information age [1]. On the other hand the technopreneurship are who make money out of existing or new technologies which may or may not have been developed by themselves [2].

Creative economy including the economy of the fourth wave, Alvin Toffler called the first wave of the economy relies on agriculture, economy the second wave in the industrial sector, and the third wave had economics in the information sector. In Indonesia, the creative economy is quite a role in national economic development though has not been touched by government interference. According to data from the Creative Economy Study Team, the creative industry sub-sector of Information

Technology in 2014 [3]. Has a Value Added Subsector (ADHB) IDR 8,610 billion or an average accounted for 0.11% of the total GDP added Subsector (ADHB) during 2010-2013. As for the labor force participation rate of the labor of the creative economy sector by 0.58% during 2010-2013. In addition, this sector has a workforce subsector 67,888 workers with national employment growth rate by an average of 0.06% in 2013 [4]. Relating to the field of software development of creative industries, these spirit of creative industries have drawn researchers' attention to the role of firm-level innovation activities in the creative industries and their economy [5], [6], [7].

Therefore, to foster the innovation it is necessary to create an environment that encourages increased capacity and capability areas create a creative economy. In this case innovation means something new for the recipient that the community concerned. Economic progress related to the level of development of 'technical change' means the stage mastery of technology. "Technical change" most are "tacit" or not codified and is built on experience. Also are cumulative (formed by "incremental" and in a certain time). Time mastery of this technology depends on the industry sector (sector specific) and the process of accumulation follow a certain trajectory typical [8]. The rapid development of information and communication technologies present huge business opportunities mainly for young entrepreneurs or students, Effect of information technology seems to open up business opportunities that could be considered unique, new, innovative, and creative for entrepreneurs. We can see so popular Microsoft Windows operating systems, the rapid sales of the iPhone, iPod, and Apple's iPad, as well as the spread of social networking sites Facebook and Twitter throughout the world. Surely the opportunity to establish a business in the information technology field is very large in the era like today. We can make products that have never been made before and will affect the development of the technology world. However, we must first recognize several key factors before starting a business in the field of information technology, among others: market feasibility, preferences, skills or familiarity, funds, raw materials. Human resources and technology as well as personality [9].

2. Method

This research begins with identifying entrepreneurs or business software technology that will be developed, training and mentoring, apprenticeship process at industrial partners, creation of business plans and monitoring and evaluation. This activity involves 30 students UPI which has the motivation to self-employment and have competence in the field of information technology.

Sustainable development is to create such sustainability (sustainability) into a form of benefits (advantages). In this regard, Barney Clark (2007) suggests VRIO models (Value, Rareness, Imperfect Imitability, and Organization). According to him, the Sustainable Competitive Advantage can be obtained by performing a continuous process of discovery that will continue to innovate. This process will run properly when first establishing the organization's core competencies are resources development capabilities and effort that relies on the potential of nature, culture and daily life of the community is a very closely in developing a sustainable business. UPI as college has organized a variety of entrepreneurial programs in an effort to increase the independence of his students that after graduating from college not be job seekers but can create jobs. Coaching methods compiled here are trying to synergize some coaching methods that already exist in the UPI.

Coaching is done by conducting a needs assessment regarding what training is needed by students to become entrepreneurs in the field of creative industries field of software technology. Then do the race simulation in the form of business ideas. This competition aims to enhance student creativity in exploring creative ideas and innovative giving solutions to the problems that harness the power of information technology. Furthermore conducted training and coaching process. To produce young entrepreneurs, a phased approach that can be mentally stimulating entrepreneurship among students, as defined in Table 1 as follows:

Table 1. Stages of Mentoring in the conduct of entrepreneurial activity in the field of creative industries in software technology-based FOSS

No	Stages of Mentoring	Action
1	First: understanding the entrepreneurial process.	identify and translate business opportunities into value-added superior transform the added value in the form of products or services, and can be consumed by the target customers. Sharpen analytical skills and entrepreneurial creativity.
2	Second: a business plan and a commitment to developing business	the operations begins with the creation of a business plan to convince a third party in order to fund its business. committed to what you want to achieve in the future to always look for ways to achieve that goal; partnership with third parties; or sell some or all of the business ownership.
3	Third: determining and adding up the value-added	Start decisive added value through market research, translates into a value added product concepts and business concepts, maximizing entrepreneurial capital as competence groups, access to funding, and social capital among other immediate family, relatives, friends, and parents so that the risk of failure when launching a business can diminimlisasikan.
4	Fourth: Consolidation of business and business development plan	For long-term goals, sustainability coaching program it is proposed formation of house innovation as a think tank to develop products and services of the creative industry entrepreneurs technology field FOSS-based software.

3. RESULTS AND DISCUSSION

This study on the activities of the steps taken to develop the skills of students in the field of creative industries FOSS-based software field. The activity starts from the preparation stage to the stage of monitoring and evaluation. More details stages are as follows:

3.1. Preparation

Community Service Program (CRP) based entrepreneurial activity started with a review of the needs required by students in developing and improving entrepreneurial competence. It was conducted with focus group discussions (FGD) and direct dialogue with the business of the students on matters concerning the business development needs of the creative industries and their understanding of the business activities and its management. The initial activity or the preparation starts from the first week to the fourth week on June 2015.

In accordance with the objectives of the PKM (Creativity Program of the Student) models Entrepreneurial Development Based Technology Creative Industries Sector Software through the use of Free Open Source Software is then at this stage the team offers a wide audience in the UPI students who already have a business or want to expand their business in order to make it more valuable. The students were mostly those who have a strong commitment to become entrepreneurs and have basic competency in developing creative industries based business based on free open source software (FOSS). At this stage chosen and netted a total of 32 students who have a rear Latra electrical engineering who are interested to participate in the creative industries PKM-based technology software field. All participants will later be given an initial understanding of the target training on how to develop a creative industry business areas of software, either on the software learning (e-learning), commerce (e-commerse), games (games) and others.

3.2. Implementation

At this stage, all participants who are interested and have the potential of entrepreneurial technology-driven creative industries are given training in how to design marketing business by utilizing e-marketing, the website designing business profile (company profile) that can be accessed widely by business partners and market more easy. The training activities conducted and guided by instructors

PKMK UPI and assisted by some experts of IT and Web Designers and Developers who have experience in the practical world.

At this stage also Entrepreneurship teams and assisted by instructors to provide guidance to students of entrepreneurship by conducting meetings face-to-face with them for 8 sessions, provide basic training as well as guidance on the business or the business of creative industries field of software based on free open source software (FOSS)
Training activities or workshops in the field of software FOSS based industry creative was held on the second Sunday in July to August 2015. As part of the next phase of activities, all participants or groups commissioned to design the website of each business profile creative and innovative in accordance with the business they are going to run the business and profile will be assessed for feasibility by the assessment team.

From the results of the training and mentoring carried out during several meetings, and guided by instructors who are experienced, participants are business enthusiasts from among the students do it seriously and enthusiastically. Enthusiast trainees (workshop) can be seen from the liveliness and seriousness they develop entrepreneurial website as a first step to publish products that want to be marketed via the Internet, social networks and other media.

In this phase of the participants are given training by the instructors how to develop the website FOSS-based entrepreneurship, these materials include:

1. Front end developing (HTML + CSS + javascript).

Figure 1. Front end developing Apperances (HTML + CSS + javascript)

Figure 1 is a display of the programing software for develop the website FOSS-based entrepreneurship.

2. Wordpress + Woocommerce Introductions (XAMMP + Database).

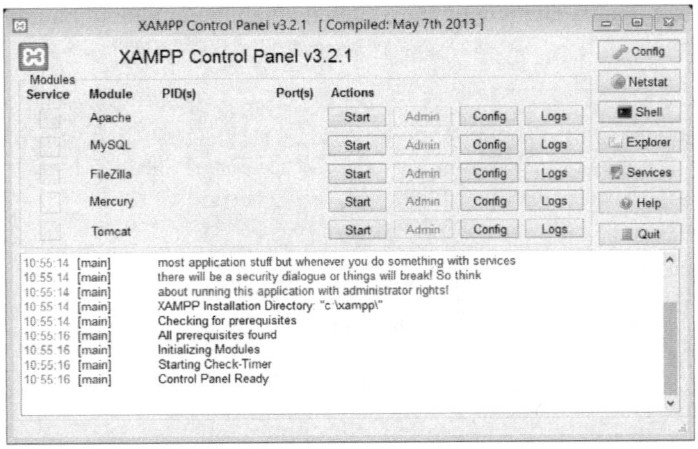

Figure 2. Wordpress + Woocommerce Apperances (XAMMP + Database)

3.3. Development

As a follow up to the stage of training the post-training mentoring process continues until participant has become considered able elaborated independently effort this website. The participants continue to be monitored progress and Entrepreneurship teams and instructors are intensive provide advisory services and technical assistance needed by the participants. In the expansion of the positive because all groups can creatively develop its business profile through the website. The stages of future activities to be implemented as a series of this program is inter connectivity among businesses student website home page into a container UPI student entrepreneur website. This website once the arena of information and communication development work done by the students UPI. The hope is FOSS-based business website can continue to be utilized even if they had graduated from his alma mater, the UPI. Here's the result of the training that has been developed by the students with various types and styles to suit their innovations.

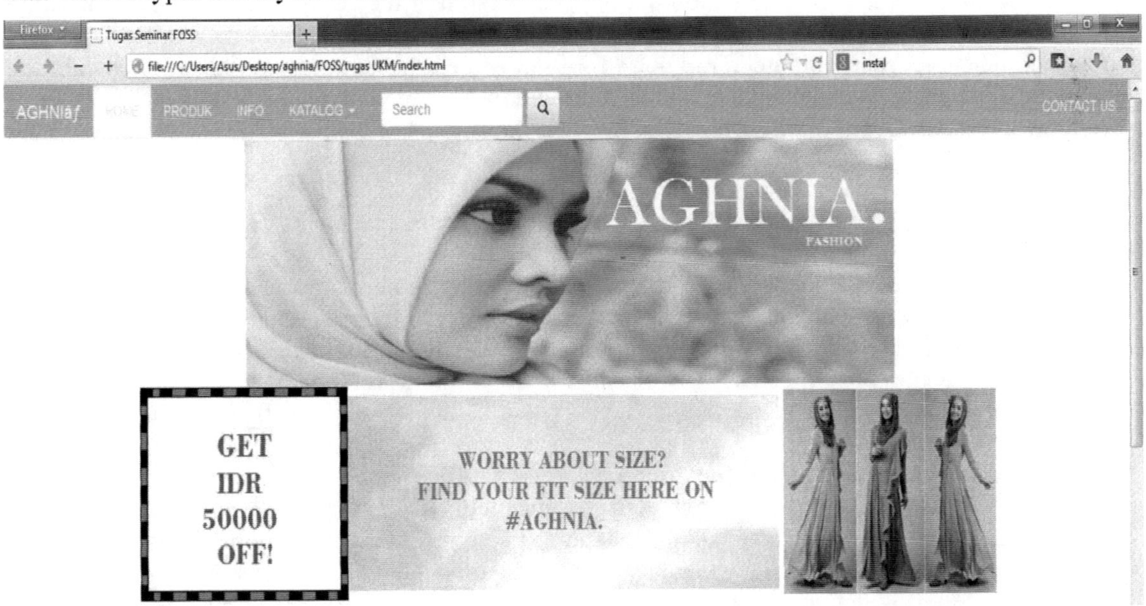

Figure 3. Blog effort students of fashion and FOSS-based Muslim fashion

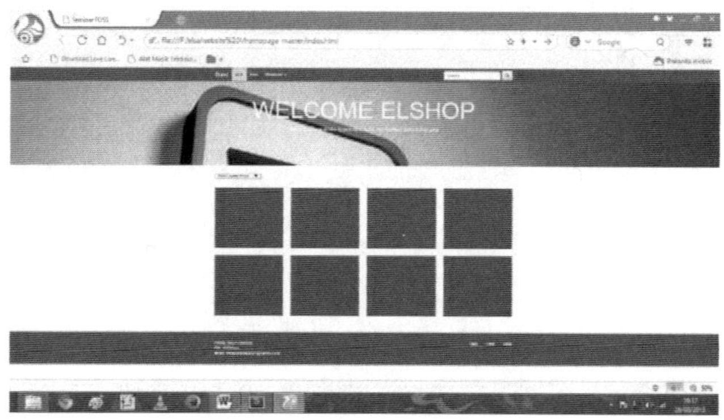

Figure 4. Blog businesses selling goods online student-based field-based FOSS

Figure 3 is the display of the home of the website of online store whereas figure 4 is the display of the home of Blog businesses selling goods online student-based field-based FOSS.

3.4. Phase Monitoring and Evaluation

Evaluation is done in Entrepreneurship PKM activity in the form of direct discussion with students who fostered and also discussions with the faculty team PKM. What are the difficulties in implementing these activities and solutions to overcome them. Students also continues to provide valuable inputs to the faculty team that PKM entrepreneurial activity goes well. Follow-up of the activities carried out are in the form of coaching/ training capacity industry for business students about the creative field of FOSS-based software that will be conducted in October 2015 in the Campus UPI. From the results of the evaluation and monitoring conducted by a team of turns of the 30 students that we provide training and mentoring only about 15 or 50% of the total number of business groups that fostered intensively and seriously develop the profile entrepreneurial venture for later entry into the communities joined in the website entrepreneurs The. Although the team PKMK already providing various facilities and instructors to guide them, but the obstacles are the reason they are not to develop a program that has been initiated, ranging from scheduled lectures are solid, coursework that accumulate and there are participants who are currently undergoing PPL, Industrial Practice, corruption and other academic activities.

4. CONCLUSION

Activity Based Community Service (PKM) entrepreneurship there are several conclusions that can be obtained include:

1. Development of entrepreneurship in the field of creative industries in software can be done in the form of stages structured activities, ranging from the mapping of the students who have asked for entrepreneurship, the selection of students who will be trained, creation of business plan and action plan, the provision of material about the development of software technology business and as well as coaching clinic and business assistance as well as help the promotion of efforts undertaken.
2. Based on the student profile is based on the observation and mentoring activities, approximately 50% UPI entrepreneurial students with various types of businesses, are already using and use the advantages of information and communication technology as a means for the promotion and marketing efforts.
3. Model fostering entrepreneurial creative industries field of software that had been developed based on the needs of entrepreneurship among students, field observations, the results of

comparative studies and internships in the industry partners who have developed and applied various software technologies.

4. Developing entrepreneurial networks between students UPI further strengthen their presence in the territory of other businesses that will awaken interconnectivity and climate entrepreneurship education is more reliable and professional.

References

[1] Chris Boulton & Patrick Turner, 2011. Mastering Business in Asia: Entrepreneurship Wiley MBA Publications Reproduced.

[2] Dina. P.T, Depositario, Aquino. A.N, Feliciano, 2011. Entrepreneur Skill Development Needs of Potential Agri-Based Technopreneurs. ISSAAS Journal Vol. 17, No.1 pp.106-120.

[3] Tim Studi Ekonomi Kreatif, 2014. Statistik Ekonomi Kreatif Subsektor Teknologi Informasi. Kementerian Pariwisata dan Ekonomi Kreatif, Jakarta.

[4] Iqbal, Ahmad, 2010. "Pengembagan Kewirausahaan Teknologi Untuk Mendorong Tumbuhnya Industri Kreatif Software Melalui Pemanfaatan FOSS". Pusat Pengkajian Kebijakan Peningkatan Daya Saing Badan Pengkajian dan Penerapan Teknologi.

[5] Green, L.,Miles, I., Rutter, J., 2007. Hidden Innovation in The Creative Industries. NESTA Working Paper. National Endowment for Science, Technology and the Arts, London.

[6] Stoneman, P., 2007. An Introduction to The Definition and Measurement of Soft Innovation. NESTA Working Paper. National Endowment for Science, Technology and the Arts, London.

[7] Miles, I., Green, L., 2008. Hidden innovation in the creative industries. NESTA Research Report. National Endowment for Science, Technology and theArts, London.

[8] Departemen Perdagangan Republik Indonesia, 2008. Pengembangan Ekonorni Kreatif Indonesia 2025, Studi industri kreatif Indonesia.

[9] Rahmawaty, Penny., Suwarto, Dyna Herlina., Endarwati, M. Lies.,2010. "Technopreneurship Course Development Program". Jurusan Manajemen FISE Universitas Negeri Yogyakarta.

IOP Conf. Series: Materials Science and Engineering **128** (2016) 012049 doi:10.1088/1757-899X/128/1/012049

Visual effects of the first ladies' Kebaya clothing on the image of Indonesian women's appearances

Suciati*

Pendidikan Tata Busana,
Faculty of Technology and Vocational, Universitas Pendidikan Indonesia
Jl. Dr. Setiabudi 207 Bandung 40154, West Java, Indonesia

*Corresponding author: suciati@upi.edu

Abstract. The image of Indonesian women on international level is partly influenced by the appearance of the First Lady. The role and position of the First Lady is the representation of Indonesian women, because basically the First Lady, as the wife who accompanies the President (head of state), has a strong background of cultural grip, high intellectuality and good personality in her daily lifestyle, including in wearing clothes, and as an ambassador of culture and design. Fashion style of the First Lady always draws praise and criticism from the public. The purpose of this study is to reveal the visualization effects of Indonesian First Ladies' kebaya clothing style in various state occasions on the image of Indonesian women's appearances. This study is a qualitative research of visual data that emphasizes the discussion of Kebaya Clothing using semiological study (connotation and denotation meaning) that bring out self-image. The results showed that the style the First Ladies' Kebaya clothing in every presidency period of their husbands had characteristics both in the style of clothing or hairstyle, indicating self-image. The conclusion of this study reveals that the First Ladies' Kebaya Clothing (National Clothing) is interpreted as having implied messages because clothing can be observed visually. Implication was done on the construction of learning patterns of clothing, national fashion design and Nusantara ethnic clothing design.

1. Introduction

Indonesia is a country that has many cultural artifacts, which visually become national identity. Identity of a nation is an important power that supports the establishment of national characteristics because, in addition to strengthening the existence, it will also provide lifting power or building power for the enforcement of security and the increase of prosperity. Characteristics are the identity that determines national direction. A nation will have an acclaimed civilization only if it can produce system of values, code of conduct and practices that are superior and are based on cultural values embraced by its people. National identity and characteristics can only be fostered through the education of good examples. The emergence of an identity crisis influenced by the progress of technology and information has been eroding local cultures, for example the declining taste of use of Kebaya clothing as National Dress.

An important role that greatly affects the dynamics of the use of National Clothing is the role of the First Lady as a figure that has great care in cultural values and national identity, and who can be made an inspiration for Indonesian women to increase their pride in Kebaya as National Clothing. The First Lady can also become a figure who is an example in wearing clothes and has high prestige and dignity. In the First Lady, there is an idealized image of Indonesian women since the presence of the

First Lady is like a mother for the country. However the multicultural culture in Kebaya Clothing (as a result of various cultural background of Indonesian First Ladies) created various designs in Kebaya clothing worn by the First Ladies. The structure of kebaya clothing of the Frist Ladies always experienced changes in their various design elements [1]. These changes gradually blur the clarity and certainty regarding visual identity of National Clothing of Indonesian Women.

2. Research Method

The diversity in public lives is highly associated with signs and marking in many things [2]. The way these signs provide information is used to provide identity guidance. One of these main identities is national identity, in this case, it is the First Ladies' Kebaya clothing as national cultural identity of Indonesian women. This method revealed that the relationship between the clothing type worn and the resulting meaning can be described by the following formula:

Table 1. Formula.

Sr (signifier) Marker Physical Form	Sd (signified) : Marked Meaning
Garment	

3. Discussion

The design of national clothing of Indonesian women should apply the diversity of Indonesia [3], which has numerous varieties in the forms of fashion artifacts [4]. The application of Indonesian cultural diversity on National clothing reflects the wisdom and respect for the diversity of Indonesian culture and its position as the clothing that reflects the personality of Indonesian women.

These conditions are summarized in the regulation of Kepres RI (Decree of the President of the Republic of Indonesia) No. 18 of 1972 dated March 7, 1972 on the National Clothing and UU (legislation) No.9 of 2010 about protocols. Kebaya clothing is a type of civilian clothing, whose position accompanies the appearance of the President as her husband on state occasions and official events. The existence of National (Indonesian) clothing originated from a workshop in 1978 in Jakarta, involving 28 (twenty eight) provinces in the country (all the provinces at that time), and the result is that long Kebaya clothing was established as consisting of a dress covering upper body and a pleated (wiru) long cloth covering the lower body. National (Indonesian) clothing model is kebaya clothing model worn by the First Lady of the Republic of Indonesia, Raden Ayu Siti Hartinah (Mrs. Tien Soeharto) in each official event both national and international.

					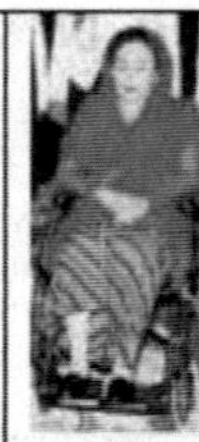	
Aspects Researched:		**Fatmawati the wife of Soekarno (1945-1966)**	**R.Ay Siti Hartinah the wife of Soeharto (1966-1998)**	**Hasri Ainun the wife of B.J Habibie (1998-1999)**	**Sinta Nuriyah the wife of Abd. Wahid (1999-2001)**	**Kristiani Herrawati the wife of Susilo B.Y (2001-1014)**
Head:	Bun	✓	✓	✓	✓	✓
Upper body:	a. Kebaya b. Baju kurung	Kebaya (short) kutubaru	Kebaya (short) kutubaru	Kebaya (short) sunda	Kebaya (long) sunda	Kebaya (long) kutubaru
Lower Body:	1. Long cloth 2. Sarong	Long batik cloth with *wiron*	Long batik cloth with *wiron*	Long batik cloth with *wiron*	Long batik cloth with *wiron*	Sarong with *tumpal*
Accessories :	a. Earrings	✓	✓	✓	✓	✓
	b. Necklace	✓	✓	-	-	-
	c. Bros	-	-	✓	✓	✓
	d. Bracelet	✓	✓	✓	✓	✓
	e. Watch	-	-	-	-	-
	f. Ring	✓	✓	✓	✓	✓
Millineris :	1. Selendang	veil	selendang	selendang	veil	selendang
	2. Bag	handbag	handbag	handbag	-	-
	3. Footwear	open slippers	Open slippers	Open slippers	Pantoffel shoes	Open slippers
	4. Fan	-	✓	✓	-	-

Figure. 1. Kebaya clothing

4. Conclusions

Kebaya clothing of the First Ladies is a sign of national identity that can be observed based on: First, Image clothing, which is two-dimensional, is the image of garment structure comprising: (a) kebaya model blouse, (b) Indonesian traditional fabric in the form of long cloth or sarong [5]. The materials of these two parts can have the basic material of either modern fabric or Indonesian traditional fabric [6][7]. (c) sewing technique that becomes the characteristic in both kebaya and sarong cloth is a traditional sewing technique, but in its development now, especially in kebaya, stitching sewing technique is generally used, (d) fashion complements in the form of selendang, slippers, handbag and fan, in addition to the bun- shaped hairstyle [8]. Then, Written clothing, which is verbal, is kebaya clothing which can give the impression of being verbally beautiful, worthy, elegant, lovely, graceful, gentle, motherly, polite, neat, clean and pretty. And Real clothing, which is 3-dimensional, is the real form of kebaya clothing.

References

[1] Zaman, Moh. Alim. (2002) : 100 Tahun Mode di Indonesia (1901-2000), Jakarta, Meutia Cipta Sarana & DPP Ikatan Penata Busana Indonesia "Kartini.
[2] Berger, Arthur S. (2010) : Pengantar Semiotika, Tanda-tanda dalam Kebudayaan Kontemporer, Yogyakarta, Penerbit Tiara Wacana
[3] Achjadi, Judi. (2003) : Permata Cita: Koleksi kain Ibu Negara, Jakarta, Penerbit Himpunan Wastraprema
[4] Sartomo, Roem. (1953) : Busana Nasional: Busana Daerah dan Modifikasi, Makalah disajikan dalam Seminar Tiara Kusuma, Jakarta, 17 September 1953.
[5] Djoemena, Nian S. (1990) : Batik dan Mitra, Jakarta, Djambatan.

[6] Anas, Biranul. (1995) : Indonesia Indah mengenai Latar belakang Kehidupan Bangsa Indonesia Adat Istiadat serta Budayanya: Kain-kain Non-Tenun Indonesia (4), Jakarta, Penerbit Yayasan Harapan kita /BP 3 TMII
[7] Hitchcock, Michael. (1991) : Indonesian Textiles, Penerbit Perplus Editions (HK) Ltd.
[8] Unajah, Uun, (2009), Pelestarian Busana dan sanggul Daerah, Dewan Pimpinan Daerah Persatuan ahli Kecantikan dan Pengusaha Salon Indonesia "Tiara Kusuma" Provinsi Jawa Barat.

IOP Conf. Series: Materials Science and Engineering **128** (2016) 012050 doi:10.1088/1757-899X/128/1/012050

Uncovering Sundanese Values by Analyzing Symbolic Meaning of Ménak Priangan Clothing (1800-1942)

M Karmila*, Suciati and I Widiaty

Study Program of Dressmaking Education,
Faculty of Technology and Vocational, Universitas Pendidikan Indonesia
Jl. Dr. Setiabudi 207 Bandung 40154, West Java, Indonesia

*Corresponding author : mkarmila@upi.edu

Abstract. This study investigates symbolic meanings found in the Sunda ethnic clothing, particularly the Menak Priangan clothing. This study aims to uncover and document those symbolic meanings found in the Menak Priangan clothing as an effort to develop Sunda cultural artefacts of West Java. This study on Menak Priangan clothing applies ethnography (visual) and aesthetic methods. The visual method is utilized in order to uncover local cultural (Sunda) values found in Menak Priangan clothing visualization, including: design, model, name, and representing colours, which then directed towards local Sundanese aesthetic concepts living within the Priangan community. Furthermore, aesthetic method is used to explore role of aesthetic values in empowering visual cultural values within certain community, particularly Sunda aesthetic values. The study results show that since the 19th century, Sunda ethnic clothing was limited to Priangan Sunda only, while traditional clothing wearing by Priangan people reflects their social strata, consisting of: a. Menak Gede (Menak pangluhurna: mayor), bearing raden title, b. Menak Leutik/Santana (mayor assistant), titles: asep, mas, agus, ujang, (Nyimas for woman), c. Somah/Cacah: ordinary people/lower class. Clothing is a cultural phenomenon within certain culture reflecting such society experiences. For Menak people, clothing and its accessories have important meanings. They wear such traditional clothing and accessories as a symbol of power they have within bureaucratic structure and as a symbol of social status they bear within traditional community structure.

1. Introduction

Menak Priangan Research on Ethnic Clothing Sunda has been pioneered by a team of writers, to initiate an independent investigation of the Sundanese cultural artifacts in 2006 with the title " Design Iket Sundanese in Bandung and Sumedang Period of 1968-2006 " and in 2007 with the title " Wedding Dress Design Shift Sunda Yang influenced Islamic element in Bandung the period of the 1980s till the 2000s ". Through this research finding that the author had the team takes a strategic effort to inventory and documentation of cultural artifacts such as the Sundanese ethnic clothing, especially clothing Menak Priangan, as a cultural heritage that should be preserved and maintained its existence. Research related to fashion Menak Sunda has been done by Ayu Septiani (2012) with the title " The Power Symbol Apparel As Menak Priangan In the 19th Century SD Early 20th Century ". The study revealed that the powers possessed by Menak is a pattern of, whereas the view that Menak identical to the king who became the guardian of God serve as a reference for the exercise of power (manifest power) . To mean that the Menak was in power , is used symbols in this case is the clothing and equipment of the marvelous greatness . From these results the authors take a common thread that there needs to be more

IOP Conf. Series: Materials Science and Engineering **128** (2016) 012050 doi:10.1088/1757-899X/128/1/012050

advanced research efforts can reveal the symbolic meanings contained in Clothing Menak Priangan through visualization analysis study fashion.

2. Research Method

In the discussion of fashion Menak Priangan used ethnographic methods (visual) and methods of aesthetic. Visual methods used to peel value kesundaan in various dimensions in marvelous fashion visualization Priangan: design, model, naming, and colors that can represent them, is then directed to the concept of aesthetic reflection on society kesundaan Priangan. Activities that first do is to find and collect a good source of primary and secondary sources. Primary sources in the form of archives relates to fashion Menak Priangan. Secondary sources used include books, scientific papers, articles in newspapers / magazines, and general reference. Secondary sources used were not all directly related to the themes discussed in this paper, but there were substantively examine the theme of clothing, and there are thematically not discuss clothes, but rather to discuss aspects of the lives of the Menak Priangan. After the data sources obtained by the interpretation of the facts is then performed (visually).

In the final stage is done through a method of aesthetic studies approach Semiotics, visual language (visualization) of an object can be observed as a system of signs. In a review of visual language, there are two important aspects of semiotics, the indices and signs (including symbols). The index is a sign that has an existential dependency relationship between the sign and the 'who indicated they were', or have a causal bond / relationship with what it represents. While the mark is a basic element in semiotics and communication, ie everything that implies, which has two categories, namely as a marker (basic form; icon, symbol) and a marker (meaning). (Sachari 2002 p.74).

3. Discussion

3.1. Priangan

According to Wahyu Wibisana (1986 : 17) in Suwardi Alamsyah P (2015 : p. 16) Priangan area covers Cianjur, Bandung, Tasikmalaya, and Sumedang districts. This area is centered in Bandung city, the capital of today's West Java province, functions as the governmental and cultural development center. History analysis of Priangan shows that its people have cultural orientation towards Pajajaran.

Since the 19th century, Sunda ethnic clothing is limited to Sunda Priangan only, in which traditional clothing wearing by Priangan people reflects their social strata, consisting of:

- Menak Gede (Menak pangluhurna: mayor), bearing raden title.
- Menak Leutik/Santana (mayor assistant), titles: asep, mas, agus, ujang, (Nyimas for woman).
- Somah/Cacah: ordinary people/lower class.

Besides these social strata, there are 2 other categories of Menak :

- Menak lama, those who have cognation with mayors of old dynasty (mayors of districts from post-traditional kingdom era).
- Menak baru, those who does not have cognation with the mayors from the old dynasty, but becomes a Menak due to his today's positional.

3.2. Menak Priangan Clothing

Papaes is the way of wearing clothes which closely related to daily manner. The way of Sundanese wearing their clothes involves their traditional clothes and its accessories as well.

The nobleman dressing manner reflects his social status at the top of social structure. Basically, the Menak dressing manner during the 19th century up to 1942 had been applied by different generations, but then the colonial government issued special regulations concerning this issue, especially for government officers.

This study is focused on the Menak Gede social status related to their dressing characteristics, including:

Table 1. Dressing characteristics

	Man	***Lady Dress***
	Formal wear of a mayor wife usually follows his husban dress	
Head	Head cover consists of an iket (a Sunda headband), made of black batik Sawunggaling/Soga Gunawijaya/Gambir Saketi	Hair is combed backward and then bun, supported with gold hairpin completed with accessories consists of diamonds or flowers
Upper Body	• Wearing white skirt with Chiang I collar, with no sleeves/with long sleeves • Wearing *bedahan (sikepan)* short cloak applying 9 shirt buttons. Golden floral pattern embroidery was applied along the edge.	• Wearing kebaya made of black velvet applying golden embroidery on the edge. • The kebaya long ends at the hips • The kebaya's neck is cut around the neck/circled • Accessories consists of: long necklace with a pendant, a serial brooch (3 layers brooch), kerabu, ring, and bracelet
Lower Body	Wearing white/ black pants • Wearing white skirt with Chiang I collar, with no sleeves/with long sleeves Belt • When wearing *Keris*, there are usually 2 or 3 belts wearing. The first layer is a rough belt (stagen), followed by special belt covered with silk materials, and the upper one is a belt called *kamus* or *epek* o *Keris* is inserted on the second belt with the handle is positioned towards the left side and should be wore when meeting the Dutch collonial officer to show respect • Daily dress of the *Menak* consists of casual cloth with white or black color	• Wearing cloth sheet (sinjang kebat) made of batik applying similar pattern with the husband batik cloth (lereng parang/doctor/parang kusuma) • A stagen belt with 4-5 meter long and 10-15 cm width is appied on the waist • Footwear consists of velvet sandals applying golden embroidery and beads Daily clothes consist of : o Kebaya made of satin/paris (plain or with flower patterns) o Sinjang sheet with lereng batik pattern o Footwear is leather sandals

3.3. Menak Priangan Clothing Accessories

The Dutch colonial government did not only regulate dressing manner of its officers, but also accessories to be used along with such clothing, as outlined in Staatsblad of 1820 No 22. Dressing accessories including tumbak ngawen (spears covered with a piece of fabric), senjata ngawen (shoulder rifle), kandaga (a safety box for official clothing, etc), padaharan, lante (a small chair made of rattan), a horse, epok (a small box to serve betel leaf), pakecohan (spittoon), gapit (artificial hands clipped with small piece of bamboo), bumbung (valuable papers storage), semambu (special belt made of silk with ornaments), a sword, a shield, songsong (royal umbrella), tumbak agem (spears used by the officer himself), and tumbak panurung (spears used by the officer's guards) (Lubis, 1998, p. 187).

IOP Conf. Series: Materials Science and Engineering **128** (2016) 012050 doi:10.1088/1757-899X/128/1/012050

Number of accessories and guards allowed was based on the officer's rank within pangreh raja structure.

Clothing accessories closely related to the Menak clothing are medals, which usually mounted on the clothes. Such honor medals were given as a reward for loyalty performed by related officer to the government. It also given to compensate treat the disappointment of the Menak suffering from government action. Such Medal of Honor, according to Bijblad no. 2500, consists of gold, silver, and bronze medals, which had to wear on the left chest using special tape. Moreover, according to the 1886 Statute no. 219, the Governor General must give a sign of loyalty and bravery medals to the regents who had shown loyalty for 30 years. When recipient officer die, his medals of honor do not need to be returned to the government, but should not be used by others as well. If the recipient of honors subjected to the death penalty or forced labor, medals should be returned to the government, while if the punishment in form of imprisonment, during prison service marks should not be used (Bijblad 4158). If a regent wearing this special medal walked past the guard of soldiers, he is entitled to respect arms of the keeper (Lopez, 1998, p 191).

3.4. Symbolic Meaning of Menak Priangan Clothing

The word "busana" is taken from the Sanskrit "bhusana". In the Java language it is known as "busono". This word refers to similar meaning in both languages, namely "jewelry". However, there was a shift in Indonesian language in which the word "busana" has been translated into "pakaian". Nonetheless, the notion of "busana" and "pakaian" are actually different, in which the word "busana" has connotation as "beautiful or nice clothes". In other words, "busana" refers to "eye catching and well suited clothes, and wore in accordance with event where it is used". This is consistent with the original meaning of the noun "busana" as "jewels", something beautiful, good, or contains artistic (aesthetic) values.

The word "busana" has several different meanings in English. Their usage depends on relevant context, such as fashion, costume, clothing, dress, and wear. The word fashion is more focused on trend, which is generally shown, certain trend during the specified time (in fashion), displayed mode (fashion show), fashion creator (fashion designer), mode book (fashion book), an image pattern of the clothes according to the latest fashion (fashion plate). Furthermore, the word costume usually associated with type of clothing, such as national dress (national costume), western clothing (west costume), Muslim clothing (Moslem costume). On the other hand, the word clothing can be interpreted as clothes, in which the term is associated with certain conditions, such as clothing for winter season (winter clothing), clothing for spring season (spring clothing). (Arifah 2003, p. 1).

Menak Priangan clothing is a manifestation of Sundanese cultural artifacts language, which full of symbolic meaning associated with the role, position and responsibilities that must be bear by a Menak. From semiotics point of view, visual language (visualization) of an object can be observed as a system of signs. From visual language perspective, there are two important aspects of semiotics exist, that are indexes and signs (including symbols). Index is a sign that has an existential dependency relationship between the sign and 'object it is signed for', or have a causal relationship with what it represents. On the other hand, mark is a basic element in semiotics and communication, i.e. everything that implies meaning, consisting of two categories, namely as a marker (basic form; icon, symbol) and as a sign (meaning). (Sachari 2002 p.74).

From visual language semiotics standpoint, anything worn by a Menak Priangan, from clothing, clothing accessories, and badge of honor as well, represent symbols of greatness/power he own, as a form of nonverbal communication of a ruler/king to his people. Menak is identical to the king who represents the God, serving as a reference in running the power (power manifestation).

From visual language semiotics perspective, analysis on symbolic meaning found within Menak Priangan clothing can be described as follows:

Table 2. symbolic meaning

Indexes		Signs, Symbols
Head	Head cover consists of an iket (a Sunda headband) or bendo, made of black batik Sawunggaling/ Soga Gunawijaya/ Gambir Saketi	Headband in the form of bendo containing a highly symbolic meaning related to the Sundanese world view, headband in form of iket/bendo related to the concept of "masagi" (4 corners), a traditional teaching tells that if someone wants to be successful he must do masagi/ngajurus/learning. So, as a king/*Menak*, one has to master extensive knowledge (masagi), because as noble person/king with power to rule, he has to master the certain knowledge enabling him to shape himself into a leader who can protect his own people
Clothes, consists of bedahan made of black velvet	Accessories from golden floral pattern embroidery, and Badge of honor	Black color implies (silow lamun teu kabuka) that a leader (*Menak*) should be the one that hideung/mature/authoritative. Meaning that becoming a leader one has to master wide range of knowledge and insight or masagi. 2. Golden ornaments symbolizes the greatness and glory of a *Menak*/king. 3. Badges of honor symbolizes loyalty, greatness and courage, as well as a distinguishing mark of social status between a *Menak* and people in the lower social strata.
	Lower clothes consists of Dodot Santana made of batik with lereng pattern (parang rusak barong, parang kusuma, kawung besar)	Dodot Santana with lereng/slopes pattern symbolizes that a *Menak*/king is on the highest strata within his society, since lowly *Menak* should not wear batik with this pattern

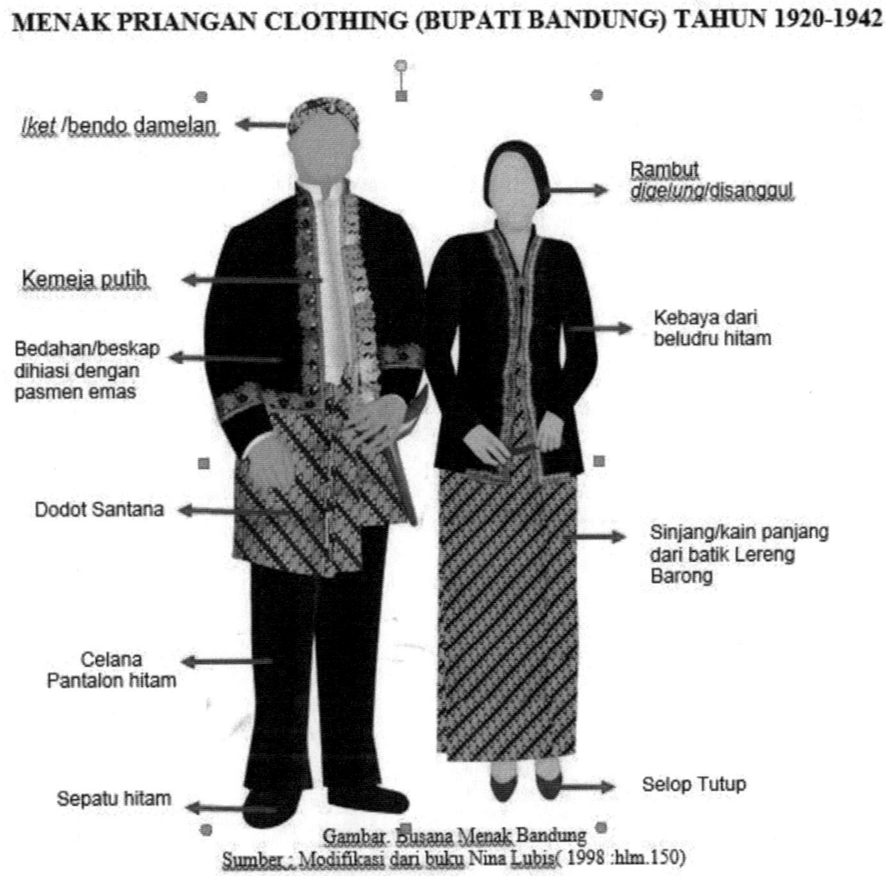

Figure 1. Busana Menaik Bandung.

4. Conclusion

Clothing is one of cultural artefacts that can communicate past events occurred in certain community to be read and studied in the future. For Menak, clothing and its accessories have a particular significance, since Menak clothing is a symbol of status in his traditional social structure, apart from being a symbol of power in the bureaucratic structure.

From clothes functions as a means of nonverbal communication point of view, the relationship between Menak as a native ruler with his people in the 19th century until the 20th can be understood as a close interdependence relationship between two different but inseparable elements. In the Javanese traditional concept, such relationship is called king- servant relationship. It indicates that the Menak life is strongly influenced by the Javanese traditions, rooted from Mataram kingdom controlled Priangan region in the past. Mataram culture influence may be seen in many aspects of Menak life, which is strongly oriented towards Javanese tradition up to two centuries after Mataram kingdom power ended. One of such tradition is symbols of power as shown in clothing and accessories used.

Menak clothing, accessories, and badges of honor represent symbols used to legitimize his power. Through dress worn a Menak/nobleman shows his authority/power to govern even to be served as just like a king.

IOP Conf. Series: Materials Science and Engineering **128** (2016) 012050 doi:10.1088/1757-899X/128/1/012050

References

[1] Achyadi, Y. (1981). Pakaian Daerah Wanita Idonesia. Jakarta : Djambatan
[2] Anonym . (2014)."BusanaTradisional Jawa Barat" .[Online].Tersedia: https://gadismuthia526.wordpress.com/2014/03/18/22/. [25 Juli 2015]
[3] Dloyana, Kusumah, Siti. Dkk. (2012).Indonesia Indah. Jakarta: Perum Percetakan Negara Republik Indonesia.
[4] Lubis, Nina H. 1998. Kehidupan Kaum Ménak *Priangan* 1800-1942. Bandung: Pusat Informasi Kebudayaan Sunda
[5] ....2003. Sejarah Tatar Sunda (Jilid 1). Bandung: Lembaga Penelitian UNPAD
[6] ....2003. Sejarah Tatar Sunda (Jilid 2). Bandung: Lembaga Penelitian UNPAD
[7] Riyanto, Arifah A. 2003. Teori Busana. Bandung: Yapemdo.
[8] Sachari, Agus.2002. Metodologi Penelitian Budaya Rupa. Jakarta: Erlangga
[9] Sumardjo, Jakob. 2003. Simbol-Simbol Artefak Budaya Sunda. Bandung: Kelir

IOP Conf. Series: Materials Science and Engineering **128** (2016) 012051 doi:10.1088/1757-899X/128/1/012051

West Java Snack Mapping based on Snack Types, Main Ingredients, and Processing Techniques

A S Nurani*, S Subekti and Ana

Home Economics Department
Universitas Pendidikan Indonesia
Jl. Dr. Setiabudi 207 Bandung 40154, West Java, Indonesia

*Corresponding author: atatsiti@upi.edu

Abstract. The research was motivated by lack of literature on archipelago snack especially from West Java. It aims to explore the snack types, the processing techniques, and the main ingredients by planning a learning material on archipelago cake especially from West Java. The research methods used are descriptive observations and interviews. The samples were randomly chosen from all regions in West Java. The findings show the identification of traditional snack from West java including: 1. snack types which are similar in all regions as research sample namely: opak, rangginang, nagasari, aliagrem, cuhcur, keripik, semprong, wajit, dodol, kecimpring, combro, tape ketan, and surabi. The typical snack types involve burayot (Garut), simping kaum (Purwakarta), surabi hejo (Karawang), papais cisaat (Subang), Papais moyong, opak bakar (Kuningan), opak oded, ranggesing (Sumedang), gapit, tapel (Cirebon), gulampo, kue aci (Tasikmalaya), wajit cililin, gurilem (West Bandung), and borondong (Bandung District); 2. various processing techniques namely: steaming, boiling, frying, caramelizing, baking, grilling, roaster, sugaring; 3. various main ingredients namely rice, local glutinous rice, rice flour, glutinous rice flour, starch, wheat flour, hunkue flour, cassava, sweet potato, banana, nuts, and corn; 4. snack classification in West Java namely (1) traditional snack, (2) creation-snack, (3) modification-snack, (4) outside influence-snack.

1. Introduction

The research was motivated by lack of literature on West Java traditional snack. West Java traditional snack is hereditary and becomes a valuable cultural asset of our nation, a liason between generations, an identity and a characteristic of each region. Therefore, it needs to be explored and preserved. Several attempts to achieve those goals are by making inventory and documentation which had been done by department of education and culture. [1] The document discusses daily foods for various age and special foods for cultural ceremony in West Java. It includes name, place, and function. Previous studies on the documentation of West Java traditional foods were conducted by Alim A. [2] who discussed rengginang (a type of cracker from glutinous rice) and Herminiati [3] who discussed opak (also a type of cracker from glutinous rice but threshed and has different flavor).

Through this research, the inventory and documentation of West Java traditional snacks were realized. The documented snacks are classified into traditional snack, snack which got outside influence, creation-snack, and modification-snack. The documentation includes names, ingredients, processing techniques, servings, and the cake functions in daily life. The documentation of traditional snack is very important since the existed inventory should be renewed along with its progress and development [4][5]. Thus, this research intends to document the more accurate and up-to-date data and information.

Sundanese is well-known as a much opened and positively responsive ethnic to any progressive transformation [6]. Therefore, we can easily find and eat snacks which get influenced from outside. Hopefully, the research is able to significantly document various traditional snack types from West Java and also the social and cultural review. The research is also a real effort in preserving traditional snack which becomes a valuable nation's cultural asset.

2. Research Method

The research was conducted in West Java which totally has 27 counties and cities. The researchers took a random sample from northern West Java, central West Java, and southern West java. Specifically, the research was conducted in 10 counties namely Garut, Sumedang, Kuningan, Cirebon, Purwakarta, Karawang, Bandung, West Bandung, and Subang. The subjects were preferred from government elements namely Department of Industry and Trade, Department of Tourism, home industry entrepreneurs, and West Java traditional food lovers and observers. Furthermore, the snacks as sample are traditional ones which are consumed daily. The researchers explored the snack names, the types, the main ingredients, the processing techniques, and the serving functions.

The primary data was collected by observations, interviews, and document collection. Observations were utilized to investigate the data of various traditional snacks from home industry entrepreneurs who produce West Java traditional snack. Besides, interviews were also conducted to government elements namely Department of Industry and Trade and Department of Tourism, local traditional food lovers and observers, public figures, elders, and other elements considered having a sufficient knowledge of local traditional foods.

The data was analyzed by using descriptive qualitative method in which qualitative analysis of various traditional foods from 10 counties including snack names, ingredients and taste, processing and serving techniques was resulted. The factor of food production and culture of each region were also described as they influence the existence of traditional snack.

3. Result and Discussion

The areas of West Java snack mapping are divided into three regions: northern, central, and southern. The northern includes 4 districts namely Karawang, Purwakarta, Subang and Cirebon. The central also involves 4 districts namely Sumedang, Bandung, West Bandung, and Kuningan. Meanwhile, the southern involves 2 counties namely Garut and Tasikmalaya. The archipelago snacks typically from West Java regions, specifically from Garut and Subang, were described based on names, types, main ingredients, processing techniques, servings, and functions. The following is picture 1 mapping of traditional food of West Java.

3.1. The Analysis of Snack Types from West Java

According to picture 1; the research on snacks typically from West Java includes snack types, ingredients, and processing techniques. The snack types from West Java are classified into four groups: (1) traditional snack, (2) popular snack (from outside influence), (3) modification snack, and (4) creation snack.

The traditional cakes are made from local comestibles, hereditary, use local tools and local cooking techniques, and relate to traditional ceremony and thanksgiving. In general, traditional snacks such as cucur, ali agrem, ongol-ongol, combro, and misro from all regions have similarity, making difficult to search for the origin. The researchers found a snack which has same name but different ingredients and processing technique. It names gemblong which in Bandung and other regions is made from glutinous rice flour and coconut, formed oval, fried then caramelized with brown sugar. However in Kuningan, gemblong is made from grated cassava and sweet potato which is formed then fried, so the texture becomes solid and crispy. Nevertheless, some snacks are the same but have different name such as the so-called apem in Bandung but in Kuningan, Cirebon, Karawang it is called cimplo which is served for rebo wekasan tradition. Other examples are kicimpring and bandros. Bandung and Garut calls kicimpring (a kind of cracker made from cassava), but Tasikmalaya names it beca. Bandros (a baked snack made from rice flour and grated coconut) is well-known in Bandung and Garut, but in Karawang and Cirebon it is called gonjin.

The popular snacks got influenced from outsiders such as Europe, China, Japan, Korea, and other Indonesian areas. The fact makes West Java snacks more various such as risoles , croquette, cake, cookies (European influence), kue mangkok, moci, moho, lumpia, siomay, kue ku, bakpau (Chinese influence), putu, getuk, carabikang, lemper, sosis solo (Javanese influence), dorayaki, moci (jepang), and toek (Korea).

The modification snack is an alteration of traditional one by adding or lessening the form, some ingredients, and color so the new snack is created and usually to targeting market. For example, initially serabi just has oncom (by-product of processed soybean production such as tempe and tofu) and egg as the topping, but currently it has hundred topping variations. In addition, simping (a flat cracker made from rice flour) initially just has one flavor – kencur (one of aromatic tropical rhizomes), but now it has cheese and sesame flavor too. Cireng (a half-fried starch snack) in the past was not popular, but now it has many variations including cireng rujak and cireng with various fillings.

Furthermore, the examples of snack as our people creativity in creating new recipes and developing foods are batagor, gehu, bala-bala, cilok, comet, moring, tahu pedas (spicy fried tofu), seblak, and cipuk.

3.2. The Analysis of West Java Snacks

According to picture 1; to make West Java snacks, various glutinous rice types are used as the main ingredient in each region. Citra, Jengkol, and Lusi glutinous rice are used to make opak, while Ranggeuyan, Lusi, Paris, Bangreng, Galuh, and Seungit glutinous rice are stored for 3 to 6 months to make rengginang. Freshly grounded Paris or Lusi glutinous rice is good to make cake such as ulen and lemper. The main ingredient of cakes ali agrem, cuhcur, burayot, and awug is freshly grounded Torondol, Pandan Wangi, Cianjur, or Rojolele rice, so cake that is fragrant, soft, and sticky can be resulted. Similarly, to make kelepon, bugis, gemblong, jalabria, and others better use freshly grounded glutinous rice. However, ready glutinous rice flour on the market less results on optimal quality because of some additives such as whitening, anticaking agent, and flour treatment agent impacting on cake texture and taste.

Chip making in each region uses various bananas such as kastroli, tanduk, bawel dan nangka muda, resulting on crispy, tasty, and yellowish product. Meanwhile, ambon banana and ripe siam banana are used to make sale pisang. Singkong mentega (a kind of cassava) is used to make combro and misro. Furthermore, to make kue obi, ripe sweet potato and cilembu sweet potato can be used.

Archipelago snacks use a mixture of palm sugar, coconut sugar, granulated sugar, and a little bit salt to get a sticky and delicious taste.

Figure. 1. Mapping of traditional food of West Java

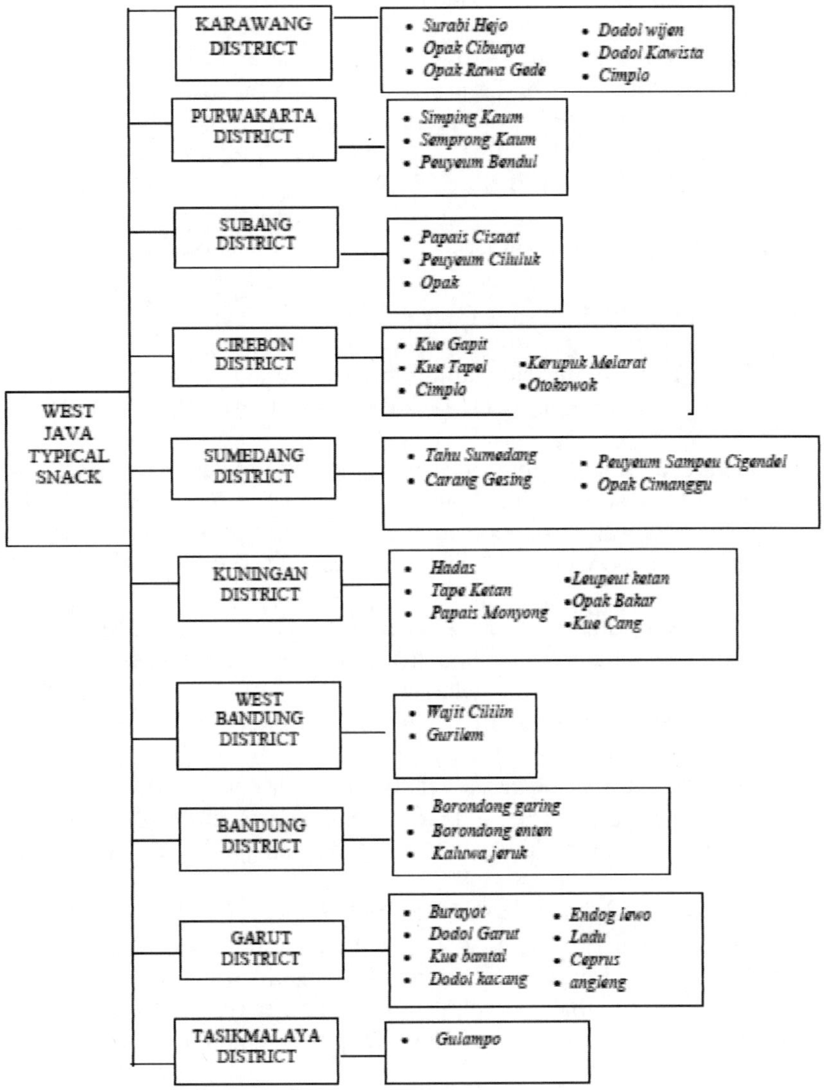

Figure. 2. West Java Snack Mapping

3.3. The Analysis of Processing Techniques of West Java Snacks

According to Figure 1, analysis of processing techniques of west java, firstly, frying is one of processing techniques of West Java snacks. It is also specifically based on the dough. For example, in making cuhcur, when the dough is put into the pan filled with hot oil, bamboo chopsticks are stick into the dough while flushing the oil onto it. Frying burayot needs dexterity and eye sharpness. After 10 seconds the dough is put into the pan filled with hot oil, it must be inverted while stick by bamboo chopsticks in order to form gelanyut (clinging form). In kremes making, when sweet potato is put into the frying pan and being 80% cooked, the sliced brown sugar is added while stirring the ingredients. The admixture is taken up soon then shaped. Furthermore, rengginang is fried within very much hot oil then the oil is flushed onto its surface. Thus, the glutinous rice cracker becomes extended. Meanwhile, frying endog londok or telur gabus is different. Before heating the oil, the dough should be put into it. If the dough is first put into the boiling oil, it will explode and the hot oil will spatter.

IOP Conf. Series: Materials Science and Engineering **128** (2016) 012051 doi:10.1088/1757-899X/128/1/012051

Steaming technique is used in making kue mangkok (Eng. bowl cake), bolu kukus (Eng. steamed cake), and kue moho. By steaming, the cakes can be extended. The steamer lid should be covered by cloth; thus, the fumes would not drop onto the cakes. Moreover, it should not be opened along the steaming process and the fire must be large. On the opposite, when steaming bugis the steamer lid should be opened occasionally in order to avoid the shape extended.

In boiling technique, duration should be noted. For example, boiling kelepon dough has no exact time but the indication of being cooked is the floating dough. If it is too long neglected, kelepon will be broken.

In caramelizing technique (Ina. membesta), sugar thickness should be noted starting from it being cooked until hairy. It means when sugar is run down, it is not broken off. In the case of making surabi, apem, kue mangkok, and carabikang, the admixture must be tapped 20 times in order to make them stringy and hollow. Furthermore, roaster technique has two ways: (1) by using sand as in making kerupuk melarat (Eng. slummy cracker) and kolontong, and (2) by direct roasting using pan or crock as in making kelapa sangan (roaster coconut), kacang sangan (roaster peanut), and borondong putih (white popcorn).

According to picture 2 ; each region of West java has typical snack. Karawang has surabi hejo which is aromatic as the result of rice flour-pandan leaves-sweet coconut milk and durian sauce mixture. Opak cibuaya and opak rawa gede are rounded, puffed-up, crunchy, and tasty. Dodol is made from glutinous rice flour mixed with sesame and kawista fruit, formed as cylinder with 15 cm long and the ends are tied.

Purwakarta has simping kaum and peuyeun bendul. The so- called simping kaum comes from kaum area and Purwakarta square. The recipe, taste, and packing are passed down from generation to generation. Simping kaum has 10 cm diameter, 3 mm thickness, kencur herb and coconut milk aroma, and savory crunchy texture. Peunyeum bendul is resulted from fermented cassava originally produced in Bendul village. It is solid, white, sweet, and sold by hanging.

The most popular snack from Subang is papais Cisaat. It comes from Cisaat area and is inherited from Dutch colonialism era. Pais Cisaat is uniqe. It is made from rice flour, coconut, and sugar, has sticky texture like dodol and sweet- savory flavor, and is covered with bamban leaves. It is not sold daily but can be ordered.

Cirebon has many typical snacks; one of them is kue tapel. Kue tapel is also inherited. It is made from glutinous rice flour which is added with grated coconut. Then, it is flattened on a thick pan, added with grated coconut and pisang raja (a type of banana) which is finely mixed with the dough, dried, taken up, and folded in a half just like crepes. Another typical snack is kerupuk melarat (Eng. slummy cracker) made from starch which is kneaded, steamed, sun-dried, and then roasted by using sand.

One of the most popular snacks from Sumedang is tahu Sumedang (Sumedang tofu). It is a product of processed soybeans, which is molded and crispy fried. The characteristic of the delicious tahu Sumedang is that the inside is empty and the outside is crispy. Besides, peuyeum Cigendel is another popular snack from Sumedang. It is made from fermented cassava, solid in the outside but soft in the inside, white, and sweet. Usually it is sold by hanging. Furthermore, Sumedang has two kinds of opak – opak cimanggu and opak oded. Opak Cimanggu has 5 cm diameter; it is puffed-up and tasty. Meanwhile, opak oded has 3 cm diameter, 0.5 cm thickness, crunchy texture, tasty flavor, and is grilled. Carag gesing is made from pisang ambon (Eng. ambon banana), added with wheat flour and coconut milk, covered by leaves, and then steamed. It is better served together with bajigur (a West Java traditional drinking made from palm sugar and coconut milk) or bandrek (a West Java traditional drinking made from ginger extract and palm sugar).

Kuningan also has many traditional snacks. Hadas is similar with awug. It is made from rice flour, grated coconut, and sugar; thus, it has sweet and savory flavor with soft texture. It is made by overlaying the admixture onto pan, given food coloring (green, red), steamed, and then cut. Commonly, it is served in a wedding ceremony. Tape ketan (fermented glutinous rice) from Kuningan is green, resulted from katuk leaves extract. Glutinous rice is fermented along one week; then, it is covered with rose apple leaves. The flavor is sweet and fresh. Tape ketan is commonly packed and sold in a black bucket. Kue cang is similar with bacang. It is made from selected glutinous rice branded Lusi or Citra but has no filling. Then, it is wrapped with bamboo leaves and served with grated coconut and sugar or kinca (palm sugar sauce). Papais monyong is made from glutinous rice flour, filled by crushed green beans or enten (a mixture of palm sugar and grated coconut) then covered with banana leaves in pyramid shape. The taste is sweet and tasty. Leupeut ketan can only be found in Kuningan. It is made from the mixture of glutinous rice and green beans poured into coconut leaf wrap then boiled. Tit has a long shape and tasty flavor.

The snacks typically from West Bandung are wajit Cililin dan gurilem. Wajit cililin is made from glutinous rice, brown sugar, and grated coconut. They are processed until well- cooked, wrapped by using corn leaves in pyramid shape. Meanwhile, gurilem is a type of cracker made from starch by roasting it with sand and formed in a long cylinder shape.

The typical snacks from Bandung District are borondong garing and borondong enten. Borondong garing is made from roasted rice mixed with thick palm sugar then molded soon after the mixture is taking up from the pan. Meanwhile, borondong enten is made from popcorn stick on enten. Another snack is kaluwa jeruk (kaluwa orange). It is made from grapefruit peeled skin which is soaked into whiting solution then boiled with sugar until the water disappears and finally sun-dried.

One of typical snacks from Garut is burayot. The snack is called burayot because of the clinging shape in which the top part is thin while the bottom is thick in a cone shape. It is made from rice flour and palm sugar sauce then fried. Another snack is the popular dodol garut which is droughty outside but watery inside. It is made from glutinous rice flour and sugar. Kue bantal (Eng. pillow cake) is made from glutinous rice flour which is directly sieved above hot pan then filled with enten and folded as pillow. Unfortunately, the snack is almost extinct. Endog lewo is made from grated cassava formed in a round shape just like marbles then fried. It is tasty and solid. Nowadays, the dough is mixed with starch in order to get crunchy texture. Ladu is originally from Malangbong area. It is made from roasted glutinous rice and added with sugar and enten (a mixture of grated coconut and brown sugar). All ingredients are cooked, molded, and then rolled into roasted glutinous rice flour so the shape is long. Angleng is another typical snack from Garut. It is similar with wajit but the main ingredient is different. Angleng is made from rice flour and wrapped with corn leaves in a pretty puffed square shape. Ceprus is produced in palm sugar producer areas. It is made by adding peeled cassava into cooked palm sugar liquid. In Sumedang, it is called wedang sampeu.

The typical snack from Tasik is gulampo which specifically comes from Cikalong area. It has a unique shape; it is small and wrapped with barangbang leaves in a square shape then tied up per 10 pieces. Gulampo is made from grated coconut and sugar. All ingredients are cooked until becoming like hair then tiny molded. The flavor is savory and sweet.

4. Conclusions

Besides the typical type of each region, the traditional snack types from West Java have similarity. They can be classified into four groups: (1) traditional snack, (2) snack which got influenced from Central Java, China, Europe, Japan, Korea, and the middle east, (3) modification snack, and (4) creation snack. The ingredients used are local agricultural products including rice, local glutinous rice such as Lusi, Bangreng, Jengkol, Citra, Paris, and Galuh, rice flour, glutinous rice flour, hunkue flour, tapioca flour, cassava, sweet potato, many kinds of banana (roid, kastroli, ambon, siam, nangka), peanut, soybean, green bean, red bean, gadung, taro, corn, palm sugar, coconut sugar, sugar, coconut, coconut milk, and other additional ingredients. The processing techniques include frying with some specific steps, steaming, grilling, roaster, sautéing, caramelizing, and boiling.

Acknowledgment

The authors would like to thank the Directorate of Higher- Education, Ministry of Education and Culture of the Republic Indonesia, which has provided funding of this research.

IOP Conf. Series: Materials Science and Engineering **128** (2016) 012051 doi:10.1088/1757-899X/128/1/012051

References

[1] Department of Education and Culture. (1985). Makanan: Wujud, Variasi dan Fungsinya serta Cara Penyajiannya daerah Jawa Barat: Proyek Inventarisasi dan Dokumentasi Kebudayaan Daerah Departemen Pendidikan dan Kebudayaan.

[2] Alim A. (2004). Mutu citra rengginang berbasis beras aromatic dengan metoda pengeringan berbeda. FT IPB.

[3] Herminiati, A.(2009). Inovasi teknologi proses pembuatan opak dari beras ketan di Jawa barat. A proceeding of workshop of grassroot innovation.

[4] Juwaedah, A. (2013). Model pembelajaran transmisi makanan tradisi berbasis indigeneous learning. A Research Report of Food and Nutrition Education Study Program of FPTK UPI.

[5] Sari, N. (2012). Alat Makan Orang Sunda Masa Kini. The Journal of Sundanese Study Center Bandung.

[6] Sari, N. (1997). Aku Cinta Makanan Indonesia. Jakarta: PT Gramedia Pustaka Utama.

Conceptual Design of Passive Safety System for Lead-Bismuth Cooled Fast Reactor

A G Abdullah, A B D Nandiyanto

Department of Electrical Engineering
Faculty of Technical and Vocational Education, Universitas Pendidikan Indonesia

Corresponding author: ade_gaffar@upi.edu

Abstract. This paper presents the results of the conceptual design of passive safety systems for reactor power 225 MWth using Pb-Bi coolant. Main purpose of this research is to design of heat removal system from the reactor wall. The heat from the reactor wall is removed by RVACS system using the natural circulation from the atmosphere around the reactor at steady state. The calculation is performed numerically using Newton-Raphson method. The analysis involves the heat transfer systems in a radiation, conduction and natural convection. Heat transfer calculations is performed on the elements of the reactor vessel, outer wall of guard vessel and the separator plate. The simulation results conclude that the conceptual design is able to remove heat 1.33% to 4.67% from the thermal reactor power. It's can be hypothesized if the reactor had an accident, the system can still overcome the heat due to decay.

1. Introduction

During steady-state of reactor operation, time-independent of temperature distribution, all of the generated heat will be removed naturally (Lamars et al., 2001). Cooling concept is one of the most critical part in the reactor design. The material selection of the fuel, cooling, cladding and reactor vessel should be the main consideration for reactor design. It must be far below its melting temperature. In case of core reactor melt down, cladding will be ruptured then fission product will be released.

Most of the nuclear power plant, i.e. 2nd generation, which operate nowadays still use active safety system to eliminate the decay heat. In this type of system, pump plays an important role to maintain reactor core temperature after shutdown. Fukushima accident have shown the minor point of the system when energy supply to pump was stop. The conceptual design of next generation reactor, i.e. 3rd generation, is relaying on passive safety system, where natural circulation is done by convection in the primary coolant. The major point of the systems: the ability to cool down reactor core without energy supply to pump, high reliability, simple and economist [1].

In this study, simulation of the design system of the passive safety system, i.e. Reactor Vessel Auxiliary Cooling System (RVACS) in 300 MWth of Pb-Bi coolant fast breeder reactor, will be analyzed. Utilization of liquid Pb-Bi allows the primary system to operate at high pressure condition, and if coupled with secondary system, it will be able to counterfeit the risk of LOCA (Loss of Coolant Accident) [1].The optimization is done to obtain high heat transfer value, thus the RVACS will give good performance on removing heat. Optimization is done on two aspect of the design: the properties aspect, which includes emissivity effect of guard vessel and perforated plate, also the effect of viscosity and air conductivity.

2. Reactor Vessel Auxiliary Cooling Systems (RVACS)

RVACS is a passive safety system in the liquid metal reactor that capable of maintaining system condition during accident [2]. The heat removal mechanism is conducted by natural circulation that passes through guard vessel air window and the wall surrounding the guard vessel. Heat from the reactor is transferred to the atmosphere via multiple stacks which connected by common inflow and outflow plane.

The RVACS performance is a function of pressure difference between airflow inlet and airflow outlet, inlet air temperature, variation air density along the path and pressure drop characteristic in system. The inlet and outlet pressure difference, and inlet temperature of RVACS are also influenced by air velocity and direction [3]. Reactor design is consist of vessel, interior and exterior of reactor vessel, which are separated by liquid metal gap. Design of RVACS can be seen in figure 1 [4]:

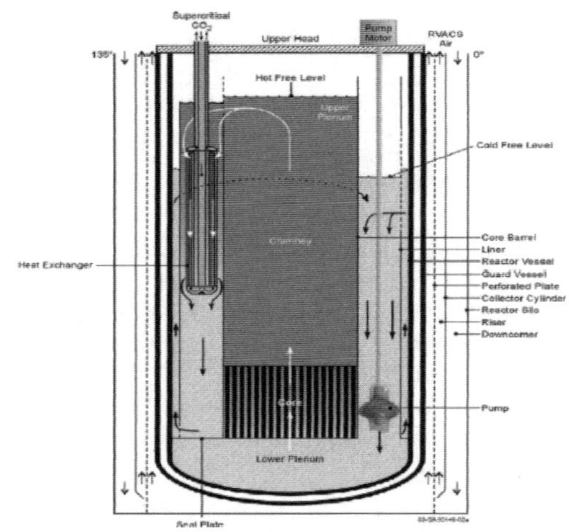

Figure 1. Pool type reactor

Heat is removed by air flow from the down comer to upper part which is passing through air riser and flow to the atmosphere via chimney. This mechanism is preceded by buoyancy force due to air density change. Heat is moved by air convection on two surfaces: outer wall of guard vessel and separator. Heat is transferred to separator from guard vessel wall by thermal radiation. And then, the heat is transferred by conduction via vessel and guard vessel.

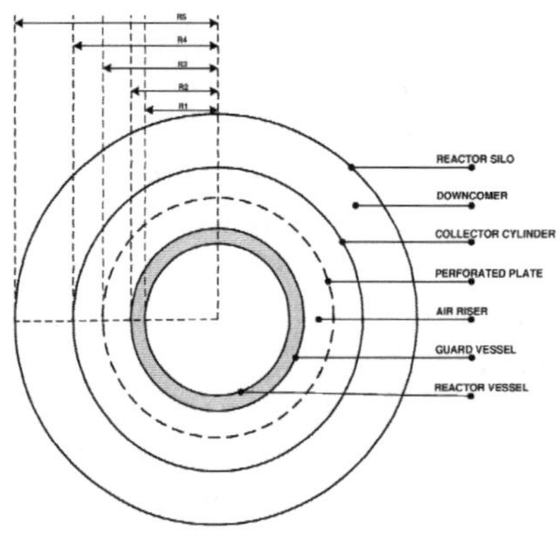

Figure 2. Cross section of RVACS

IOP Conf. Series: Materials Science and Engineering **128** (2016) 012052 doi:10.1088/1757-899X/128/1/012052

Figure 3 shows the RVACS geometry. In RVACS design, separator element is added on the surface of guard vessel (do not directly contacted). The purpose of the separator addition is to decrease heat on convection process. However, in some calculation the addition separator element is sometimes neglected.

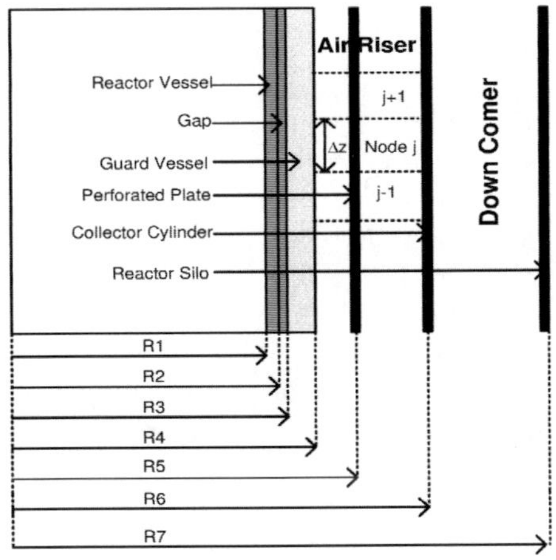

Figure 3. RVACS Geometry

3. Computation Methods

The RVACS design uses several assumptions: Analysis is conducted under steady state; Vessel and guard vessel temperatures are assumed to be constant, azimuthally; Separator is perfectly isolated; Heat convection coefficient (h) in riser is azimuthally constant; All of the inlet and outlet pipes are perfectly isolated; The heat convection between coolant and reactor wall are constant; Coolant temperature is linearly varies on axial direction; Heat that transfer via liquid metal gap is modeled in conduction model; Axial conduction and radiation is neglected.

According to fig. 3, calculation of heat transfer from guard vessel to separator is the radiation calculation:

$$\dot{Q}_{i\,us} = C_G A_G (T_G^4 - T_S^4) \tag{1}$$

A_G is the area of guard vessel surface, T_G and T_S are guard vessel temperature and separator temperature, respectively, while C_G is define as follows:

$$C_G = \sigma\left[\frac{1}{\varepsilon_G} + \frac{A_G}{A_S}\left(\frac{1}{\varepsilon_S} - 1\right)\right]^{-1} \tag{2}$$

Where σ d is boltzman constant, ε_G and ε_S are the emissivity of guard vessel and separator, respectively. The heat transfer from separator to air was modeled by the pure convection process:

$$\dot{Q}_{o\,us} = h_j A_S (T_S - T_{b\,u\,l})_i \tag{3}$$

Heat convection coefficient, h_j, is calculated based on [5]:

$$h_j = 0{,}0 \quad R2^{0,8}\, P\, 2e^{0,4} \frac{k_a}{D_h}\left(\frac{T_{Gr}}{T_a}\right)_i^{-0,4}\left[1 + \left(\frac{z_j}{D_h}\right)^{-0,3}\right] \tag{4}$$

Thus, total of heat transfer that pass through guard vessel is:

$$\dot{Q}_{out,G} = h_j A_G (T_G - T_{bu}) + \dot{Q}_{ia,G} \quad (5)$$

The heat transfer to surface of outer guard vessel from primary cooling system was modeled by conduction process. The calculation was done by serial resistance between cooling temperature and guard vessel outer wall temperature.

$$R_{TO} = R_C + R_V + R_G + R_A \quad (6)$$

Where:

$$R_C = (h_C \pi D_1 \Delta z)^{-1}$$
$$R_{cylinder} = \frac{\ln(D_{ou}/D_{in})}{\pi k \Delta x} \quad (7)$$

k is thermal conductivity of the material that fill the gap between reactor vessel and guard vessel. D_{in} and D_{in} outer diameter and inner diameter of the cylinder, respectively. Thus, the heat transfer that enter the guard vessel is:

$$\dot{Q}_{inG} = R_{TO}^{-1} (T_C - T_G) \quad (8)$$

4. Result and Discussion

The referred design that was used in simulation is shown in table 1. Analysis was covered most of the coolant temperature. If the result of coolant shows temperature below maximum of cladding temperature, then the design is assumed to be done successfully. This because power level decay that passes through cladding is very small and there will be a few thermal resistances between cladding and primary coolant.

Tabel 1. Main parameters of RVACS design

Reactor power	225 MWth
Coolant	Pb-Bi
Fuel	UO_2-PUO_2
High of RVACS	40 m
Downcomer gap	5,86 m
Radius of reactor vessel	4,62 m
The Thickness of reactor vessel	0,4 m
The Thickness of guard vessel	0,1 m

The simulation was started by calculating total of heat transfer on referred design. It was found that RVACS is capable of eliminating heat of about 4.39 MWth or about 1.95% of total thermal power of reactor. Heat is eliminated in RVACS system by natural circulation from the static atmosphere around reactor vessel. The result of the simulation is shown in figure 4.

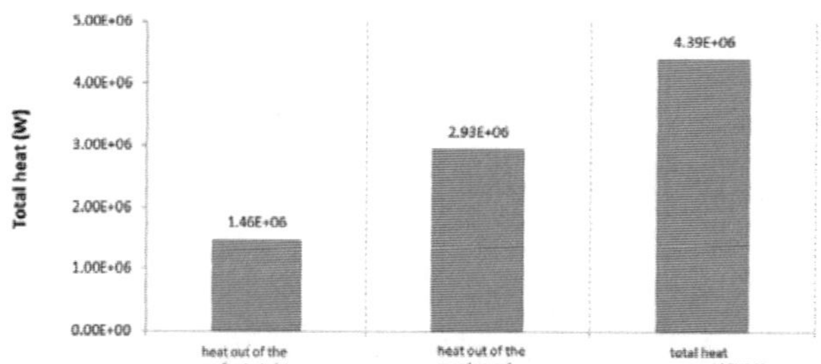

Figure 4. Total heat removal of RVACS reference design

Figure. 5 and 6 show change of the RVACS performance due to optimization of guard vessel emissivity, perforated plate, viscosity and air conductivity. Performance of RVACS increased when guard vessel emissivity and perforated plate values were close unity. On the other hand, optimizations of viscosity and air conductivity indicated insignificant contribution to the power of RVACS heat transfer.

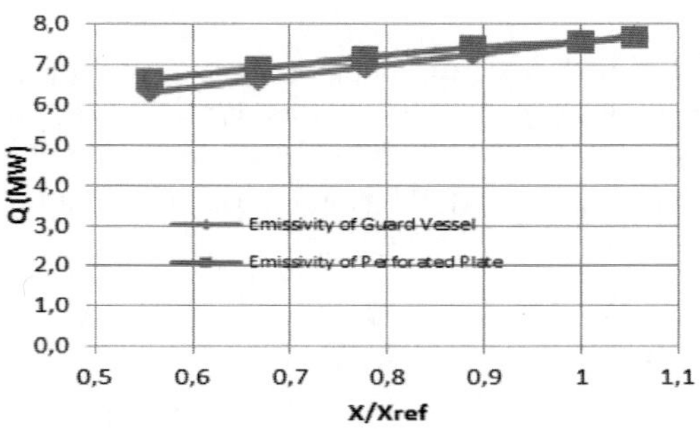

Figure 5. The effect of changes in emissivity value of guard vessel and the perforated plate

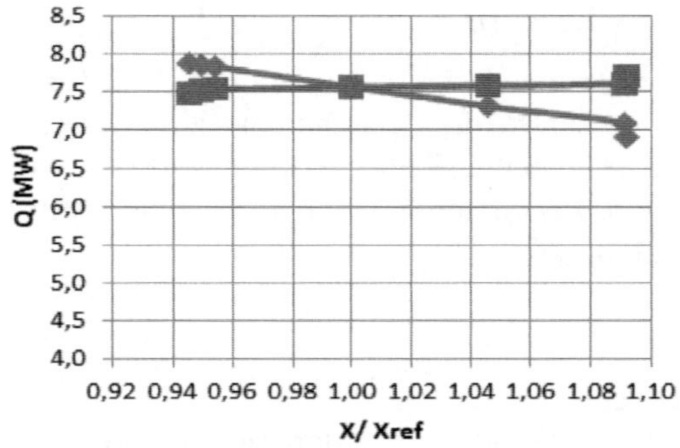

Figure 6. The effect of changes in viscosity and air conductivity

Figures 7, 8 and 9 show optimization results on RVACS geometry. Performance of RVACS was highly affected by optimization of RVACS height. While diameter optimization or gap channel gave less influence, this means that power of heat transfer is similar with simulation result on reference design.

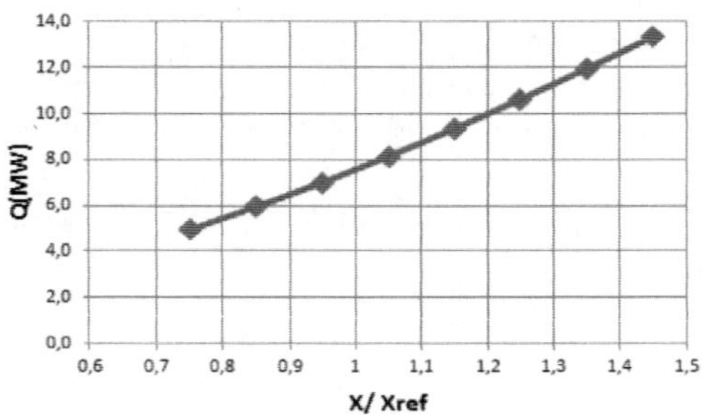

Figure 7. The change height effects of RVACS

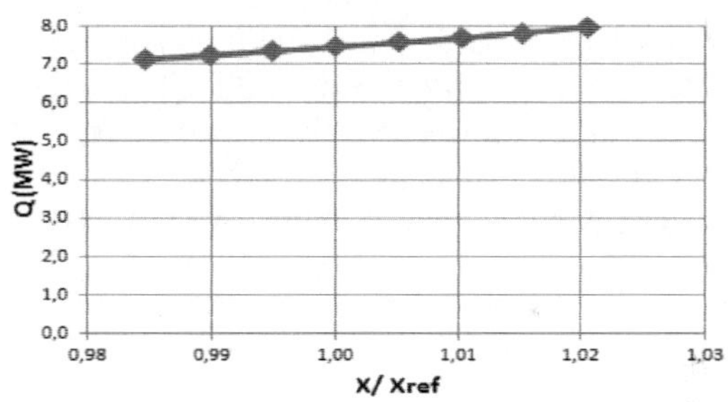

Figure 8. The change diameter effect of RVACS

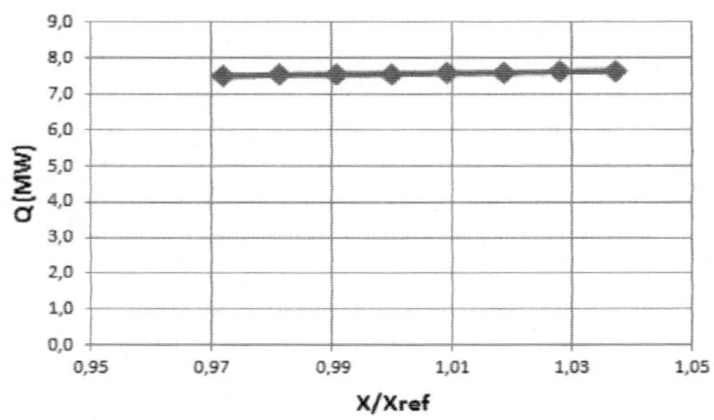

Figure 9. The change channel gap effect

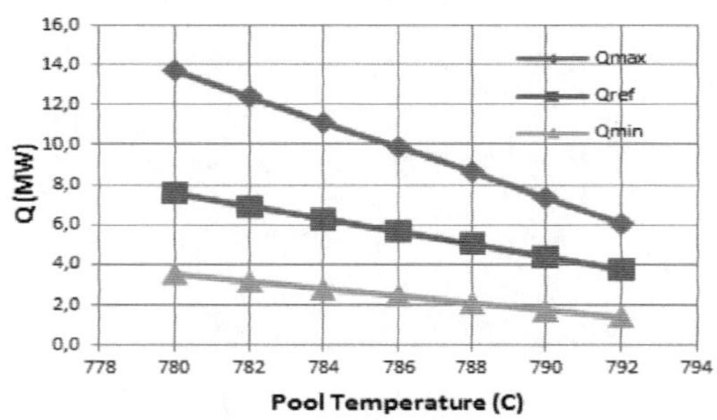

Figure 10. Heat removal capabilities of RVACS

IOP Conf. Series: Materials Science and Engineering **128** (2016) 012052 doi:10.1088/1757-899X/128/1/012052

Figure 10 shows combination of result of all optimization, i.e. minimum design, reference design and maximum design. Simulation result concludes that this theoretical design was able to eliminate heat from 1.33% to 4.67% of power thermal reactor.

5. CONCLUSSION

The conceptual design generates heat transfer value that met the requirement of natural circulation reactor design by the environment. The passive safety system of RVACS will be activated to remove decay heat under accident condition. The calculation results show that the level of heat removal is good enough to prevent temperature to reach very high value, when the RVACS has good and efficient heat removal system.

References

[1] Joshua Whitman, Design of Passive Decay Heat Removal System for The Lead Cooled Flexible Conversion Ratio Fast Reactor, Nuclear Science and Engineering, Massachusetts Institute of Technology, (2007).

[2] M. M. Razzaque, T. H. Ramin and B. I. Spinrad, A Passive Emergency Cooling System For An Advance LMR- The Trench Reactor, Ann. Nuclear Energy, Vol. 16, No 7, pp. 345 – 351, (1989).

[3] Hejzlar P., Buongiornoj., Macdonald P., And Todreas N.,Design Strategy and Constraints for Medium-Power Lead-Alloy-Cooled Actinide Burners, NuclearTechnology, 147, pp.321-343, (2004).

[4] Robert C. Petroski, Design of a 2400MW Liquid-Salt CooledFlexible Conversion Ratio Reactor, Nuclear Science and Engineering, Massachusetts Institute of Technology, (2008).

[5] J. Heineman, M. Kraimer, And P. Lottes, Experimental And Analitycal Studies of A Passive Shutdown Heat Removal System for Advanced LMR's, Argone National Laboratory, Illinois, (1988).

[6] Lamars, J. R. and Baratta, A. J., 2001, Introduction to Nuclear Engineering Third Edition, Prentice-all New Jersey,ISBN-13: 978-0201824988, pp: xx

The effect of modified potato flour substitution on the organoleptics characteristics of toddler biscuit

Dewi Cakrawati*1, Puji Rahmawati2

Agro-industrial Technology Education Program, Faculty of Technology and Vocational Education, Universitas Pendidikan Indonesia

E-mail : [1]dewicakrawati@upi.edu, [2]pujirahmawati@upi.edu

Abstract. Toddler biscuit is complementary food given to infants to help meet their nutritional needs. This research was undertaken to develop toddler biscuit with subtitution of physically modified potato flour. There were two puposes of the research, first to know the characteristics of physically modification on potato flour; secondly to know biscuit characteristics with modified potato flour substitution. There were two factor analysis in the development of biscuit; first factor was pre heating mehods and substitution rate. The research was conducted with experimental method using split plot design. The functional properties on modified potato flour as swelling capacity, water absorption capacity, solubility and viscocity were analyzed. Organoleptic analysis using quality hedonic test showed no interaction between potato starch modification and concentration of modified potato flour. Quality hedonic test showed all toddler biscuit socred in the range of "slightly like" to "like moderately". Modifying potato starch by boiling and steaming with flour concentration of 30% producing toddler biscuit with organoleptic characteristics acceptable according to the panelists.

1. Introduction

Toddler biscuit is complementary food (solids) given to children from the age of 6 months to help meet the needs of energy and other nutrients. The biscuits should contain the nutrients needed by toddlers, including carbohydrate, proteins and fats, and micronutrients such as vitamins and minerals. Sources of carbohydrates are usually obtained from brown rice or green beans. Wheat has not been widely used for fear of babies' allergic reaction to the gluten contained in it. Sources of protein can be obtained from milk, while the source of fat from margarine or butter. The addition of milk also helps meet the needs of minerals in infants. Some experts say toddler biscuit can help train toddlers' motor nerves, especially their ability to hold something and to stimulate the process of chewing.

Biscuit is a food product made of flour with a diameter of 6-10 cm with a square or round shape, which is made by roasting [1]. Biscuit dough does not have to be inflated, so that the flour used as a raw material can be substituted by other materials to add nutritional value or even reduce the consumption of wheat. Some toddler biscuit manufacturers in Indonesia have begun to use materials other than arrowroot flour such as tubers, because it has better digestibility, or flour banana to give flavor and increase the fiber content of food.

One of the potential ingredients from Indonesia is potato tuber, whose production has continued to increase year by year. Potato production in Indonesia in 2010 reached 1,060,805 tons. Potatoes have many sources of species diversity, consisting of a variety of local species and some superior varieties.

Potato starch contains more than 12.5% resistant starch. According to reference number [2], potatoes are a source of resistant starch type 2, i.e. starch that is naturally present in plants. Resistant starch cannot be digested and absorbed in the small intestine of healthy individuals and is resistant to hydrolysis enzyme amylase.

According to reference number [3], the potato is one food that contains phenolic compounds so susceptible to enzymatic browning when experiencing physical treatments such as peeling, slicing, or bumping. Some ways to prevent browning is to avoid contact with oxygen, using a metal chelating compound or inactivatingpolifenoloxidaseenzymes by means of heating. According to reference number [4], enzymatic browning occurs when food containing phenolic compounds undergoes enzyme polyphenol oxidase wounding, so the substrates of the phenolic compounds produce brown quinone compound. The presence of oxygen and metal ions accelerates the enzymatic browning. Reference number [5] shows that the potato variety of granola that is extracted using distilled water by 10% contains resistant starch of approximately 56% of the amount of the starch. According reference number [6], warming causes changes in the type of potato resistant starch in potatoes of type 2 and type 3. In terms of quality, resistant starch type 3 is more resistant to digestive enzymes that act better as a prebiotic.

Based on the above explanations, the researchers intend to conduct research on the effect of pre-cooked potato starch on potato biscuit-making through experimental design method using Split Plot Design, focusing on two things: modification of potato starch and the substitution rate of potato flour used.

2. Research Method

This research was carried out for 8 months since the signing of the contract in the Laboratory of Study Program Education on Agro-Industry Technology, Faculty of Technology and Vocational Education, UPI. Materials used are potato that bought from local farmer, distilled water, butter, milk, sugar, flour were procured from local market.

2.1. Research Procedure

1. Modified potato flour
   The first phase of this research was making modified potato flour using pre heating. The treatment consisted of boiling, steaming and baking. The potato then was mashed into small particle and put in the refrigerator with a temperature of 40°C for 24 hours. After about 2 hours, the potato wasthendried using electric oven with a temperature 50°C until the weight became constant.
2. Making Potato biscuit
   The making of potato biscuit was a modification of the method proposed by Herudiyanto and Hudaya (2009).
3. Organoleptic Analysis
   Organoleptic analysis of modified potato biscuit was performed using hedonic quality test with a scale of 1-5, where a value of 1 is very good, 2 is good, 3 is rather good, 4 is rather good, and 5 is very good. The panelists were 15 semi-trained people in the field of toddler biscuits.

2.2 Research Design

The study design employed a Split Plot Design with 2 treatments and 3 replications. The first treatment was the modification of starch consisting of three levels: (a1) boiling (a2) steaming, and (a3) baking. The second treatment was to find the level of substitution modifed potato flour covering 4 levels, namely 10% (b1), 20% (b2), 30% (b3), 40% (b4). If the calculated F value is greater than the interaction of the F value in the distribution table at the level of 5%, then Test Least Square design will be conducted. However, there is no interaction Duncan's multiple range test will be carried out.

3. Results and Discussion

New product development is a strategic area of the food industry, where currently consumers desire food products with two functions: To meet the nutritional needs and to have added value for health. Such food product is known as functional food [7]. According reference number [1], the definition of biscuits can refer to the definition given by Samuel Johnson, in which biscuit means a food made from good quality flour with the addition of sugar that has a hard texture with a longer shelf life.

3.1 Variance Analysis of Potato Biscuit Colour

Results of analysis of variance showed that there was no interaction between the substitution rate of starch and modification of starch. Test results of the influence of substitution rate of starch and modification of starch individually on toddler cookies' colour presented in Table 1.

Table 1. The effect of flourmodification method and substitution rate on Potato Biscuit color

Treatment		
Modification methods	Average	Notation
Boiling	3.74	a
Steaming	3.71	a
Baking	1.914027	b
Substitution rate	Average	notation
10%	3.55	a
20%	3.17	a
30%	2.87	a
40%	2.69	b

Description: The average treatment marked with the same lowercase letters is not significantly different at the 5% significance level according to Duncan's multiple range test.

One of the organoleptic characteristics that influence consumer acceptance is the color. Golden brown color indicates that the biscuits are well-cooked. According to reference number [4], brownish yellow color is a result of caramelization process because of the sugar and flour in biscuit dough. The addition of potato starch affects the color of biscuits because, as Eskin (2000) has noted, the potato is one commodity with a high content of phenolase enzyme so susceptible to browning. Reference number [7] added that one way to prevent enzymatic browning in potato is the provision of heat that would inactivate the phenolase enzyme. In addition, according to reference number [6], warming will change the potato's resistant starch into resistant starch type 2 type 3, that is more resistant to treatment. Toddler biscuit has moisture level of 4.02±0.06%. This is consistent with slow water content of resistant starch digested from banana cookies based on research by reference number [8] which has a water content of 7.7%. According to reference number [4], the color can also be affected by the biscuit dough characteristics, including pH, water content, the presence of sugar and amino acids, as well as temperature and baking time.

According to reference number [1], biscuits have a low water content so as to provide a crunchy texture effect and inhibit the growth of microorganisms, so that the biscuits have a long shelf life. Ash content of food is affected by the mineral content of the food material. The results show the level of flour substitution was higher than the ash content. It is thought to be influenced by the mineral content in potatoes and wheat used. To produce good and even colour, it is most preferable to use an electric oven that has a stable and measurable temperature. According to reference number [7], the process of heat transfer during roasting takes place by convection, conduction and radiation, with a small size biscuit, the heat on the surface will be the same as that in the core of biscuits. This will lead to the formation of brown colour almost evenly on the biscuit with the highest colour intensity on the outer part of the cookies.

3.2 Variance Analysis on Potato Biscuit Taste

Potato flour does not have a distinct flavor, so the addition of butter and milk helps improve the flavor of the biscuit. Butter and milk provide the savory flavor for the biscuits and provide additional nutrients of fat and protein. Differences in the composition of the biscuit dough affect the fat content and proteinof the biscuits produced. The addition of butter and milk causes potato biscuits to have higher levels of fat and protein content, which is higher than the slowly digested cookies in the research of Aparacio-Saguilan et al. (2006), where the fat content was 12.7 ± 0.1% and protein levelof 5.09 ± 0.1%. The addition of milk may also cause the Maillard reaction which will affect the resulted color biscuits.

Results of analysis of variance showed that there was no interaction between the substitution rate of starch and modification of starch. Test results of the influence of substitution rate of starch and modification methods of starch individually on the flavor of toddler cookies are presented in Table 2.

Table 2. The effect of flour modification method and substitution rate on Potato Biscuit Taste

Treatment		
Modified methods	Average	notation
Boiling	3.638596	a
Steaming	3.610621	a
Baking	2.540871	a
Substitution rate	rata-rata	notation
10%	3.75543	a
20%	3.26761	a
30%	3.087653	a
40%	2.896013	b

Description: The average treatment marked with the same lowercase letters is not significantly different at the 5% significance level according to Duncan's multiple range test.

Table 2 shows that the substitution of modified potato flour and flour level substitution effect was not significantly different in terms of their effects on potato biscuit taste. This is in line with the finding of Aparacio-Saguilan et al.(2006) which showed that the flavor of cookies with and without the addition of resistant starch was not significantly different. Cookies with the addition of35% resistant starch was still acceptable for the panelists, although the highest level of acceptance of the cookies is an addition of 15% resistant starch.

3.3 Variance Analysis on Potato Biscuit after Taste

Results of analysis of variance showed that there was no interaction between the way the flour is modified and the potato flour substitution rate. Test results of the influence of flour modification methods and flour substitution rate on toddler biscuits after-taste are presented in Table 3.

Table 3 The effect of flour modification method and substitution rate on Potato Biscuit after-Taste

Treatment		
Modified methods	Average	notation
Boiling	3.418946	a
Steaming	3.412443	a
Baking	2.633714	a
Substitution rate	Average	notation
10%	3.651002	a
20%	3.174184	a
30%	3.011018	a
40%	2.773019	a

Description: The average treatment marked with the same lowercase letters is not significantly different at the 5% significance level according to Duncan's multiple range test.

Table 3 indicates that the effects of substitution of modified potato flour and the level of substitution on potato biscuit after-taste were not significantly different. The after-taste can affect the level of consumer acceptance. According to Waring (2005),resistant starch content of the potato can help strengthen the butter flavor; it can also affect the after-taste of potato biscuits. This buttery nature is often used in the manufacture of low-fat food products, in which the resistant starch is used to reduce the use of fat.

3.4 Variance Analysisof Potato Biscuit Crispiness

Results of analysis of variance showed that there was no interaction between the ways starch is modified and the substitution rate of the starch. In addition, the results of the analysis of each factor showed no significant results that further analysis is not needed.

Crispiness is one of the characteristics that greatly affect the level of consumer acceptance of cookies. The level of biscuit crispiness in terms of the substitution of modified potato starch ranged from grades 3-4, meaning that the panelists could still receive the crispiness of the biscuits. According to Waring (2000), one of the ways to ensure the public are getting enough dietary fiber intake is to forficate dietary fiber in foods, where dietary fiber resistant starch groups can help improve the texture and crispiness of the biscuit. Crispy biscuits are influenced by the raw materials used.As put forward by Manley (2000), the use of butter, in addition to affecting the flavor, increases crispiness. The butter to be used should be frozen, so that during the mixing the batter will spread evenly and melt when baked. This method will make a crunchy biscuit. The process of rolling and folding the dough biscuits will also affect the crispiness of biscuits.

4. Conclusions

1. Potato biscuits with 30% substitution of flour modification characteristic shave acceptable organoleptic characteristics according to the panelists.
2. Potato flour biscuits with the addition of physical modification at30% substitution level contained fat and protein for 28%and9%, respectively.

5. Acknowledgment

The work is funded by Dana DIPA (Bantuan Operasional Perguruan Tinggi Negeri - BOPTN) Universitas Pendidikan Indonesia Budget Year 2014 by virtue ofRector's Decree No. 3414/UN40/LT/2014

References

[1] Manley, D. 2000. Technology of Biscuits, Cracker and Cookies.Woddhead Publishing Company, Cambridge

[2] Herawati, H. 2010. Standardisasi Pati Termodifikasi untuk Produk Pangan. Prosiding PPI Standardisasi 2010 –Jakarta, 11 November 2010 (on-line) http//www.bsn.go.id (downloaded 18 Januari 2013)

[3] Eskin, N.A. 2000. Biochemistry of Food. Academic Press Inc, San Diego

[4] Villamiel M., M. D. del Castillo, and N.Corzo. 2006. Browning Reaction: Biochemistry and Food Processing. Y.H.Hui (ed). Iowa: Blackwell Publishing

[5] Sari, F.T.,Nurhayati, Djumarti. 2013. EktraksiPatiresistendaritigavarietaskentang yang berpotensisebagaikandidatprebiotic (Extraction of resistant starch of three potato varieties potential to be prebiotic candidates). *BerkalaIlmiahPertanian*, Vol 1, Nomor 2, pp. 38-42. on-line) diunduh http//www.pdf.unej.id (accessed 18 January 2013)

[6] Wulan, SN., E. Saparianti, SB Widjanarko, N. Kurnaeni. 2006. Modifikasi Pati Sederhana dengan Metode Fisik, Kimia dan Kombinasi Fisik-Kimia untuk Menghasilkan Tepun Pra Masak Tinggi Pati Resisten yang Dibuat Dari Jagung, Kentang dan Ubi Kayu (Modification of simple starch with physical, chemical, and combination of physical-chemical methods to produce pre-cooked high resistant starch made from corn, potato, and tuber). *Jurnal Teknologi Pertanian*, Vol 7 No. 1 pp. 1-9 (on-line) http//www.pustaka-litbang.go.id (downloaded 21 July 2012).

[7] Alsaffar, AA. 2011. Effect of food processing on the resistant starch content of cereal dan cereal product- a review. *International journal of food science and technology* 46 pp. 455-462.

[8] Aparicio-Saguilan A., S G Sayago-Ayerdi. A Vargas-Torres, J Tovar, T E Ascencio-Otero, LA Bello-Perez. 2006. Slowly Digestable Cookies Prepared From Resistant Starch-Rich Lintnerized Banana Starch. Journal of Food Composition and Analysis Volume 20, Issues 3-4 May 2007. Pp. 175-181 *periodicals* http://www.researchgate.net (retrieved 10 November 2013).

Contrast enhancement for satellite image segmentation with fuzzy cluster means using morphological filtering

Harsiti[1], T A Munandar[2], A Suhendar[1], A G Abdullah[3*], D Rohendi[4]

[1]Information System Department, Universitas Serang Raya, Banten, Indonesia

[2]Informatics Engineering Departement, Universitas Serang Raya, Banten, Indonesia

[3]Electrical Engineering Education Department, Universitas Pendidikan Indonesia, Bandung, Indonesia

[4]Mechanical Engineering Education Department, Universitas Pendidikan Indonesia, Bandung, Indonesia

*Corresponding author: ade_gaffar@upi.edu

Abstract. Image segmentation is a stage in image processing, responsible for dividing an image into regions homogeneous, based on the similarity. Suppose the grey level of a pixel by pixel gray neighbors. The quality of image segmentation is generally influenced by the characteristics and the handling of images to be processed. This paper presents the results of satellite image segmentation using a fuzzy cluster region means (FCM). To improve the contrast of the image, conducted morphological filtering techniques. Satellite image analysis performed on five districts in the province of Banten Indonesia. Difference's segmentation results evident when non-negative parameter value is converted to a 2 and 4. The higher the value of a non-negative parameter is given, then the details of the edges of objects clearer segmentation results. The combined use of a top-hat and boots-hat filtering on objects before satellite imagery analysed by FCM, indicating that it merges with the background object. Background object in the original image is the object of rice fields and is not part of observation in this study. It was identified to have the same gray level similarity with the object of building.

1. Introduction

Image segmentation is the process of pre-processing the image in order to perform the introduction of an object image. Image segmentation divides the image into regions homogeneous in accordance with its neighboring color pixel gray level [1]. Many methods are used for image segmentation among other intensity-based, discontinuity, similarity, graph, Pixon, hybrid methods and clustering [2]. In the discussion of other methods of image segmentation can be realized with some other approaches such as the histogram threshold, clustering (fuzzy and hard), region growing, region splitting and merging, edge-based, physical model- based, fuzzy approaches, neural networks and genetic algorithm [3], [4]. Previous studies have used cluster technique for image segmentation requirements [3], [5], [6].

Cluster technique is used because of its ability to discover the structure of a set of data that does not have a label [3]. In this case, the structure is known as the color pixels that have a different gray level with its neighboring pixels. In addition, the cluster technique also produces image segmentation that is capable of providing an adequate sense, uses little memory and is faster in performing image segmentation, especially for natural images [4]. On the other hand, the quality of the image will be segmented play a very important role to get better results. One technique for improving image quality is to modify the contrast level before segmental. The development of science in the field of image processing has produced many important approaches for improving the image contrast. Some methods such as histogram equalization, logarithmic image processing (LIP) [7], [8] is used to enhance the contrast in the image quality before entering the stage of image processing. However, the use of histogram's equalization not provide better image quality compared to LIP. Same thing with LIP, despite having results with good image quality, but takes a very long time to be able to produce better image quality [7].

To overcome this, it is proposed another approach to improve image quality by using morphological filtering (MF). The concept of MF has been widely used by some researchers to increase the level of contrast of the image before image processing step. The MF can improve image quality of the detection signal from the non-linear frequency modulated (FM) unknown, although with high noise [9]. Able also eliminates noise and interference in the process of retaining the basic geometric shape of the image [10], [11]. Has a fast process [12], filter out noise remaining [13]. MF can analyze an object's image perfectly [14].

2. Morphological filtering

The MF is a technique used to analyze the shape of the image. An MF works based on two main concepts namely erosion and dilation. Dilation works by adding pixels to the boundaries of objects in the image, while erosion work otherwise. In practice, the concept MF dilation and erosion are often used simultaneously. Morphological opening is the image that is based on the concept of erosion, followed by a dilation process. The Morphological closing is an image analysis process that begins with dilation, followed by erosion. This is then called by morphological filtering [15]. In the development of MF technique has several methods, including the top-hat and boots-hat filtering.

Tot-Hat technique is a method to filter the object image. This technique can reduce the results of the morphological opening (MO). Filtering methods other objects are Bot-Hat techniques. This method can show MO by reducing pixel of the original image of the input. Top-Hat filtering is defined as in equation (1).

For example: $f: E \rightarrow R$ (1)

An image to grayscale formatting, has a function to map a number of points based on the Euclidean distance, or discrete space E (example R^2 aorZ^2) to in the form of a line that is more real. If given b(x) as a structural element grayscale, then the top-hat transformation is given by the following equation:

$T_w(f) = f - f \circ b$ (2)

The operation is called white top-hat, where o is a morphological opening operation. Operation black top-hat is the opposite of white top-hat which is defined as follows:

$T_b(f) = f \bullet b - f$ (3)

Operator • is a morphological closing operation.

3. Fuzzy Cluster Means (FCM)

Fuzzy C-Means (FCM) was first introduced by Dunn and later modified by Bezdek [16] [17]. The algorithm works iteratively group data so as to form a C cluster membership by minimizing weight and distance data is based on the objective function.

$$F_{FCM} = \sum_{k=1}^{n} \sum_{i=1}^{c} (\mu_{ik})^q d^2(x_{k,} v_i) \quad (4)$$

$X = \{x_1, x_2, \ldots, x_n\} \subseteq R_p$ is the set of p-dimensional vector data, n is the number of data, c is the number of clusters with the provisions of $2 \leq c \leq n$, μ_{ik} is degree of membership x_k for cluster , and q is the weight of reappointment for each value of fuzzy membership, v_i is a prototype cluster center, and $d^2(x_k, v_i)$ is measured the distance between the data objects x_k and cluster center v_i. Here are the stages of FCM algorithm:

a. Determine the value of each parameter c, q and ϵ
b. Determine the partition initial matrix $U = [\mu_{ik}]$
c. Determine loop counter $b = 0$
d. Calculating of cluster center $\{v_i^{(b)}\}$ to $U^{(b)}$ by the equation :

$$v_i^{(b)} = \frac{\sum_{k=1}^{n} (\mu_{ik}^{(b)})^q \, x_k}{\sum_{k=1}^{n} (\mu_{ik}^{(b)})^q}$$

e. Determine the membership $U^{(b+1)}$. For $k = 1$ to n, count the value of $I_k = \{i \mid 1 \leq i \leq c,\ d_{ik} = \| x_k - v_i \| = 0\}$, $/I$; to matrix column k, determine the new membership function by the equation:
   i. If $I_k = \phi$ then

$$\mu_{ik}^{(b+1)} = \frac{1}{\sum_{j=1}^{c} \left(\frac{d_{ik}}{d_{jk}}\right)^{\frac{2}{(q-1)}}}$$

   ii. Else $U_{ik}^{(b+1)} = 0$ for all $i \notin I$ and $\sum_{i \in I_k} U_{ik}^{(b+1)} = 1$; next k;

f. If $\| U^{(b)} - U^{(b+1)} \| < \epsilon$, then clustering is completed. Other than , set $b = b + 1$ and repeat steps (d)

4. Research Methods

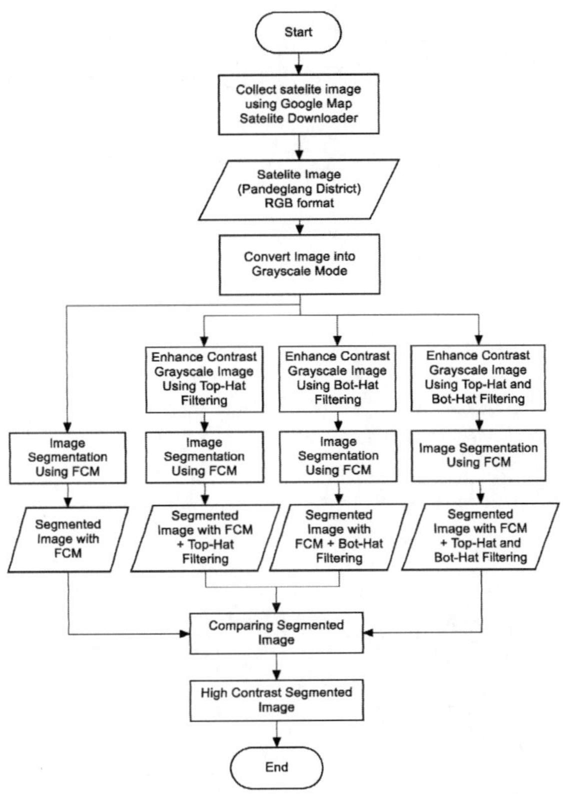

Figure 1. The research procedure

This study begins with the collection of satellite imagery with Google Map Satellite Downloader application. This study uses five sample districts in Pandeglang, Banten province. The image of the five samples are then converted from RGB format into grayscale mode. Four stages image segmentation: image segmentation with FCM original, then do a top-hat techniques and bot-hat, after that do image segmentation with FCM algorithm. MATLAB software is used in this study as an analysis tool (Figure 1).

5. Result and Discussion

The segmentation process for this study captivated samples in the Pagelaran district. Figure 2 shows a sample image that is still original (before segmentation). The segmentation process is done through the following four steps: (1) determination of parameter FCM, (2) input and reading of data image, (3) conversion satellite image of the YCbCr color (grayscale), and (4) grouping satellite image to segment color image by using FCM.

Figure 2. Sample of satelite image

Before performing segmentation analysis of satellite images, first set the parameters: number of clusters (c) for image segmentation is 2, the weight of reappointment fuzzy (q) is 2, the smallest error to be expected (ϵ) is 0.0001 and the maximum number of iterations is 100. Satellite imagery will be entered in Matlab, using the format * .jpg / *. JPEG, with a size of 50x50 pixels. Image processing steps initiated by using FCM. Then, to improve the quality to the image is done by using a top-hat filtering. Furthermore, to improve image quality, is used bot-hat filtering. The results of image processing shown in figure 3.

Figure 3. Satellite images after the improved quality using FCM, Top-Hat and Bot-Hat filtering.

Image processing followed by processing the image size 1116 x 659 pixels, as shown in Figure 4. Figure 4 (a) is derived by processing the image using FCM. Segmentation pattern still looks rough. Each image at the contour edges of buildings and other areas is, even so, not clearly visible on form.

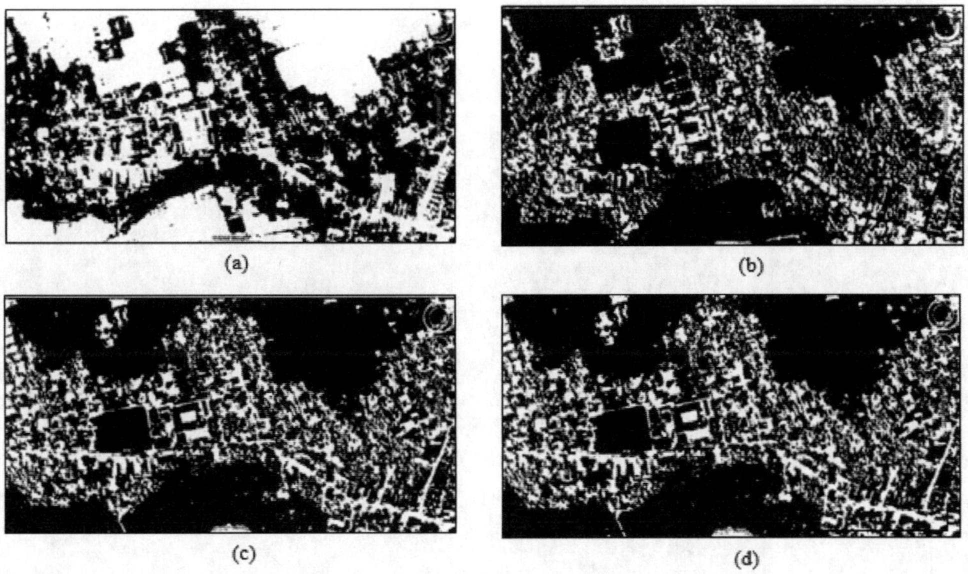

Figure 4. Image processing result for Pagelaran district.

To clarify the contours of the edges of the image, then, the analysis of morphological filtering combined into FCM method. Analysis of the top-hat, bot-hat and a combination of both, is used as a method of analysis to strengthen the results of satellite imagery using FCM The use tool strel aimed to clarify the structure of the object contour image, so that the edges of objects to be a segmented image became clear. A disk shape is used to clarify the edges of objects by taking the midpoint of the image as dot-forming discs. To make the structure more detail, use non-negative parameter values by 8. It is because of that, the smaller the value parameter disk is used; the shape of the object image edge structure will be increasingly unclear. The parameter value> 0 will make the operation faster analysis of morphological filtering.

The second stage during the process of segmentation is to connect the FCM with top-hat filtering techniques. This technique is used prior to the FCM analysis. Filtering the results shown in Figure 4(b). The use of top-hat technique can extract the smallest element to the image. The results of image segmentation indicate that the object edges within the image can be detected more clearly than just using FCM only. Top-hat can give the object brightness level higher on the satellite image.

In the next step, the result of image segmentation is done by combining FCM and bot-hat filtering techniques. As with the previous process, bot-hat filtering technique is used prior to analysis by FCM. At this stage, the object in the image that appeared dark conditioned so easy to analyze. The results of image processing on bot-hat technique is not much different when compared with the results of a top-hat filtering. However, there are some objects that escaped from a top-hat and then detected by filtering bot-hat, so the segmentation results using bot-hat shows the results of a more detailed and clearer than the top-hat. The result of the segmentation shown in Figure 4 (c).

The next segmentation techniques are combined the techniques of top-hat and bot-hat filtering. The result from the segmentation shown in Figure 4 (d). The object of the satellite images which have been segmented very difficult to recognize, why the object on the image is given two handling simultaneously, namely the provision of light from the object (top-hat) so that the brightness level is increasing. The brightness level is increased again after bot-hat techniques are also applied to the object to the image. Background objects and objects from the image tended to appear on segmentation results that make interpretation of the object with the background. Satellite image segmentation followed by lowering non-negative parameter. The parameter values are varied replaced by a value of 2 and 4, and the results can be seen in Figure 5.

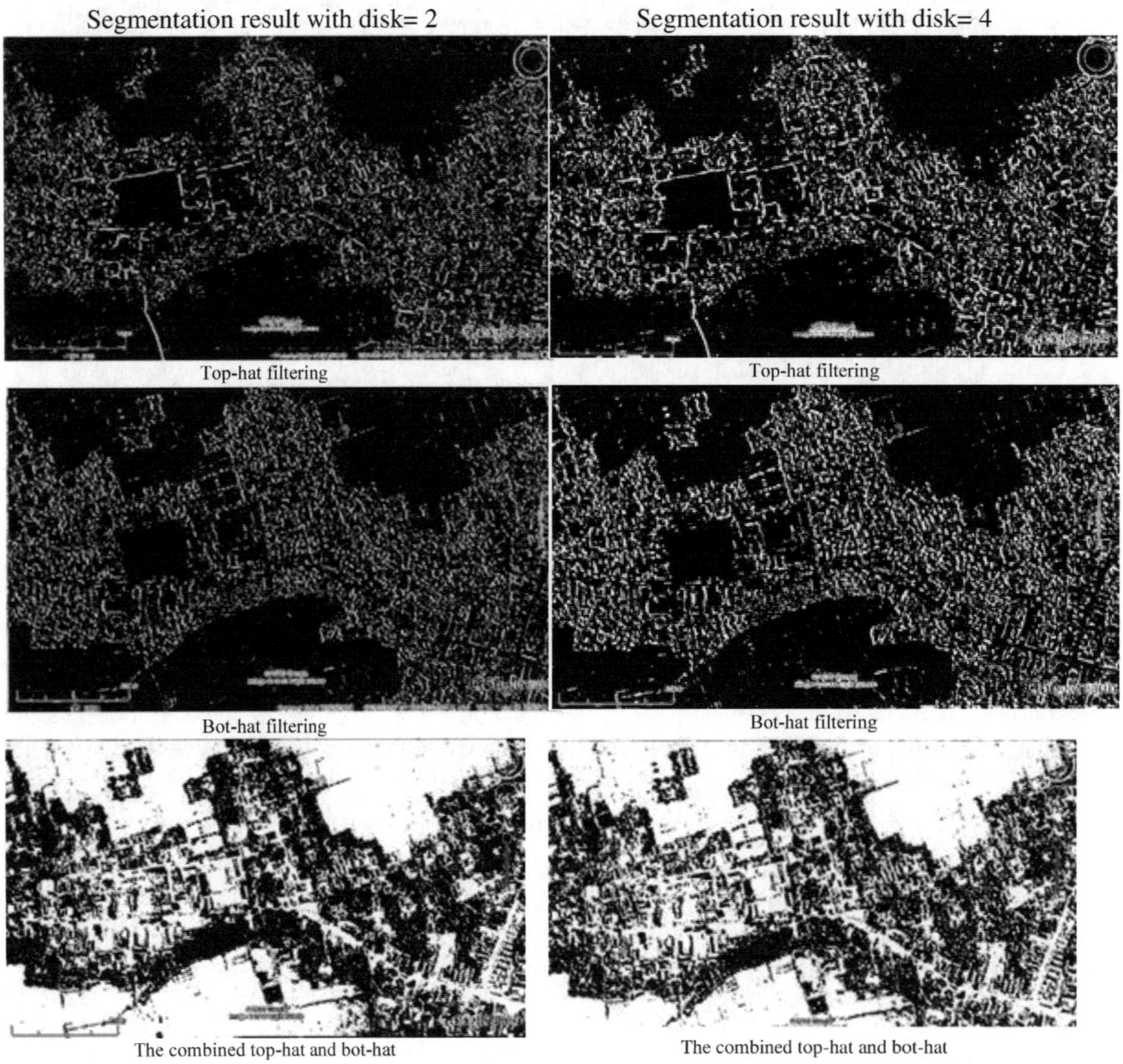

Figure 5. Image processing result using non-negative parameter

Differences in the satellite image segmentation results showed a significant difference after lowering the value of a non-negative parameter. Experiments with the value to the parameter = 2, the result from the segmentation can divide very clearly between the observed object in the background object.

Nevertheless, the detail with the object is generated not as the result of segmentation by value parameter = 4. Visually, the difference can be observed in Figure 5. The object of observation on a satellite image can be seen clearly, but objects precisely fused with a background object. Background object on the original image is the object of paddy (not a part of observation for this study), but identifiable resemblance gray level equal to building objects. Segmentation results for four other districts also showed similar results. The higher the value of this parameter is given a non-negative, then the level of detail edge detected image of the object more clearly.

6. Conclusion

The results of this study indicate that the use of morphological filtering in general can boost the image of the object contrast. The object image can be clarified using a combination of a top-hat and bot-hat filtering before analyzed using FCM. Despite irregularities in two-sample areas. Object image contrast enhancement is affected by non-negative parameter values given to morphological filtering techniques. The greater the value of the disk is given. The edges of objects detailed segmentation results became clear.

References

[1] Andyana, I,M,B., 2015, Segmentasi Citra Berbasis Clustering Menggnakan Algoritma Fuzzy C-Means dan Cat Swarm Optimization, tesis, Magister Teknik Elektro, Universita Udayana, Bali.

[2] Khan, A.M. dan Ravi, S., 2013, Image Segmentation Methods: A Comparative Study, International Journal of Soft Computing and Engineering (IJSCE), Volume-3, Issue-4, September 2013

[3] Sathya, B., dan Manavalan, R., 2011, Image Segmentation by Clustering Methods: Performance Analysis, International Journal of Computer Applications, Volume 29– No.11

[4] Bhanu, B., Lee, S., Ming, J., 1995, Adaptive image segmentation using a genetic algorithm, IEEE Transactions on Systems, Man and Cybernetics, (Volume:25 , Issue: 12)

[5] Islam, S., dan Ahmed, M., 2013, Implementation of Image Segmentation for Natural Images using Clustering Methods, International Journal of Emerging Technology and Advanced Engineering, Volume 3, Issue 3, March 2013

[6] Suri, P, R., dan Mahak, 2012, Image Segmentation With Modified K-Means Clustering Method, International Journal of Recent Technology and Engineering (IJRTE), Volume-1, Issue-2

[7] Murinto, Putra, W,P., dan Handayaningsih, S., 2008, Analisis Perbandingan Histogram Equalization dan Model Logarithmic Image Processing (LIP) untuk Image Enhancement, Jurnal Informatika, Vol. 2, No. 2

[8] Ahmad, N., dan Hadinegoro, A., 2012, Metode Histogram Equalization Untuk Perbaikan Citra Digital, Proceeding of Seminar Nasional Teknologi Informasi dan Komunikasi Terapan 2012

[9] Haiyan, S., Penglang, S., dan Shouhong, Z., 2006, Detection of Unknown Nonlinier FM Signals By Time-Frequency Morphological Filtering, Proceeding of 2006 8th International Conference on Signal Processing.

[10] Li, H., Xiao, De-Yun, dan Zhao, X., 2009, Morphological Filtering Assisted Field-Pipeline Small Leakage Detection, Proceedings of the 2009 IEEE International Conference on Systems, Man, and Cybernetics, San Antonio, USA

[11] Mingshan, G., Hong'e, R., dan Yan, M., 2009, Multi-sale Morphological Filtering Method for Preserving the Details of Images, Proceeding of 2009 Second Asia-Pacific Conference on Computational Intelligence and Industrial Applications

[12] Liu, Z., Wang, J., dan Liu, B., 2011, ECG Signal Denoising Based on Morphological Flitering, Proceeding of 5th International Conference on Bioinformatics and Biomedical Engineering, (iCBBE) 2011.

[13] Verma, R., Mehrotra, R., dan Bhateja, V., 2013, An Integration of Improved Median and Morphological Filtering Techniques for Electrocardiogram Signal Processing, Proceeding of 2013 IEEE 3rd International Advance Computing Conference (IACC)

[14] Prakash, K, S., Prakash, R, P., dan Binu, V, P., 2015, A Study of Morphological Filtering Using Graph and Hypergraphs, Proceeding of 2015 IEEE International Advance Computing Conference (IACC).

[15] Thapar, S., dan Garg, S., 2012, Study And Implementation Of Various Morphology Based Image Contrast Enhancement Techniques, International Journal of Computing & Business Research, Proceeding of 'I-Society 2012' at GKU, Talwandi Sabo Bathinda (Punjab)

[16] Yang, Y., dan Huang, S., 2007, Image Segmentation By Fuzzy C-Means Clustering Algorithm With A Novel Penalty Term, Computing and Informatics, Vol. 26, 2007, 17–31

[17] Cannon, R.L., Dave, J.V., dan Bezdek, J.C., 1986, Efficient Implementation of the Fuzzy C-Means Clusteng Algornthms, IEEE Transactfions On Patern Analysis And Machine Intelligence. Vol. PAMI-8, NO. 2, March 1986